CO 1 35 80324 5C   £45.00

BOOKSTORE LOAN

NEWTON AYCLIFFE
1/88
25. JAN 92
12. APR. 1988 TRIM
13. JUN. 1988
-2. AUG. 1988
06. MAY 89
29. JUN 89
20. JUL. 1989
8/89
08. MAR 90

3580324 3

BOOKSTORE LOAN

COUNTY COUNCIL OF DURHAM

COUNTY LIBRARY

The latest date entered on the date label or card is the date by which book must be returned, and fines will be charged if the book is kept after this date.

STAR WARS AND
EUROPEAN DEFENCE

Also by Hans Günter Brauch

In English:
DECISIONMAKING FOR ARMS LIMITATION – ASSESSMENTS AND PROSPECTS (*editor with Duncan L. Clarke*)
ALTERNATIVE CONVENTIONAL DEFENSE POSTURES IN THE EUROPEAN THEATER – THE FUTURE OF THE MILITARY BALANCE AND DOMESTIC CONSTRAINTS (*editor with Robert Kennedy*)
MILITARY TECHNOLOGY, ARMAMENTS DYNAMICS AND DISARMAMENT (*editor*)

In German:
STRUKTURELLER WANDEL UND RÜSTUNGSPOLITIK DER USA (1940–1950) – ZUR WELTFÜHRUNGSROLLE UND IHREN INNENPOLITISCHEN BEDINGUNGEN
ENTWICKLUNGEN UND ERGEBNISSE DER FRIEDENSFORSCHUNG (1969–1978). EINE ZWISCHENBILANZ UND KONKRETE VORSCHLÄGE FÜR DAS ZWEITE JAHRZEHNT
ABRÜSTUNGSAMT ODER MINISTERIUM? AUSLÄNDISCHE MODELLE DER ABRÜSTUNGSPLANUNG MATERIALIEN UND REFORMVORSCHLÄGE
KERNWAFFEN UND RÜSTUNGSKONTROLLE – EIN INTERDISZIPLINÄRES STUDIENBUCH (*editor*)
CHEMISCHE KRIEGFÜHRUNG – CHEMISCHE ABRÜSTUNG, DOKUMENTE UND KOMMENTARE (*editor with Rolf-Dieter Müller*)
GIFTGAS IN DER BUNDESREPUBLIK – CHEMISCHE UND BIOLOGISCHE WAFFEN (*with Alfred Schrempf*)
DER CHEMISCHE ALPTRAUM ODER GIBT ES EINEN C-WAFFENKRIEG IN EUROPA?
DIE RAKETEN KOMMEN! VOM NATO-DOPPELBESCHLUSS BIS ZUR STATIONIERUNG
PERSPEKTIVEN EINER EUROPÄISCHEN FRIEDENSORDNUNG
ANGRIFF AUS DEM ALL. DER RÜSTUNGSWETTLAUF IM WELTRAUM
SICHERHEITSPOLITIK AM ENDE? EINE BESTANDSAUFNAHME, PERSPEKTIVEN UND NEUE ANSÄTZE (*editor*)
VERTRAUENSBILDENDE MASSNAHMEN UND EUROPÄISCHE ABRÜSTUNGSKONFERENZ (*editor*)
MILITÄRISCHE NUTZUNG DES WELTRAUMS – EINE FORSCHUNGSBIBLIOGRAPHIE (*editor with Rainer Fischbach*)

Star Wars and European Defence

Implications for Europe: Perceptions and Assessments

Edited by
Hans Günter Brauch

*Chairman, Study Group on Peace Research
and European Security Policy
and Lecturer in International Relations
Institute of Political Science
Stuttgart University*

Foreword by
Denis Healey

Preface by
Raymond L. Garthoff

MACMILLAN
PRESS

© Hans Günter Brauch 1987
Foreword © Denis Healey 1987
Preface © Raymond L. Garthoff 1987

All rights reserved. No reproduction, copy or transmission of this publication may be made without written permission.

No paragraph of this publication may be reproduced, copied or transmitted save with written permission or in accordance with the provisions of the Copyright Act 1956 (as amended), or under the terms of any licence permitting limited copying issued by the Copyright Licensing Agency, 33-4 Alfred Place, London WC1E 7DP.

Any person who does any unauthorised act in relation to this publication may be liable to criminal prosecution and civil claims for damages.

First published 1987

Published by
THE MACMILLAN PRESS LTD
Houndmills, Basingstoke, Hampshire RG21 2XS
and London
Companies and representatives
throughout the world

Typeset by Acorn Bookwork, Salisbury, Wiltshire

Printed in Hong Kong

British Library Cataloguing in Publication Data
Star wars and European defence: implications
for Europe: perceptions and assessments.
1. Strategic Defence Initiative
I. Brauch, Hans Günter
358'.8'0973   U393
ISBN 0-333-41286-9

For Colonel Jonathan Alford
Deputy Director
The International Institute for Strategic Studies
a highly regarded colleague
and former military man
who died
unexpectedly on 14 August 1986
at the age of 53
His impressive personality and his work
for a better understanding
of the Military Balance and of confidence building measures
will be remembered
by his many admirers and friends

and

For Stephan Tiedtke
a German peace researcher
a respected colleague
and a
good human being
who died
suddenly on 18 February 1986
at the age of 43
His work for a better understanding
of Soviet foreign policy and for a relaxation
of tensions in Europe
will be remembered
by his friends and colleagues
in East and West

Contents

List of Figures	x
List of Tables	xi
Notes on the Contributors	xiii
Foreword by Denis Healey	xix
Preface by Raymond L. Garthoff	xxxiii
Acknowledgements	xli
List of Abbreviations	xliii

Introduction
Hans Günter Brauch   1

PART I   AMERICAN AND SOVIET SPACE AND BMD PROGRAMMES

1   SDI – A Global Defence Against Ballistic Missiles
Simon Peter Worden   9

2   Offensive Nuclear Forces, Strategic Defence and Arms Control
Gary L. Guertner   26

3   SDI – A Reaction to or a Hedge Against Soviet BMD Projects: Soviet Military Space Activities and European Security
Hans Günter Brauch   50

PART II   EUROPEAN PERCEPTIONS OF AND REACTIONS TO SDI

4   SDI – The British Response
Trevor Taylor   129

5   French Political Reaction to SDI – The Debate on the Nature of Deterrence
Alain Carton   150

6   SDI – The Political Debate in the Federal Republic of Germany
Hans Günter Brauch   166

Contents

7   Perceptions and Reactions to SDI in the Benelux
    Countries
    Robert J. Berloznik   234

8   SDI – A Topic of Transatlantic Consultation and Debate
    in International Parliamentary Bodies
    Hans Günter Brauch   256

9   EUREKA – A West European Response to the
    Technological Challenge Posed by the SDI Research
    Programme
    Alain Carton   311

PART III   ASSESSMENTS OF THE STRATEGIC CONSEQUENCES OF SDI FOR EUROPE

10 The Implications of BMD for Britain's Nuclear Deterrent
    Paul B. Stares   331

11 The Implications of SDI for French Defence Policy
    Alain Carton   341

12 The Geostrategic Risks of SDI
    Peter A. Wilson   352

PART IV   THE ABM TREATY – FROM SDI TO EDI

13 The ABM Treaty – Its Evolution, Interpretation and
    Grey Areas and an Official Attempt at Reinterpretation
    John B. Rhinelander   373

14 From SDI to EDI – Elements of a European Defence
    Architecture
    Hans Günter Brauch   436

15 Compliance with the ABM Treaty – Questions Related to
    Soviet Missile Defence and to the United States SDI
    *Thomas K. Longstreth, John E. Pike and John B.
    Rhinelander*   500

16 Closing the Window of Vulnerability
    John E. Pike   537

APPENDIX

A   *Treaty Between the United States of America and the
    Union of Soviet Socialist Republics on the Limitation of
    Anti-Ballistic Missile Systems (ABM-Treaty)*   555

Contents

B Agreed Statements, Common Understandings, and
   Unilateral Statements Regarding the Treaty Between the
   United States of America and the Union of Soviet Socialist
   Republics on the Limitation of Anti-Ballistic Missiles   562
C Protocol to the Treaty Between the United States of
   America and the Union of Soviet Socialist Republics on the
   Limitation of Anti-Ballistic Missile Systems   568
D Standing Consultative Commission Agreements   571
Bibliography   572
Index   588

List of Figures

1.1	Ballistic missile trajectories	20
2.1	Strategic nuclear offensive–defensive linkages for United States systems	31
2.2	Layered defence concept	38
2.3	Countermeasures to advanced BMD	40
14.1	The Soviet tactical ballistic missile threat for Western Europe	442
15.1	Estimated 1990 coverage of fans of Soviet LPARs	530
15.2	Estimated 1990 coverage of fans of American LPARs	531

List of Tables

1.1	Possible kinetic energy defensive system	17
3.1	Milestones in the development of military space systems	82
3.2	Relative United States/USSR standing in the 20 most important basic technology areas	108
3.3	Relative United States/USSR technology level in deployed military systems	109
6.1	Attitudes of 5 West German parties on 4 defence issues	198
14.1	Changes in Soviet ballistic missile launchers aimed at targets in NATO Europe from 1970–80 according to data from IISS	444
14.2	Survey of conceptual frameworks and of systems components for an EDA	452
15.1	SDI compliance issues	522
16.1	Comparison of endoatmospheric interceptors	541

Notes on the Contributors

Robert J. Berloznik (Belgium) has been a research associate at the Polemological Centre of the Free University of Brussels since 1983; prior to that he worked for the University Centre for Development Cooperation in Brussels. Publications (in Dutch): *Politieke, Sociale en Kulturele Aspekten van de Patroon-Cliënt verhouding in Latijns-Amerika* (1981); *Vrede en Ontwikkeling* (1984); *Militaire Aspekten van het Amerikaanse Space Transportation System* (1985); *De militarisering van de ruimte* (1985).

Hans Günter Brauch (West Germany) is director of the Study Group on Peace Research and European Security Policy (AFES) at the Institute of Political Science at Stuttgart University, a lecturer for international relations and a research associate working on weapons technologies. He was a student of political science, modern history, international law and English language and literature at Heidelberg and London University (University College) and he holds a Dr phil. degree from Heidelberg University (1976). He has been a research associate at Heidelberg, Harvard and Stanford Universities and he has taught at the Universities of Darmstadt and Tübingen. He has written and edited fifteen books in German and English on arms control, defence policy and international security and he has contributed articles for readers in the United States, Great Britain, Denmark, Finland, South Korea and West Germany and for scientific and political journals. He is co-editor of two other English readers (with Duncan L. Clarke), *Decisionmaking for Arms Limitation – Assessments and Prospects* (Cambridge, Mass.: Ballinger, 1983) and (with Robert Kennedy), *Alternative Conventional Defense Postures in the European Theater – The Future of the Military Balance and Domestic Constraints* (1987). He has published two books on military space matters, in German: *Angriff aus dem All. Der Rüstungswettlauf im Weltraum* (Berlin-Bonn: Dietz, 1984) and as co-editor (with R. Fischbach), *Militärische Nutzung des Weltraums – Eine Forschungsbibliographie* (Berlin: Berlin-Verlag Arno Spitz, 1987).

Alain Carton (France) is a fellow at the Institute for East–West Security Studies in New York and he was a Parliamentary assistant (for defence) in the Socialist Group in the National Assembly till July 1986. He received his M.A. in International Relations from the Sorbonne (1981). From 1981–5 he attended the Institut de Sciences Politiques in Paris. He has been a Parliamentary Assistant to Senator R. Pontillon, a research assistant at CIRPES working on infranuclear deterrent and French–German relations. He is author of: 'Perceptions Allemandes de la Doctrine Rogers', *Revue de Défense Nationale*, July 1983; *L'ecole Allemande de Techno-Guerilla, Cahier d'Etudes Stratégiques* (Paris: CIRPES, 1984).

Raymond L. Garthoff (United States) has been Senior Fellow in Foreign Policy Studies at the Brookings Institution, Washington, DC since 1980. A graduate of Princeton University, he holds a Ph.D. from Yale University. He served as a specialist on Soviet political and military affairs at the Rand Corporation (1950–7), in the Office of National Estimates, CIA (1957–61); as special Assistant for Soviet Bloc Politico-Military Affairs in the United States Department of State (1961–8); Counselor in the United States Mission to NATO (1968–70); Deputy Director, Bureau of Politico–Military Affairs, Department of State (1970–3); Senior Advisor on the United States SALT Delegation (1969–73); Senior Foreign Service Inspector (1974–7) and United States Ambassador to Bulgaria (1977–9). He is the author of six books and contributor to over 30 other books, and has published over 100 articles in professional journals. Among his major publications are: *Détente and Confrontation: American–Soviet Relations – From Nixon to Reagan* (Washington: The Brookings Institution, 1985); *Perspectives on the Strategic Balance* (Washington: The Brookings Institution, 1983); *Soviet Military Policy* (New York: Praeger, 1966); *Soviet Strategy in the Nuclear Age* (New York: Praeger, 1958, 1962); *Soviet Military Doctrine* (Glencoe, Ill.: Free Press, 1953).

Gary L. Guertner (United States) is Professor of International Relations at California State University, Fullerton. He was an officer in the Marine Corps, held the Henry Stimson Chair of Political Science at the United States Army War College (1981–4); and served as scholar-in-residence at the United States Arms Control and Disarmament Agency in Washington. He has published numerous articles on defence and foreign policy in professional journals, for example: 'Strategic Defense: The Burdens of Proof', *Foreign Policy*, June 1985; 'Strategic Defense: New Technology and Old Tactics', *Parameters*,

Autumn 1985; 'Nuclear War in Suburbia', *Orbis*, Spring 1982; 'Soviet Strategic Vulnerability: Deterring a Multi National State', *Political Science Quarterly*, Summer 1981; 'Offensive Doctrine in a Defense Dominant World', *Air University Review*, Fall 1985.

Denis Healey, leading foreign policy spokesman of the British Labour Party (Shadow Foreign Minister). Born on 30 August 1917, he has been a member of the House of Commons since 1952, specialising in foreign affairs and defence matters. In 1952–4 he was a member of the Parliamentary Assembly of the European Council and from 1952 to 1955 he was a member of the Parliamentary Assembly of the European Council and from 1963 to 1965 of the Assembly of the Western European Union. In the Cabinet of Prime Minister Harold Wilson he served as Defence Minister from 1964–70 and in the cabinets of Prime Ministers Wilson and Callaghan he was Chancellor of the Exchequer from 1974–9. From November 1980 until May 1983 he was the Deputy Chairman of the British Labour Party.

Thomas K. Longstreth (United States) is a foreign policy advisor to Senator Edward Kennedy. He was Associate Director of Research and Analysis with the Arms Control Association (1982–5) and a senior analyst with the Center for Defense Information. He received his B.A. from Wesleyan University (Connecticut) and his M.A. in security policy studies from George Washington University. He has authored numerous studies and articles on strategic and arms control issues. He is co-author with John E. Pike and John B. Rhinelander of *The Impact of US and Soviet Ballistic Defence Programs on the ABM Treaty* (Washington, DC: National Campaign to Save the ABM Treaty, 1985).

John E. Pike (United States) has been associate director for Space Policy with the Federation of American Scientists in Washington, DC since 1983. He initiated the Space Policy Working Group, consisting of staff of Congressional offices and public interest organisations concerned about space weapons issues. He authored studies and articles on space weapons, arms control, verification and national security issues. He was a consultant to Universal Studios, science writer (1981–3), a political consultant in Atlanta, Georgia and Nashville, Tennessee (1975–80) and an energy policy analyst with the Tennessee State Energy Agency, Nashville. He holds a B.S. in

Technology Assessment from Vanderbilt University, Nashville. He has authored, among other publications, 'Criteria for Verification', *Bulletin of Peace Proposals*, 16,3(1985) pp. 255–62; 'Anti-Satellite Weapons', *Public Interest Report*, November 1983; 'Charges of Soviet Treaty Violations', *Public Interest Report*, March 1984; *The Strategic Defence Initiative – Budget and Program* (Washington: Federation of American Scientists, 10 February 1985); 'Anti-Satellite Weapons and Arms Control', *Arms Control Today*, December 1983; 'Space Weapons – History and Current Debate', *Bulletin of the Atomic Scientists*, May and August 1984 (with Richard Garwin); 'The Soviet ABM Program – Is there a Gap?', *Arms Control Today*, September 1984; he is co-author with Thomas K. Longstreth and John B. Rhinelander of *The Impact of US and Soviet Ballistic Missile Defense Policy on the ABM Treaty*.

John B. Rhinelander (United States) is partner of Shaw, Pittman, Potts & Trowbridge and a member of the Board of Directors of the Arms Control Association. He is a graduate of Yale University (1951); University of Virginia Law School and was Editor-in-Chief of Virginia Law Review (1961); law clerk to a Supreme Court Justice; Special Civilian Assistant to the Secretary of Navy; Chief Counsel and Deputy Director of the Office of Foreign Direct Investments in the Department of Commerce; Deputy Legal Advisor, Department of State; Legal Advisor to United States SALT I delegation; General Counsel at the Department of Health, Education and Welfare; Undersecretary at the Department of Housing and Urban Development; Lecturer on nuclear weapons and arms control at the University of Virginia Law School. His major publications are: with Mason Willrich (co-editor): *SALT – The Moscow Agreements and Beyond* (New York: Free Press, 1974); co-author with Thomas K. Longstreth and John E. Pike: *The Impact of US and Soviet Ballistic Missile Defense Programs on the ABM Treaty*; 'Implications of US and Soviet BMD Programs on the ABM Treaty', in SIPRI (ed.), *Space Weapons and International Security* (Stockholm: SIPRI, forthcoming).

Paul Brian Stares (Great Britain) is a research associate at the Brookings Institution. He holds a Ph.D. from the University of Lancaster (1979–82) and a B.A. from North Staffordshire Polytechnic. He is the author of: *Space Weapons & US Strategy Origins & Development* (London; Sydney: Croom Helm, 1985); and co-editor with Michael Schwarz: *The Exploitation of Space – Policy Trends in the military and commercial uses of outer space* (London: Butterworth, 1985); 'A comparative Assessment of US and Soviet Space Program,' *Daedalus*, Spring 1985; (with John Pike) 'The "Star Wars" Initiative: Problems and Prospects', *Space Policy*, Summer 1985; 'Space and US National Security', in W. Durch (ed.), *National Interests and the Military Use of Space* (Cambridge: Ballinger 1984).

Trevor Taylor (Great Britain) is the Principal Lecturer in International Relations, North Staffordshire Polytechnic. He holds a Ph.D. from the London School of Economics (1972), a M.S. from Lenigh University, Pennsylvania (1968) and a B.Sc. from the LSE (1967). He is the author of: *European Defence Cooperation* (London: Routledge and Kegan Paul, 1984); *Defence, Technology and International Integration* (London: Frances Pinter, 1982); and editor of *Approaches and Theory in International Relations* (London: Longman, 1972).

Peter A. Wilson (United States) has worked since 1981 for a number of research firms dealing with national security issues including the RAND Corporation and the M/A-COM Linkabit Corp. Prior to joining the United States government in the mid-1970s, he was a consultant to the RAND Corporation, the Hudson Institute and a member of the staff of the Center for Naval Analyses. During the late 1970s he was a member of the Policy Planning Staff of the Department of State specialising in nuclear arms control, European security and rapid deployment issues. He has published among other publications, 'A New Strategic Arms Agreement: Taking the Long View', *SAIS Review*, Winter 1984 (with Theodre H. Moran); 'Battlefield Guided Weapons: The Big Equalizer', *US Naval Proceedings*, March 1975.

Simon Peter Worden (United States) was an advisor to the United States Delegation to the Negotiations on Nuclear and Space Arms with the Soviet Union. Colonel Worden holds a Ph.D. in Astronomy. He worked for the Strategic Defence Initiative Organisation (SDIO) in the United States Department of Defense.

Foreword[1]

AN ALTERNATIVE REALISTIC PERSPECTIVE BEYOND NUCLEAR DETERRENCE

by Denis Healey

In his seminal speech in March 1983 when President Reagan launched his Star Wars concept, he said: 'The human spirit must be capable of rising above dealing with other nations by threatening their existence'; and he made it clear that he did not believe that peace could rest much longer on the threat of mutual suicide. I agree with him.

Secretary Weinberger carried the logic of Reagan's remarks further in a speech he made on 9 October 1985 when he said: 'The world has changed so profoundly since the 1950s and 1960s when most of our strategic ideas were formulated, that many of these concepts are now obsolete'. He specifically rejected NATO's current strategy of flexible response by saying 'our position on the uses of military power represents a rejection of received wisdom about limited war and gradual escalation'. And he described what he saw as the Reagan Administration's commitment to make conventional deterrence work.

The insights and objectives thus described by the President and his Defence Secretary are very much those on which the Labour Party bases its policies for defence and disarmament. Unfortunately the policies developed over the last few years in Washington, far from fulfilling these objectives, move in precisely the opposite direction. For example, the deadlock at the Geneva Summit over President Reagan's Star Wars programme now threatens the world with an accelerating arms race in both offensive and defensive weapons. And this is something which the President warned his hearers very firmly against in his initial speech. In March 1983 he said: 'If defensive systems were paired with offensive systems they could be viewed as fostering an aggressive policy and nobody really wants that'.

Yet the United States is now committed to pursuing its Star Wars defensive programme simultaneously with introducing new offensive nuclear weapons into its arsenal. The Administration has said that the period in which 'defensive systems will be paired with offensive systems' may last for several decades or longer. Indeed no one has yet suggested how it could be brought to an end once it gets under way – short of a nuclear holocaust. Mr Gorbachev has made it absolutely clear that the Soviet Union sees no alternative but to follow suit in both defensive and offensive systems.

I believe nevertheless that it is both possible and necessary to carry the logic of the American statements I have quoted to a conclusion in practical policies for defence and disarmament.

It may be useful to put the problems we face today in the perspective of the history of nuclear strategy and diplomacy over the last 40 years.

Nuclear weapons have not prevented war outside Europe, even wars in which the superpowers were involved, like the Soviet war in Afghanistan, the American war in Vietnam and the war between China, North Korea and the United Nations in the 1950s.

Nevertheless it is difficult to believe that the post-war settlement in Europe would have survived so long without the deterrent effect of nuclear weapons. Yalta and Potsdam divided this historic continent into two against the will of most of its peoples along a line which ran through the middle of Germany, its most powerful state. Yet that settlement has already lasted twice as long as the 1918 settlement. The deterrent effect of nuclear weapons must have contributed to this result.

Despite this, there have been growing doubts, not just in the unofficial peace movements but right at the heart of the NATO institutions, both about the strategies on which nuclear deterrence is based and about the role of nuclear weapons in relations between Russia and the West. These doubts have a long history, reaching back to the end of the Second World War.

Perhaps we can learn something from the mistakes of those early years. As it was, by the end of 1946 the pattern of the cold war had set and strategic considerations dominated nuclear thinking on both sides.

When the war ended, all the Red Army needed to reach the Rhine was boots. The United States was pulling its troops out of Europe very fast. Britain was demobilising. No other Western European country was providing significant forces and Germany was disarmed. Four years later, the United States was persuaded to extend its nuclear umbrella over Western Europe, through NATO. That has been the foundation of Western European security ever since.

During the early post-war years, the possibility of actually dropping nuclear weapons on the Soviet Union before Russia had built up a nuclear stockpile of its own was discussed in Washington. The Joint Chiefs of Staff produced plans for attacking the Soviet Union which, thanks to the Freedom of Information Act, are now available to the public. In 1948 they had a plan for dropping nuclear weapons on twenty Soviet cities. In 1954 they had a plan for dropping a thousand nuclear bombs on the Soviet Union. Yet the American government never adopted these plans, even at a time when it had practically a monopoly of nuclear weapons.

Although by 1960 the Americans still had 20 times as many nuclear weapons as Russia, Christian Herter said in public that the United States would never actually use its nuclear forces against the Soviet Union unless its own survival was directly at stake. In other words the nuclear deterrent was already unreliable so far as America's allies were concerned. Until President Reagan and Secretary Weinberger, no American statesman since then has expressed such doubts while in office although many have done so after retiring – notably Henry Kissinger and Robert McNamara.

Meanwhile both sides have enormously increased the number of their nuclear weapons, especially in the ten years between 1970 and 1980. The United States increased the number of its strategic warheads in that decade from 4,000 to 10,000, and the Soviet Union from 1,800 to 6,000. In 1985 America had 11,000 strategic nuclear weapons, the Soviet Union 9,000. On top of that both sides have a great number of tactical nuclear weapons. The total nuclear stockpile of the two great powers together is now over 50,000 – the equivalent of well over a million Hiroshimas.

There has never been a period during this long miserable story in which either side seemed likely to have a meaningful superiority over the other in the strategic nuclear field. All the experts including even Richard Perle, the Pentagon's prime hawk, agree that there is now effective strategic parity between the United States and the Soviet Union. Nevertheless the arms race continues unabated. By 1990, unless there is agreement on stopping the race, each side will have over 13,000 strategic nuclear weapons, if they observe the SALT II agreement; if they do not observe it, each side will probably have about 20,000 strategic nuclear weapons.

Many years ago Winston Churchill asked what was the point of buying more bombs simply to make the rubble bounce. We must ask today, why on earth are both sides acquiring these colossal arsenals when they could not actually use them without committing suicide.

Yet the reason for the continuing arms race is all too obvious. Whatever the original reason for the increase in stockpiles it has led some people on each side to shift from the idea of having nuclear weapons in order to deter a war to the idea of having nuclear weapons in order to fight a war – in which, to use the American word, it can 'prevail'.

An absolute precondition of victory in a nuclear war would be the ability to destroy the enemy's retaliatory forces in a surprise attack – or 'first strike'. The technical feasibility of a successful first strike may appear closer today in theory.

On top of the miniaturisation, the multiple independent re-entry vehicles (or MIRVs) are quite extraordinarily accurate. An increase of 10 per cent in accuracy is equivalent to an increase of a 100 per cent in destructive power against a hard target. In fact it is now technically possible for each side to plan for carrying out a first strike against its enemy's fixed bases on land.

Thus each side thinks it has an incentive to increase the number of its nuclear missiles so as to have too many targets for an enemy first strike to cover, or so as to be able to destroy the increasing number of enemy nuclear weapons in its own first strike. And each side can avoid the vulnerability of fixed bases on land by making its land-based missiles mobile or by putting its missiles at sea or in aircraft. Moreover the moment either side thinks its enemy might be on the point of acquiring the capability for a first strike it may be tempted to pre-empt that first strike.

In my opinion the idea that any government would authorise a first strike even with the weapons which may soon be available, never mind with its existing weapons, is a fantasy. A first strike at present would require the explosion of at least 1,000 warheads on land, with a high probability of causing a nuclear winter. Even so, it would leave undamaged sufficient enemy missiles to wreak intolerable retaliation. The whole of his submarine-based missiles as well as most of his airborne missiles would escape entirely; those alone would suffice to blow up the world several times over.

Unfortunately there is a tendency for the military on each side grossly to exaggerate the enemy capability because it is the best way of getting money for themselves. The United States Air Force first sought to deceive public opinion on this issue as far back as Eisenhower's time by claiming that Russia had 300 ICBMs, when the Samos satellite showed it had only 60. President Eisenhower, however, had sufficient experience to be deeply distrustful of the 'military–industrial complex'. He was not taken in.

Unfortunately nuclear strategy in the West today tends to be determined by tiny elites of middle-ranking bureaucrats and staff officers who have no personal experience of world war and are obsessed by esoteric theories. These elites are predominantly civilian and are under no effective political control, partly because of the enormously rapid turnover of defence ministers in most countries.

Effective political control of nuclear strategy requires not only that the minister should work at the problem himself. He must also engage the interest of his Prime Minister and other key members of the Cabinet, which is not always easy. President Reagan has often shown ignorance of the most fundamental facts on which nuclear strategy must be based. My impression is that the Soviet leaders keep themselves far better informed.

So far this bizarre black comedy has not had any catastrophic effects in the real world because till now the strategic nuclear balance between Russia and the USA has been invulnerable to quite large variations in their relative capability, and to big differences in the composition of their forces. But the new weapons already under development on both sides, and in some cases already deployed, could upset that stability.

Each side depends for knowledge of its enemy's capacity and for early warning of attack largely on spy satellites; but each side is trying to develop systems for destroying spy satellites – 'ASATS' – which would rob the enemy of his eyes and ears and greatly increase the chances of carrying out a surprise attack. Secondly, many of the new missiles can hit their targets so fast that the decision how to react to them will have to be taken not by human beings but by computers; this would be particularly true of the United States Star Wars systems.

The first time a possibly ambiguous warning that missiles were on their way hit the monitor screens there would be no time for the base commander at Molesworth to consult his own President, never mind for President Reagan to consult Mrs Thatcher. A situation in which the survival of the human race depends on the microcircuits of the computers rather than the human brain is a worrying one for anyone who knows how often computers can go wrong.

Both sides are also producing a large arsenal of new weapons such as cruise missiles which are designed to carry both nuclear and conventional warheads. If either side detected 100 cruise missiles on its monitors it would have to assume they were carrying nuclear warheads rather than conventional ones and react in kind, because if it waited to find out it might be dead. Another danger of cruise missiles is that they are easily hidden – particularly at sea – and would make it much more difficult to verify a disarmament agreement.

The third dangerous new development is the attempt to produce a comprehensive defence against strategic nuclear attack. The only man in the world who believes that Star Wars might make nuclear weapons 'impotent and obsolete' is President Reagan himself. None of his officials who are working on the Star Wars project believe that. What they think it may be possible to do – perhaps within ten years – is to produce some defence for most of America's land-based ICBMs against a Soviet first strike, although that defence would probably be based at first on land rather than in space.

The State Department's official pamphlet about Star Wars says that its purpose is not, as President Reagan claims, to replace the nuclear deterrent, but to strengthen or 'enhance' it. The possibility that Star Wars may have this limited ability for defending missile sites is an incentive for the other side to increase the number of its own missiles. Mr Gorbachev has made that very clear.

In any case, if America is really worried about the vulnerability of its ICBMs it would be cheaper simply to scrap them and follow the advice of the poet: 'Put these missiles out to sea/Where the real estate is free/And they're miles away from me.'

The main reason why they do not take this obvious step is that American defence policy is still dominated by interservice rivalry. The United States Air Force does not want to give up part of the strategic triad in favour of the United States Navy. Star Wars is only one of the malign consequences of this rivalry.

Unfortunately some Russians seriously believe – as President Reagan warned they might – that the purpose of Star Wars is to protect America's land-based missiles against a ragged response by the Soviet missiles which survived an American first strike. In other words they suffer from a mirror image of American fears of a Soviet first strike. Moreover American scientists now suggest that the sort of space-based systems America is trying to develop could be used as easily for attack as for defence. Indeed if the laser weapons were used to incinerate cities, they could produce the equivalent of a nuclear winter no less than nuclear weapons. That is why Russia insists so firmly on implementing the Shultz–Gromyko agreement to prevent the arms race in space.

If you take the risk of instability seriously, by far the most important task in the field of disarmament is to stop the nuclear arms race in its tracks immediately by halting the modernisation of nuclear forces.

Stupendously excessive as they are, the existing arsenals of nuclear weapons are not likely in themselves to produce a war. But the weapons on the way could well destabilise this situation. That is why in the Labour Party we so strongly support a freeze on nuclear weapons as a first step to their reduction. We want a freeze on the testing and deployment of all new systems both offensive and defensive. We think it could be achieved by some fairly simple technical methods if the two sides agree on the objectives and on the cutoff points.

The first step would be a comprehensive test ban treaty which would prevent either side from testing new types of nuclear warheads. Scientific advances have made it possible to detect and measure nuclear tests even down to a few tons in yield, especially if you can put some of the new sensing devices in Soviet and American territory, where they could be manned by neutrals. Six non-aligned countries, including Sweden and India, have already offered to man such stations. The other way you could guarantee the freeze is by banning all tests of the components of new nuclear delivery systems which can be observed by so-called 'national means', for example spy satellites.

If you once got an agreed freeze, backed by a ban on the testing of new delivery systems and a ban on all nuclear tests, then it would be much easier to attack the problem of cutting existing arsenals.

Both sides have already agreed a 50 per cent cut in existing strategic weapons, although each side's specific proposals are heavily slanted to favour its own particular interests. The job of negotiations would be to reconcile these differences.

This brings us to a problem which directly affects Britain's security. NATO is likely to continue to need to possess nuclear weapons to deter nuclear attack so long as the Soviet Union possesses them. But, supposing you got a ban on the modernisation of nuclear weapon systems and on strategic defence together with a big cut in existing arsenals which institutionalised nuclear parity at a much lower level, could America's remaining nuclear forces continue to deter a purely conventional attack on her European allies?

It is worth noting that the allies seem less worried than the United States about the effect of deep cuts in nuclear weapons on their own security. And all the European governments secretly share President Mitterrand's hostility to Star Wars. Indeed, the Europeans take the prospect of another world war less seriously than the Americans – something which is already causing trouble in the United States Congress.

In fact there has been no time since the end of the Second World War when Western intelligence believed that the Russians were planning to launch an all-out invasion of Western Europe. NATO has always believed that the real danger of war with the Warsaw Pact lies not in an attack out of the blue in Central Europe, but from the spillover of a conflict between Russia and the West in some other area, like the Middle East, or of internal fighting inside Eastern Europe like the Berlin rising or the invasion of Hungary. In such a situation nuclear deterrence is of limited value, because once the fighting has begun, deterrence has failed in its main purpose. The risk of relying on nuclear deterrence is that you might be involved in a conflict which nuclear weapons have not deterred, and to which a nuclear response is not appropriate.

Recent students of Soviet strategic thinking, like C. N. Donnelly of the Royal Military Academy, Sandhurst, reach the same conclusion by a different route. He argues that Russia is pursuing its aims in Europe by all means short of direct armed conflict but if war were to break out unexpectedly Russia is determined to win quickly without using nuclear weapons 'before sufficient time has elapsed for the United States to commit itself to a strategic nuclear war'.

The possibility of war in Europe is there, however, small. NATO must have a military strategy for reducing that possibility to the minimum and for stopping the fighting without a nuclear holocaust if war should break out.

Over twenty years ago when I became British Defence Secretary, NATO was committed to a 'tripwire' strategy under which the first significant movement of Soviet troops across the dividing line would trigger all-out nuclear war. My American colleague Robert McNamara and I sought to find a strategy which would be morally more acceptable and politically more credible.

We persuaded NATO to adopt a strategy of 'flexible response' under which if Western forces faced defeat in conventional fighting NATO would introduce nuclear weapons in discrete steps, giving the enemy the chance at each stage to stop fighting rather than invite escalation to the next rung on the ladder which led to all-out strategic nuclear war. At the time this was at least an advance on the tripwire strategy, which Germany was reluctant to abandon. Moreover the nuclear threshold had already been raised much higher by very substantial conventional forces of which by far the majority were provided by the European allies.

Over the last two decades, however, several factors have undermined the feasibility of flexible reaponse. We now know that the electromagnetic pulses emitted by nuclear explosions could make communications between the battlefield and the high command difficult if not impossible. There is no evidence that NATO governments have yet agreed any guidelines for the use of nuclear weapons under flexible response. For both these reasons it would be impossible for NATO to control escalation as flexible response requires.

Statements by leading Americans have cast doubt on the readiness of the United States government to authorise even the first use of nuclear weapons. Other American statements appear to contemplate a nuclear war which is limited to European soil. For Germany even the limited use of nuclear weapons would mean the nuclear holocaust. And since the Warsaw Pact now has parity with NATO at every level of nuclear warfare it cannot be assumed that it would pay NATO to initiate the use of nuclear weapons at any level.

It is not surprising that many of those generals who have had responsibility for planning to fight a war in Europe have become highly sceptical of existing NATO strategy. General Rogers, the Supreme Allied Commander in Europe, has made it clear that he is doubtful whether the NATO governments would ever authorise any use of nuclear weapons in a European war. He has complained that they appear to want him to take the fateful decision and he is rightly unwilling to assume a responsibility which must belong to governments. He has also recently stated that he does not think it would be possible to keep a nuclear war in Europe limited; on the contrary, he believes it would escalate very fast into a general nuclear exchange.

On the other hand he says he would be forced to use nuclear weapons in the first few days of a large-scale conventional war because NATO is so inferior to the Warsaw Pact in conventional forces.

This view is widely disputed. The International Institute for Strategic Studies says again in its latest annual survey of the military balance that Russia's conventional superiority is not sufficient to tempt her to risk an all-out conventional attack on Western Europe. New estimates by the CIA of Soviet defences are much less frightening than they were a few years ago. And NATO's new estimates of the ready forces on both sides are much more optimistic than they used to be.

However, to judge the relative capability of the opposing forces in an actual war it is necessary to carry out a far more sophisticated analysis than a simple count of weapons and manpower. Experts who have attempted this, notably Kaufman, Mearsheimer, Mako and Cordesman, all suggest that NATO forces could provide an effective defence against even an all-out Warsaw Pact attack if comparatively small and inexpensive changes were made in their organisation, equipment, deployment and strategy.

NATO itself has recently been discussing important changes in its conventional strategy. I believe, however, that the approach favoured by NATO officials of striking deep into Eastern Europe, perhaps with very expensive and sophisticated new weapons, is inappropriate, both because it would provoke a pre-emptive attack and because Soviet strategy and deployment are changing so as to provide fewer targets for such Western weapons to hit.

In any case, there is not much point in being able to hit targets 300 miles behind the front line if the Red Army can puncture the front line and then spread out widely in West Germany. Moreover, the AirLand battle strategy which the American forces in Germany have adopted unilaterally involves the use of nuclear and chemical weapons as well as conventional weapons, so it is quite inconsistent with a strategy of conventional deterrence or defence.

There is now a growing feeling among military experts that NATO must look in a different direction – towards a non-provocative strategy of conventional deterrence which could protect NATO territory without using nuclear weapons if deterrence should fail.

The first step would be one which already has wide support – from the Palme Commission, for example. All nuclear weapons should be withdrawn from a strip, say, 150 kilometres deep on both sides of the dividing line. NATO commanders have long been worried by the presence in the front line of nuclear weapons which are inconsistent with any attempt at conventional defence and might be overrun in the first hour or so of a conflict; they talk of the 'use or lose' dilemma.

Beyond this precondition for moving to a non-nuclear strategy lie three main fields for action to make better defensive use of NATO's conventional forces – reserves, barriers and equipment. Each of these would require some change in NATO's present tactical doctrines.

The most important would be to make better use of NATO's existing reserves of trained manpower. If NATO's European reserves were organised and equipped even as well as those of neutral countries like Sweden, Switzerland and Finland the European allies could double their present contributions on the Central Front. Andrew

Hamilton, a leading American analyst, calculates that Britain could in this way double its ground combat power in Central Europe for the cost of the Trident submarine programme alone.

Other analysts calculate that the defensive capability of NATO's existing forces could be increased some 40 per cent by the preparation of defensive positions in peacetime. I recently discussed with a Soviet General in Moscow General Rogers's proposal for laying pipes underground on West German territory which could be filled with an explosive slurry to create wide and deep tank traps in case of war. The Soviet General opposed it on the grounds that it would provide NATO forces with 'an inviolable sanctuary' – the best recommendation possible, I would have thought!

The third main area for action would be the exploitation of the new technologies to improve defensive weapons. This would be far better than to develop expensive new weapons which may not work for deep strike against targets which may not be there.

In the coming months we are likely to see a flood of proposals along these lines for providing NATO with an adequate non-provocative conventional strategy. They will come from men on both sides of the Atlantic with personal experience of warfare and deep knowledge of the problem.

No one like me who, after six years as a soldier in war and six years as Defence Secretary in peace, has had the opportunity to discuss these problems with Russians and neutrals as well as with our allies can fail to conclude that security in the nuclear age will depend on working with one's political opponents as well as with one's friends.

This obvious fact is almost universally accepted so far as disarmament and arms control are concerned. Every negotiation on arms control implies a readiness by each side to limit its own defence efforts for the sake of co-operation with its opponent.

I believe that the same insight should also be applied to defence itself. Indeed the Stockholm agreement on Confidence Building Measures involves mutual notification and observation of military manoeuvres.

Even the slightest acquaintance with the nuclear problem rams home the fact that each side is driven by fears which are the mirror image of the other's. Would it not be possible to exorcise those fears by openly exchanging knowledge of one another's defence preparations – which is largely available in any case by satellite photography and signals interception? Then defence policies could be adjusted to minimise unnecessary fears.

Let me end as I began with another quotation from the speech by Secretary Weinberger on 9 October 1985. He then revealed that there should be regular talks between the military leaders of each nation and regular meetings at the highest levels of the Departments of Defence and State with their Soviet counterparts. It seems an excellent idea. On his side Mr Gorbachev has proposed that NATO and the Warsaw Pact should have such contacts.

Surely this is an area where an honest exchange of views can do nothing but good. If at present neither Moscow nor Washington is prepared to respond to the other's invitation, Britain and Western Europe should take the lead in pressing for military discussions between the two alliances. Mutual confidence is the only basis for real security in the nuclear age.

As far as Mr Gorbachev's disarmament proposals of early 1986 are concerned at least two major obstacles exist on the Western side.

First, President Reagan's continuing conviction that his Star Wars programme can make nuclear weapons impotent and obsolete – a conviction which none of his own advisers share and which is rejected by his allies. Star Wars now blocks agreement with the 50 per cent cut in strategic nuclear arsenals on which Moscow and Washington have both agreed in principle.

Second, the determination of the British and French governments to increase up to ten times the destructive power of their existing nuclear forces.

On Star Wars, recent Soviet statements offer the prospect of both sides continuing research providing they agree to ban testing and deployment of space-based weapons or their components. Surely Moscow and Washington could now agree on this, which is in any case compulsory under the ABM Treaty they have both ratified!

On Intermediate Nuclear Forces NATO agreed not to deploy Cruise and Pershing missiles if the Russians agreed to arms control.

Now that the Russians have gone further and accepted President Reagan's zero option some European governments are saying they want to keep Cruise and Pershing missiles even if the Soviet Union dismantle all their SS-20s aimed at Europe. Yet the only conceivable reason for keeping these vulnerable American missiles on European soil would be the belief that America would be more ready to use them than its strategic forces because it might then hope to keep a nuclear war limited to the European side of the Atlantic. It would in fact decouple America from the defence of NATO. It is obvious that the British government is now using these delaying tactics in the hope of talking the new Soviet offer to death. We must not let this happen.

Yet the press tells us that some European governments, including the British, want to torpedo an agreement on INF by insisting that America keeps Cruise and Pershing missiles on their soil, whatever Russia may do. Apparently they believe this is the only way of 'coupling' America's strategic forces to their defence. Exactly the opposite is the case. The only conceivable reason for putting these vulnerable missiles on European soil would be the belief that Washington might be more prepared to authorise their use in war because it could then keep the United States a sanctuary and limit the nuclear holocaust to Europe. We must now speak up for the survival of our continent and get all those missiles out of Europe. The present British government can only gain from Mr Gorbachev's agreement to let it keep Polaris until a later stage in the disarmament negotiations, provided we cancel the Trident programme. Meanwhile the Labour Party is confident that in office it could secure a Soviet commitment to decomission equivalent Russian nuclear forces when it fulfils its long-standing commitment to decomission the Polaris fleet for good.

House of Commons, London   Rt Hon. DENIS HEALEY
C.H. M.B.E. M.P.

Notes

1. This foreword is based on an abridged version of Mr Healey's Fabian Autumn lecture delivered on 26 November 1985. The long text has also been published as Fabian Tract 510, 'Beyond Nuclear Deterrence' (London: Fabian Society, March 1986).

Preface

by Raymond L. Garthoff

As weapons extend beyond this world – literally as well as figuratively – old problems assume a new aspect, and new issues arise. Especially significant is the new interest by the United States in strategic ballistic missile defence (BMD) utilising space-based systems or components. This re-opens a dimension of the strategic situation quiescent since the Outer Space Treaty of 1967 and the Anti-ballistic Missile (ABM) Treaty of 1972. The implications for Europe are multifold. Decisions on strategic defence weapons programmes of both the United States and the Soviet Union have potentially direct, and inescapably indirect, impacts on European security – in terms of the Soviet threat, the American deterrent, alliance solidarity, alliance strategy, prospects for arms control, and the future of East–West détente. Among the more specific aspects of this range of consideration is the area of theatre (or tactical) BMD, inasmuch as it relates directly to possible defence of Europe itself. The implications for Europe of space-based and other potential strategic BMDs of the United States and the Soviet Union are, however, no less significant.

What is new and of special significance about Star Wars and BMD at the present time? We face a conjunction of strategic, technological and political developments that create a new context as well as introducing new content. The establishment of a rough strategic parity or balance between the Soviet Union and the United States – while long anticipated and reflected in the first strategic arms limitation talks and the strategic arms limitation treaty (SALT) agreements of 1972 and 1979 – has raised questions about the efficacy of American deterrence that have only gradually come to the surface. In my judgement this issue is more grounded in Western concerns than in Soviet perceptions, and thus poses more a question for reassurance of the members of the alliance than a problem for deterrence of Soviet pressure or attack. Nonetheless, the state of the United States–Soviet strategic relationship is one source of new doubts in the West as to the continued effectiveness of the American deterrent extended to protect Western Europe.

The technological development of greatest moment is the widespread perception that scientific research has now opened new prospects for possible advanced weaponry far more effective than was envisaged when strategic defence was last seriously considered in the late 1960s. Moreover, the United States appears to be well in the lead in developing the range of scientific–engineering abilities required to make such systems attainable. Many thus now see technology as a possible means to alter the strategic balance to the advantage of Western deterrence and defence.

Strategic uncertainties and technological promise propel interest in strategic defence to the fore just as political confidence between East and West is at a new nadir. The modicum of wary confidence between the Soviet Union and the Western powers – in particular in the belief of a shared interest in arms control to reduce the risks of war and the costs of an arms race – that flourished in the climate of détente in the early and mid-1970s is now almost entirely gone. Political tensions have again become predominant, particularly between the United States and the Soviet Union, but also in general East–West relations. There is (again especially in influential circles in the United States) a strong scepticism as to the feasibility and even desirability of arms control – and of détente.

The European-centred East–West détente that began in the late 1970s established deeper roots than the more short-lived and spectacular American–Soviet détente of the 1970s. It has therefore survived the collapse of the latter in the aftermath of Afghanistan. While Western European (and American) views are by no means uniform, and a fundamental community of interests remains, there is a general divide in assessments of the threat and the appropriate policy response. The more confrontational stance of the Reagan administration (particularly evident from 1981 through 1983), compounded by its efforts to line up European support for economic sanctions, led many Europeans to distance themselves from its embrace. Yet there has remained a strong desire on both sides of the Atlantic to maintain the alliance.

The strongest demonstration of alliance solidarity was its successful effort to implement the 1979 decision on deployment of American intermediate-range nuclear forces (INF) in Europe. Whether in fact the INF deployment has enhanced the security of the alliance is debatable, but at least the alliance persisted in implementing its deployment decision despite strenuous Soviet efforts to derail it. Moreover, the NATO INF deployment has been seen as reinforcing the coupling of shorter-range tactical nuclear weapons to American strategic forces, and thus denying the Soviet Union a feared advantage for political pressure through a monopoly of intermediate-range missiles in Europe. This rationale, strong in the minds of the European proponents of the original INF decision, reflected above all a fear of weakened credibility of the American deterrent under strategic parity. The Carter Administration, in turn, agreed to the deployment – and indeed championed it – not because it was seen as necessary to reinforce deterrence of Soviet pressure, but to reassure the European allies (and to make amends for the mishandled 'neutron bomb' affair). The original rationale continued to be the mainstay of European support for the INF, although it was obscured by the adoption by the Reagan Administration of an initially non-negotiable but politically effective position of seeking a 'zero option' that, even if accepted, would have left longer- and shorter-range Soviet missiles in Europe and no countervailing American 'coupling' force. Nor did many seem to notice that the Soviets, rather than attempting to use their SS-20s to decouple Europe from the United States, consistently and loudly insisted that no limited nuclear war in Europe was possible. The *Soviets* thus insisted on 'coupling' any NATO use of nuclear weapons with strategic war with the United States.

The unexpected 'Star Wars' speech by President Reagan in March 1983, and subsequent American elaboration of a massive investment in a new Strategic Defence Initiative (SDI), as it came to be called, set the central element in a new agenda for East–West relations, and to a lesser extent for the Western alliance as well, following the INF deployment struggle.

There had been no alliance consultation or even notification before the Star Wars initiative. This was in stunning contrast to the joint alliance study and decision on the INF, and to the pattern of consultation long established in the SALT negotiations. When in 1967 the United States had first proposed SALT to the Soviet Union, and also began contingent preparations for a possible decision to deploy anti-ballistic missiles (ABM – the then-current term for ballistic missile defence, BMD), it had engaged in detailed consultations. One consequence was the initiation in April 1967 of a NATO study of the whole ABM issue, including the pros and cons of an ABM defence for Europe. This study – conducted under the auspices of the new Nuclear Planning Group (NPG) – concluded (in April 1968) that the reasons leading the United States in 1967 to decide to deploy a light ABM defence against China were not relevant to a European defence, and that ABM deployment in Europe was not warranted.

Even more important, perhaps, than this conclusion was the very fact of a joint study – one with active (but not dominating) American participation, permitting the European NATO members to reach their own conclusions. Subsequent alliance support for decisions by the United States on ABM deployment, and an ABM curtailment in SALT, were greatly aided by this active consultation and joint study.[1]

The question of the strategic relationship between the United States and the Soviet Union has been the subject of extensive consultation in the NPG and the North Atlantic Council in Brussels. The strategic relationship, key military technological developments, strategic arms control, and ballistic missile defence were well established elements of a long-standing transatlantic dialogue from the beginning of SALT and ABM to Star Wars. Hence the shock of President Reagan's unilateral initiative. (The fact that, as soon became apparent, the subject had not even been discussed with most interested parties *within* the United States government may have mollified some European feelings about being left out, but it did not increase confidence in the decision itself.)

President Reagan's Star Wars speech did include explicit, if enigmatic, reference to protection of Europe as well as the United States, speaking of defence of 'our own soil or that of our allies'.[2] But he did not explain how such a defence could be established or what it would mean, especially given the large and varied arsenal of Soviet weapons within range of Western Europe other than 'strategic ballistic missiles'. (The statement was reportedly a last-minute insertion at the urging of the Department of State, when it was belatedly informed of the impending initiative.) The even greater improbability of the feasibility of such defence of Europe reinforced the impression that reference to the 'soil of allies' was a cosmetic political addition. The President also promised 'closer consultation with our allies' even as he demonstrated that his initiative would not be subject to review.[3]

The Star Wars initiative is challenging to the entire bedrock of alliance strategy: deterrence through threat of retaliation. If the constraints of the ABM Treaty were removed – from the Soviet Union, of course, as well as the United States – the prospect of strategic BMD of the two superpowers would raise disquieting questions as to the continued effectiveness of extended deterrence. Whatever contribution the INF deployment had made to coupling was more than overshadowed by the prospect of Soviet and American protection and a concomitant *decoupling* of Western Europe. While arguably the protection of BMD could make the United States more ready to risk an escalation to limited attacks on the Soviet Union, that possibility was more than outweighed by the fact that BMD would provide protection to the Soviet Union, negating the deterrent value of the prospect of such attacks. Moreover, the President had raised the prospect of *substituting* strategic BMD for nuclear retaliatory forces, referring to reliance on 'the specter of retaliation, of mutual threat' as a 'sad commentary on the human condition'.[4] 'Wouldn't it be better to save lives rather than to avenge them?' And he set as the goal of his initiative 'to give us the means of rendering these nuclear weapons impotent and obsolete ... to achieve our ultimate goal of eliminating the threat posed by strategic nuclear missiles'. His aim of replacing interactive deterrence by unilateral protection was underscored by his admission that 'defensive systems ... if paired with offensive systems ... can be viewed as fostering an aggressive policy, and nobody wants that.'[5]

The effective protection of a nation against nuclear attack had never been a serious prospect (and, in the view of most scientists, is not today). 'Damage limitation' has always been a secondary aim to deterrence, and as the Soviet Union acquired strategic parity with the United States by the 1970s mutual deterrence was born of mutual vulnerability. A resurrected active pursuit of strategic defence, even apart from its possible attainment, thus challenges the prevailing strategic order.

During the late 1960s and early 1970s this reality of mutual vulnerability and mutual deterrence came to be recognised in Moscow and led to important modifications in Soviet military doctrine, arms control, and broader policy objectives. Soviet acceptance (and, indeed, advocacy) of severe constraints on BMD development and deployment as embodied in the ABM Treaty in fact reflected this changed outlook. Contrary to widely-held assumptions in the West, the Soviet interest was (and is) in mutual deterrence, and not merely in a short-term neutralisation of American disadvantage in ABM development.[6]

The Star Wars initiative thus challenged a common basis for seeking stability through arms control established since the start of SALT fifteen years earlier, and has had a far more serious impact in shaking Soviet confidence in American interest in arms control than is generally realised. While this fact is not sufficiently appreciated in the West, the more concrete challenge of the SDI to the existing arms control regime is.

As the discussion in this volume shows, many Europeans were thus alarmed not only at the prospect of a strategically decoupling development, but also by a politically decoupling unilateral initiative by the United States. The SDI appeared to reflect an American desire to retreat from the complex realities of politics, strategy, and arms control in the nuclear age. The United States seemed to display an interest in drawing back into a 'Fortress America' rather than engaging with its allies – and its adversary.

Not only would the prospects for new steps in arms control be burdened (if not, indeed, hopelessly prejudiced) by such a move, but the principal existing achievement of arms control – the ABM Treaty – is not compatible with full testing (to say nothing of deployment) of comprehensive BMDs. The prompt hostile Soviet reaction raised further concerns. In short, the initiative itself, its motivations, its pursuit, and its consequences have been recognised to have had a highly negative impact on East–West Relations.

Nonetheless, the United States has persisted with the SDI, and has sought European support. It sought support for–and gained grudging acquiescence in – a strategic arms control negotiating position that was predicated on maintaining its strategic defence programme – and other space weapons programmes – unconstrained. That this approach was manifestly non-negotiable is widely recognised in Europe, but this was glossed over and the matter not permitted to become an issue with the United States. An American campaign to stress Soviet research in the field, while perhaps effective in countering Soviet efforts to constrain research, was irrelevant to the main issue: while concern over Soviet BMDs justified hedging by comparable research, the only way to head off Soviet BMD deployment was by maintaining and strengthening arms control, not by ending it in pursuit of a programme that required abandoning even the existing arms control restraints.

The United States also sought to elicit European support by dangling the prospect of technological sharing. By enlisting European governments and industry in aspects of the SDI programme, a wider constituency of political support could be built (as was done within the United States).

The general European reaction has been to try to maintain an alliance consensus by emphasising those aspects of the United States approach that could be supported – research, and inclusion in arms control – while remaining publicly silent (and privately very cautious in any criticism) of the basic initiative itself and its most objectionable features, including the prospect of stalemating arms control and further burdening East–West relations.

Two specific aspects of the SDI of direct European concern remain. One is the threat that Soviet BMD, unleashed if an open competition in strategic defence displaces the ABM Treaty, will negate the British and French nuclear deterrents. The second is the question of what the Europeans want to do (and can do) about ballistic missile defence of Europe itself.

Theatre or tactical defence, and the relationship of tactical BMD to strategic missile defence, raise many of the same problems as strategic missile defence, but with some important differences. For one thing, the question of relationship to air defence and cruise missile defence is heightened. For another, the question of 'decoupling' is not posed, and could even be allayed, by an effective theatre defence. Finally, the ABM Treaty covers strategic BMD, and the area of tactical BMD is not directly limited. In practice, this means that some tactical BMD is allowed, but that there is also an ill-defined grey area overlapping constrained strategic BMD. Moreover, while BMDs in Europe might be 'tactical' for the United States, they might also be 'strategic' for the Europeans, and in any case tactical BMDs in the USSR might also have strategic defence capabilities. The question of tactical theatre defence is thus a complex one requiring wide-ranging consideration.

BMD – from the ABM debates of the 1960s and ABM Treaty of 1972 to the Star Wars SDI of the 1980s – has been a persistent central element in the East–West strategic picture, in arms control, and ultimately in political relations. It has also, both for that and for other reasons, been a perennial theme in the translatlantic dialogue and alliance relationship. Yet while the implications of the subject for Europe are widely recognised to be major, there has been no comprehensive study. The present work seeks to fill that gap.

RAYMOND L. GARTHOFF
The Brookings Institution,
Washington, D.C.

Notes

1. See Raymond L. Garthoff, 'BMD and East–West Relations', in Ashton B. Carter and David N. Schwartz (eds), *Ballistic Missile Defense*

(Washington, DC: Brookings Institution, 1984) pp. 280–5; and see Paul Buteux, *The Politics of Nuclear Consultation in NATO, 1965–1980* (London: Cambridge University Press, 1983). I participated in the NATO SALT consultations of 1967 (and subsequently through 1972), and in the 1967–68 NPG study of ABM.

2. President Ronald Reagan, 'National Security: Address to the Nation, March 23, 1983', *Weekly Compilation of Presidential Documents* (hereafter *Presidential Documents*) 19, 28 March 1983 (Washington, DC: United States Government Printing Office) p. 447.
3. Secretary of Defence Caspar Weinberger had met with the other NATO ministers of defence in a session of the NPG ending the very day of the President's speech – but he had not been in a position even to prepare the way for it or advise them of the initiative.
4. Reagan, *Presidential Documents* 19 (1983), p. 447.
5. Reagan, *Presidential Documents*, p. 448.
6. Garthoff, *Ballistic Missile Defense*, pp. 286–314; 323–8.

Acknowledgements

The concept of this book grew out of a panel called 'From Star Wars to the Strategic Defence Initiative: European Perceptions and Assessments' the editor had organised and chaired at the 26th Annual Convention of the International Studies Association on 8 March 1985 in the Washington Hilton & Towers, Washington, DC with Paul Stares, Alain Carton and Hans Günter Brauch as the European speakers and Walter B. Slocombe and John Pike as the American discussants.

The Berghof Foundation for Conflict and Peace Research has provided the financial support for the research project, 'Destabilising strategic weapons technologies – Their implications on deterrence, arms control and on the policy for peace and security in Europe', which has been conducted by the editor at the Institute for Political Science at Stuttgart University. The Berghof Foundation, the German Research Association and the International Studies Association have provided funding for essential travel.

The editor appreciates the permission of Lexington Books, D. C. Heath and Co to use three illustrations in chapter 2 from the book by Gary L. Guertner and Donald M. Snow, *The Last Frontier: An Analysis of the Strategic Defense Initiative* (Lexington, Mass., 1986). Chapter 15 is an updated and footnoted version of an article previously published in a report for the National Campaign to Save the ABM Treaty by Thomas K. Longstreth, John E. Pike and John B. Rhinelander, *The Impact of US and Soviet Ballistic Missile Defense Programs on the ABM Treaty*, Washington, DC, March 1985. The permission to use material from chapters VIII, IX and XI of this study is appreciated.

The editor is grateful for both the encouragement and the many suggestions from the members of the Study Group on Peace Research and European Security Policy (AFES) (Institute of Political Science, Stuttgart University), who provided an intellectual framework for the discussion of space-related issues during three conferences the editor organised for students and for youth officers of the Bundeswehr in February and July 1985 and in February 1986 with the financial support of the Landeszentrale für Politische Bildung Baden-Württemberg.

Last but not least the editor is grateful to Mr Delbert Andrews and his wife Helga Andrews for their assistance in language editing. They have developed together a true German–American partnership. Herbert Beck, a graduate of the University of Zürich and of the LSE, helped with proofreading. Keith Povey, an editorial services consultant, with the vital assistance of Mrs Barbara Docherty, spent many long days and evenings at his farm in Devon polishing the language, especially that of the non-native English speakers, and checking for errors. Mr Povey did the final proofreading, and Ms Sue Ramsay prepared the index. Many others, not mentioned individually, in and out of the US and of European governments, have discussed the problems related with SDI with the editor and have provided useful insights. Their kind assistance is appreciated. However, the sole responsibility for any remaining errors rests with the editor. The cutoff date for most chapters is early 1986. Only the major subsequent developments were incorporated in the proofs in November 1986.

Last but not least, the editor expresses his gratitude to Mr Tim M. Farmiloe and to Mr Simon Winder of The Macmillan Press in London for the smooth bridge-building across the Channel and for the pleasant co-operation, as well as to Kermit Hummel of St Martin's Press in New York for co-publishing this book in the United States. Acorn Bookwork in Salisbury, England, and the printers in Hong Kong helped to transform an idea into reality within a few months' time.

Stuttgart University   HANS GÜNTER BRAUCH

List of Abbreviations

ABM	anti-ballistic missile
ACDA	United States Arms Control and Disarmament Agency
ACIS	Arms Control Impact Statement
AEG	Allgemeine Elektricitäts-Gesellschaft
AF	US Air Force
AFES	AG Friedensforschung und Europäische Sicherheitspolitik
AGALEV	Anders Gaan Leven (Belgian Ecologist Party)
AGARD	Advisory Group for Aerospace Research and Development
AIM-9L	American Air Defence System
ALCM	air-launched Cruise missile
AMRAAM	advanced medium-range air-to-air missile
AOA	airborne optical adjunct
AOS	airborne optical system
APVO (Strany)	Soviet Aviation Units
ARIANE	name of a European rocket for launching satellites
ASAT	anti-satellite
ASBM	air-to-surface ballistic missile
ASEAN	Association of South-East Asian Nations
ASRAAM	advanced short-range air-to-air missile
ASW	anti-submarine warfare
ATAM	anti-tactical aerodynamic missile
ATBM	Anti-tactical ballistic missile
AT(B)M	Anti-tactical (ballistic) missile
ATM	Anti-tactical missile
AWACS	airborne warning and control system
AWS	advanced warning system
BDI	Bundesverband der Deutschen Industrie
BENELUX	Belgium, Netherlands, Luxembourg
BMD	ballistic missile defence
BMEWS	ballistic missile early warning system
BRITE	Basic Research in Industrial Technology for Europe

List of Abbreviations

BSTS	boost surveillance and tracking system
C^3	command, control and communication
C^3I	command, control, communications and intelligence
CD	Conference on Disarmament
CDA	Christen Demokratisch Appel (Dutch Christian Democratic Party)
CDI	conventional defence improvement
CDU	West German Christian Democratic Party (Christlich Demokratische Union)
CEP	circular error probability
CERN	European Organisation for Nuclear Research
CESTA	Centre d'Etudes des Systèmes et de Technologies Avancées
CIA	Central Intelligence Agency
CITES	Consortium of Italian Aerospace and Electronic Firms
CMF	conceptual military framework
CND	Campaign for Nuclear Disarmament
CNES	Central National d'Etudes Spatiales
COCOM	Consultative Group Coordinating Committee (for multilateral export controls for trade with socialist countries)
CONUS	Continental United States
CPN	Communist Party of the Netherlands
CPSU	Communist Party of the Soviet Union
CSU	West German Conservative party in Bavaria (Christlich Soziale Union)
CVP	Flemish Christian Democratic Party in Belgium
D1	German Space shuttle mission (Deutschland 1)
DARPA	Defense Advanced Research Projects Agency
DDR&E	Director of Defense Research and Engineering
DESY	Deutsches Elektronen Synchroton in Hamburg
DFG	Deutsche Forschungsgemeinschaft
DGB	Deutscher Gewerkschaftsbund (German Trade Union Congress)

List of Abbreviations

DHIT	Deutscher Industrie- und Handelstag (German Industrial and Trade Council)
DIA	Defense Intelligence Agency – a US intelligence service
DoD	US Department of Defense
DOT	designating optical tracker
DP	Demokratisch Partei – a Dutch liberal party
DSAT	defensive satellite
EAD	extended air defence
EADI	European Aerospace Defence Initiative
EC	European Communities
ECU	European Currency Unit
EDA	European Defence Architecture
EDI	European Defence Initiative
EDIG	European Defence Industrial Group
EEC	European Economic Community
EFTA	European Free Trade Association
EIB	European Investment Bank
EML	electromagnetic rail gun
EMP	electro magnetic impulse
ENDO-NNK	endoatmospheric non nuclear kill technology
EORSAT	electronic ocean reconnaissance satellite
EPC	European Political Co-operation
ER	Enhanced radiation
ERHIT	extended range version of SRHIT (see: FLAG)
ERIS	Exoatmospheric re-entry vehicle interception system
ESA	European Space Agency
ESM	electronic support mission
ESPRIT	European Strategic Programme for Research and Development in Information Technologies
ET	emergent technologies
EURATOM	European Atomic Agency
EURECA	(a) ESA Research Programme; (b) European Research Co-Ordination Agency – later EUREKA
EUREKA	European high technology programme
EURO BMD	European ballistic missile defence

EUROBIOT	European Research Project on Biological Technologies
EUROBOT	European Research Project on Robotics
EURO COM	European Research Project on Communication Technology
EUROMAT	European Research Project on Material Science
EUROPA	European observation and civil technology rocket project.
EVI	Europäische Verteidigungsinitiative
EXO-NNK	exoatmospheric non-nuclear kill technology
FAAD	forward area air defence
FABMD	field artillery ballistic missile defence
FBIS	Foreign Broadcast Information System
FCO	Foreign and Commonwealth Office
FDF	Front Démocratique des Francophones (a Belgian political party)
FDP	Freie Demokratische Partei (West German Liberal Party)
FEBA	forward edge of the battle area
FLAGE	flexible lightweight agile guided experiment
FOBS	fractional orbital bombardment system
FOFA	follow-on-forces attack (Rogers Plan)
FPS	an American radar system
FRG	Federal Republic of Germany
FSSS	Future Security Strategy Study of October 1983
FY	fiscal year
GAC	General Advisory Committee of the US Arms Control and Disarmament Agency
GAP	Greng-Alternative Partei – a Dutch ecologist party
GBL	ground-based laser
GLCM	ground-launched Cruise missile
GNP	gross national product
GRIP	Groupe de Recherche et d'information sur la Paix (Belgium)
HEDI	high endoatmospheric defence interceptor
HEL	high-energy laser
HERMES	French project for a manned space plane
HOE	homing overlay experiment

List of Abbreviations

IBC	integrated communication networks
ICBM	intercontinental ballistic missile
IEDSS	Institute for European Defence and Strategic Studies
IEPG	Independent European Programme Group
IFF	friend or foe identification system
IFRI	Institut Français des Relations Internationales
IISS	International Institute for Strategic Studies
IKV	Interkerkeligk Vredesberaad (Dutch Inter-Church Peace Council)
INF	intermediate-range nuclear forces
IR	infra-red
IRBM	intermediate-range ballistic missile
ISMA	International Satellite Monitoring Agency
ITT	International Telephone and Telegraph Corporation
JCS	Joint Chiefs of Staff (in USA)
JET	Joint European Torus
JTACMS	joint tactical missile system
JSTARS	joint surveillance and target attack radar system
KH-12	Keyhole-12 (United States military reconnaissance satellite)
KP/CP (also KP/PC)	Communist Party in Belgium
LANDSAT	land satellite (earth resources technology satellite)
LEDI	low endo-atmospheric defence interceptor
LEP	long endurance platform
LNO	limited nuclear options
LoADS	low altitude defence system
LODE	large optics demonstration experiment
LPAR	large phased-array radar
LRACM	long-range air-delivered Cruise missiles
LRINF	long-range intermediate nuclear forces
LRTNF	long-range theatre nuclear forces
LSAP	Social Democratic Party, Luxembourg
LWIR	long-wave infrared
MAD	mutual assured destruction
MARV (also MaRV)	Manoeuvring re-entry vehicle
MAS	mutual assured survival

List of Abbreviations

MBB	Messerschmitt-Bölkow-Blohm
MIRV	multiple independently-targetable re-entry vehicle
MITI	Ministry of International Trade and Industry (Japan)
MLRS	multiple-launch rocket system
MOBS	multiple orbital bombardment system
MOL	manned orbital laboratory
MOU	Memorandum of Understanding
MP	Member of Parliament
MPR	military photo-reconnaissance
MPRSAT	military photo-reconnaissance satellite
MRACM	medium-range air-delivered Cruise missile
MRBM	medium-range ballistic missile
MRINF	medium-range intermediate nuclear forces
MRTNF	medium-range theatre nuclear forces
MT	megaton
MX	missile experimental (new American ICBM)
NAA	North Atlantic Assembly
NADC	NATO Aerospace Defence Command
NASA	National Aeronautics and Space Administration
NATO	North Atlantic Treaty Organisation
NAVSAT	navigation satellite
NAVSTAR	navigation satellite timing and ranging
NEC	new strategic concept
NNK	Non-nuclear kill
NPG	Nuclear Planning Group (of NATO)
NSDD	National Security Decision Directive
NSC	National Security Council
OAMP	optical aircraft measurement programme
OCA	offensive counterair
OCV	Overlegcentrum voor de Vrede (a Belgium peace organisation)
OSD	Office of the Secretary of Defense (in the USA)
OTA	Office of Technology Assessment (United States Congress)
PAR	perimeter acquisition radar
PARCS	perimeter acquisition radar attack characterisation system
PBV	post-boost vehicle

List of Abbreviations

PBW	particle beam weapon
PGM	precision guided munitions
PKO Strany	organisation in the Soviet military responsible for anti-satellite systems
PRC	People's Republic of China
PRL	Parti Réformateur Libéral – French speaking Belgian liberal party
PRO Strany	organisation in the Soviet military responsible for anti-ballistic missiles
PS	Parti Socialiste (French Socialist Party)
PVDA	Partij van de Arbeid (Dutch Social Democratic Party)
PVO Strany	national air defence forces of the homeland (Soviet Union)
RACE	Research and Development in Advanced Communication Technology for Europe
R&D	research and development
RDTE	research, development, testing and engineering
RF	radio frequency
RIIA	Royal Institute of International Affairs (Chatham House in London)
RORSAT	radar ocean reconnaissance satellite
RPR	Rassemblement pour la République (French Gaullist Party)
RPV	remotely piloted vehicle
RUSI	Royal United Services Institute
RV	re-entry vehicle
RVSN	Soviet Strategic Rocket Forces
SACEUR	Supreme Allied Commander Europe
SALT	Strategic Arms Limitation Treaty
SAM	surface-to-air missile
SAMRO	Satellite Militaire de Reconnaissance Optique
SBAMS	space-based anti-missile system
SCC	Standing Consultative Commission
SDA	Strategic Defence Architecture
SDI	Strategic Defence Initiative
SDIO	Strategic Defence Initiative Organisation
SDP	Social Democratic Party (Great Britain)
SEL	Standard Elektrik Lorenz (A former German subsidiary of ITT)
SHAPE	Supreme Headquarters Allied Powers Europe

List of Abbreviations

SICBM	small intercontinental ballistic missile
SIPRI	Stockholm International Peace Research Institute
SLBM	submarine-launched ballistic missile
SLCM	sea-launched Cruise missile
SME	Serpent Monétaire Européen
SNF	short-range nuclear forces
SNLE	Sous Marin Nucleaire Lanceur d'Engin
SP	Flemish Socialist Party
SPAS	Name of an MBB satellite
SPD	Sozialdemokratische Partei Deutschlands (West German Social Democratic Party)
SPOT	name of a French satellite for remote sensing
SRBM	short-range ballistic missile
SRHIT	short-range homing interceptor technology
SRINF	short-range intermediate nuclear forces
SRTNF	short-range theatre nuclear forces
SS	surface-to-surface ballistic missile
SSBN	ballistic missile submarine, nuclear-powered
SSN	submarine, nuclear-powered
SSTS	space surveillance and tracking system
START	Strategic Arms Reduction Talks
SYRACUSE	Système de Radiocommunication Utilisant un Satellite
TABM	tactical anti-ballistic missile
TACM	tactical missile
TBM	tactical ballistic missile
TDI	Tactical Defence Initiative
TGE	Technology, Growth and Employment Committee
TIR	terminal imaging radar
TNF	Theatre Nuclear Forces
UDF	Union pour la Démocratie Française (French Liberal Party)
UK	United Kingdom
USA	United States of America
USCENTCOM	United States Central Command
USIS	United States Information Service
USSR	Union of Soviet Socialist Republics
VAKA	Vlaams Aktiecomité tegen Atoomwapens (Belgium peace group)

VPKO	Soviet military command
VPRO	Soviet military command
VPVO	Soviet military command
VVD	Volkspartij voor Vrijheid en Demokratie (Dutch Liberal Party)
WEU	Western European Union
WP	Warsaw Pact
WRK	Westdeutsche Rektorenkonferenz (Conference of West German University Presidents)
WTO	Warsaw Treaty Organisation
ZMT	Zenith Missile Troops

Introduction
Hans Günter Brauch

Star Wars and European Defence is the outcome of a truly transatlantic effort, both to *analyse the European perceptions* of President Reagan's vision and the SDI research programme and to assess *the implications of SDI* for the security policy of the two European nuclear states, Great Britain and France, and for European security in general. By drawing together a distinguished group of American and of younger European authors, this book covers the SDI debate in Europe until December 1985 and in West Germany until the end of February 1986. As a representative sample of the fourteen European NATO countries, the reaction in six nations will be analysed in detail – in the two nuclear weapon states (Great Britain and France) and in the four non-nuclear NATO countries (the Federal Republic of Germany, Belgium, the Netherlands and Luxembourg). A much broader survey of the debate within the defence committees of the sixteen NATO countries is offered in chapter 8, which focuses both on transatlantic consultations on the SDI and on the debate within three transnational parliaments – the North Atlantic Assembly, the Western European Union and the European Parliament.

President Reagan's 'Star Wars' speech of 23 March 1983 was initially interpreted by the European NATO allies as a purely tactical move – primary aimed at disarming the anti-nuclear rhetoric of the American Catholic bishops, who were to meet in April 1983. In the intermediate-range nuclear forces (INF) deployment countries – Great Britain, West Germany, Belgium and the Netherlands – President Reagan's anti-nuclear undertones were perceived with dismay and concern. However, both the governments concerned and the European anti-nuclear peace movement were too concerned with the deployment of Pershing II and Cruise missiles to pay much attention to the proclaimed long-term shift in the American military strategy from 'offence dominance' to 'defence dominance'.

After NATO's Nuclear Planning Group meeting at Cesme in Turkey (April 1984) where the European Defence Ministers received the first extensive briefing on theSDI programme from their American counterpart, Mr Weinberger, the second major security debate of the 1980s after INF) gradually emerged about the role and the implications of strategic and tactical ballistic missile defence (BMD) systems for European security, for the future of East–West relations and for transatlantic and West European security policy. During 1985, SDI (and increasingly a European Defence Initiative, EDI) were the primary divisive security issues within European cabinets, parliaments, and affecting bilateral and multilateral relations. It is disputable whether SDI was instrumental in forcing the Soviet Union back to the arms control negotiations in Geneva in March 1985 or in the realisation of the first summit between President Reagan and the newly-appointed Secretary General of the CPSU, Michael Gorbachev. Nevertheless, SDI left NATO Europe even more divided than the INF dispute, with many prominent INF supporters joining the SDI critics, most noticably French President Mitterrand and the former West German Chancellor, Helmut Schmidt.

As of February 1986, SDI has left Western Europe divided into three political camps – the opponents, the reluctant supporters and those who are still undecided. At least five European NATO countries (France, Norway, Denmark, Netherlands, Greece) and Canada have rejected any official participation in the SDI research programme, while three nations (Great Britain, the Federal Republic of Germany and Italy) have moved towards a policy of conditional support, on conditions which might not be acceptable to the United States Department of Defense and could thus produce major bilateral disputes in the future. The remaining six NATO countries have either little to expect in terms of SDI contracts (e.g., Turkey, Portugal, Luxembourg and Iceland) or they are focusing their energies on other major security debates (e.g., in Spain, the referendum on its continued NATO membership and in Belgium, the deployment of Cruise missiles).

The twelve contributors to this book have varying professional backgrounds: of the seven American authors two were directly involved in the negotiation of SALT I and the ABM Treaty: Raymond L. Garthoff was a senior advisor to the United States SALT Delegation and John B. Rhinelander was its legal advisor. Peter Wilson was a member of the Policy Planning Staff of the Department of State during the Carter Administration, working on nuclear arms control and European security issues, and Colonel Simon Peter Worden was the representative of the Strategic Defence Initiative Organisation in the United States Delegation to the Negotiations on Nuclear and Space Arms with the Soviet Union. Thomas K. Longstreth is presently foreign policy advisor to Senator Edward

Kennedy and John E. Pike is one of the most quoted American critics of the SDI. As the Associate director for Space Policy of the Federation of American Scientists, Mr Pike initiated the Space Policy Working Group in Washington. Gary L. Guertner is the only purely academic contributor. He taught at the United States Army War College and was a scholar-in-residence at the United States Arms Control and Disarmament Agency before returning to the California State University in Fullerton.

Of the five European contributors (ten chapters altogether), one, Alain Carton was a Parliamentary Assistant specialising in defence matters in the French National Assembly, two are working as research associates (Paul Stares at the Brookings Institution in Washington and Robert J. Berloznik at the Centre for Polemology in the Free University of Brussels). Trevor Taylor and Hans Günter Brauch are teaching international relations at Technical Universities, Taylor in Stoke-on-Trent in Great Britain and Brauch in Stuttgart in the Federal Republic of Germany. The first three European authors, Stares, Carton and Berloznik, are still in their late 20s and early 30s. This volume brings together arms control experts with long experience both in government and academia and new talents.

The book is organised into four parts: the first deals with the global strategic balance, with the American and the Soviet BMD programmes and with the fundamental relationships between offence, defence and arms control. The second part offers detailed analyses of the perceptions and of the reactions to the SDI in Great Britain, France, the Federal Republic of Germany and in the Benelux countries as well as analyses of the transnational parliamentary debates and the joint European reaction to the technological challenge posed by the SDI–EUREKA. Part III analyses the implications of SDI on the national nuclear forces of the United Kingdom and France and on geopolitical aspects in Europe and Asia from a British, French and American perspective. Part IV focuses on military conceptual thinking and hardware planning for an EDI, on the ABM Treaty, its evolution, interpretation, grey areas, the compliance issue and the rationale behind the schemes to 'close the stated window of vulnerability' by a shift towards a 'defence dominant' world.

In chapter 1, Colonel Worden, who holds a Ph.D. in astronomy, sets the stage for a discussion of the global and the regional or tactical BMDs from the vantage point of the American administration. In chapter 2 Gary Guertner concentrates on the relationship between nuclear forces, strategic defence and arms control. In his conclusions,

Guertner warns against the technical follies of the unilateralists who 'advocate deployment of strategic defences even in the absence of arms control', and he pleads instead for close co-operation of science, arms control and military strategy to avoid 'SDI risks becoming America's technological Vietnam'.

Chapter 3, by Hans Günter Brauch, provides a detailed overview of the shifting official and unofficial interpretations of Soviet BMD and military space activities prior to and after President Reagan's speech of 23 March 1983. Brauch, in analysing the gradual shift in the selling of BMDs from an 'initiative' to a 'response' to the Soviet BMD and space threat, interprets the American efforts to justify it as the most recent manifestation of what has been described by Theodore J. Lowi as 'oversell mechanism'. The 'oversell of the remedy' – the remedy being the 'revolutionary' SDI – was followed by an 'oversell of the threat'.

In chapters 4–7, Taylor, Carton, Brauch and Berloznik analyse the evolution of the official position of the governments of Great Britain, France, the Federal Republic of Germany and the three Benelux countries as well as the different party positions and the roles of industry and academia in these six countries during 1984 and 1985 (and for West Germany until April 1986).

In chapter 8, Hans Günter Brauch deals with the lack of consultation prior to the two US unilateral decisions to implement BMDs, the first in 1967 and the second in 1983. During the first ABM debate (1967–72) the European NATO allies opposed a European BMD system. However, in the second BMD debate in the three transnational parliaments, the North Atlantic Assembly, the Western European Union (seven NATO members) and in the European Parliament (with MPs from nine NATO members and Ireland), a tendency in favour of conditioned support for the SDI research programme within the confinements of the ABM Treaty has prevailed. EUREKA, as a West European response to the technological challenge posed by the SDI, is addressed by Alain Carton in chapter 9, in which he focuses on the global competition in the field of high technology, on the problems of technology transfer and on European co-operation in the field of advanced technology between the Versailles Summit in 1982 and the Paris Conference on Technology in July 1985. The origins had led to EUREKA and the results of the first four EUREKA meetings are also reviewed.

Part 3 opens with chapter 10 by Paul Stares, in which he focuses on the relationship between SDI, Soviet BMD and Britain's nuclear modernisation programme. In chapter 11 Carton focuses on the French defence policy and the ABM debate, on the modernisation of the French nuclear forces, on the role of space technology in French defence policy and on the official French position relating to the concept of an EDI. In chapter 12, Peter Wilson analyses the geostrategic risks of the SDI, and the potential political gain the Soviet Union may have from a neutralisation of the nuclear forces of third countries.

Part 4 opens with chapter 13 by John B. Rhinelander. Based on his experience as the legal advisor to the US SALT delegation at the time when the ABM Treaty was being negotiated, Rhinelander reviews its evolution, offers an authoritative interpretation of the existing grey areas and discusses the efforts of the Reagan Administration to reinterpret the ABM Treaty.

In chapter 14 Hans Günter Brauch concentrates on the political and military background of the evolving debate on an EDI (or a Tactical Defence Initiative, TDI), on the Soviet Theatre and Tactical Nuclear Forces targeted against Western Europe and on potential Soviet countermeasures to SDI and EDI. Brauch also develops the conceptual framework in which diverse components of a European Defence Architecture (EDA) are presently being discussed in NATO circles and surveys the American weapons projects which are potential components of an EDI such as SDI-related programmes, antitactical systems and deep-strike systems (e.g., joint tactical missiles (JTACMS) or Army joint tactical missile systems (TACMs). Brauch concludes with a preliminary evaluation of an EDI for European security and with a plea for arms control – a political alternative to both SDI and EDI.

Longstreth, Pike and Rhinelander, in chapter 15, review the present ABM Treaty compliance debate in the United States in view of the questions related to Soviet missile defence activities and to the US SDI.

John E. Pike concludes the book with chapter 16, in which he analyses prior attempts to defend the ICBM, potential SDI components for site defence, the goals for less-than-perfect intermediate defences, the problem of preserving assured destruction and the related problem of escalation stability.

The authors from both sides of the Atlantic hope to contribute to an informed debate on the political, military, strategic, economic, technological and arms control aspects of President Reagan's SDI. If future transatlantic conflicts within NATO are to be avoided, a thorough understanding both of the long term-goals associated with the SDI by the present American Administration and of the detailed conditions on which the British, the German and the Italian governments have based their support for the SDI research programme is required. Any United States effort to undermine the 'firewall' dividing research and testing as incorporated in the ABM Treaty may very well provoke a transatlantic crisis of major proportions.

Part I
American and Soviet Space and BMD Programmes

1 A Global Defence Against Ballistic Missiles
Simon Peter Worden

I INTRODUCTION

Deterrence of war has been at the heart of Western security for the past forty years. Since the advent of nuclear weapons, Western leaders have sought to minimise the risk of destruction by maintaining effective nuclear-capable forces to deter aggression and by pursuing complementary arms control agreements. This approach appears to have worked.

The completion of the ABM Treaty in 1972 between the United States and the Soviet Union marked what was believed to be an understanding and agreement between the United States and Soviet Union for the basis for deterrence. The preamble to that Treaty summarised these considerations:[1]

> Considering that effective measures to limit anti-ballistic missile systems would be a substantial factor in curbing the race in strategic offensive arms and would lead to a decrease in the risk of the outbreak of war involving nuclear weapons.
>
> Proceeding from the premise that the limitation of anti-ballistic missile systems, as well as certain agreed measures with respect to the limitations of strategic offensive arms, would contribute to the creation of more favorable conditions for further negotiations on limiting strategic arms.
>
> Declaring their intention to achieve at the earliest possible date the cessation of the nuclear arms race and to take effective measures towards reductions in strategic arms, nuclear disarmament, and general and complete disarmament.

This agreement embodied the understanding that 'all-out' nuclear exchanges would be catastrophic for all sides, and that measures to encourage reductions in strategic nuclear forces were of top priority. At that time it was believed that the pursuit of strategic defensive capabilities would complicate the process of reducing nuclear arsenals.

The basis for the ABM Treaty and its virtual prohibition against ballistic missile defences was embodied in four strategic understandings which are now in question.

First, the West believed that limitations on defences would lead to deep reductions in strategic arsenals. These hopes have been illusory. Since 1972, the number of warheads on Soviet land-based strategic missiles has increased by nearly a factor of four, to about three times Western levels, or about 6,000 warheads. In contrast to this buildup, the United States reduced during the last two decades its total number of nuclear warheads by 8,000 and – more importantly – the destructive power of its nuclear forces by 75 per cent.

Second, the Western powers understood in 1972 that threats to the survivability of respective retaliatory forces would be reduced. This was viewed as such an important point that the United States made a unilateral statement during the 1972 ABM Treaty negotiations that it would view a failure to achieve reductions in threats to retaliatory force survivability as the basis for withdrawal from the ABM Treaty. The Soviet Union has not only expanded its offensive forces, but it has expanded precisely those forces which threaten the survivability of Western retaliatory assets. The operational Soviet ICBM force consists of some 1,400 launchers. Two-thirds of these have been built since 1972, and consist of the heavy MIRVed (multiple independently-targetable re-entry vehicles) SS-17, SS-18, and SS-19 ICBMs. The SS-18 and SS-19 carry more and larger MIRV warheads than the United States Minuteman III, the most modern US ICBM. The Soviet ICBMs carry at least six to ten warheads, as opposed to three for the Minuteman III. The SS-18 was specifically designed to destroy US ICBM silos and other hardened targets. The Soviet SS-18 force alone can destroy more than 80 per cent of the US ICBM silos using two nuclear warheads against each.[2]

Third, the United States entered into the ABM Treaty with the understanding that its provisions would not be circumvented or violated. The Treaty drafters recognised that the long-lead item to a deployment of a nationwide ABM capability was a nationwide radar net. They also recognised that there are legitimate reasons for a comprehensive radar system to provide early warning of a missile attack. For this reason the ABM Treaty very carefully constrains the placement of early warning radars. Article VIb of that Treaty enjoins the signatory parties 'not to deploy in the future radars for early warning of strategic ballistic missile attack except at locations along the periphery of its national territory and oriented outward'.[3] The Soviet Union is currently constructing a large phased array radar near Krasnoyarsk, in central Siberia which violates this ABM Treaty provision.[4] This radar is physically indistinguishable from other radars in the Soviet Union which the Soviet government has identified as ballistic missile early warning radars. Significantly, its location fills an obvious gap in the Soviet Union's early warning coverage. The Krasnoyarsk radar, however, is not located near one of the Soviet Union's borders and is oriented toward national borders several thousand kilometres away. The Soviet Union has claimed that this radar is designed not for early warning or ABM-related purposes, but for space tracking. However, the location and orientation of the radar is not suitable for space tracking or any other allowed functions. But it will have the inherent capability to track ballistic missiles in flight trajectory and perform additional ABM functions in a manner contrary to the Treaty.

The Krasnoyarsk radar violation, taken in concert with other ABM-related activities – some also in violation of ABM Treaty provisions – has led the United States to conclude that the Soviet Union could be preparing to deploy a prohibited nationwide BMD.

Fourth, the acceptance of the ABM Treaty limitations on defences reflected technical realities true during the late 1960s. As both the United States and the Soviet Union developed missile defence systems, it became clear that those systems could be more easily and cheaply overwhelmed by new offensive missile warheads, particular MIRV warheads, then could the defensive systems be enhanced to meet increased threats. Moreover, the ABM systems of that period relied on large engagement radars which were vulnerable to relatively low-cost countermeasures. However, the technology of the 1980s has become sufficiently advanced that the technical undesirability of missile defence may no longer exist.

With respect to Western Europe, Soviet strategy and the potential role of missile defences has also changed. It is clear that the Soviet Union is prepared for the decisive use of force to achieve military purposes. But they are also aware that the unselective use of force could result in a catastrophic nuclear exchange. The Soviet Union has prepared a wide variety of options for use of nuclear weapons to achieve military objectives against Western forces and facilities. A possible – and some believe likely – Soviet use of nuclear weapons would be in the event of unexpected reverses in the centre of Europe, on the lightly-armed flank of NATO, or in the Persian Gulf. The risk of a global nuclear conflagration would be a strong incentive to keep such nuclear weapons use confined to the theatre of operations in which they are first used. It would therefore be likely that nuclear strikes would be selective, controlled, restricted to military targets and sustainable over the course of a prolonged conventional conflict. These considerations imply that an initial pre-emptive strike may consist of a few tens of weapons, followed by continued selective use of a number of small nuclear weapons through the duration of a conflict. Perhaps no more than a few hundred weapons would be involved, in contrast to a simultaneous launch of many thousands of warheads likely in global nuclear war. This limited use of nuclear weapons provides a framework for discussing strategic defences as they relate to Europe.[5]

If the use of nuclear weapons are to be avoided in the context of conventional conflicts which could affect Europe, then it must become clear to any potential aggressor that either the consequences of using nuclear weapons will be unacceptable, or that the military gain expected from nuclear weapons use would not justify the cost of the systems to deliver such weapons. These two aspects of deterrence are mutually supporting and have never been completely independent. Strategic defences help deter when they can clearly limit and perhaps prevent military gains through an aggressor's offensive actions.

II TECHNICAL REQUIREMENTS FOR AN EFFECTIVE MISSILE DEFENCE

The defensive systems which the United States investigated during the 1960s, and which the Soviet Union are currently deploying, were based on performing missile intercepts in the last few minutes of the missile's flight. This intercept scenario has the advantage that, as the missile re-enters the atmosphere – in the so-called 'terminal phase' of its flight – any penetration aids designed to confuse or overwhelm the defence are stripped away by air friction in the earth's atmosphere.

The objective of the Strategic Defence Initiative (SDI) is to find technologies which would make it possible to intercept the missile soon after it is launched, and to have many different intercept opportunities throughout the missile's flight. This multiple-layered defence concept is the key feature of an effective defence. Even if each layer has limited effectiveness, overall performance can be very good. Three individual layers with 80 per cent intercept success in each layer combine to give better than 99 per cent overall effectiveness. Moreover, efforts to defeat or counter a given layer are unlikely to succeed against other layers based on different defensive technologies.

In each layer, systems must be defined which would first detect a missile attack, define missile targets and identify individual threatening objects. Next, defensive weapons must be pointed and directed at the ballistic missile or warhead targets. Finally, a battle management system is needed to manage each phase of the battle reliably and hand unintercepted targets on to subsequent layers.

The first part of a ballistic missile flight is the boost phase, during which the main rocket engines are burning to thrust the warheads out of the atmosphere towards their targets. It is relatively easy to see the booster's hot rocket exhaust, even from distances very deep in space. Since the boost phase lasts five minutes or less, and perhaps as short as one minute, a boost-phase defence has the difficult task of intercepting the boosters during this short period.

The next phase exists only for MIRVed missiles. A small 'post-boost vehicle' is released when the booster burns out which directs the multiple warheads to different targets and can also deploy hundreds of decoys, balloons and other penetration aids designed to confuse and overwhelm later defensive layers. Both boost and post-boost phases offer an opportunity for weapons and other defensive systems to destroy warheads and decoys while they are still in one package. Clearly, the later intercept occurs in the post-boost phase, the less leverage there is in firing at and destroying the post-boost vehicle.

The longest phase is mid-course, which lasts up to thirty minutes for ICBMs but only a few minutes for the shortest-range missiles. During mid-course, the warheads and any penetration aids travel above the earth's atmosphere on a ballistic trajectory. Since the flight paths of light and heavy objects are identical, tracking alone cannot tell the difference between a warhead and a lightweight decoy. The primary technical problem during mid-course is thus to discriminate warheads from decoys and other objects designed to confuse the defence.

The terminal phase begins when the warheads and penetration aids begin to re-enter the atmosphere at about 100 kilometres above the earth's surface.

When President Reagan launched the SDI in his 23 March 1983 speech, he challenged the scientific community to determine how effective defences could be built. During the summer of 1983, about fifty of the United States' top experts in defensive technologies met and produced the 'Defensive Technologies Study', better known as the 'Fletcher Study' (after James C. Fletcher, who was the former head of NASA and who directed this study effort).[6] The study identified new technologies which might make effective defence possible. The study also outlined a roadmap for a research programme to pursue these technologies.

The Fletcher Study identified 'directed energy weapons' as an important new technology. Directed energy weapons include lasers (which could destroy a missile by burning or punching a hole in a warhead or missile) and particle beams (which contain individual atoms or charged particles moving at velocities near to the speed of light). The particles can penetrate deep inside a target and destroy or disrupt its internal workings. Weapons based on these technologies could stand off thousands of kilometres in space, or even be located on the ground on the other side of the planet to destroy missiles soon after they are launched.

New technologies are being developed for infrared and visible light sensors which could detect the heat, the shape and even produce an actual image of a warhead in space. From several different types of such data it will be possible to ensure that decoys which credibly pass for warheads would be as costly to produce and launch, and nearly as heavy, as real warheads.

There have been significant advances in computers. The Fletcher Study concluded that compact computers capable of nearly a billion operations per second would be needed to handle a multilayered defence against tens of thousands of objects. Computers with this capacity are now approaching the state-of-the-art position.

A combination of new sensor technologies and new computer capabilities now make possible a 'kinetic energy interceptor'. These are small homing warheads which could seek warhead or missile targets and destroy them by physically hitting them. Technology has already proceeded to the point where such interceptors weigh a few tens of kilogrammes, considerably less than even a warhead target, and thereby potentially much less expensive to proliferate than offensive missiles and warheads.

The Fletcher Study also identified reasons why an intensive research programme is necessary before a decision to proceed to deployment can be made. Five critical questions were identified. First, it is necessary to understand how, in detail, a ballistic missile could be intercepted in the boost phase. Second, the means confidently to discriminate between warheads and decoys must be developed and validated. Third – and extremely important – the means to make potential defences survivable so that the defences are not themselves an appealing target must be developed. The fourth requirement is equally critical: once a defensive system is in place, it must be possible to upgrade it to intercept additional missiles and warheads more cheaply than it would be for the opponent to proliferate those offensive forces. Moreover, this cost–exchange advantage must be maintained even against specially redesigned offensive threats. Last, although billion operation per second computers exist, better means to programme these computers are needed. The Fletcher Study estimated that ten million lines of computer instruction would be needed, as large as any computer system ever constructed. The Fletcher Study has stood the test of time. Today's SDI programme is focused on technologies for a defence which would be lethal, survivable, and cost-effective.

III EVOLVING GLOBAL DEFENCE CONCEPTS

While it is premature to consider the details of a possible future global defence system, representative concepts do exist. The applicability of these concepts to the European missile problem may be examined in the context of these representative system constructs.

The SDI is designed to formulate options which would deter unilateral Soviet moves to implement defences against ballistic missiles, and which could also form the basis for agreements jointly to pursue a more stable strategic posture based on defensive capabilities. It is premature to propose a specific multilayered defensive system in any detail. Nonetheless, certain precepts and general outlines for systems may be examined now.

The SDI Programme is formulated to provide system options of two basic types. It is believed that some technologies will be sufficiently developed to be incorporated into systems feasible during the 1990s. Through a combination of arms control and carefully planned defensive deployments, these systems could retain their effectiveness indefinitely. However, the possibility of improved offensive capabilities – either offensive systems with increased penetrability through the defensive system or capable of providing a credible threat to the survivability of the defensive system – could eventually provide an incentive for breakout from a defence-reliant strategic posture. A second objective of the SDI is thus to provide technical options for maintaining and improving defence effectiveness against threats specifically designed to defeat defences.

Only one possible multilayer 'global' BMD system will be discussed here. Although less than half of the SDI funding is devoted to weapons research, discussion generally centres on the type of technology which might be used for the actual missile or warhead destructive system. Credible possibilities exist for *both* kinetic energy (interceptor) and directed energy (laser or particle beam) technologies. However, the basic three-layer system discussed here is based on kinetic energy technologies such as were demonstrated in the successful Homing Overlay Experiment in June 1984. In this experiment, a mock warhead was fired from Vandenberg Air Force Base in California to the Kwajalein Missile Range almost 8,000 kilometres away in the Pacific Ocean. When radar sensors at Kwajalein detected the incoming warhead, a missile was fired which contained a non-nuclear interceptor. After leaving the atmosphere this device opened up an infrared (heat-seeking) sensor which detected the incoming warhead. A computer controlled a rocket engine on the interceptor to manoeuvre it into the direct path of the warhead and execute a direct collision over 100 miles up in space. This experiment demonstrated for the first time in history that a 'bullet could hit a bullet'. A specimen system architecture based on this technology is detailed in Table 1.1. The detailed numbers and functional subsystem types are only one of a great number of alternatives.

The intercontinental offensive strike force is assumed to consist of 800 ICBMs of the Soviet SS-18 class, each MIRVed with eight warheads and 24 decoys weighing 10 per cent of the warhead weight. The intermediate range offensive forces consist of 400 SS-20 class missiles with three warheads and three decoys each and 400 SS-22 class missiles with one warhead each.

The defensive system is divided into three phases along with associated battle management systems. For the boost phase, thousands of rocket-powered homing interceptors are carried on several hundred satellite carriers, each holding tens of interceptors. An interceptor capable of diverting from its course sufficiently to intercept all missiles within its range might weigh in the range of 100 kilogrammes. Each satellite carrier could intercept the long-burning

Table 1.1 Possible kinetic energy defensive system

Offensive threat	
Offensive missiles	800 SS-18 (8 warheads each)
	400 SS-20 (3 warheads each)
	400 SS-22 (1 warhead each)
Total warheads	8,000
Decoys (10% warhead wt)	20,800
Defensive elements	
Space-based interceptors	10,000
Carrier satellites	500
Boost surveillance satellites	5
Mid-course sensor satellites	10
Ground-based interceptors	3,200
5 sites in United States	
3 sites in Europe	
Airborne optical system	18
15 based in United States	
3 based in Europe	
Terminal radars	37
30 based in United States	
7 based in Europe	
Terminal interceptors	600
150 sites in United States	
30 sites in Europe	

SS-18 up to 2,000 kilometres away and an SS-20 up to 1,000 kilometres away if the interceptors were fired soon after a launch was detected. These interceptors cannot operate inside the atmosphere and have no potential against ground targets. A boost-phase surveillance and tracking system deep in space would provide attack warning and booster track information. These same systems could operate in the boost phase and also the post-boost phase for the multiple warhead systems.

A space-based kinetic energy system has at least one additional intercept opportunity during the mid-course. However, there are two complicating factors during this phase. First, the interceptor must attack a 'cold' target rather than the 'hot' booster. Since infrared sensors 'see' the heat of the target, cold objects are much harder to detect and track than hot ones. In order to intercept such cold objects, the interceptor must be guided fairly near to the target in order for its heat-seeking sensor to take over and perform final homing on to the target. Second, decoys may be present in large numbers during the mid-course. These decoys can be constructed so that they look very much like the warhead targets and could confuse and exhaust the defence's interceptors. For these reasons, a network of infrared sensor platforms are needed in addition to those sensors which would detect the boosters during the boost phase. These sensors would be able precisely to locate targets, discriminate decoys from warheads and track potential targets. Current sensor technologies are adequate accurately to identify at least half of a large number of sophisticated decoys. Less massive decoys may be discriminated much better. An interceptor system must thus be sized to handle some (but probably not a prohibitive number of) decoys.

The mid-course provides additional opportunities for both warhead interception and discrimination of warheads from decoys. Ground-based interceptors can be committed to attack a target based on accurate tracks and target identification while potential targets are still in early mid-course flight. A first wave of intercepts would thus occur while the warheads are still thousands of kilometres from their targets. There appears to be time in the mid-course for at least two full 'shoot-look-shoot' cycles. After the first mid-course intercept attempt, mid-course sensors can assess the outcome and commit another wave of interceptors for those cases where the first intercept failed. Since each ground-based interceptor site can defend an area of several thousand kilometres diameter, a handful of sites would be sufficient for either Europe or the United States. Each site would contain several hundred interceptors. Since interceptors weighing about 1,000 kilogrammes or less appear feasible (due in large part to the very small sizes possible in a non-nuclear interceptor), beach interceptor missile could be relatively inexpensive to deploy.

A terminal layer provides an additional intercept layer within the atmosphere at altitudes between 10–30 kilometres. At these high altitudes even attacking warheads which have been 'salvage-fused' to detonate when intercepted will be unable to damage unhardened sites on the ground such as cities or large military bases. Intercept inside the atmosphere or 'endo-atmospheric' intercept has the advantage that the atmosphere serves to strip away lightweight decoys. Two different sensor concepts exist for this phase. In the first, a set of 'optical' subsystems consisting of various infrared and laser-ranging sensors would be carried on an airborne platform to perform tracking and discrimination functions. Such 'Airborne Optical Systems' (AOSs) might ultimately be carried on high-altitude, unmanned aircraft. Each airborne sensor system could perform the required sensor function for an area of 1,000 kilometres diameter or more.

Consequently, only a few would be needed for all of Western Europe. The second sensor option, which may ultimately be used in conjunction with the first, might be a series of ground-based 'Terminal Imaging Radars' (TIRs). These radars would begin discriminating warheads from decoys outside the atmosphere by forming an image of a potential threat object. The ground-based radars may be small enough to be mobile to enhance their survivability. Each radar could cover a circle several hundred kilometres across. The smaller 'footprint' of coverage for the radars makes them best suited for defence of high-value areas such as military staging areas, battle areas, or densely populated civilian targets.

Terminal interceptor missiles would be distributed over many sites, with only a few missiles per site. Both the distribution and possibly mobility of sensors and interceptors could make terminal systems highly survivable. Non-nuclear terminal interceptors guided to their targets by a combination of external commands and on-board infrared homing sensors are the current focus of SDI investigation. The interceptor is guided to within a few metres of the target, which is then destroyed by a small pellet cloud explosively launched from the interceptor.

In addition to the 'strategic' possibilities described here, work is proceeding on an 'underlay' defence for use on the battlefield against the shortest-range ballistic missiles (SRBMs). These interceptors would operate below 15 kilometres altitude and may either use a 'hit-to-kill' warhead like the terminal interceptors described above or could use a small nuclear charge. The latter option would be an effective response against salvage-fused attackers. The single kiloton-class nuclear warheads would be small enough to cause little or no damage on the ground, but would effectively disrupt the large attacking warhead before it could detonate.

IV ISSUES RELATED TO THE DEFENCE OF EUROPE

An interesting result of SDI analyses to date is that the boost-phase and space-based mid-course systems postulated for the defence of the United States appear to be *oversized* for the defence of Europe.

The dominant requirement for mid-course defence is decoy discrimination. Figure 1.1 illustrates why discrimination is simplified for shorter-range missiles. Residual atmospheric drag prevents the

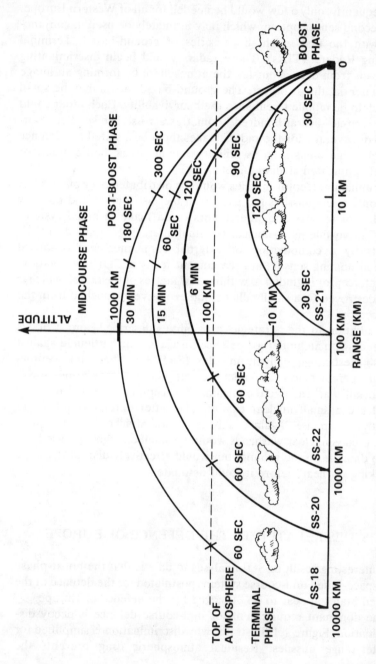

Figure 1.1 Example trajectories of typical ballistic missile threats; Boost, post-boost, midcourse and terminal phases are noted for missiles with trajectories similar to Soviet SS-18, SS-20, SS-22 and SS-21 systems

deployment of credible decoys, even at very high levels within the atmosphere. Since only the longest-range missiles threatening Europe – the SS-20s – spend the predominant share of their flight time outside the atmosphere, the decoy problem is non-existent for many threats to Europe. Even for the SS-20, payloads are much less than for ICBMs, leaving little excess payload for decoys. Additionally, sensors in Europe can observe most of a missile's trajectory – unlike ICBMs, where the curvature of the earth hides much of the trajectory. The longer relative observation time enables the European defence better to watch the critical warhead and decoy deployment phase, thus greatly facilitating discrimination.

Systems sized for boost-phase defence of the United States against ICBMs appear to be more than adequate for the European defence problem. Although the shortest-range missiles never leave the earth's atmosphere and are thus denied to currently feasible space-based kinetic energy interceptors, these missiles have much lower velocities and can be detected and intercepted throughout most of their flight by terminal interceptors. The longer-range SS-20s and SS-22s do spend a substantial time outside the atmosphere and have a boost phase (although shortened by a factor of two) during which they are vulnerable to space-based and boost-phase weapons. The launch area for these missiles is spread out over nearly the same territory as for ICBMs. The limitations on intercepting shorter-range missiles are offset by the greatly reduced number of warheads and missiles which would be involved in an attack against Europe. The use of nuclear weapons in Europe would likely occur over a much longer time than an ICBM attack on the United States, giving the defence time to bring far more of its assets into play than in a mass ICBM attack on the United States. SDI calculations have shown that space-based kinetic energy systems such as those discussed in this chapter are oversized for the postulated European threat by a factor of three. Indeed, the ground-based systems alone appear to be adequate to defeat the current and near-future threats to Europe.

V COST-EFFECTIVENESS, SURVIVABILITY AND STABILITY

The 'cost–exchange' issues appear greatly to favour the defence based on current information. Offensive warhead costs do not vary significantly from the $30 million per warhead between ICBMs, submarine-launched ballistic missiles (SLBMs), or intermediate-range ballistic missiles (IRBMs). It is not possible to establish defensive system costs at this time. However, the cost goals of the SDI are consistent with scaling estimates based on current missile and space systems. These cost estimates suggest that future defensive systems could have a substantial cost-effectiveness margin over offensive systems. One of the highest priority goals of the SDI is to establish definitive cost data for possible defence systems.

Strategic defence stability criteria are not based primarily on system costs, although a future system must be affordable. More important than basic costs are the 'marginal costs' needed to enhance the defensive system capability to counter increased offensive forces. Between a third and half of most defensive system cost estimates are in the sensor portion of the system. These sensor systems would require minimal enhancement for even large changes in threat size. Marginal cost exchange between offence and defence would thus be primarily based on enhancements to only the weapon portion of the defensive system.

Survivability is an essential requirement for an effective defence. Although survivability of both ground-based and space-based defensive elements must be pursued, the issue of space element survivability is new and has not yet been fully investigated. There is no way of knowing whether a Soviet attack on Western Europe would involve attacks on space-based strategic defence systems; however, this possibility must be taken seriously. Three types of threats to space-based elements might exist: kinetic energy anti-satellite weapons (ASATs), nuclear explosive ASATs and directed energy threats such as space-based particle beams or ground-based lasers. It appears that satellites may be hardened against nuclear attack to the degree that essentially a direct nuclear hit would be needed to destroy the satellite. Shielding appears feasible against lasers and particle beams. Such shielding appears effective against even advanced directed energy threats that might not exist for many decades. Both nuclear and non-nuclear ASATs may be countered through a combination of manoeuvring and enforcement of a keep-out zone using the same kinetic interceptors envisioned for missile defence weapons as self-defence weapons. A similar cost-exchange advantage as would exist against ballistic missiles should exist for ground-based ASAT attackers, since a ground-launched ASAT would require as nearly as heavy and sophisticated a booster as a nuclear warhead. Sensor satellites with vulnerable optical elements may be placed at sufficiently high altitudes that they would be both difficult to detect and difficult to attack. Moreover, these sensor satellites could also be defended by escort kinetic energy interceptors.

Ground- and air-based system survivability is also essential. The mobility of air-based and possibly ground-based sensors makes it very difficult to locate and destroy these elements. The dispersal and possible mobility of interceptors similarly enhances their survival. Since a multilayered defensive system has boost-phase and mid-course defensive layers, it would be difficult for an aggressor to calculate the force levels necessary to destroy those ground-based defensive elements which lie behind the first defensive layers. The fact that a successful pre-emptive attack must first destroy each layer of a defensive system in turn greatly decreases the aggressor's confidence in a successful attack. Deterrence is thereby enhanced greatly by the very existence of multilayered defensive systems.

To ensure long-term stability a defensive system must evolve to counter new types of offensive threats which might have been specifically designed to defeat an initial system. The addition of advanced technologies – such as powerful directed energy devices – offers options of this kind. It appears that moderate power lasers or particle beams may be added in the 1990s to aid in actively discriminating decoys from warheads. This all but eliminates the utility of decoys – for example, moderate power laser pulses can strike a potential threat object, causing it to recoil. Heavy objects – presumably warheads – would recoil only slightly compared to a lightweight decoy, thereby identifying the heavy warhead as a threatening object.

Within 15–20 years it appears likely that directed energy technology will have advanced sufficiently so that lasers and particle beams will be powerful enough to serve as primary defensive weapons. High-energy lasers can reach deep within the atmosphere to attack even boosters with very short burn times. An opponent seeking to defeat a boost-phase defence by developing 'fast-burn' boosters which burn out while still within the atmosphere would thus achieve little from this development. Lasers and particle beams appear feasible with effective ranges of many thousands of kilometres and re-target times a small fraction of a second. Space-based directed energy weapons or directed energy weapons with space-based elements could thus counter even extremely large numerical threats.

VI OTHER ISSUES

A multilayered BMD might also be able to counter other threats. For example, conventionally armed or chemically and biologically armed missiles would be vulnerable to the same defensive system. There is already considerable concern that such non-nuclear missiles pose unacceptable threats to NATO installations or critical nodes. A multilayered BMD of the type discussed in this chapter could eliminate these threats as well as nuclear-armed missile threats.

Bombers and Cruise missiles also present a serious threat to NATO. While space- and some ground-based kinetic energy defences appear to have little potential against these 'airbreathing' threats, directed energy weapons capable of penetrating the atmosphere do. A constellation of space-based laser weapons capable of countering a large ICBM attack may consist of 100 or less laser platforms. These constellations are not only oversized against missile threats to Europe but are even more capable against airbreathing threats. The critical requirement for airbreathing defences is not weapons, but rather the need for advanced sensor systems. Many of the infrared and radar systems being considered within the framework of SDI, particularly space-based sensors, have considerable potential for identifying and tracking airbreathing threats.

VII CONCLUSION

Defences based on successful SDI research programmes might be feasible in the 1990s which are also extremely effective against ballistic missile threats to Europe. Such defences would be part of a global multilayered defence system which would enhance deterrence by denying military gain to an aggressor from a ballistic missile attack. The technology level and defensive force requirements for European defence appear relatively moderate in comparison to those necessary for ICBM defences.

The global BMDs envisaged have great potential to be survivable and cost-effective at the margin. Both criteria are necessary to foreclose the possibility of an offence–defence arms race spiral and to serve as a valid inducement to arms control. The addition of directed energy technology can further close options for new offensive countermeasures which might degrade the effectiveness of initial defences.

These same directed energy technologies also appear to have considerable potential for negating airbreathing threats to Europe.

The United States has offered to pursue its SDI in full co-operation and collaboration with its Allies to ensure that the defence needs of the entire Western Alliance are met. Through such joint efforts the day when President Reagan's vision of a world free from the threat of nuclear war will be brought within our reach.

Notes

1. US Arms Control and Disarmament Agency, *Arms Control and Disarmament Agreements*, 1982 edn (Washington, DC: United States Government Printing Office, 1982) pp. 137–47.
2. US Department of Defence, *Soviet Military Power 1985* (Washington, DC: United States Government Printing Office, 1985).
3. See note 1.
4. US Arms Control and Disarmament Agency, *Arms Control: US Objectives, Negotiating Efforts, Problems of Soviet Compliance* (Washington, DC: ACDA, May 1984); 'The President's Unclassified Report to the Congress on Soviet Noncompliance With Arms Control Agreements', *White House Press Release*, 1 February 1984 (Mimeo).
5. The discussion of European defence requirements follows Gregory H. Canavan, 'Theater Applications of Strategic Defense Concepts', (Los Alamos: National Laboratory Publication P/AC:85-149, LA-UR-85-2117, 1985).
6. US Department of Defense, *The Strategic Defense Initiative: Defensive Technologies Study* (Washington, DC: United States Government Printing Office, 1984).

2 Offensive Nuclear Forces, Strategic Defence and Arms Control
Gary L. Guertner

Offensive and defensive weapons with their supporting command and control networks form an integrated system for the conduct of nuclear war. Each part depends to some degree on the functioning of the other parts, and these interactions influence both the choice of weapons we deploy and the strategies for their employment. In the past, reducing the numbers of nuclear weapons and maintaining strategic stability required co-ordinated and parallel offensive–defensive constraints to minimise unilateral advantages and prevent unrestrained competition.[1]

The Reagan Strategic Defence Initiative (SDI) has fundamentally altered previous efforts to engage in co-ordinated offensive–defensive arms control negotiations, and instead views strategic defence and offensive arms reductions as twin goals. In his 'Star Wars' speech (23 March 1983) the President directed the military, scientific, and industrial communities to undertake a long-term research programme to 'achieve our ultimate goal of eliminating the threat posed by strategic nuclear missiles'.[2] Following six months of intense study by three panels, the Administration proposed spending $25 billion dollars from FY85 to FY89 to permit informed decisions in the early 1990s on whether to initiate deployment of a multilayered defence capability after the year 2000.[3]

Shifting from a strategic posture of 'offence domination' to 'defence domination' is fraught with technical, fiscal, strategic and political uncertainties. This chapter examines strategic offensive–defensive relationships in the context of strategic doctrine and nuclear weapons employment policies, formal treaty commitments, negotiating strategies and the President's SDI, which cuts across the entire spectrum of strategic arms control issues.

I STRATEGIC DOCTRINE

Classically, deterrence of war and strategic nuclear weapons employment policies have a paradoxical relationship in that deterring nuclear war has required policies and credible plans and strategies for fighting and, if not winning, at least assuring that potential adversaries could not win. While 'winning' a nuclear war has little meaning in view of the major destruction that would accompany the use of nuclear weapons, it still seems clear that, to be deterred, potential aggressors must be denied confidence that they could achieve their war aims.

United States and Soviet strategic doctrines have evolved from this paradox with important differences in emphasis.[4] American force structure and declared employment policies have evolved to deter Soviet execution of their war plans through assured retaliation by survivable American nuclear forces. Current strategy increasingly includes 'damage limitation' through preferential attack options against Soviet strategic, theatre, or conventional forces before these forces can be launched or fully deployed, and the threat of 'assured destruction' through escalation to urban/industrial targets if aggression continues. The most salient feature of United States doctrine has been the evolution of graduated and flexible responses that incorporate limited nuclear attacks to maintain options for intrawar bargaining, escalation control, and prompt conflict termination.

Soviet doctrine places greater emphasis on warfighting and damage limitation through large-scale, pre-emptive attacks against military targets. These threatening features are not in themselves a rejection of deterrence. Soviet force structure and declaratory policy emphasise that the better their armed forces are prepared to fight a nuclear war, the better their society is equipped to survive its effects, and the more clearly the adversary understands this, the more he will be effectively deterred. This doctrine is sometimes called 'deterrence through denial' – that is, seeking to deny the opponent the prospect of military victory. It covers all of the Soviet Union's strategic bases since it rests on well-established warfighting doctrines and capabilities in the event that deterrence fails.

While it is true that Soviet and American strategic doctrines are converging in their emphasis on hard-target counterforce and damage limiting capabilities, potentially destabilising doctrinal asymmetries remain. The most obvious is the apparent Soviet rejection of United States limited nuclear war concepts including escalation control and intrawar bargaining. The Soviets view these concepts as attempts at political intimidation rather than elements in a strategy conceived by those who take war seriously. For the Soviets, denial of military victory requires a strategy of damage limitation through robust pre-emption (e.g., launch on tactical warning), credible defensive systems and attacks of greater magnitude than those prescribed by United States limited nuclear war strategy.

II OFFENSIVE AND DEFENSIVE RELATIONSHIPS TO SOVIET–AMERICAN DOCTRINAL ASYMMETRIES

The credibility of both Soviet and United States strategic doctrine is sensitive to the evolving relationships between offensive and defensive forces. This relationship is complicated by the fact that nothing in nuclear strategy is purely defensive, in the sense that it does not directly support or lend credibility to offensive operations. Any calculation of a first strike or pre-emption is conditioned in part by active (e.g., ABM) and passive (e.g., civil defence) defensive capabilities to absorb residual second-strike forces.

Any potentially dangerous and destabilising realignment of offensive–defensive strategic capabilities should be regulated by arms control strategies that formally shape opposing force structures so as to preclude plausible pre-emptive options. Arms control agreements that create greater levels of strategic stability (i.e., mutually survivable forces) combined with the evolving United States strategy of limited attack options increases the level of Soviet uncertainty as to the nature and scope of an American response to aggression. A strategy incorporating strategic defence may or may not add to stability or to the evolving denial function of United States forces. Those outcomes will depend on the success and reliability of developing technologies and on the Soviet Union's willingness to negotiate offensive limitations rather than embark on new strategic initiatives of its own.

The technological issues of strategic defence and deployment decisions will be more a matter of judgement than proof, since neither United States nor Soviet planners are likely to have a full, confident understanding of how their systems would operate in a wartime environment. Full-scale testing of a layered defence is even more impractical than full-scale testing of offensive systems. Only components can be tested. Because we see only isolated pieces, we thus cannot discern the entire puzzle. Systemic reliability will be a matter of faith and inference, since only a full-scale attack could demonstrate the reliability of sensors, communications, and weapons. Successful tests of individual weapons against single or even multiple targets will not constitute proof of systems reliability. Battle management – always war's least controllable or predictable variable – will require reliable and survivable warning sensors; tracking capabilities for thousands (or even tens of thousands) of targets; instant (and therefore pre-delegated) communications, command and control systems; as well as demonstrable weapons.

The primary effect of so much uncertainty is likely to push both sides toward worst case judgements about the effectiveness of their own and each other's defence system. The same logic that drives one to doubt one's own capabilities must drive one to expect the enemy to be effective. Future stability in the offensive–defensive relationship may, therefore, be measured primarily by the constraints both sides are willing to accept on future systems. Precisely which general combinations of offensive–defensive constraints would degrade Soviet capabilities most is debatable because of operational uncertainties. On balance, offensive constraints would affect the Soviet's robust style of pre-emption/damage limitation more than they would the evolving United States strategy of limited attack options and escalation control. Defensive constraints affect both American and Soviet strategic doctrine. When combined with offensive limits, however, they degrade Soviet forces more than those of the United States, since defensive constraints make the execution of United States limited nuclear options more credible than a Soviet strategy based on massive pre-emption to achieve 'acceptable' levels of damage from a retaliatory strike.

Defensive advantages by either side will greatly enhance the credibility of its strategic doctrine. Neither side is therefore likely to accede to a posture of defensive inferiority. Failing arms control remedies, the disadvantaged party will seek to re-establish its strategic position through offensive countermeasures, defensive countermeasures, or both. In the Soviet case, these measures could also include doctrinal modifications. For example, the Soviets could seek compensation for perceived offensive shortfalls by moving toward a 'softer' strategic target set, including greater emphasis on countervalue targets to compensate for the rapidly declining penetrability of their strategic forces.

III OFFENSIVE–DEFENSIVE RELATIONSHIPS

Arms control strategies aimed at achieving equal ceilings, equal sublimits, or offsetting advantages in offensive weapons (e.g., United States bomber or SLBM advantages for Soviet ICBM advantages) may satisfy domestic political requirements and public perceptions of the strategic balance, but they do not necessarily support the operational effectiveness of nuclear forces if deterrence fails. This is not to suggest that warfighting plans and strategies should drive arms control policy. Nevertheless, strategic force levels codified by treaty will shape warfighting options for the future, and the two must be related. Their credibility to deter war will depend to a large degree on the relationship between offensive trade-offs and defensive systems that may or may not be constrained by arms control agreements.

Figure 2.1 illustrates possible relationships of offensive systems to defensive and offensive threats, defensive systems to offensive systems, and in the case of SDI technology, potential offensive threats from 'defensive' systems.

Examples from Figure 2.1 include:

1. Strategies designed to negotiate higher United States bomber limits to trade against Soviet ICBMs must take into account offensive threats to bomber bases and defensive (air defence) threats to bomber penetration.
2. Submarines must be able to survive offensive threats to their home ports, anti-submarine warfare (ASW) at sea, and their missiles must be able to penetrate Soviet missile defences.
3. Space-based BMDs can be attacked by anti-satellite weapons (ASAT) and possibly ABMs. Space-based BMD systems must be able to defend themselves and therefore must have the ability to destroy ASATs. Limitations on ASATs may enhance the survivability of space-based defences, but either side could circumvent treaty limitations by labelling an ASAT weapon as a BMD system or component. Conversely, BMD constraints could be circumvented by labelling a BMD weapon as an ASAT system or component. Because of their dual capabilities, both or neither should be constrained by treaty, but not one or the other.
4. Similarly, space-based defences could attack other space-based defences. War in space could, therefore, begin with pre-emptive attacks by 'defensive' systems against defensive

Offensive System	Defensive Threat	Offensive Threat	Arms Control Remedies	Unilateral Remedies
Bombers	• Air Defense	• SLBMs and SLCMs • ICBMs	• SLCM Ban/Limitation • SSBN Patrol Area limits • Ban SLBM Depressed Trajectories • Air Defense Constraints	• ALCMs • Penetration Aids/Stealth Technology • Tactics • Higher Alert Rates • Basing Changes
ICBMs	• ABM Ground BMD-Space	• ICBMs SLBMs (Future)	• Launcher Limits • Missile Limits • Warhead Limits • Throw-weight Limits • ABM/BMD Constraints	• High Alert Rates • Hardening • C^3I Improvements • Launch Under Attack • Mobile ICBMs • Penetration Aids/MARVs
SSBNs/SLBMs (Submarines/Sub-launched ballistic missiles)	• ASW (Anti-Sub warfare) • ABM/BMD	• SLCMs • ICBMs (Threatens non-alert SSBNs) • SLBMs	• ABM/BMD Constraints • ASW Constraints • SSBN Sanctuaries • ICBM, SLCM, SLBM Constraints	• High Alert Rates • Technology/Stealth (Silence) • Penetration Aids/MARVs • Tactics • C^3

Defensive System	Defensive Threat	Offensive Threat	Arms Control Remedies	Unilateral Remedies
ABM-Ground	• BMD-Space (Potential)	• MARV • Penetration Aids • Stealth Tech. • Cruise Missiles • EMP-Electro Magnetic pulse • Tactics/Mass	• BMD Constraints • MARV Constraints • Offensive Force Reductions	• Increase Assets • Harden System
BMD-Space	• ABM-Ground • BMD-Space • DSATs (Defensive Satellites)	• ASAT • Tactics/Mass • EMP, Blackout • ALCMs • GLCMs • SLCMs	• Ban BMD • Ban ASAT • Constrain: SLCM, GLCM, ALCM • Reduce Ballistic Missile Forces	• Increase Assets • Harden System • Tactics/Maneuver

Figure 2.1 Strategic nuclear offensive–defensive linkages for US systems

systems. Under such circumstances, battle stations would require escort vehicles or defensive satellites (DSATs) that could proliferate like components of a naval battle group around an aircraft carrier.

New technologies that may emerge from an unconstrained SDI could further obscure the offensive-defensive relationships depicted in Figure 2.1. If, for example, space-based 'defences' acquired a dual capability to destroy offensive weapons in flight and surface-based targets (e.g., ICBMs, ABMs, ships), then 'defensive' systems could not only support the offence indirectly by limiting a retaliatory attack, but also directly through pre-emptive attacks against all targets. Explorations of new strategic frontiers promise a shield, but could deliver a sword as well.

These examples illustrate why nuclear arms control negotiations require a comprehensive, long-term approach to Soviet–American strategic capabilities. Treaties cannot embrace every possible threat or contingency, but neither should they result in vulnerable force structures because negotiators failed to comprehend the offensive–defensive relationships among strategic forces. It may also be worth noting potential unilateral remedies (Figure 2.1, Column 5) or countermeasures that can be taken outside the context of an arms control treaty to shore up United States defences against evolving vulnerabilities or to strengthen United States deterrent capabilities independent of treaty constraints. No treaty can lock all the doors to potential countermeasures.

Threats to strategic stability further complicate the distinction between offensive and defensive forces. If a large percentage of silo-based ICBMs, for example, could confidently ride out an attack, their offensive second-strike missions would be credible. But as fixed-based ICBMs become increasingly threatened, a 'use them or lose them' or 'defensive' pre-emption option becomes more attractive to a side with a countermilitary strategy.[5] When pre-emptive, damage limiting attacks seem rational, strategic stability vanishes, deterrence fails and distinctions between offensive and defensive roles and missions become almost totally obscured. Ideally, arms control agreements can promote strategic stability by reducing offensive forces to levels sufficiently low that deterrence can be achieved and maintained through unilateral programmes (e.g., mobile ICBMs, bomber alert rates, more submarines at sea).

The maintenance of strategic stability is a major objective for arms control. The American approach to the problem has correctly been concentrated on reducing Soviet ICBMs. The insistence of distinguishing between fast-flying (ballistic missles) and slow-flying (bombers and cruise missiles) strategic forces was initially a good strategy at START (the strategic arms reduction talks). The United States attempted to negotiate major reductions in ballistic missiles, arguing that they were the most destabilising systems given their short flight time, accuracy, high yields and constant state of readiness. The Soviets, not unexpectedly (since approximately 75 per cent of their strategic warheads are deployed on land-based missiles) argued that all nuclear weapons were equally dangerous. Ballistic missiles rely on speed, while bombers and cruise missiles rely on stealth. Treaty limits should therefore be aggregated in such a way that each party would retain the freedom to mix strategic forces in its own way.[6] For the Soviets, that meant protecting their considerable investment in land-based missiles.

The United States should continue to seek the reduction of Soviet ICBMs to their lowest possible levels but without clinging to a non-negotiable position that overdraws the saliency of the fast-flying–slow-flying distinction. Stability is largely a function of the ratio of hard-target capable warheads to vulnerable counterforce targets in a pre-emptive attack. The smaller the ratio, the greater the stability. Targeting ICBMs with ICBMs is not a solution to the problem of strategic stability. Bombers and cruise missiles can perform second-strike, counterforce missions if it is agreed that the only rationale (compatible with strategic stability) for strategic counterforce targeting is to preclude reloading of silos and retention of reserves. If the Soviets strike first, there is no reason why they could not be as prepared to launch any withheld missiles on short warning of retaliatory ICBMs as they would be on longer warning of approaching bombers armed with cruise missiles. Moreover, if the United States reduces its own ICBMs (and/or deploys mobile ICBMs) to expand bomber and cruise missile forces, it would reduce Soviet counterforce capability by trimming the target base against which ICBMs are useful.[7]

In summary, employment policies that are compatible with strategic stability could be achieved if the United States can exact favourable bomber for ICBM trade-offs. This could be done either through unequal but off-setting limitations on bombers/bomber weapons and missiles/warheads or through aggregating total counterforce capability with less emphasis on fast- *v.* slow-flying delivery vehicles. The Soviets may find such trade-offs to be an acceptable way of protecting their investment in heavy ICBMs from sudden, arms control-dictated changes. 'Sudden' is the key word in this context, since over time the vulnerability of fixed, land-based missiles will drive the Soviets to more survivable mobile ICBMs. The technological requirements of mobile ICBMs will indirectly move Soviet forces toward the United States objective of reducing the numbers of heavy missiles and the numbers of warheads each is capable of carrying.[8]

IV FORMAL TREATY LINKAGES BETWEEN OFFENSIVE AND DEFENSIVE SYSTEMS

SALT I formally linked offensive and defensive forces in the arms control process. The ABM Treaty and the *Interim* Agreement or SALT I (emphasis on 'interim') rest on the assumption that limiting strategic defence reduced first-strike incentives and contributed to a more stable deterrence. The United States expected that limitations on defensive forces would reduce the requirements for Soviet offensive forces. The inability to limit MIRVs in SALT I, however, guaranteed that attempts to drive down Soviet offensive forces through defensive limits would fail. Whether through strategic design or imitation, it was certain that in the absence of treaty constraints the Soviets would follow the American lead with MIRV deployments of their own. The results created, for the first time, a convergence of Soviet strategic doctrine with a theoretical capability to execute it.

The permissive nature of SALT I and the slow pace of follow-on negotiations leave reductions in offensive forces as the most enduring objective of arms control. The ABM Treaty codified the need for progress in the reduction of offensive nuclear weapons. Ambassador Smith's unilateral statement of understanding of this requirement added a specific time frame (five years) and stressed:

> The US Delegation believes that an objective of the follow-on negotiations should be to constrain and reduce on a long-term basis threats to the survivability of our respective strategic retaliatory forces. If an agreement providing for more complete strategic offensive arms limitations were not achieved within five years, US supreme interests could be jeopardized. Should that occur, it would constitute a basis for withdrawal from the ABM treaty.[9]

SALT II placed a ceiling on the growth of offensive systems, but at levels higher than those anticipated by SALT I strategists who attempted to gain Soviet offensive constraints indirectly through the ABM Treaty. As Ambassador Smith's statement makes clear, the Soviets were put on notice at SALT I that the failure of this strategy would constitute grounds for abrogation of the ABM Treaty. This long-standing position of linking defensive constraints to offensive reductions, combined with the President's SDI, will play a crucial role in future bargaining strategy.

V NEGOTIATING STRATEGY AND SDI

The Soviets are obviously concerned about future US programmes that may emerge under the rubric of SDI, and this could provide leverage for the United States if it is willing to accept constraints at reasonably early stages in the development of BMDs. The Soviets have made it clear in a variety of fora that the President's initiative would create a fundamental change in the Soviet–American strategic relationship. A succession of Soviet leaders have charged that the purpose of the 'new conception', in combination with strategic modernisation programmes, is to give the United States strategic superiority.

If operational linkages exist between offence and defence, then offensive and defensive weapons should not be decoupled for purposes of arms control. For the United States in the near term, the broader the negotiating agenda and the greater the linkage between offensive–defensive forces, the greater the negotiating leverage. Americans may be in a position to trade SDI constraints for both Soviet offensive force reductions and matching Soviet ABM constraints. Arms control leverage and bargaining strength, however, are time-sensitive commodities with a short shelf-life. If the Soviets match United States SDI 'initiatives' or breakout of the ABM Treaty, the United States will find the bargaining parameters narrowed to trading defensive constraints for defensive constraints. The opportunity for rolling offensive missile reductions into the bargain will have been substantially reduced, if not lost.

The issue of offensive–defensive trade-offs in the negotiations has been made more complex by the larger issue introduced by the United States in the form of a 'new strategic concept'. Reportedly drafted by Paul Nitze, the new American concept links the goal of deep cuts in offensive weapons with the development of strategic defences over a long, carefully-phased, transition period. During the next ten years the United States will seek a radical reduction (builddown) in offensive nuclear arms, followed by a period of mutual transition to effective non-nuclear defence forces as technology makes such options available. In a final 'ultimate period' strategic defences could make it possible to eliminate all nuclear weapons.[10]

The Soviets have made it clear that they reject the new strategic concept, and are not interested in turning negotiations into protracted seminars or tutorials on the virtues of strategic defence. The real issue behind the debate over sequential or simultaneous progress in the three Geneva negotiating fora (START, INF, and Defence–Space) is whether the United States is willing to bargain future SDI constraints on testing and development in exchange for deep reductions in offensive forces.

The START–SDI relationship may be the most difficult to reconcile because of the efforts required to explore new technological frontiers in the offensive–defensive relationship and the impact of those explorations on existing arms control treaties. In the near term, extensive research may increase Soviet incentives to negotiate new agreements and comply with existing ones. More certainly, R&D could eventually provide insurance against a sudden Soviet strategic breakthrough or long-term unilateral advantages gained through non-compliance or treaty abrogation. These issues are complicated by divisions within the Administration. Some officials want to use SDI as negotiating leverage; others seek to protect it from all forms of arms control-dictated constraints.[11] The debate is complicated even among the pro-arms control faction by questions of how far you can go in 'demonstrating' a new technology before violating existing agreements, and how one gets newly-demonstrated technologies to the bargaining table without precipitating Soviet countermeasures.

By the late 1980s pressure will mount (in proportion to funds expended) to test or demonstrate the new technologies that are expected to provide defence against ballistic missiles. At that point, decisions will have to be made about future ABM Treaty compliance.

Three major options seem the most likely:

1. Hold off testing BMD technologies until all efforts to negotiate reductions in Soviet offensive forces have been exhausted.
2. Negotiate ABM Treaty modifications with the Soviets which would allow testing of BMD components and new technologies.
3. Exercise the right to abrogate the ABM Treaty and move ahead with full-scale testing and deployment of BMD.

Option 1 is the most desirable for the future of a stable arms control regime. However, uncertain progress in both United States and Soviet development of BMD makes it difficult to assess United States leverage for SDI–START trade-offs.

The Soviets are not likely to agree on ABM Treaty amendments (option 2) if they think that the United States research programme is further advanced than their own or if they believe that those programmes can be constrained by domestic politics. The Soviets could drag out the negotiations for such amendments in the attempt to delay and kill United States programmes. They could (instead of or in addition to such delays) refuse to agree to such amendments, giving Washington the unhappy choice of either halting its programmes under Soviet pressure or withdrawing from the ABM Treaty (and thus being subject to criticism that the United States had torpedoed the Treaty). The latter could fit Soviet interests if they were ready to begin deployment of a large-scale traditional ABM system, while the United States was still some years away from an advanced BMD, or even state-of-the-art deployment, or possibly even a decision to deploy.

Abrogation of the ABM Treaty (option 3) has the same serious political consequences, and would make it difficult for the President to achieve the major goal of his SDI – to decrease reliance on offensive nuclear forces.

In the long term, if BMD becomes technically feasible and cost-effective, the Soviets may find that the value of ballistic missiles is reduced, which may increase the chances of negotiated reductions. Alternatively – and more likely in the short- term – they may develop active (offensive) and passive (defensive) countermeasures to future defensive systems.

VI SDI AND SOVIET COUNTERMEASURES

SDI and its related research and development programmes are aimed at producing a multilayered, multitechnological approach to BMDs. Attacking ballistic missiles in each phase of their flight with weapons that destroy them in different ways forces the offence to attempt the difficult task of overcoming various threats (see Figure 2.2.).[12]

The earliest or boost phase of a missile's flight segment starts from launch and extends only a few minutes (2–5) into flight. During this time a missile produces an extremely intense heat or infrared signal that can be picked up by space-based detection and warning systems. Because the missile is carrying all its nuclear warheads, decoys and penetration aids, it is a very high-value target for the defence. The capability of interception is critical for a credible layered defence system. Unfortunately, during the first 200,000 feet of a missile's ascent it is relatively immune from attack either because United States defences may not be able to react quickly enough or because many defensive technologies cannot penetrate deeply into the earth's atmosphere from their space-based orbits. For these reasons many scientists are not optimistic that reliable boost-phase intercepts are possible with any technology visible on the horizon.

The second or post-boost phase occurs after the missile's first stages have burned out and fallen away from the post-boost vehicle (PBV) or 'bus' which carries the warheads and decoys. The warheads

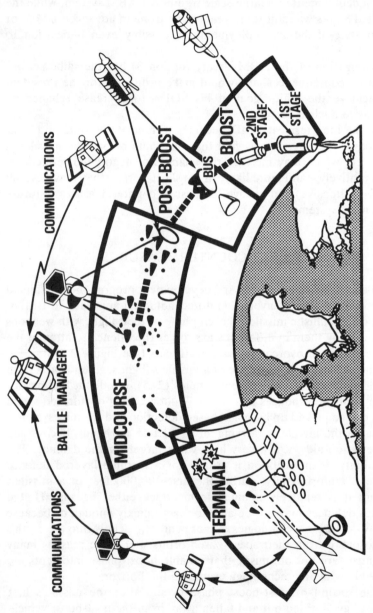

Figure 2.2 Layered defence concept

and decoys are released during a sequence of small, controlled manoeuvres. The 'bus' generates small infrared bursts of energy as it manoeuvres, and it is therefore still detectable from space-based sensors. However, once warheads, chaff, decoys and other objects are released, detection becomes far more difficult. The PBV is therefore a high-value target that quickly declines in value as it launches each of its warheads on separate trajectories to their targets.

During the third or mid-course phase, lasting 15–20 minutes, the released warheads and decoys follow their predetermined trajectories. Discrimination between threatening and non-threatening objects by defensive sensors is made difficult by the potentially large number of decoys that could be deployed. Detection is further complicated by the fact that warheads and decoys cool rapidly to a temperature very close to that of the surrounding space, making it difficult for infrared sensors to acquire and hold any target – much less distinguish nuclear warheads from thousands of other objects, and fire on them in the relatively short period of time involved.

During the final re-entry phase the warheads re-enter the atmosphere. Detection again becomes comparatively easy because of the heat generated during re-entry and because lightweight decoys and chaff burn up. Unfortunately, re-entry lasts only 30–100 seconds. Detection and interception must take place almost instantly to avoid offensive–defensive duelling literally overhead.

These technical challenges will face defence advocates even if the Soviets played the role of co-operative adversary. The latter is not likely, and prudent planners will have to anticipate a variety of Soviet countermeasures. Several of these have already been identified in United States studies and by the Soviets themselves.[13] Figure 2.3 summarises many tactics that strategic defences will be required to overcome, including:

1. Proliferation of missiles and warheads to exhaust defences in each phase or layer.
2. Reflective coatings and spin boosters to serve as protection from lasers.
3. Increased thrust and/or reduced payload to reduce boost phase to less than 1 minute.
4. Deployment of ASATs, including space mines, to degrade space-based defences. (As Robert DeGrasse and Stephen Daggett have pointed out, the ultimate cost and complexity of a space-based system may be affected by the need to

TRAJECTORY PHASE	HARDENING	DECOY	DATA DENIAL	EXHAUST	EVADE	ATTACK DEFENSE
Boost	•Coatings: Ablative Reflective •Spin Booster	•Dummy Boosters •Smoke/Flare Generators	•Jam with ground-based systems •Camouflage •Smoke	•Mass Attacks •Increase Payload •Mobile Basing	•Fast Burn Boosters •Depressed trajectories	•ASAT •Space mine
Post Boost	•Harden bus, •Single RV missiles		•Mask with Aerosols	•Multiple buses	•Fast burn buses	•ASAT •Space mine
Midcourse	•Harden RVs against X-rays	•Chaff •Baloons •Optical & LWIR Decoys (Long Wave Infrared)	•Aerosols •Radar & optical jamming	•MiRVs		•ASAT •Space Mine
Terminal		•Optical, radar & LWIR decoys	•Optical & radar jamming	•MiRVs	•MaRV •Salvage Fusing •Cruise Missiles •Stealth Technology	•Precursor Attacks

Figure. 2.3 Countermeasures to Advanced Ballistic Missile Defenses

defend itself from pre-emptive attack. Like aircraft carriers, the cost of protecting the primary assets may exceed the price of the firepower the system delivers.)[14]
5. Deployment of thousands of decoys or chaff to avoid mid-course detection. (This includes the use of masking aerosols and hiding warheads inside specially constructed balloons to make them indistinguishable from decoys.)
6. Manoeuvrable warheads (MaRVs) to evade terminal defences.

Additional countermeasures worth mentioning in greater detail include traditional military tactics of mass and dispersal. Defences must be dispersed to protect against Soviet deployments of offensive missiles across the full breadth of the Soviet Union and at sea. The offence has a tactical advantage, since it can disperse its forces at sea while massing them on land through additional deployments of silo-based and mobile ICBMs (both road- and rail-mobile ICBMs are probable) to concentrate attacks in the traditional style of Soviet military doctrine.

Soviet attempts to exhaust defences by maintaining a superior balance of offensive forces would be costly, and Soviet leaders would undoubtedly harbour grave doubts about their adequacy to execute Soviet strategic doctrine. Saturation attacks against a portion of the satellite fleet that is over the target area at any given time, however, is more feasible and would require defensive responses in the form of additional battle stations or technical breakthroughs that would allow quick orbital changes and the corresponding capability to mass defensive forces. The fuel requirements and response time for orbital flexibility make this option unattractive from both cost and battle management perspectives. Additional battle stations could drive costs closer to the estimates of critics and further weaken political support for strategic defence.[15]

SDI's concentration on BMD could also encourage the Soviet Union to circumvent space-based defence through massive deployments of cruise missiles carried on long-range bombers, submarines and surface ships. No defence would be complete without multibillion dollar programmes for air defence to meet these new threats.[16] Just as offensive nuclear forces have been structured around the triad of land-based missiles, sea-launched missiles, and long-range bombers, so too will an era of defence dominance be forced to deal with all three legs of the strategic defence triad – missile defence, air defence and civil defence.

Conventional force postures may also be affected by strategic defence. If defences succeed in limiting the threat from offensive nuclear forces, what would success look like for NATO? Would the world be safe for conventional warfare, and if so how might that translate into additional requirements for procurement, personnel and conventional weapons that would be needed to deter the Soviet's massive capabilities on the ground? How would the United States strategy of extended deterrence be affected if NATO's nuclear umbrella disappeared in a 'defence dominant' world? NATO allies could face the prospect of increased demands on resources to deter conventional aggression in Europe at a time when an abrogated ABM Treaty exposes them to rapid Soviet deployments of state-of-the-art ABM systems that would be effective against limited British and French nuclear deterrents. In a 'defence dominant' world the deterrent value of these forces would give way to the conventional balance and battlefield nuclear weapons. Short-range, low-yield, but predictably large numbers of tactical nuclear weapons would ensure the continued high cost of fielding and protecting the full spectrum of conventional forces required to fight Europe's air–land battle.[17]

The viability of each Soviet countermeasure is, like SDI technologies, subject to debate and the assumptions of various studies and advocates. Nevertheless, if only a few of the many responses available to the offence are feasible, deployment of defensive systems will have to be very cost-effective at the start to overcome domestic opposition.

In summary, a vigorous SDI programme is not only likely to be expensive, but it could undercut efforts to get reductions in ballistic missiles, and could also trigger new and unconstrained competition in both offensive and defensive systems. Certainly the United States could not risk neglecting its offensive forces in the face of massive Soviet offensive countermeasures to BMD programmes. The evolution of American technologies to counter those countermeasures may be possible, but only at correspondingly increased system complexity and cost. And even these defensive counters will be just one more step in the never-ending contest between offence and defence.

Attempts to constrain the offensive–defensive competition forecast by critics will be complicated by many uncertainties. Efforts to limit defensive weapons and related counting rules required by both present and future arms control regimes will be obscured by as yet unanswered technical questions. For example, how many satellites will be required to defend against a growing Soviet offence?[18] What are their rates or volume of fire, their sustainability, their forms of energy regeneration, and/or the reload/reconstitution capabilities of the various space and ground-based systems under R&D? Answers to these basic questions are critical to any assessment of the systems cost-effectiveness or its capability of staying ahead of offensive countermeasures.[19] The technical and economic feasibility of the transition to a 'defence dominant' world cannot be known until technical developments make clear choices of systems possible. The ABM–START–SDI dilemma is that answers to the questions of systems, force levels and capabilities probably cannot be learned in an environment where previous arms control agreements have survived or where new agreements are being negotiated.

As these and other SDI issues spill into public debate, offensive–defensive interactions will become increasingly visible and divisive. As they do, they are likely to undermine the President's major selling point – that defensive systems can decrease reliance on offensive nuclear weapons. This is especially true in the face of conflicting statements on SDI's mission in strategic policy. The President has repeatedly described SDI as a step leading to the total defence of the American people. Secretary Weinberger has gone even further in describing SDI as 'morally right' since 'it is an attempt to devise a system that protects our people instead of avenging them'.[20]

No one closely associated with defensive technologies, however, believes that a leakproof system required for defending the American urban population is likely. Strategic defence can protect the deterrent force and decrease incentives for an attack on the United States. This is the only conceivable and credible mission for SDI, and it should be clearly stated as part of the debate. Fred Iklé, Under Secretary of Defence, has defined the mission of a defensive system in precisely these terms when he revealed that the Administration might build an 'interim' system that would protect nuclear missiles, but not cities.[21] When so stated, defensive technologies can be more readily contrasted with arms control as an alternative to reducing the Soviet offensive threat.

Finally, it becomes essential to view American SDI and strategic modernisation programmes as part of a single strategy that feeds Soviet strategic planning. As long as our strategic modernisation programme includes unconstrained hard target systems – MX, D-5, and possibly cruise missiles – the Soviets will have few incentives to reduce their offensive weapons. The ultimate 'layered defence' includes the capability to destroy a high percentage of Soviet ICBMs in their silos or support bases (in the case of mobile ICBMs). The deployment of 'interim' defences makes the threat from offensive systems all the more credible. As long as the Soviets confront the dual threats of strategic modernisation programmes and SDI with interim stages, there will be little chance of achieving the President's goal of moving toward a 'defence dominant' world while negotiating offensive arms reductions. These twin goals are hopelessly in conflict.

VII SDI AS AN ARMS CONTROL DILEMMA

The strategic and political difficulties facing SDI are as complex as the technical barriers. Domestically, timely and adequate funding from Congress will confront a maze of contentious economic, technical, strategic and arms control debates during which SDI must enjoy technological successes and a sustained political consensus to maintain costly, multi-year programmes. Internationally, the viability of SDI will depend on United States ability to limit offensive weapons. SDI and arms control must, therefore, be carefully co-ordinated rather than made to appear antithetical. An arms control strategy that co-ordinates offensive and defensive systems requires the preservation of the negotiating process to provide a forum for reducing the level of offensive nuclear weapons. This will almost certainly require the continuation of interim restraint (abiding by SALT I and II ceilings) on offensive arms and non-abrogation of the ABM Treaty. This, in turn, will require a clear commitment to accept future constraints on SDI testing and deployments.

Offensive–defensive co-ordination would also require major revisions in previous START proposals. Despite considerable rhetoric about deep cuts, both sides have attempted to protect very large numbers of offensive weapons. The initial ceilings proposed at START in 1983 were so high and the prohibitions on specific weapons so few that both sides' positions were designed to support an offence dominant world and thus were inconsistent with more recent movement towards strategic defence. If the United States wishes to encourage effective defences, for instance, it must not encourage:

1. MaRVs as a principal means of penetrating terminal ABMs.
2. Air- and sea-launched cruise missiles (SLCMs) which would require large and costly air defence systems.

3. The proliferation of new types of ballistic missiles that would facilitate the development of countermeasures to SDI (fast-burn boosters, reflective coatings, spin boosters).
4. High ceilings for sea-launched ballistic missiles, since the position uncertainty and short flight times make defence against them more difficult.
5. ASAT testing and deployments, since they are major threats to C^3I, (command, control, communications and intelligence), battle management and satellite battle stations.

Limiting offensive–defensive competition requires mutually balanced constraints. Ascendancy on one side of the equation by either party is certain to provoke short-term countermeasures on the other side and long-term competition in both offensive and defensive technologies. In summary, Negotiating leverage and willingness to accept constraints in both offence and defence are two sides of the same coin. Failure to compromise will surely result in a Soviet effort to match United States defences. Mutual fear that the other side may develop offensive means to penetrate defences will fuel competition in both. These are the strategic hurdles to reaching the President's goals of offensive reductions in a defence dominant world. Arms control that includes both defensive and offensive constraints and unilateral basing and deployment solutions to the problems of weapons vulnerability offer the most cost-effective and stable solutions to the twin problems facing strategic forces – survivability and penetrability.

This, in essence, was the conclusion of the Scowcroft Commission Report endorsed by the President on 29 April 1983.[22] The bipartisan report recommended the deployment of a small, mobile ICBM, smaller and more numerous submarines, improved communications and increased penetration effectiveness of retaliatory forces. These steps are more likely to lead to successful arms control negotiations and strategic stability than the offensive–defensive arms competition that is likely to grow out of The President's SDI.

VIII CONCLUSION

Long-term strategic stability as defined here requires arms control stability. Arms control stability grows out of an environment of formal constraints and well-defined force structures in both offensive and defensive systems sufficiently secure that neither side feels compelled to deploy new systems to match or offset the deployments of the other. Strategic stability and arms control stability are indivisible.

The proponents of SDI must avoid what Freeman Dyson has described as the 'technical-follies future'.[23] The 'technical-follies' are the open-ended offensive–defensive arms race that surely will be precipitated by unilateralists who advocate deployment of strategic defences even in the absence of arms control. The unilateralists who would attempt to create a security regime based solely on technology could never be at rest. Soviet responses would, sooner or later, negate the best efforts of those who abandon diplomacy in their search for security through technology alone. Science, arms control and military strategy must succeed together, or SDI risks becoming America's technological Vietnam.

Notes

1. 'Strategic stability' is defined here as a condition of the military balance that, in a crisis, offers the Soviet Union no incentive to initiate a nuclear attack. Neither is the United States under pressure to do so. By contrast, 'strategic instability' would be a condition in which either the United States or the Soviet Union or both believed that victory could be achieved and defeat averted only by striking the other side pre-emptively.
2. *Weekly Compilation of Presidential Documents* 19, 28 March 1983 (Washington, DC: United States Government Printing Office, 1983) p. 448.
3. The three studies were: (1) The Fletcher Panel (Defense Technology Study), (2) The Hoffman Study Group (Future Security Strategy Study), and (3) The Miller Study (Department of Defense study on policy and strategic stability). Only the Hoffman study is unclassified and available to the public. The weapons to be researched for their feasibility include high-energy lasers, particle beams, microwaves, and conventional technologies (for example, kinetic energy). Layered defence could include space-, air-, ground- and sea-based deployments of all or combinations of these weapons.
4. Studies of Soviet and United States strategic doctrine are too numerous to cite. Those that compare and contrast the Soviet and American approach to strategic doctrine include Fritz Ermarth, 'Contrasts in American and Soviet Thought', *International Security* 3 (Fall 1978); Benjamin Lambeth, *Selective Nuclear Options in American and Soviet Strategic Policy*, R-20034-DDRE (Santa Monica, California: Rand Corporation, 1976); and Dennis Ross, 'Rethinking Soviet Strategic

Policy: Inputs and Implications', *Journal of Strategic Studies* 1 (May 1978). For the evolution of United States Strategic doctrine see Fred Kaplan, *The Wizards of Armageddon* (New York: Simon & Schuster, 1983); Leon Sloss and Marc Dean Millot, 'US Nuclear Strategy in Evolution', *Strategic Review* (Winter 1984) pp. 19–28; and Desmond Ball, *Targeting For Strategic Deterrence, Adelphi Paper* 185 (London: International Institute for Strategic Studies, 1983).

5. The Department of Defense specifically rejects such a strategy for the United States. The Soviets through former Minister of Defence, Dimitri Ustinov and Giorgi Arbatov, head of the Institute for the Study of the United States and Canada, have stated they might move towards a strategy of launch on warning if their land-based missiles become vulnerable to a pre-emptive strike. See *Washington Post*, 11 April 1982, p. 5, and *Los Angeles Times*, 18 July 1982, p. 3E.

6. Discussed in Strobe Talbot, *Deadly Gambits* (New York: Alfred A. Knopf, 1984) p. 282.

7. This point is made by Richard K. Betts in 'Elusive Equivalence: The Political and Military Meaning of the Military Balance', in Samuel Huntington (ed.), *The Strategic Imperative* (Cambridge, Mass.: Ballinger, 1982) p. 126. This also assumes that in a protracted nuclear war, the United States could maintain an intelligence capability to 'see' Soviet targets and distinguish empty from unfired silos.

8. United States bomber and cruise missile advantages that offset Soviet ICBM advantages will, as Figure 2.1 indicates, depend on the continued penetrability of United States forces. Unconstrained Soviet modernisation of air defences will, at some point, make this trade-off unattractive. The reverse could also be true with unconstrained United States BMDs.

9. *Arms Control and Disarmament Agreements* (Washington, DC: United States Arms Control and Disarmament Agency, 1982) p. 146.

10. The 'new strategic concept' was first made public in a little-noticed speech to the Foreign Policy Association by Deputy Secretary of State, Kenneth W. Dam, *Geneva and Beyond: New Arms Control Negotiations, Current Policy 647* (US Department of State, Bureau of Public Affairs, 14 January 1985). Ambassador Nitze further developed the concept in a speech before the Philadelphia World Affairs Council on 20 February 1985, and in his testimony before the Senate Foreign Relations committee on 26 February 1985. Cynics have already noted that Nitze's 'ultimate period' in which nuclear weapons wither away is matched in its idealism only by the Marxist theory of the 'withering away' of the state.

11. The President's Science Advisor, George Keyworth, called space-based defences 'inevitable', *Baltimore Sun*, 18 September 1984, p. 1. Undersecretary of Defense Fred Iklé has stated, 'the US should not trade its "star wars" research program for reductions in Soviet nuclear missiles'. Quoted in *Washington Post*, 27 October 1984, p. A16. Ambassador Rowny has favoured demonstrating the technical feasibility of Star Wars before attempting to trade it against Soviet offensive reductions. Lt General James Abrahamson, head of the strategic defence organisation at the Department of Defense has suggested a 'three-year'

deployment delay in exchange for a reduction of 500 Soviet warheads. Quoted in *Christian Science Monitor*, 29 October 1984, p. 3. In a *New York Times* interview on 12 February 1985, and during his send-off speech to the United States Delegation on 8 March 1985, the President stressed that the United States would not negotiate limits on SDI research or testing. Deployment of strategic defences has never been specifically ruled out of current or future negotiations.

12. The layered defence concept is described in *Directed Energy Missile Defense In Space – A Background Paper* (Washington, DC: United States Congress, Office of Technology Assessment, 1984); John Tirman (ed.), *The Fallacy of Star Wars* (New York: Vintage, 1984), chapter 5 and 6; Sidney Drell *et al.*, 'Preserving the ABM Treaty: A Critique of the Reagan Strategic Defense Initiatives, *International Security* 9 (Fall 1984) pp. 67–83; *Defense Against Ballistic Missiles: An Assessment of Technologies and Policies Implications* (Washington, DC Department of Defense, 6 March 1984); and Leon Sloss, 'The Return of Strategic Defense', *Strategic Review* (Summer 1984) pp. 37–44.

13. For example, *Directed Energy Missile Defense*, pp. 45–54, and Tirman (ed.), pp. 21–6. The Soviets have also openly discussed countermeasures. See, for example, Committee of Soviet Scientists for Peace Against Nuclear War, 'Strategic and International Political Consequences of Creating a Space-based Anti-Missile System Using Directed Energy Weapons' (paper presented at the Meeting of Delegates of the United States National Academy of Sciences and the Academy of Sciences of the USSR on Problems of International Security and Arms Control, Moscow, 8–11 May 1984) pp. 21–6. It should also be noted that Soviet sources which emphasise countermeasures against SDI have great propaganda value and should not be taken as official Soviet views on the real technological potential of SDI research.

14. William D. Hartung *et al.*, *The Strategic Defense Initiative: Costs, Contractors, and Consequences* (New York: Council on Economic Priorities, 1985) p. 18.

15. See note 18.

16. Major General John A. Shaud, Director of plans for the Air Force stated, 'If you're going to fix the roof, you don't want to leave the doors and windows open'. Air Defence requirements against Soviet cruise missiles have been discussed in *Washington Post*, 25 August 1984, p. A4, and in 'Air Defense: Protecting America's Skies', *Heritage Foundation Backgrounder* (Washington, DC: Heritage Foundation, 13 September 1984) (mimeo). Secretary of Defense Weinberger has also stated that space defence will have to be backed up by radars and planes to protect North America from bombers. See *New York Times*, 17 January 1985, p. A1.

17. For these and other reasons the European press has been largely negative toward the idea of a strategic defence for the United States. For a thorough discussion of NATO political and military reaction see Paul E. Gallis, Mark Lowenthal, and Marcia Smith, *The Strategic Defense Initiative and United States Alliance Strategy* (Washington DC: Congressional Research Service, 1 February 1985). The Chinese are

also opposed since unconstrained Soviet defences could negate their strategic forces.

18. No single issue in the strategic defence debate is more divisive than satellite numbers. Estimates vary widely, depending on often unstated assumptions about technical performance, Soviet countermeasures, and the type of defence being described. For example, satellites required against currently deployed Soviet ICBMs:

   1. Union of Concerned Scientists   300
   2. Office of Technology Assessment   160
   3. Livermore Laboratory   90
   4. Drell, Farley, and Holloway   320
   5. High Frontier   432
   6. Brzezinski, Jastrow, and Kampelman   114

   Resolution of these differences is essential to the future of SDI since, as Robert Jastrow estimates, each satellite could cost as much as an aircraft carrier. The methodological battle over calculations of these numbers remains intense.

19. According to testimony by James A. Thomson, Director of National Security Strategies Program at the RAND Corporation, the offence will have an advantage under all circumstances. A strategy for deterrence that relies solely on the defence's ability to deny damage to a determined Soviet attacker is beyond United States reach. Testimony before the Defense Appropriations Sub-Committee of the House Appropriations Committee, 9 May 1984.

20. Speech before the Pittsburgh World Affairs Council, Department of Defense Press release, 30 October 1984.

21. *Washington Post*, 27 October 1984, p. A16. 'Interim' defence means defence of ICBM sites against the terminal phase of an ICBM attack. See Department of Defense, *Defense Against Ballistic Missiles: An Assessment of Technology and Policy Implications* (Washington, DC: Department of Defense, 1984) p. 8.

22. President's Commission on Strategic Forces, *Report of March 21 1984* (mimeo) pp. 9, 11, 20, 21.

23. Freeman Dyson, *Weapons and Hope* (New York: Harper & Row, 1984) chapters 5 and 7.

3 SDI – a Reaction to or a Hedge Against Soviet BMD Projects: Soviet Military Space Activities and European Security
Hans Günter Brauch

The Strategic Defence Initiative (SDI) of the United States and improved Soviet capabilities for ballistic missile defence (BMD) will have dramatic implications for European security in general, for the future of East–West relations and in particular for NATO's strategy of 'flexible response'.

In his address on 23 March 1983 President Reagan offered a vision beyond the presently valid doctrine of mutual vulnerability or mutual assured destruction (MAD), which appeared to have been the common strategic rationale of both the United States and the Soviet Union since the initiation of the SALT process in 1969. His answer to the nuclear dilemma was: 'We embark on a program to counter the awesome Soviet missile threat with measures that are defensive'. After calling on the scientific community 'to give us the means of rendering these nuclear weapons impotent and obsolete', President Reagan concluded by characterising his decision as one 'which holds the promise of changing the course of human history'.[1] Did President Reagan *initiate* a strategic revolution, did he merely *react* to past and present Soviet BMD and military space activities, or did he intend to provide a *hedge* against Soviet violations of the 1972 ABM Treaty or potential BMD 'breakout' options in the future?

President Reagan's 'Star Wars' address obviously supports Theodore J. Lowi's thesis that 'Presidential behaviour since World War II can be summarised as "oversell": The President has been forced to (1) oversell the crisis and (2) oversell the remedy. These are the continuities in the formulation of foreign policy'.[2] President Reagan's effort to oversell the remedy was followed by official and unofficial efforts to oversell the threat posed by Soviet BMD and other military space projects, and by accusations of Soviet violations of the ABM Treaty.

Any assessment of Soviet military space projects and BMD capabilities is confronted with the dilemma, as Michael Deane notes, that Soviet military literature normally 'reveal[s] very little' about Soviet operational capabilities and weapons, but there is an 'absolute void with regard to Soviet anti-missile and anti-space weaponry'.[3] Therefore the analyst who wants neither to neglect observable Soviet military space activities nor to become an apologist of Soviet diplomatic and scientific rhetoric to stop the 'militarisation of outer space' has to rely almost exclusively on Western (and predominantly American) sources: official governmental reports, testimony in the United States Congress, intelligence leaks to or speculations in the trade press and on the interpretations of this limited and often biased evidence by different schools of Soviet military experts. These analyses often play a great role in policy formulation, or in its subsequent legitimisation. Any effort to assess Soviet military space activities and potential Soviet reactions to the American SDI must therefore be of a tentative nature. Methodologically, the date of President Reagan's 'Star Wars' speech in March 1983 may assist in the interpretation of official United States pronouncements about Soviet military space activities. May drastic increases in the nature of the threat posed by Soviet BMD and military space activities be interpreted as an inherent part of the 'oversell mechanism'? The more the effort of the Reagan Administration to 'oversell the remedy' of a vision beyond deterrence has been challenged in the United States, the more the traditional technique 'overselling the threat' has re-emerged. Has SDI – which was praised as a major unilateral political and strategic decision to redirect the longer-term strategic goals by changing today's research priorities[4] – become another necessary reaction to a unilateral Soviet 'edge in active and passive defences'?[5] This chapter will focus on the following two questions:

1. What is the status of Soviet space weapons projects (ASAT, laser and exotic beam weapons), of ground- and space-based BMD and of ATM capabilities?
2. May President Reagan's SDI be interpreted as a *unilateral decision* unrelated to Soviet military space activities, as an *unavoidable reaction* to Soviet BMD capabilities and programmes or as a *hedge* against Soviet violations or future breakouts of the 1972 ABM Treaty?

I SOURCES AND INTERPRETATIONS OF SOVIET MILITARY DOCTRINE

Four major sources about Soviet space-related weapons will be distinguished here:
- (a) Soviet official pronouncements by its political and military leaders and articles by Soviet military and civilian defence and foreign policy experts;
- (b) United States official assessments in Congressional testimony and in governmental reports, for example in *Soviet Military Power*;
- (c) official leaks or speculations that have been reported in military trade journals, for example in *Aviation Week & Space Technology*;
- (d) assessments and interpretations of Soviet military activities in the academic literature by independent or Pentagon-sponsored civilian and military experts.

(a) Soviet sources on military doctrine and military space activities

Soviet publications distinguish between military doctrine and military science. In the Soviet conception, as interpreted by David Holloway:

> Doctrine embodies the fixed positions of the state on questions of war and military policy; military science, on the other hand, consists of the study of war and of the methods of waging it and is thus constantly developing. Doctrine has to take into account the economic and political conditions of the state, while military–scientific research is not constrained in this way. Doctrine is likely to remain stable for some time, being revised only in response to major political or military developments; military science, on the other hand, is constantly advancing as it tackles new problems. Doctrine is defined by the Party leadership, while military science is largely the prerogative of the General Staff and the military academies. Doctrine expresses the political character and purpose of the state, but draws on military science in the formulation of its military–technical side.[6]

In analysing the relationship between doctrine and technology in the publications of Soviet authors, in Soviet history and in the context of the weapons acquisition process, Holloway found that this 'relationship is seen as one of complex interdependence' whereby 'technological change exerts a determining influence on the methods of warfare, but doctrine has a significant role in adapting those methods to new weapons and in discerning at an early stage the military utility of new technologies.[7] In a historical perspective Holloway interprets doctrine not as 'a function of technology alone' but as one that 'is deeply rooted in political, cultural and economic conditions. Military technology, on the other hand, is not merely a function of doctrine, but rests on industrial and technological power'.[8] And in the weapons acquisition process according to Holloway 'doctrine' is the 'organising principle' in the formulation of a project. Given these complex relationships between military doctrine formulated by the political leadership and military science developed by the professional military and military technology which relies on the research, development and industrial capabilities of the Soviet Union, an analysis of Soviet military space activities has to rely on the following six sources:

1. general political statements of the Soviet political leadership, of the members of the Politbureau and of the Central Committee of the CPSU who have an impact on the formulation of military doctrine;[9]
2. publications in the military journals for professionals who may contribute either to the development of military science or to the education and indoctrination of the troops;[10]
3. statements by delegates to the major arms control negotiations;[11]
4. publications of Soviet foreign and defence experts in the major institutes of the Academy of Science, who often act as interpreters of Western developments to the Foreign Ministry and to Party bodies;[12]
5. the demonstration of Soviet hardware capabilities at military parades and the testing of new strategic and space weapons as perceived by the national technical means of the United States.[13]
6. the analysis of publicly available remote sensing data, for example from American (LANDSAT) and European (SPOT) satellites.[14]

However, as the case of the United States demonstrates, declaratory politics is hardly identical with the operative military doctrine or with the nuclear employment policy of the Soviet Union, and Soviet negotiators have been extremely reluctant to discuss aspects of their operative doctrine with their American counterparts.[15] Given the tight secrecy relating to Soviet military space programmes and R&D projects, all evidence relating to Soviet military hardware is based exclusively on Western sources (primarily United States official publications, unofficial leaks and independent assessments by the IISS[16] and SIPRI)[17].

(b) United States official assessments of Soviet military capabilities

Most details about Soviet military hardware programmes have been published by both the executive and the legislative branches of the United States government, for example in:

1. the Annual Defence Posture Statements of the Secretary of Defence and of the Joint Chiefs of Staff, as well as the Chiefs of Staff of the individual services;[18]
2. in testimony to the Armed Services, Appropriations and Intelligence Committees of both the Senate and the House of Representatives,[19]
3. in published assessments of the CIA, the Defence Intelligence Agency (DIA), of other intelligence services as well as the Government Accounting Office, the Congressional Budget Office, the Office of Technological Assessment and the Congressional Research Service.[20]

In order to focus specifically on Soviet military activities the United States Department of Defense has published four editions of *Soviet Military Power*, two of them before President Reagan's Star Wars speech (in September 1982 and March 1983) and two thereafter (in April 1984 and April 1985).[21] In October 1985 the United States Department of Defense and the State Department released a joint publication: 'Soviet Strategic Defense Programs'.[22] One method available to the analyst is to review these official United States sources for their consistency or for the major changes in the background of the independent expert discussions in the United States and in other Western and neutral countries.

(c) Reports in the media and in the trade press on Soviet developments

Details about Soviet military hardware programmes and research and development projects are often reported in United States dailies, news magazines, in newsletters for the defence industry and in the so-called trade press. These publications often refer to secret United

States intelligence assessments or to secret reports of branches of the United States Department of Defence which have been leaked deliberately to the press during budget debates as an additional public boost for their own defence projects in the United States Congress. Sometimes they originate from former high-level intelligence officials. One analytical method is to confront alarmist projections of the past with available evidence of the present. However, Soviet secretiveness about its own hardware programme assists suspicion and alarmism in the defence-minded Western trade press which cannot be easily refuted by an independent expert.

(d) Western academic opinion about Soviet military policy, Soviet military doctrine and Soviet hardware projects

The interpretation of the role of the Soviet Union in world politics, according to Klaus von Beyme, has oscillated 'between two poles: primacy of domestic politics or primacy of foreign policy'.[23] While adherents of the totalitarian model have emphasised the change in domestic power structures proponents of the convergence model of Soviet foreign and defence policy have increasingly stressed the impact of international relations on Soviet activities in world affairs, especially since 1945.[24] With the gradual shift from the totalitarian to the convergent paradigm in the 1960s, many Western analysts have also downgraded the role of ideology as a major motivating force in relation to the pursuit of national interests of the Soviet state.[25]

In discussing the perceived contradiction between ideology and pragmatism in detail, Adomeit has argued that 'the ideological content of Soviet foreign policy is equivalent to the degree of Soviet support to world revolution, more specifically, the extent to which the Soviet Union is willing to employ military force on behalf of local communists in various areas of the world'.[26] For Adomeit the main contradiction is 'between *Rechtfertigungsideologie*' and '*Antriebsideologie*', the argument being that the Soviet state is indeed an ideology in power but that ideology is merely providing legitimacy to action (*Rechtfertigung*) and can no longer be regarded as a guide to action, i.e. furnishing motivation (*Antrieb*). Proof of this thesis is derived from the undoubtedly valid observation that 'Marxist–Leninist doctrine has served to justify all sorts of policies'.[27]

In the present debate on Soviet military doctrine the political leadership has focused primarily on the peaceful coexistence between different social systems, admitting that a victory in a nuclear war is unlikely,[28] while those military writers primarily concerned with the education and the motivation of the Soviet troops have repeatedly stressed the goal of winning and victory if the imperialists should start a major war against the Soviet Union or the socialist camp.[29]

One reason for the controversial Western assessments of Soviet military doctrine may be seen in the different focus of the two major schools of analysts. While the *hawks*[30] prefer to discard most statements by the Soviet political leadership and by the civilian 'interpretors' as pure propaganda aimed at the erosion of motivation for defence in the West and focus primarily on that part of the Soviet literature that is to develop 'military science' and to create motivation of the troops, many *doves*[31] take Soviet political rhetoric and Soviet political behaviour during the détente period more seriously while downgrading the political relevance of the literature, which plays a major role in the indoctrination of the troops.

While the former school – which has had a major impact on the Reagan Administration, and on many conservative politicians in Western Europe as well – argues that the Soviet Union never accepted the strategic assumption of mutual vulnerability (MAD), the latter school – which has provided the intellectual foundation for a policy of co-operation with the Soviet Union – rejects the thesis that the Soviet Union believes that it can fight and win a nuclear war, and argues instead that the Soviet Union has adopted a strategic policy based on the assumption of mutual vulnerability. While the *hawks* generally are supportive of President Reagan's SDI proposals and have often contributed to its legitimisation,[32] the *doves* (or the proponents of a co-operative policy with the Soviet Union) have tended to be rather critical of SDI.[33]

Hart distinguished among six different schools of thought on Soviet military doctrine, which are 'currently vying for decision-makers' attention' in the United States:

(a) primitivism;[34]
(b) convergence;[35]
(c) neoClausewitzians;[36]
(d) Talmudic;[37]
(e) imperial;[38]
(f) eclectic.[39]

He concluded that 'the present configuration of views on the Soviet Union consists basically of a coalition of primitivists and advocates of convergence [representing the doves] arrayed against the neo-Clausewitzians [the hawks]'.[40]

In order to avoid 'all-inclusive, single factor, interpretative models' Hart suggests 'less ambitious mechanisms that focus on specific aspects of Soviet military thought'.[41] As far as Western interpretations of Soviet doctrinal writings on nuclear weapons, ballistic missile defence and military space activities are concerned, we may distinguish two major schools of interpretation:

1. those who stress *'nuclear warfighting'*[42] and *'damage limitation'*[43] as the primary focus of Soviet military doctrine (hawks or neoClausewitzians); and
2. those who refer to *'mutual vulnerability'*[44] and *'deterrence'*[45] as the dominant Soviet strategic premises (doves or those who represent the convergence, imperial, primitivist and eclectic schools).

While the latter school – based on careful historical and systematic analyses of the changes which occurred in Soviet doctrinal writings and justifications since the mid-1950s after Stalin's death – favours a policy of co-operative arms management, the former questions Soviet treaty compliance and favours a policy of unilateral United States procurement decisions.

II BMD, SOVIET MILITARY SPACE ACTIVITIES AND SOVIET MILITARY DOCTRINE

(a) Soviet views and interpretations of Soviet military doctrine

The Soviets distinguish among the following three terms: military doctrine (*doktrina*), military science (*nauka*) and military art (*voyennoye iskusstvo*). According to the *Soviet Dictionary of Basic Military Terms* (1965), 'military doctrine' is the officially accepted system of scientifically founded views of a nation on 'the purposes and character of possible war, on the preparation of the country and the armed forces for it, and also on the methods of waging it'.[46] In the Soviet view, military doctrine has two aspects: political and military–technical.[47] While the former – determined by a nation's political and military leadership – describes the purpose and character of war, the latter 'deals with the methods of waging war, the organization of the armed forces, their technical equipment and combat readiness'.[48]

'Military science' is defined as 'a system of knowledge concerning the nature, essence and content of armed conflict, and concerning the manpower, facilities and methods for conducting combat operations by means of armed forces and their comprehensive support . . . Soviet military science is based on Marxist–Leninist teachings and is guided by the method of materialistic dialectics and historical materialism'.[49] 'Military art', as the main field of military science, comprises the 'theory and practice of engaging in combat' and it includes 'tactics, operational art, and strategy, which constitute an organic unity and are interdependent'.[50]

As far as Soviet thinking on nuclear weapons is concerned, Holloway stressed two themes: 'the prevention of such a war, and preparation to wage it. In Soviet thinking, these two aspects are not conflicting, but complementary.' Holloway rejects the drawing of 'a sharp contrast between the war-fighting policy of the Soviet Union and the war-deterring policy of the United States'.[51] While the prevention of nuclear war is seen as the responsibility of the Party leadership, military science and strategy have remained the domain of the military. In Marshal Sokolovskii's book, a major collective work of military science, the problem of how to fight and win a nuclear war is thought through.[52] Holloway has observed a shift in Soviet military doctrine since the 1950s; while in the late 1950s and early 1960s 'Soviet military theoretists were concentrating on the study of a future war and how to wage it, since the mid-1970s the Party leaders have laid more stress on the political side of military doctrine in an apparent attempt to adapt the doctrine to the relationship of strategic parity with the United States. However, 'the military stress on preparing to fight and win a nuclear war has been reinforced by the ideological belief that, if nuclear war did take place, it would be the decisive contest between socialism and capitalism, and that socialism would emerge victorious'.[53]

Both the Party leadership and major contributors to military science – as for example Marshal Sokolovskii – have referred to the terrible potential destruction of such a war, and to the responsibility of the CPSU to prevent it and to further peaceful coexistence. While surprise played a major role in the Soviet strategic debate in the mid-1950s, Holloway argues that since the late 1960s Soviet thinking 'has placed less emphasis on the idea of pre-emption . . . Perhaps pre-emption is now seen to be inherently problematical in nuclear war too'.[54]

The retired Soviet General Milstein has praised the SALT I treaties and he has indirectly accepted mutual vulnerability as a fact of life:

> First, the agreements stopped the race in the most destabilising field of strategic arms – in the field of ABM systems – and thus strengthened the stability of the situation as a whole. Second, they have halted the build-up in the most modern strategic offensive arms – ICBMs and ballistic missiles mounted on submarines. Third, concrete obligations were undertaken on preventing a nuclear war.[55]

Before the conclusion of the ABM Treaty all Western experts agreed that Soviet military doctrine emphasised 'the importance of strategic defence, rather than to rely upon a strategic offensive capability for deterrence as the United States has done'.[56] Referring to the historical legacy of 1941, Major General Nikolai Talensky in 1964 provided a clear doctrinal rationale for BMD: 'the creation of an effective anti-missile system enables the state to make its defence dependent chiefly on its own possibilities, and not only on mutual deterrence'.[57] Three years later Major General I. Anureyev[58] argued in the official journal *Military Thought* that 'a most important factor which makes it possible to accomplish the task of changing the correlation of forces in one's own favour is anti-air defence (anti-missile and anti-space)'.[59] The third edition of Marshal Sokolovskii's book *Military Strategy* (published in 1968)[60] had several photographs on BMD, including the statement that 'one of the cardinal problems for Soviet military strategy is the reliable protection of the rear from nuclear strikes – anti-ballistic missile defence'. Though he conceded that 'means of nuclear attack undoubtedly predominate over methods and means of defence against them', the BMD mission, along with anti-air and anti-space (and civil defence) remained'.[61]

After the 1972 ABM Treaty, the BMD mission 'virtually disappeared from Soviet military writings',[62] just as Soviet military uses of outer space have not been discussed since the Soviet Union joined the Outer Space Treaty of 1967.[63] Only generals of the Air Defence Forces continued to refer in general terms to BMD and 'to a need for defence against ballistic missiles as well as aircraft'.[64] While references to Soviet doctrinal statements on BMD and on the military use of space disappeared in the Soviet literature, official interpretations of Soviet military doctrine in the official Pentagon publication *Soviet Military Power* have become more detailed.

(b) Official American assessments of Soviet military doctrine in the four volumes of *Soviet Military Power* (1981–85)[65]

While the first edition (published in September 1981) was limited to a 'factual portrayal of Soviet Armed Forces', the second edition published in March 1983 left 'no doubt as to the USSR's dedication to achieving military superiority in all fields'. Under 'protection of the homeland' in a nuclear war there were only minor references to Soviet BMD and space missions, while the primary emphasis was on Soviet nuclear warfighting capabilities, with a special reference to the Soviet theatre nuclear force capabilities confronting Western Europe.[66] In the third edition, strategic defences were interpreted as 'vital to the overall Soviet strategy for nuclear war ... The operations of Soviet defensive and offensive forces are closely coupled: attack strategies are geared in large part to the reduction of the defensive burden. In the Soviet concept of a layered defence, effectiveness is achieved through multiple types of defensive capabilities compensating for shortcomings in individual systems and for the likelihood that neither offensive strikes nor any one layer of defence will stop attacking weapons. The Soviets have made major improvements in their deployed strategic defences and have invested heavily in ABM-related developments'.[67]

Soviet Military Power (1985) devoted a full chapter to Soviet strategic defence and space programmes, emphasising the close coupling of Soviet defensive and attack forces.

A comparison of the authoritative official United States assessment in *Soviet Military Power* indicates a major increase of the perceived Soviet BMD activities and projects and their military space activities after President Reagan's Star Wars speech and after the establishment of the SDI research programme.

(c) Nuclear warfighting and rejection of mutual vulnerability: the hawkish interpretation of Soviet military doctrine and of Soviet theatre nuclear, BMD and military space activities

The official interpretation of the Soviet military doctrine and policy has been influenced by a group of Soviet experts in the United States who agree on at least most of the following aspects:

1. the Soviet Union does adhere to its aim of world revolution

therefore ideological references in Soviet military doctrine should not be downgraded;[68]
2. the political statements of the Soviet leadership perform a propagandistic purpose, and Soviet peace rhetoric intends to undercut the defence motivation in Western Europe and to drive a wedge between the United States and its European allies;[69]
3. the writings of Soviet military officers should be taken very seriously;
4. United States and Soviet strategic thought differs in the interpretation of the major concepts of deterrence and mutual assured destruction;[70]
5. the Soviet leadership (and especially the military) have not accepted strategic parity and mutual vulnerability as major premises for their defence planning;[71]
6. Soviet military planners believe that they can fight and win a nuclear war, and they plan accordingly;[72]
7. the Soviets pursue an aggressive military space policy.[73]

Richard Pipes – a Soviet specialist and one of the major concept formulators of the Committee on the Present Danger – has challenged the view of many American strategists and officials that in the real world of strategic nuclear weapons 'nobody could possibly win',[74] claiming: 'By the mid-1960s [the Soviet Union] adopted . . . a "war-fighting" and "war-winning" doctrine.'[75]

For Pipes, SALT cannot solve the problem of quantitative limitations and qualitative constraints:

> as long as the Soviets persist in adhering to the Clausewitzian maxim on the function of war, mutual deterrence does not really exist. And unilateral deterrence is feasible only if we understand the Soviet war-winning strategy and make it impossible for them to succeed.[76]

While Pipes, Van Cleave and others[77] used Soviet military writings and their interpretation of Soviet military doctrine to attack the very premises of the SALT process, Joseph D. Douglass, Jr, Amoretta M. Hoeber and Lewis Allen Frank have, by the same method, legitimised the nuclear force modernisation process.[78]

Another group of the SALT critics has gradually shifted their primary focus from 'nuclear warfighting' in Soviet military doctrine to *The Role of Strategic Defense in Soviet Strategy*.[79] Michael Deane, after a detailed analysis of Soviet writings on air and BMD since the conclusion of the 1972 ABM Treaty, stated:

> In a political sense, mutual deterrence is deemed unacceptable by the Soviets because it suggests (1) an approval of a stable international world order based on the status quo and (2) a denial of the peacetime political utility of Soviet military power. In a military sense, mutual deterrence is irrational because, if deterrence should fail and nuclear war break out, each side courts its own destruction if it has refrained from establishing the forces for war-winning and war-survival. . . . In an operational sense, mutual deterrence is considered destabilising if at all practical in the long run . . . Thus, instead of lowering the possibility of nuclear war, the Soviets perceive mutual deterrence as raising the likelihood of conflict.[80]

For Deane, Soviet anti-air, anti-ballistic and anti-space defence activities are part of the Soviet 'war-fighting and war-winning strategy'. From his reading of the publications of leading officers of the Soviet Air Defence, Deane concluded: 'Soviet spokesmen seem to address the issue of air defence as if the ABM Treaty did not exist'.[81] And he added: 'The Soviet objective appears to be the development of a long-term strategic advantage through the significant technological breakthrough or a short-term advantage during a crisis situation wherein the sudden deployment of a large number of ABM systems might constrain the opponent's freedom of action . . . It can be expected that the Soviets will continue to improve their anti-ballistic missile and anti-space defences until they can indeed construct an "impenetrable" system, and that their emphasis on point defence will continue as long as it appears that the United States will deploy an effective cruise missile'.[82]

David Yost – a somewhat more sophisticated representative of the damage limitation school, which maintains that the Soviets never endorsed the principle of MAD – referred to several possible motives why the Soviet Union signed the 1972 ABM Treaty. Their goal is to be able to:

(a) obtain ceilings on United States ICBM and SLBM launchers inferior to those allowed the USSR in the SALT I Interim Agreement;

(b) leave United States ICBMs and other hardened targets unprotected so that Soviet counterforce targeting objectives could be pursued;

(c) slowdown and hamper United States BMD research and development efforts;
(d) gain time for Soviet BMD technology and deployed capabilities to equal and surpass those of the United States; and
(e) fund Soviet BMD activities at less cost than such efforts would require if the United States were competing on a more intensive basis.[83]

Other adherents to the hawkish interpretation of Soviet military doctrine have referred to pragmatic considerations which may have influenced the Soviet Union to conclude the ABM Treaty: 'cost considerations, the low performance capabilities of the existing Soviet ABM teachnology, and a perception that it was the correct moment to make political gains through arms control'.[84] Payne and Stroude have interpreted the ABM Treaty as an example of the successful use of arms limitation to promote Soviet ends.[85] According to Hensel, the proponents of the hawkish interpretation of Soviet military doctrine 'view Soviet agreement to the limitations embodied in the ABM Treaty as consistent with the tenets of that doctrine and reject the proposition that it represented Soviet acceptance of the principle of mutual societal vulnerability as the basis for superpower deterrence.[86]

Lt General James A. Williams, director of the Defence Intelligence Agency, recently argued that according to Marshal Sokolovskii the Soviet space doctrine holds 'that space – a fourth military environment – is critical to the success of forces in the other three environments – land, sea, and air. The Soviets envision using space for offensive operations and have the domination of it as a goal. They intend to deny any enemy the use of space for military purposes while using it for their own military operations'.[87] Williams warns that the Soviet leadership has given high priority to its military space programme, based on the belief that it could make the following gains:

1. enhanced prestige and world recognition as a technological superpower;
2. global support to the Soviet military forces under their combined arms approach; and
3. sufficient military power to deny the use of space to potential adversaries.

He concludes that 'the Soviets' military capabilities in space pose a serious and growing threat. Should the current trends in spending and system development continue unabated, it could ultimately affect the strategic deterrent posture of the United States'.[88]

Uri Ra'anan, in a brief presentation of the few doctrinal references to the role of space in Soviet military doctrine, concluded that Soviet space programmes are 'supervised predominantly by Party and state personnel in charge of military industry or security'. Based on unclassified Soviet sources up to the 1960s and on restricted material in the 1970s, Ra'anan describes the components of Soviet space doctrine. The five goals are to:

(a) create space weapons systems ancillary to the existing military services;
(b) neutralise space systems (including reconnaissance and communications systems) of other countries;
(c) develop strategic offensive space systems, both to conduct warfare in space itself and to attack targets in the planetary surface;
(d) in this general context, particular emphasis is placed upon achievement of surprise, through destroying the adversary's capacity to engage in electronic intelligence-gathering and through jamming of his communications systems, stressing the space aspects of these objectives;
(e) diminution of the vulnerability of the Soviet space systems and the enhancement of the vulnerability of the Soviet space systems and the enhancement of the vulnerability of adversary space weapons systems and anti-space defence systems.[89]

This interpretation of the role of theatre nuclear weapons, of BMD, and especially the interpretation of the Soviet position on MAD (which reflects the official United States assessment) has been challenged by a school of Soviet experts, most of whom still support a co-operative strategic policy *vis-à-vis* the Soviet Union.

(d) Deterrence and mutual vulnerability: the interpretation of Soviet military doctrine and of Soviet BMD and military space activities from a détente perspective

The second group of American and other Western experts on Soviet military policy and doctrine are in broad agreement on four premises:

1. The Soviet leadership has accepted mutual vulnerability as a fact of life by signing the ABM Treaty in 1972.[90]

2. Though the Soviet military as well as the American military is planning to fight a nuclear war, both the American and the Soviet leadership have accepted deterrence as a guiding principle.[91]
3. The ABM Treaty has been a major achievement of a cooperative arms control policy which has contributed to strategic and arms race stability. It should therefore be maintained and strengthened.[92]
4. The statements of the Soviet political leadership should not be discarded as pure propaganda, and in analysing the publications of the Soviet military the element of indoctrination of the troops should be taken into account.[93]

Raymond Garthoff and David Holloway, the most prominent American and British authors representing this point of view, have analysed the changes in Soviet attitudes towards SALT, BMD and MAD before and after the initiation of the SALT negotiations in 1969. Garthoff described the Soviet military approach to SALT as

> constructive, if conservative ... In SALT the Soviets have accepted mutual deterrence, both through advocacy of equal security for the USA and USSR and even more tellingly through sponsoring an ABM limitation specifically precluding a defence of the country against overwhelming startegic missile attack and thus ensuring mutual vulnerability. They have accepted strategic parity as reflected in the SALT agreements, and as a goal for further agreements ... SALT led Soviet military and political leaders both to understand as never before the indivisibility of strategy and arms control ... SALT has also led to recognition of the importance of both strategic and 'political–military' interaction between the USA and the USSR and indeed has enhanced considerably the importance of that interrelationship.[94]

Referring to the writings of authoritative Soviet military writers, Garthoff pointed out that General Talensky argued in 1965 that there 'is no more dangerous illusion than the idea that thermonuclear war can still serve as an instrument of politics, that it is possible to achieve political aims by using nuclear weapons and still survive'.[95] For Garthoff, the writings by Marshall Sokolovskii 'show Soviet military recognition of the emergence of mutual deterrence'. However,

> mutual deterrence in Soviet writings is usually expressed in terms of assured retaliatory capability which would devastate the aggres-

sor, because this formulation . . . is more responsive to ideological sensitivites over the idea that the USSR could be considered a potential aggressor and thus needs to be deterred. In addition, this formulation avoids identification with the specific content of the US concept of 'mutual assured destruction', often expressed in terms of countervalue capability for destroying a specific percentage of the opponent's industry and population. This US interpretation is more limited than the Soviet recognition of mutual deterrence.[96]

Since the 24th Party Congress in 1971, Garthoff noted, 'Soviet military and political leaders [have] ceased to call for strategic parity . . . Instead, mutual deterrence, a balance, parity and equal security are advocated'.[97] In 1977 Leonid Brezhnev, in his speeches at Tula in January and at the 60th anniversary of the Bolshevik Revolution in November, stated that the Soviet Union 'does not and will not seek military superiority over the other side. We do not want to upset the approximate balance of military strength existing at present . . . between the USSR and the United States'.[98]

Garthoff and many other Western experts have also noted a growing Soviet concern about United States modifications of its nuclear declaratory politics and its force posture modernisation since the mid-1970s, which have often been described by Soviet commentators as a United States attempt 'to destabilise global strategic parity in its own favour' and as a failure to recognise 'an axiom of contemporary international relations: security in the age of nuclear parity is based on stability, and stability is based on the mutual acknowledgement of equality, and on abandoning the aspiration for superiority'.[99] Garthoff also rejects the argument of the hawkish interpretation

> that Soviet statements on such propositions as mutual deterrence and the unacceptability of nuclear war are 'for export', and they are contrasted with selected open Soviet military discussions. The evidence from such sources as the confidential USSR Ministry of Defence organ *Military Thought* dispels such erroneous conclusions. The record indicates that the Soviet political and military leadership accepts a strategic nuclear balance and parity between the USSR and the USA as a fact, and as the probable and desirable prospect for the future . . . In Marxist–Leninist eyes, the military power is not and should not be the driving element in world politics.[100]

When the United States first approached the Soviet leadership to negotiate constraints on their respective strategic nuclear forces, both

Soviet 'military doctrine and "disarmament" doctrine favoured BMD and other forms of defence'[101] but by November 1969 the views of General Talensky 'had been superceded by the concern over the "arms race stability" of an ABM–offensive missile interaction in the arms race'.[102] Since the early 1970s several 'Soviet spokesmen [have] stressed the particularly destabilizing effect of the combination of ABM and MIRV deployments'.[103] Though the Soviet National Air Defence forces played down the implications of the ABM Treaty for its own mission, Garthoff concludes after a careful analysis of Soviet writings and BMD activities since 1972 that:

> The Soviet Union has adjusted its military doctrine and concrete military programs to accommodate the major arms control restraint on the offensive–defensive arms competition established in the ABM Treaty of 1972. This treaty, in turn, represents not only a military–political decision, but also a political investment in a détente policy that features arms control as one of its principal elements ... While Soviet proposals – not unlike our own – are fashioned to their advantage, the ABM Treaty stands as witness to the fact that the Soviet Union is prepared to support balanced arms control agreements embodying substantial constraints on the military programmes of both sides. The political dimension of the ABM Treaty should not be ignored in evaluating both Western stakes and Soviet reactions in considering the impact of possible future initiatives on BMD on East–West relations'.[104]

In analysing the Soviet response to SDI Garthoff and Holloway quoted many official statements and commentaries in the Soviet press referring to the common understanding reached in the ABM Treaty 'that attempts to create ABM systems destabilize Soviet–American relations and would impede agreement in limiting the strategic capabilities of both sides'.[105]

Drell, Farley and Holloway – in agreement with Garthoff – however cautioned:

> The Soviet leaders apparently accept that mutual vulnerability to devastating retaliatory strikes is an objective condition, for the time being at least, but their preparations for war arouses fears that they either do not recognise the reality of this condition, or are trying to escape from it. There is a tension between these two elements of Soviet thinking. Both elements have to be recognized if Soviet policy on ABM systems is to be understood.[106]

Soviet political and military leaders and scientific spokesmen have portrayed President Reagan's SDI 'as part of a drive to achieve superiority' which would intensify the arms race and would have a destabilising effect on the Soviet–American strategic relationship. All the Soviet commentary, Drell *et al.*, conclude, 'points to the inevitability of Soviet measures to counter an American effort to build an ABM system'.[107]

Between these two interpretations of Soviet military doctrine a third school has emerged that has tried to combine the two opposite assessments of Soviet doctrine and military activity.

(e) A centrist perspective on Soviet military doctrine

A centrist perspective on Soviet military doctrine integrates two facets of Soviet military doctrine: the emphasis of the political leadership on parity and elements of deterrence in the United States – Soviet relationship and the concern of the Soviet military with warfighting if deterrence should fail. Berman and Baker have stressed that:

> Despite Soviet interest in avoiding nuclear conflict and maintaining a credible deterrent in peacetime, if war should occur, the USSR would seek ... militarily [to] prevail over the enemy as the best means of assuring national survival. Some Soviet military writings refer to victory as the goal in wartime, but few define its meaning in the context of all-out nuclear war. Nevertheless, Soviet nuclear strategy continues to adhere to the traditional objectives associated with military victory of physically defending the homeland against military attack, decisively defeating the enemy's forces, and occupying enemy territory.[108]

Stanley Scienkiewicz[109] and Fritz Ermath[110] have analysed the many differences in both the formulation and content of strategic doctrine of both superpowers. According to Scienkiewicz:

> Soviet doctrine does not dismiss deterrence. On the contrary, most of the open literature seems preoccupied with enhancing it. It is a theory of deterrence, however, which is substantially at variance with the US formulation ... Seeking doctrinal convergence as a formal goal in SALT thus seems impractical at best ... [However,] the Soviets have shown themselves able to reach SALT agreements at variance with their military doctrine. The obvious example is the ABM treaty.[111]

While United States doctrine, in the view of Fritz Ermath, plans to deter nuclear war at the lowest cost ('arms race stability') and strategic risk ('crisis stability'), 'Soviet strategic doctrine stipulates that Soviet strategic forces and plans should strive in all available ways to enhance the prospect that the USSR could survive as a nation and ... defeat the main enemy should deterrence fail – and by this striving both help to prevent nuclear war and help the USSR attain other strategic and foreign policy goals.'[112] While in the United States perspective, 'strategic deterrence tended to become the only meaningful objective of strategic forces', Ermath has noted a shift in the Soviet concept under the conditions of parity 'from primary emphasis on defensive themes of war prevention and protection of prior political gains to more emphasis on themes that include the protection of dynamic processes favouring Soviet international interests'.[113] Ermath draws three conclusions from the lack of understanding of the 'fundamental differences between the US and the Soviet strategic thinking, both at the level of objectives and at the level of methods':

(a) the 'hawk's lament' of underestimating the competitiveness of Soviet strategic thinking;
(b) the 'dove's lament' of projecting United States view on to the Soviets and of overselling the SALT achievements; and
(c) the excessive United States confidence in strategic stability.[114]

In agreement with David Holloway, Sayre Stevens pointed to two distinct aspects of Soviet strategic thinking and policy:

> One constitutes an effort to prepare to fight and to survive a nuclear war; this is the domain of the military and produces the 'military science' ... The other, the domain of the political leadership, is the conduct of a peace policy intended to prevent war and to limit the threat to Soviet national security through political means. It is the responsibility of the political leaders to preserve peace and that of the military to ensure the capability to punish the aggressor and survive war. Viewed in this light, the often-conflicting statements about strategic policy that emerge from the USSR are easier to understand.[115]

In the narrower military perspective, damage limitation becomes an inherent complement of a warfighting capability. However, according to Stevens:

Soviet strategic thinking cannot be explained in terms of the effects of BMD alone, but these effects must be seen in conjunction with strategic counterforce strike capability and with passive defenses that play their own part in reducing the effects of nuclear attack. This view allows the Soviet Union to recognise the predominant significance of offensive forces in the nuclear era and at the same time to establish the need for active defenses as part of its overall strategic force capabilities.[116]

Hensel, in a discussion of 'Soviet media images of superpower space policy', applied a pluralist interpretation to the Soviet strategic debate by assuming 'two opinion groups within the Soviet political and military leadership' relating to 'the appropriateness of mutual societal vulnerability versus damage-limitation as the principle underpinning Soviet–American strategic nuclear deterrence' as well as two other 'opposing coalitions concerning the question of whether to develop and extensively deploy space weapons and ballistic missile defences'.[117] Hensel assumes that 'within the military–defence industrial establishment, there would be a coalition of interests which feel that both offensive and defensive space weapons, as well as traditional antiballistic missile systems should be incorporated into the Soviet arsenal as they become technologically feasible'. Hensel assumes that the Soviet Air Defence Forces – the responsible command for Soviet space weapons – represent the principal supporters of space weapons within the Soviet military leadership, while other components of the Soviet military–defence industrial complex 'would be less enthusiastic and, perhaps, hostile to the idea'. Those within the Soviet Party leadership 'who accept the principle of mutual societal vulnerability to nuclear retaliation' might prefer 'to avoid the destabilizing effects associated with the development of [space and BMD] weapons'. Hensel concludes:

> This opinion group collectively is likely to be united with those elements of the military–defence industrial opinion group opposing ambitious programmes designed to develop and extensively deploy space weapons and [BMD]. It is this coalition we hear when we receive that portion of the Soviet media's message which stresses the technical non-feasibility, expense, and destabilising impact of space weapons and ballistic missile defences . . .
>
> Other civilian elements, foremost the defence industrial bloc, opt for the warfighting, damage-limitation approach. Some compo-

nents of the latter bloc would profit from an ambitious program designed to develop and extensively deploy space weapons and BMD ... It is they who would prefer a comprehensive ban on space weaponry. Finally, at present, this coalition appears to remain dominant within the Soviet political/military hierarchy, despite the presence of an increasingly powerful counter-coalition which would like to overturn the ABM Treaty and ambitiously develop and extensively deploy BMD and space weapons'.[118]

These three conflicting interpretations of Soviet military doctrine in general, and BMD and space doctrine in particular, focus on different aspects of the Soviet debate, and they provide only partial answers. Nevertheless, the centrist and the détente perspective reflect more closely dominant perceptions in Western Europe as to Soviet behaviour, while the hawkish United States interpretation appears to be too closely linked with providing legitimacy for the often contradictory strategic rationales of the Reagan Administration as it relates to the defence of Western Europe.[119] Declaratory politics, however, both in the United States and in the Soviet Union may hardly provide objective criteria for evaluating and interpreting Soviet BMD and military space efforts.

III THE SOVIET MILITARY ORGANISATION FOR AIR DEFENCE, BMD AND SPACE OPERATIONS

Political, military, industrial and research and development components contribute to the Soviet BMD and space effort. Relatively little is known about the specific organisations and about the decision-making process among them. While some Western analysts have tried to apply the model of the military–industrial complex with modifications to explain this interrelationship,[120] Holloway has reached the opposite conclusion: 'The build-up of military power since the 1950s must be seen as the product of conscious political choices, and not as a result of pressures from a military–industrial complex'.[121] While several studies exist about the central policy-making bodies dealing with military affairs,[122] Stephen M. Meyer has frankly admitted that:

> the structure of Soviet space activities remains a riddle. There is some evidence, however, suggesting that a State Committee on

Space Exploration oversees Soviet non-military activities in space, and co-ordinates long-term space planning between the Ministry of Defence and other state organs. Regardless of the exact character of the State Committee on Space Exploration, the command and control of Soviet military activities in space is firmly in the hands of the Ministry of Defence, and administered by the various services of the armed forces. There is no Soviet space command.[123]

Within the Soviet military, two commands are responsible for nuclear, BMD and space activities: the Strategic Rocket Forces (RVSN) and the Air Defence Forces. Both commands stress maintaining 'constant combat readiness'.[124] Soviet theatre nuclear forces are deployed with all six armies of the RVSN, five of which may target Western Europe.

According to Berman and Baker:

> the majority of the divisions of some rocket armies are assigned for theater strike missions against NATO or China while the remaining divisions may be mainly oriented toward covering intercontinental military and industrial targets. Importantly, the deployment of variable-range ballistic missiles makes it possible for each rocket army to cover both regional and intercontinental targets.[125]

All Soviet strategic defensive forces are under the command of the National Air Defence Forces of the Homeland (PVO Strany), which is responsible for early-warning satellites, anti-ballistic missiles (PRO Strany), anti-satellite systems (PKO Strany), aviation units (APVO Strany) and surface-to-air missiles (Zenith Missile Troops, ZMT), which can be used for anti-tactical (ATBM) missions. According to Meyer:

> Control of ground-based early warning radars is probably split between the Radio-Technical Troops and the VPRO. Therefore, it is possible that either the VPRO subordinate command or the VPVO command itself has operational control over Soviet space-based early warning systems. The VPRO may also control some aspects of Soviet research programs on space-based beam weapons for anti-missile defence.
>
> Another branch of the VPVO, the VPKO, controls the Soviet ASAT program and most likely operates much of the Soviet ground-based space tracking network. It is also believed that the VPKO controls the Soviet ground- and space-based laser research program for ASAT applications.[126]

While the Strategic Rocket Forces control the nuclear threat confronting Western Europe, the Air Defence Forces could administer many of the potential countermeasures against both the SDI and EDI, which are intended to protect Western Europe against regional Soviet nuclear forces.

IV SOVIET COUNTERMEASURES AGAINST SDI AND EDI

Any decision about European participation in SDI or in favour of an EDI will have to provide answers to the following questions:

1. Can SDI and/or EDI protect Western Europe against existing Soviet nuclear forces (intercontinental forces, LRINF, MRINF and SRINF)[127]
2. With which offensive and defensive countermeasures could the Soviet Union neutralise Western defence efforts?

Both American[128] and Soviet[129] analysts have referred to various conceivable countermeasures against a multilayered, space- and ground-based BMD system as outlined in the context of the SDI research programme.

A report of a working group of the Committee of Soviet Scientists for Peace, Against Nuclear Threat headed by R. Z. Sagdeyev and A. A. Kokoshin distinguished between active and passive countermeasures against a space-based anti-missile system. 'The former include various ground (sea)-, air- or space-based weapons using either missiles or lasers', for example 1) ground (sea)-, air- or space-based ballistic missiles, 2) space mines, 3) high-power ground-based lasers, 4) obstacles put up on the trajectories of the combat stations, and 5) false missile launches. Among the passive countermeasures the Soviet scientists include:

> camouflaging missile launchings in the laser-working optical range (all types of smoke screens), multilayer missile casings and ablating coatings. Coatings with a high reflection factor at the working wavelength (including retro-reflectors) may also have a part to play as countermeasures against a number of types of lasers (operating in the visible and infrared spectrum) . . . An effective network of the means of countermeasures can be set up within a very short time with the use of the already available technology. As indicated in Western sources, corresponding units and components are much better tested and more reliable than the elements and subsystems

to be used for SBAMS based on directed-energy weapons. Estimates show that a system of the means of countermeasures may also be much cheaper than a large-scale SBAMS. The cost of a highly efficient countermeasures system (counter-SBAMS) with regard to the SBAM itself is most likely to make up 1 or 2 per cent.[130]

As far as a European-based AT(B)M as part of an EDI is concerned, the Soviets could initiate the following additional countermeasures:

1. increase the number of forward deployed SRINF and MRINF (e.g., the SS-21, SS-22 and SS-23);
2. increase the nuclear-capable aircraft in the European theatre in case of a conflict;
3. shift from ballistic to theatre-wide land-, air- and sea-based Cruise missiles.

In order to counter these potential Soviet offensive countermeasures against an EDI, a European defensive system would require a capability to destroy Soviet

1. short- and medium-range ballistic missiles with an *AT(B)M system*;
2. nuclear- and dual-capable aircraft with an *improved air defence*;
3. dual-capable ground-, air- and sea-based Cruise missiles with an *anti-Cruise missile defence system*.[131]

Following this consideration of Soviet countermeasures against SDI and EDI, Soviet space weapons and BMD capabilities will be surveyed.

V SOVIET PASSIVE MILITARY SPACE SYSTEMS AND ACTIVE SPACE WEAPONS AGAINST SATELLITES (ASAT)

Ever since the Soviet Union launched the first Sputnik into orbit, outer space has increasingly been utilised by the military of both superpowers in launching *passive space systems* (satellites for reconnaissance, electronic intelligence, command, control, and communication (C^3) early warning, navigation, meteorology and geodesy) and *active space weapons* (multiple and fractional bombardment systems: MOBS and FOBS, as well as anti-satellite weapons (ASAT)). Since the early 1980s spokesmen for the United States Department of Defense, defence-minded Congressmen and the trade press have claimed that 'the Soviets have the world's most active military space program, [which] dominates the Soviet Union's overall space effort';[132] that they outspent the United States in space by about $3,000 million in, for example, FY83;[133] that the gap with the United States in space weaponry would become wider;[134] and that by 1990 a manned military space station would enable the Soviet Union to attack ground sea and air targets on earth.[135] Has the Soviet Union been the pace-setter and has it become the front-runner in the military use of space?

(a) Passive Soviet military systems (satellites)

Proponents of the 'space gap' thesis, most recently Paul Nitze, have claimed that 'in 1984 the Soviets conducted about 100 space launches. Of these, some 80 were purely military in nature, with much the remainder serving both military and civil functions. By way of comparison, the total number of United States space launches in 1984 was about 20'.[136] This data are quite misleading, as Jasani,[137] Meyer and Smith[138] have shown, given the short lifetime of most Soviet satellites compared to the American ones.

In the 1981 official *Soviet Military Power* report, the Soviet military space activities were described: 'In the past ten years they have been launching spacecraft ... at the rate of four to five times that of the United States. The annual payload weight placed into space by the Soviets ... is ten times that of the United States ... We estimate that 70 per cent of Soviet space systems serve a purely military role, another 15 per cent serve dual military/civil roles, and the remaining 15 per cent are purely civil'.[139] In the second edition of March 1983 the Pentagon claimed: 'The Soviet quest for military supremacy has expanded into space [ranging from] ... extended manned missions, to meteorological, communication, navigational, reconnaissance, surveillance, targeting and anti-satellite missions ... The USSR currently has under development a heavy-lift space launch system ... that will be able to place payloads at least six to seven times the weight of those of the US space shuttle into orbit'. In the view of the United States Department of Defense, 'the Soviet military space program also reflects an ever-increasing use of space for world-wide surveillance and warning ... The Soviets are increasing their use of space systems for command, control, and communications ... [They] are emphasising the development of communication networks using satellite systems that will be placed in geostationary orbits [and] over the next ten years, the Soviets will develop and deploy an even more advanced series of communication satellites'.[140]

In April 1984 the third Pentagon report claimed that 'the Soviets maintained the world's most extensive early warning system for both ballistic missile and air defence ... The current Soviet launch detection satellite network is capable of providing about 30 minutes' warning of any US ICBM launch, and determining the area from which it originated'.[141] After repeating the functions of Soviet satellites it predicted that over the next ten years:

> the Soviets are likely to develop primarily for military purposes:
> 1. a permanently manned SKYLAB-size space station to be operational in the next 2–3 years with a 6- to 12-person crew, and
> 2. a very large modular station, to be operational by the early to mid-1990s, which could house as many as 100 personnel.

By all measures, the Soviet level of effort devoted to space is increasing significantly over the activities noted in the 1970s. The projected yearly rate of growth of the Soviet space programme is expected to outpace both the annual rate of growth in overall Soviet military spending and that of the Soviet gross national product (GNP) for a number of years to come. The Soviets have embarked upon a long-term, broad-based effort to expand their operational military capability in space. A major Soviet objective is to expand warfighting capability in space and achieve a measure of superiority in that arena. It is clear the Soviets are striving to integrate their space systems with the rest of their Armed Forces to ensure superior military capabilities in all arenas.[142]

The most extensive description of the Soviet 'strategic defence and space programs' was published in 1985, when it was claimed that 'Soviet military doctrine establishes requirements for the military space program ... Space assets play a major role in [the combined arms concept of warfare] in the areas of anti-satellite warfare; intelligence collection; command, control, and communications; meteorological support; navigational support; and targeting ... Their reliance on these systems is growing. Space weapons also play an important role in their strategic operations'.[143] The fourth Pentagon report also states that the United States has no counterpart to Soviet ocean reconnaissance satellites, to the EORSAT for electronic intelligence and to the RORSAT for radar purposes.

> Their mission is to detect, locate, and target US and Allied naval forces for destruction by anti-ship weapons launched from Soviet platforms ... The Soviets have recently employed a new radar-carrying satellite system. Designed for mapping ice formations in polar regions, these satellites will greatly enhance the ability of the Soviet Navy to operate in icebound areas ... The launch rate of satellites to geostationary orbits has risen in recent years. In the period 1974–8, one to two launches per year were conducted. In 1979, the rate increased to five per year, and eight launches occurred in 1984.[144]

This alarmist perspective of the Soviet military space effort is in contrast with the factual presentations by Nicholas L. Johnson in the *'Soviet Year in Space'*,[145] with the comparative reports by Marcia S. Smith about the *'Space Activities of the United States, the Soviet Union and other Launching Countries/Organizations'* for the Congressional Research Service,[146] and with the annual chapters by Bhupendra Jasani in the SIPRI Yearbooks.[147]

In a recent article Stephen M. Meyer pointed to many severe shortcomings in the Soviet space support missions. For him the 'high annual launch rate of Soviet military photo-reconnaissance satellites is a poor and misleading indicator of Soviet "presence" in space or for the Soviet MPR capabilities from space because the Soviet Union has purposefully adopted a high launch rate/low mission longevity engineering approach to the MPR mission'.[148] Meyer also doubts the military usefulness of space-based MPR missions (given the normal weather conditions over NATO Europe) over air and ground MPR systems. With their low orbits (under 400 kilometres) Soviet MPRSATs would also be vulnerable to attack by the planned United States ASAT system and the 'vulnerability of the two main Soviet launch complexes, Pletetsk and Tyuratam, raises doubts about the efficacy of reconstituting the space-based MPR mission during a major war. These would be easy targets for United States ICBMs, SLBMs, and bomber-launched Cruise missiles, as well as British and French nuclear forces'.[149] The Soviet electronic intelligence satellites with orbits at 600 kilometres would also be vulnerable to the United States F-15-ASAT. Soviet C^3 satellites, especially the Molniya satellites with highly variable orbits from 500 kilometres to 40,000 kilometres would be less vulnerable to United States ASAT attacks. The same would also apply to Soviet early warning satellites. The two types of Soviet ocean reconnaissance satellites could, according to Meyer, be easily disrupted:

> RORSATs orbit at fairly low altitudes and their radars act as beacons. A US ASAT system with a radar homer would have little trouble locating its prey. Though Soviet EORSATs orbit at somewhat higher altitudes and do not emit intense signals, they would be quite vulnerable to ASAT attack using infrared seeking. It should also be noted that electronic countermeasures can be employed to jam, deceive, and otherwise confuse Soviet ocean reconnaissance satellites.[150]

While current Soviet navigation satellites orbit within the operational limits of the United States ASAT system, the next generation of NAVSATs (Glonass) will orbit at about 18,000 kilometres, well beyond the reach of any ASAT system.

If the Soviet passive military satellites are compared with the comparable United States systems and if their relative vulnerability against the ASAT capabilities of the other side are taken into account, the Soviet space threat as portrayed by the United States Department of Defense appears to be less convincing.

(b) Active Soviet weapons in space

The Soviet Union tested in the early 1960s a multiple orbital bombardment system (MOBS) and up till 1971 a fractional orbital bombardment system (FOBS), both of which have been prohibited in the meantime – the MOBS by the Outer Space Treaty of 1967 and the FOBS by the SALT II Treaty of 1979.[151] Therefore the only remaining Soviet space weapon system, which has been tested twenty times between 1968 and 1982, is an ASAT System. How has the Soviet ASAT system been evaluated since 1981 by the United States Department of Defense, in the press and in academic and factual analyses by civilian experts?

In 1981 the Pentagon said about the Soviet ASAT:

> The demonstrated Soviet non-nuclear low-altitude orbital ASAT interceptor poses a known, if presently limited, threat to some US satellites. It is anticipated the Soviets will continue to work in this area with a goal of negating satellites in high orbit, as well as

developing more effective kill mechanisms, perhaps using a laser or some other type of directed energy weapons.[152]

In 1983, the second Pentagon report described the ASAT capability as 'the capability to seek and destroy US space systems in near-earth orbit', and then speculated about future developments:

> One direction of the Soviet Union's space weapons program is toward the development and deployment of a space-based laser system. The Soviets could launch the first prototype of a space-based laser anti-satellite system in the late 1980s or very early 1990s. An operational system capable of attacking other satellites within a few thousand kilometers range could be established in the early 1990s.[153]

In 1984, the Pentagon claimed for the first time that the Soviet ASAT interceptor

> can reach targets orbiting at more than 5,000 kilometres, but it is probably intended for high-priority satellites at lower altitudes. The anti-satellite interceptor is launched from Tyuratam, where two launch pads and storage space for additional interceptors and launch vehicles are available. Several interceptors could be launched each day from each of the pads. In addition the Soviets have two ground-based test lasers that could be used against satellites. The Soviets also have the capabilities to conduct electronic warfare against space systems.[154]

In April 1985, the United States Department of Defense repeated all the previous claims about potential ASAT capabilities in the late 1980s and in the 1990s.[155] More recently, Paul Nitze again repeated these claims by stating:

> The Soviets [have] had the [ASAT] capability since 1971 . . . Using a radar sensor and a pellet-type warhead, the interceptor can attack a target in orbit at various altitudes during the interceptor's first two revolutions. An intercept during the first orbit would minimize the time available for a target satellite to take evasive action . . . In addition to the orbital interceptors, the Soviets could also use their operational ABM interceptors in a direct-ascent attack against low-orbiting satellites'.[156]

Less reluctant in describing Soviet ASAT capabilities have been unnamed intelligence and Pentagon sources who leaked horror stories about Soviet satellite killers in order to build up support for the United States ASAT programme after the election of President Carter. In late November 1976, *Newsweek* and *Penthouse* claimed that two United States satellites 'were hit by a single Soviet laser over the Indian Ocean within the last year'. These reports were denied by a Pentagon spokesman while the then Secretary of Defence Rumsfeld 'did not deny that Soviet lasers could have caused the damage'.[157] Four years later, in May 1980, Richard Burt quoted from a CIA intelligence estimate 'produced by the Carter Administration [which] concluded that the Soviet Union has developed a ground-based laser weapon that could be used to destroy American space capabilities'.[158]

In October 1981 *Aviation Week & Space Technology* reported: 'The Soviet Union is operating in low-earth orbit an anti-satellite battle station equipped with clusters of infrared homing guided interceptors that could destroy multiple US spacecraft'. Once again, the Pentagon denied this alarmist claim.[159] However, for weeks later the same trade journal repeated its claim, quoting from an intelligence report that the data on the new killer satellite system were now 'very hard from a variety of sources and methods; harder than anything we've seen in a long time'.[160] In April 1982 the *Sun* quoted from a Joint Chiefs of Staff report to Congress that 'The Soviets have an operational anti-satellite system deployed. The system has the potential to destroy satellites in high orbit. The Soviets have no operational space-based weapons system'.[161] Two months later Lt General Kelly H. Burke, chief of Air Force research and development, claimed publicly that 'the Soviet Union could launch a laser anti-satellite weapon in the next five years, but he contended 'that it would be "ineffective" except as a political and propaganda weapon'.[162] During budget time in March 1983, DARPA director Robert S. Cooper reported 'that the Soviet Union could launch "the first prototype of a space-based laser antisatellite weapon in the late 1980s or early 1990s" [and] have an operational system capable of attacking other satellites within a few thousand kilometers in the 1990s'.[163]

Both the official Pentagon reports and the alarmist claims in the press omitted any reference to the performance record of the Soviet ASAT system since 1968. During a SIPRI hearing, the then director of the weapons Laboratory in Los Alamos Donald Kerr admitted in 1983 that the picture about Soviet ASAT capabilities is 'still hazy'.[164]

Clarence A. Robinson, the senior military editor of *Aviation Week & Space Technology*, maintained in 1983 that of twenty Soviet ASAT tests 'in 16 ... the intercept distance was close enough to be ruled a success ... The intercept altitudes for the killer satellites are believed to range from 160 to 1,500 kilometers'.[165] Marcia Smith has stated that the Soviet Union 'has demonstrated the capability to intercept a target within one orbit, up to an altitude of approximately 2,300 kilometers'.[166] One possibility of measuring success is for Smith 'to consider a test in which the interceptor manoeuvres to within one kilometre of the target as being sufficient to have inspected the target, or to have destroyed it if that action has been desired. Using this measure and media accounts of the tests, one can conclude that 13 of the 20 Soviet ASAT tests have been successful'.[167]

Nicholas L. Johnson stated in his annual reports on *The Soviet Year in Space* for 1982 and 1983 that eleven of the twenty Soviet ASAT tests have been failures.[168] This evaluation has been supported by Paul Stares.[169] For Stephen M. Meyer, twelve of the twenty Soviet ASAT tests failed. While the first test series from 1968–71 involving a radar homing satellite was rather successful (five of seven tests), the second test series from 1976 to 1982 was not. It incorporated two changes – a 'pop-up' intercept technique (four of seven tests failed) and a new homing system 'believed to employ optical tracking' (all six tests failed.[170] According to Meyer the co-orbital approach technique of the Soviets is:

> vulnerable to direct ascent interception (that is, to the US F15/ASAT) because of the time they spend catching their target. The switch to the 'pop-up' interception by the Soviets, however, casts serious doubt on the prospects for destroying Soviet ASAT interceptors by conventional means. The use of radar homing in Soviet ASAT interceptors presents a number of opportunities for countermeasures, including target satellites designed to change orbits upon sensing radar illumination. Yet, here again, the possible move to optical tracking by the Soviets would greatly complicate the task of developing satellites that could 'defend themselves'.[171]

(c) Comparison of Soviet and United States passive military satellites and ASAT weapons

The contrast of official Pentagon claims and capability projections for the Soviet ASAT and the poor performance record of the primitive operational Soviet ASAT is evident. All experts agree that the fifth generation of the United States ASAT – consisting of an F-15 fighter and a two-stage homing missile – which is presently being tested is far superior to the demonstrated Soviet test performance. According to Stares, of the thirty-three ASAT tests the United States conducted between 1959 and the end of 1984, six were failures, twenty-three were successful and for four no data have been made available.[172]

Not only has the performance of United States systems and the applied technology in most cases been superior, after the Sputnik shock in 1957, the United States has also been the pace-setter in using satellites for military purposes.

Table 3.1   Milestones in the development of military space systems

Type of the vehicle	Date of deployment USA	USSR	Pacesetter
First satellite in orbit	1958	1957	USSR
Reconnaissance satellite	1959	1961/1962	USA
First ASAT test	1959	1968	USA
Navigation satellite	1960	1970	USA
Communications satellite in synchronous orbit	1966	1974	USA
Early warning satellite in synchronous orbit	1966	1975	USA
First operational ABM system	1975	1970	USSR

Source: This table is partly based on Bhupendra Jasani (ed.) *Outer Space – A New Dimension of the Arms Race* (London: Taylor & Francis, 1972) p. 93.

VI   SOVIET BMD – OLD SYSTEMS AND NEW DEVELOPMENTS

Both in the United States and in the Soviet Union interest in BMD dates back to the 1940s.[173] Ever since the United States recognised the existence of a Soviet ABM test site in U-2 reconnaissance pictures of Sary Shagan in April 1960, there has been an intensive debate within the United States intelligence and strategic community.

(a) From Griffon to Galosh: Soviet BMD prior to the ABM treaty

After the recognition of the first ABM test site at Sary Shagan and of 'pieces of at least three systems that were later to have BMD applications' in April 1960, United States intelligence observed a series of missile tests in 1961 and 1962 launched from Kapustin Yar to impact at Sary Shagan.[174] These were the Leningrad system with the Griffon missile, the Tallinn Line with the SA-5 missile and the Moscow system with the exoatmospheric Galosh missile.

Though some United States observers claimed that the surface-to-air missiles (SAM) sites which were built around Leningrad by 1962 had a marginal BMD capability, many doubted it. After the Griffon programme was dismantled in 1964 public attention shifted to the so-called Tallinn Line, which consisted of the initial versions of the SA-5 air defence missiles which, according to Stevens, were probably intended 'for defence against aircraft at long ranges up to very high altitudes'.[175] The third system around Moscow with its nuclear-armed Galosh interceptor missile (which was first shown at a military parade in 1964) 'consisted of the Hen House, Dog House, and Cat House search and target acquisition radars: mechanically-steered dish antennas for tracking targets and for tracking the interceptor missiles and guiding them to their targets.'[176] According to David Yost 'aside from its vulnerability to nuclear effects, the system had two key shortcomings. First its only credible capacity was against small attacks, but United States ICBMs and SLBMs were soon to be equipped with multiple independently-targetable re-entry vehicles (MIRVs): 'while the Galosh system may have been marginally effective against the American missile force of the 1960s, it would be virtually useless against the MIRVed force of the 1970s"'.[177] Second, the systems radars were vulnerable to deception by exoatmospheric decoys and chaff'.[178]

Obviously because of these severe limitations the initial twelve complexes with a projected 128 Galosh missiles were curtailed to 96 in 1968 and further cut back to 64 missiles in 1969 and reduced to 32 missiles in 1980.[179] In 1974 Paul Nitze recalled United States estimates of 'the early 1960's' that the 'Soviets would ... deploy some 2,000 long-range exoatmospheric interceptors, and, in addition, some 6,000 to 8,000 short-range endoatmospheric interceptors'.[180] Efforts at overselling the Soviet BMD in the 1960s to overcome the deep scepticism of then Secretary of Defence McNamara were obviously disproved both by the low performance of Galosh and by the ABM Treaty constraints.[181]

(b) From support for BMD to silence about and public opposition to a multilayered BMD system – changes in Soviet policy and conflicting Western assessments

Since the 1950s there has been a limited public debate in the USSR on the potential implications of BMD on disarmament. Although Pyotr Kapitza,[182] who later won the Nobel prize, had called in 1956 for an 'obligatory exchange of information concerning experimental work on defence measures', General Talensky argued in 1964 that an ABM system would enable 'the state to make its defences dependent chiefly on its own capabilities, and not on mutual deterrence, that is, the goodwill of the other side'.[183] When Soviet Prime Minister Aleksei Kosygin was asked on 9 February 1967 about the Soviet view on ABM, he replied: 'I think that a defensive system which prevents attack is not a cause of the arms race ... Its purpose is not to kill people, but to save human lives.'[184] However, as Garthoff has carefully documented, even the Soviet military was divided about the effectiveness of a BMD. While Marshal Malinovksy and Marshal Grechko stressed the limited effectiveness of BMD, the commander of the Air Defence Forces General Batitsky speaking for his service claimed that a Soviet BMD could 'reliably protect the territory of the country against ballistic missile attack'.[185] In November 1969, the leader of the Soviet SALT delegation, Mr Semenov, acknowledged a change in the Soviet position on ABM or BMD stressing that ABM systems could stimulate the arms race and could be strategically destabilising.[186] During the SALT I negotiations, Garthoff has pointed out 'it was the Soviet side that took the initiative in proposing Article I [of the ABM Treaty] "not to deploy ABM systems for defence of the territory of the country". The Soviets were ready to consider "a complete ban of ABM" and to accept "qualitative limitations"'.[187]

In the aftermath of the conclusion of the ABM treaty in 1972, several prominent Soviet commentators praised the ABM Treaty because it 'introduced into the arms race for the first time some elements, albeit minimal, of restraint, certainty, and predictability',[188] while others stressed the 'destabilising effect of the combination of ABM and MIRV deployments'.[189] On the other hand, Marshall Grechko – while stressing the contribution of this Treaty to 'prevent a competition between offensive and defensive nuclear missile arms' – stated also that the Treaty 'does not place any limitations on the conduct of research and experimental work directed toward the solution of problems of the defence of the country from nuclear missile strikes'.[190]

Both the relevance of the Treaty for Soviet military doctrine and its subsequent military research and development activities in the BMD field has become the major area of contention in the United States strategic debate between the hawks, the supporters of détente and the centrists.

While Garthoff admitted that since 1972 'the Soviet Union has . . . carried forward an active research and development programme on BMD, within the limitations of the treaty',[191] the hawks and spokesmen of the intelligence community claimed that the BMD activities were 'considerably greater than that of the United States[192] and that 'the known Soviet programmes appear aimed at achieving deployable weapon systems'.[193]

Yost has recently grouped the potential BMD capabilities the Soviets are acquiring through their research and development efforts under four headings: 'an improved Moscow BMD system including transportable components for what has been designated the ABM-X-3-system; potentially BMD-capable SAMs, including SA-5s, SA-10s and SA-X-12s; new large phased-array radars that could support the new interceptors; and directed-energy weapons'.[194] How have these developments been interpreted in the Pentagon's reports on *Soviet Military Power*, in the trade press and in the academic literature?

(c) Soviet BMD developments as seen by American officials, journalists and security experts

In 1981 the Pentagon provided a brief factual account about the components of the Soviet ABM: 'The Moscow defences currently include the ABM-1B/Galosh interceptor missiles, battle management radars and missile engagement radars. The Soviets . . . also continue to engage in an active and costly ABM research and development effort, which they are permitted to do under the ABM Treaty of 1972. Their main concentration appears to be on improving the performance of their large phased-array detection and tracking radars and developing a rapidly deployable ABM system . . . Improving the Moscow defences is allowed by the 1972 ABM Treaty as long as the 100 interceptor launcher limit is not exceeded.'[195] In 1983, a few days before President Reagan's Star Wars speech, the United States Department of Defense acknowledged:

> The system cannot presently cope with a massive attack, however, and the Soviets have continued to pursue extensive ABM research and development programs, including a rapidly deployable ABM system and improvements for the Moscow defences ... They are building additional ABM sites and are retrofitting sites with new silo launchers. To support these launch sites, the Soviets are building the new ABM radar in the Moscow area ... It is a phased-array radar with 360 degree coverage ... These improvements to the Moscow defences, which could be completed by the mid-1980s, are allowed by the 1972 ABM Treaty.[196]

The Pentagon report speculated: 'Space-based ABM systems could be tested in the 1990s, but probably would not be operational until the turn of the century'.[197]

One year after President Reagan initiated SDI with his vision beyond deterrence there was a drastic shift in the Pentagon evaluation of Soviet BMD capabilities and programmes:

> Since 1980, the Soviets have been upgrading and expanding this system within the limits of the 1972 ABM Treaty ... When completed, the new system will be a two-layer defence composed of silo-based long-range modified GALOSH interceptors designed to engage targets outside the atmosphere; silo-based high-acceleration interceptors designed to engage targets within the atmosphere; associated engagement and guidance radars; and a new large radar at Pushkino designed to control ABM engagements. The silo-launchers may be reloadable. The new Moscow defences are likely to reach operational status in the late 1980s.

For the first time the Pentagon pointed to potential Soviet breakout capabilities:

> The USSR has an improving potential for large-scale deployment of modernized ABM defences well beyond the 100-launcher ABM Treaty limits. Widespread ABM deployment to protect important target areas in the USSR could be accomplished within the next 10 years. The Soviets have developed a rapidly deployable ABM system for which sites could be built in months instead of years ... The Soviets seem to have placed themselves in a position to field

relatively quickly a nationwide ABM system should they decide to do so.

For the first time, in the third edition of *Soviet Military Power*, it was stated that two Soviet air defence missiles (SA-10, SA-X-12) 'may have a potential to intercept some types of US strategic ballistic missiles as well. These systems could, if properly supported, add significant point-target coverage to a wide-spread ABM deployment'.[198]

In the fourth edition of *Soviet Military Power*, Soviet BMD activities were portrayed in even grimmer terms, claiming that

> developments aimed at providing the foundation for widespread ABM deployments beyond Moscow are underway. The new SA-X-12 surface-to-air missile, which incorporates BMD capabilities, is nearing operational status, while research on directed-energy BMD technology continues apace ... The first new launchers are likely to be operational this year, and the new defences could be fully operational by 1987.

Then follow various force projections which refer to potential Soviet ABM breakout options:

> They have been testing all the types of ABM missiles and radars needed for widespread ABM defences beyond the 100-launcher limit of the 1972 ABM Treaty. Within the next 10 years, the Soviets could deploy such a system at sites that could be built in months instead of years ... The new, large phased-array radars under construction in the USSR, along with the HEN HOUSE, DOG HOUSE, CAT HOUSE, and possibly the Pushkino radar, appear to be designed to provide support for such a widespread ABM defence system. The aggregate of the USSR's ABM and ABM-related activities suggest that the Soviet Union may be preparing an ABM defence of its national territory.[199]

This obvious BMD threat upgrade since 1984 is in the United States tradition; after overselling the remedy they oversell the threat alike to justify the tremendous financial resources needed for SDI.

In the United States press, various reports have been published indicating that the 'Soviet Union had a lead over the United States in developing weapons to repel nuclear attacks from space'.[200] In early December 1983 a CIA report was leaked to *Aviation Week & Space Technology* about Soviet BMD activities, including 'vigorous research and development for directed-energy weapons – laser, particle beam and microwave devices', and White House officials were quoted as saying: 'We are concerned about a breakout from the ballistic missile treaty by the USSR'.[201] And in January 1984, a few days after President Reagan had signed a decision memorandum initiating SDI, the same trade journal reported: 'President Reagan has been given evidence by the Central Intelligence Agency that the Soviet Union is producing components for and has in place or under construction the major elements of a nationwide BMD system'.[202] In this alarmist treatment a White House official was quoted as saying: 'The US could be witnessing a Soviet move to place itself in a position to abrogate the Anti-ballistic Missile Treaty and rapidly deploy a system to defend key areas such as intercontinental ballistic fields'. The new evidence would show 'that the mutual assured destruction concept may be on the way out'. Therefore, the report continues 'the President has concluded that the Soviets are doing everything they can with the BMD to bring the US to its knees'.[203] And on 9 March 1984 a Reagan defence official was quoted in the Washington Post saying 'that the Soviet Union is a decade ahead of the United States in developing high-technology defence against strategic missiles'.[204] However, on 22 March 1984 DARPA director Robert Cooper backtracked, admitting in a secret session of the Senate Armed Services Committee: 'I don't think that the Soviets are far advanced as to where we stand in many, if not most, of these technologies'. When he was confronted with the statement Richard de Lauer had made two weeks earlier at a public meeting, Mr Cooper appeared to back off.

The 'overselling the threat' strategy was harshly criticised by Senator Jeff Bingaman of New Mexico: 'It strikes me that we are taking very speculative information with regard to potential Soviet capability, and we are building it up as a likelihood for purposes of public consumption. The public is out there thinking these guys have all kind of capabilities'.[205]

During the budget debates for FY 85, the *Washington Times* headlined on 10 May 1985: 'CIA:Soviets are developing their own "star wars" system'. According to this press account: 'The CIA assessment of Soviet efforts that are counterparts to the Reagan Administration's SDI ... found Soviet programs on a par with and well in advance of US research. It indicates the Soviet Union is on the verge of launching a major directed energy weapons development program'.[206] On 21 June 1985 the same conservative daily claimed:

'The Soviet Union is believed to be ahead of the United States in research on three key beam weapon technologies considered critical to a space-based nuclear missile defence system, a senior Defence official said'.[207] And on 25 June 1985 the *Baltimore Sun* quoted a new intelligence estimate of the CIA which claimed: 'The Soviet Union now has the "potential" to deploy a countrywide defence against ballistic missiles, "breaking out" of the 1972 Treaty limiting such arms, White House officials said'.[208] However, on 26 June 1985 the National Intelligence Officer, Lawrence K. Gershwin, testified before a joint session of two Senate subcommittees that the CIA has not 'judged it likely that the Soviets would in fact move to such a deployment in the near term.[209]

Soviet hawks, especially those of the Air Defence Forces, unnamed intelligence sources, White House, Pentagon officials and alarmist press accounts all provided the ammunition for a strong contingent of hawkish analysts with which they were to base their explicit or implicit efforts to gather support for United States BMD programmes as called for in SDI.

In a highly speculative account of Soviet policy options relating to space-based lasers for BMD, Rebecca Strode concluded:

> With this strategic promise on the technological horizon, Soviet compliance with the 1972 ABM Treaty could well become problematic *whether or not* the United States seeks a change in the current BMD *status quo*. In the absence of Soviet BMD deployment, the USSR may find the survivability of its own time-urgent, hard-target capable ICBM force threatened. The pressure to reconstitute a secure and accurate second-strike ICBM capability through BMD deployment would be especially strong if Moscow believed it could attain an advantage well before the US could attain a comparable capability.

For this reason, Strode argues, 'a substantial US research effort in the area of laser BMD would be prudent as a hedge against and deterrent to an attempted Soviet "break-out" from the ABM Treaty, regardless of whether or not the US ultimately chooses to develop and deploy such a system'.[210]

Given this clear political message it is not surprising to be informed that the available evidence suggests 'that Soviet acceptance of ABM limitations represented a short-run effort to compensate for technological inferiority, not an abandonment of a damage-limiting, warfighting strategy'.[211] And if this interpretation of Soviet motives for the ABM Treaty were correct, 'one could expect that future Soviet advances in BMD technology could lead the USSR to reconsider the Treaty – either by relaxing the Treaty's restrictions through renegotiation, or by outright withdrawal from the agreement'. However, the argument continues, 'US development of a superior BMD technology could bring the Soviet Union to the negotiating table for serious arms control talks'.[212] Strode concludes with the hope that the economic difficulty of the USSR 'of pursuing large-scale offensive and defensive deployments simultaneously might however, be so severe as to force greater reliance in the long run on one option (probably BMD) at the expense of the other'.[213]

Sayre Stevens, a former deputy director of intelligence at the CIA who had been responsible for all the CIA's intelligence analysis, described the 'steady, unfrenzied progress toward defined development program in the Soviet BMD effort since 1972'. For Stevens, the 'recitation of theoretically derived implications of the ABM-X-3 [a main feature in Pentagon and hawkish statements] is simply that – theoretical'.[214] In his assessment, 'the removal of half of the sixty-four Galosh launchers at the Moscow BMD sites marked the beginning of an upgrade program' that appeared 'to be overdue'. Stevens downgraded the exaggerated performance projections in US publications by stating: 'Although these improvements to the Moscow system will significantly enhance its limited capability, these weapons cannot seriously hinder a US attack on Moscow, given the very large weapons inventories that currently exist'.[215] Stevens sees few powerful incentives to abandon the ABM Treaty in the near term, because the USSR 'has only now achieved the level of technology that was available to the US ten years ago. The major difference now is that Soviet technology is much closer to application'. Stevens stressed 'that areas in which the US has more or less superiority (for example, microelectronics in general, large-scale integrated circuits, phased-array radars, compact and high speed data processors, and the like) are particularly important to the development of advanced BMD systems'.[216] Stevens does not think it likely that the Soviet Union will break out of the ABM Treaty but rather 'that the US initiative might trigger a Soviet deployment response'.[217] Given the Soviet secretiveness, the dispute between the hawkish and the moderate perceptions of Soviet BMD activities cannot be refereed by an independent expert. However, extreme doubts may be justified towards the most recent hawkish efforts to 'oversell the threat'.

VII SOVIET AIR DEFENCE MISSILES WITH POTENTIAL BMD CAPABILITIES – THE SAM UPGRADE IN THE AT(B)M CONTEXT

Among the allegations of Soviet Treaty violations[218] only one dealt with the problem of the upgrading of SAM systems, claiming 'concurrent testing of ABM and SAM components'. Interestingly, the Reagan Administration did not charge that the Soviet Union was violating the ABM Treaty by introducing two new air defence missiles, the SA-10 and SA-X-12, both with potential anti-tactical ballistic missile (AT(B)M) capabilities.

During the negotiation of the ABM Treaty[219] and during the SALT I hearings in the United States Congress[220] considerable attention was devoted to the relationship between the Soviet and the American air defence systems and their ABM potential. Both American defence and some intelligence officials and members of Congress were concerned that Soviet aircraft defence systems – in the early 1970s especially the SA-2 and SA-5[221] – might be upgraded to obtain some ABM capabilities thereby circumventing the ABM treaty.

Once again a shift can be observed in the evaluation of the most recent Soviet SAMs for air defence (SA-10 and SA-X-12) after President Reagan's 'Star Wars' speech. In its first report on *Soviet Military Power* in September 1981 the Pentagon stated: 'The SA-10 system is the latest Soviet strategic SAM system and is designed for increased low-altitude capability. With radars which are more advanced than previous systems, the SA-10 was designed to counter low-altitude manned aircraft, *although it may have some capability against cruise missiles*' (my emphasis). In the same edition, the old SA-5 was described as 'a long-range interceptor designed to counter the threat of high-performance aircraft'.[222] In March 1983 the Pentagon maintained that the primary emphasis in the modernisation of the Soviet SAM force was 'a counter to low-altitude targets. Newer systems demonstrate longer range, particularly at low altitude; improved mobility; increased target handling capability and increased firepower. Deployment of the SA-10 system, which can engage multiple aircraft and *possibly cruise missiles at any altitude*, has steadily increased. In addition to deployment around the USSR, the system is replacing the 30-year-old SA-1s around Moscow. Development of a mobile SA-10 is under way' (my emphasis).[223]

On 9 March 1983, Department of Defense Assistant Secretary T.

K. Jones made the following remarks about the SA-X-12, downgrading its performance:

> The SA-X-12 is considerably in the future, fortunately. Moreover, the maneuvering capability of the Pershing front end would probably give that system a hard time. It is something we are looking at carefully as the 12 is developed. We don't have all the information on it. We will just have to keep watching it.
>
> As for the ground-launched cruise missile, we see the deployment of the better quality Soviet air defence system centering around Moscow and Soviet cities – not Warsaw Pact areas – therefore we believe the GLCM going into targets it would be applied to will remain effective for many years.

When asked about the potential AT(B)M capability, Mr Jones replied:

> It appears the Soviets are developing that to counter the shorter range ballistic missiles and the ABM Treaty was drafted so that is a legal development and they could deploy it fully. However, at the margin, a system that has good capability against something like the Pershing II would also have reasonable capability to defend reasonable areas against our ICBMs and submarine-launched ballistic missiles.[224]

In its third report, the United States Department of Defense repeated its prior cautious statements, adding: 'The mobile SA-10 could be operational by 1985'.[225] However, in its fourth edition of April 1985 a major change in the evaluation of the Soviet SAM occurred:

> The first SA-10 site reached operational status in 1980. Nearly 60 sites are now operational and work is underway on at least another 30. More than half of these sites are located near Moscow. This emphasis on Moscow and the patterns noted for the other SA-10 sites suggest a first priority of terminal defence of wartime command and control, military, and key industrial complexes ... The SA-X-12 has good low-altitude air defence capabilities *as well as the ballistic missile defence capabilities*. (my emphasis)[226] The SA-X-12 is both a tactical SAM and anti-tactical missile. *It may have the capability to engage Lance and both the Pershing I and Pershing II ballistic missiles. The SA-10 and SA-X-12 may have the potential to intercept some types of US strategic ballistic missiles as well*. These systems could, if properly supported, add significant

point-target coverage to a widespread ABM deployment (my emphasis).[227]

In 1984 the potential ABM capability of the SA-X-12 was only implied in general terms:

> The SA-X-12 is both a tactical SAM and anti-tactical ballistic missile. Both the SA-10 and SA-X-12 may have the potential to intercept *some types of US strategic ballistic missiles as well* (my emphasis). These systems could, if properly supported, add significant point-target coverage to a wide-spread ABM deployment.[228]

The first alarmist accounts about the new 'Soviet super-weapon' (the SA-12) also appeared in April 1983, only a few days after President Reagan's 'Star Wars' speech. 'Tests monitored by the CIA', revealed Jack Anderson on 5 April 1985 in the *Washington Post*, 'showed that the SA-12 successfully shot down Soviet missiles roughly equivalent to US Pershing II missiles . . . That means the Pershing IIs . . . may be obsolete before they're put in place . . . What is hair-raising about the SA-12 is that it may be capable of knocking our Poseidon missiles out of the sky . . . This means the SA-12, designed originally to ward off US short-range tactical missiles, might be capable of defending the Soviets' land-based ICBMs against our subbased long-range missiles'.[229]

In the same vein, a British source reported on 14 April 1985 about 'New Soviet missile defences':

> The CIA believes the SA-12 can successfully defend Moscow against the American Pershing-2 missiles, and submarine-launched Poseidon ballistic missiles. The first Soviet ATBM was tested in 1974 . . . Since then, the SA-5 and the SA-12 have been tested as both an ATBM and an ABM. But they seem to be aimed mainly at the American Lance and Pershing . . . Another Soviet missile, the relatively new SA-10 . . . is a weapon which the Russians probably plan to use against the American cruise missile.[230]

Although T. K. Jones had downgraded the SA-X-12 in 1983 in Congressional testimony, DARPA director Robert Cooper claimed a year later that 'the SA-12 mobile air defence missile system [could] . . . be used by the Soviets as an ABM system when used in conjunction with the large phased-array radar the Soviets have built in violation of the ABM Treaty . . . [They] could build and store thousands of the ABM missiles for the SA-12 that could be used in a rapid breakout of the ABM Treaty'.[231] And on 2 April 1984 Clarence A. Robinson quoted in *Aviation Week & Space Technology* an unnamed United States official as claiming that the SA-12 'clearly has ballistic missile intercept capability ... We estimate that within four years or so at the present production rates the Soviets would stand a good chance of intercepting approximately 17 of the US strategic missile force in the terminal regime. When you approach the 20 line, this causes grave uncertainty of penetration and will force alteration of the single integrated operational plan'.[232]

The authoritative Brookings study on BMD made the United States strategic Cruise missile modernisation programme 'responsible for the vigor with which the Soviet Union has sought increasingly capable air defences'. According to Stevens, in responding to these improvements Soviet designers:

> have built into new SAM systems the attributes required to counter them. These SAM improvements are in substantial measure – though not entirely – applicable to the BMD intercept problem ... All of these points suggest that the SAM upgrade is substantially more credible than it was ten years ago. The Soviets could have, with its new SAMs, a BMD capability able to enhance damage limitation that is not controlled by the ABM Treaty, whereas the United States, with no strategic SAMs, has none.
>
> Reports that the Soviet Union is now developing an ATBM system give cause for even greater concern ... Soviet development of such systems constitutes a qualitative change in the nature of SAM upgrade concerns, for although these weapons systems are not specifically constrained by the ABM Treaty, they will almost surely possess some significant capability against long-range strategic ballistic missiles ... The widespread deployment of ATBMs to Soviet tactical forces in the future could only produce a serious concern about their possible employment by the PVO in a strategic role ... Their rapid deployment in large numbers (possibly from covert storage) would constitute another means whereby the Soviet Union could extend its defensive forces, very possibly within the ABM Treaty. Although the treaty does prohibit giving non-ABM systems the capability to intercept strategic ballistic missiles, it would be extremely difficult to make an airtight case that it occurred if the Soviet Union denied the allegations.[233]

Within the United States government obviously no agreement could be reached whether the SA-12 tests against a SS-12 MRINF could be considered a violation of the ABM Treaty.

The ATBM problem indicates a major grey area of the ABM Treaty which legitimise Western development of an ATBM capability in the context of an EDI. The Soviet SA-10, SA-X-12 and the American Patriot ATM could all contribute to an erosion of the ABM Treaty and to a new level of the arms competition in Europe.

VIII SOVIET LASER-, PARTICLE BEAM- AND OTHER SPACE WEAPONS-RELATED R&D PROGRAMMES

In a detailed Report to Congress the United States Department of Defense described five major research areas for the SDI with a special emphasis on directed energy weapons technologies (laser, particle beams and nuclear-driven directed energy concepts).[234]

To what extent has a shift in the public assessment of similar Soviet research efforts occurred since March 1983 which could support or undercut the traditional effort of overselling the threat after overselling the remedy?

(a) Directed energy weapons as seen by the United States Department of Defense (1981–5)

In the first Pentagon report it was stated that the Soviet Union had reduced the technology gap from '10-to-12 years behind US capability' in 1965 to 'three-to-five years with a few outstanding developments following US technology by only two years and some problem areas lagging by as much as seven years'. As far as the area of directed energy weapons is concerned the 1981 report claimed:

> The Soviets have devoted substantial resources to high technology developments applicable to directed energy weapons. Their knowledge of radio frequency weapons ... and the fact that they are developing very high peak-power microwave generators, gives rise to suspicions of possible weapons intent in this area as well. The Soviets have been interested in particle beam weapons (PBW) concepts since the early 1950s ... The Soviet high energy laser program is three-to-five times the US level of effort and is tailored to the development of specific laser weapon systems. The Soviet laser-beam weapons program began in about the mid-1960s ...

Available information suggests that the Soviet laser weapon effort is by far the world's largest.[235]

In March 1983 the United States Department of Defense pointed to a known future development of space weapons systems . . . involving:

1. the present generation of anti-satellite vehicles;
2. a very large, directed-energy research programme, including the development of laser-beam weapons systems which could be based either in the USSR, abroad the next generation of Soviet ASATs or aboard the next generation of Soviet manned space stations.

For the first time the report claimed in its brief section on directed energy: 'Their high energy laser programme is three-to-five times the US effort. They have built numerous classified facilities dedicated to the development of these weapons.' As far as the potential applications in weapons systems are concerned, the report stated:

> The Soviets are committed to the development of specific laser weapon systems. Soviet development of moderate-power weapons capable of short-range ground-based applications such as tactical air defence and anti-personnel weapons, may well be far enough along for such systems to be fielded in the mid-1980s. In the latter half of this decade, it is possible that the Soviets could produce laser weapons for several other ground, ship and aerospace applications.[236]

After President Reagan's 'Star Wars' speech, the reference to the weapons applications of directed energy in the Soviet Union became more specific in the Pentagon's third report of April 1984:

> Soviet directed energy development programmes involve future ABM as well as anti-satellite and air-defence weapons concepts. By the late 1980s, the Soviets could have prototypes for ground-based lasers for ballistic missile defence. The many difficulties in fielding an operational system will require much development time, and initial operational deployment is not likely in this century. Ground- and space-based particle beam weapons for ballistic missile defence will be more difficult to develop than lasers. Nevertheless, the Soviets have a vigorous program underway for particle beam development and could have a prototype space-based system ready for testing in the late 1990s.[237]

As far as Soviet ASAT applications are concerned, the report states:

> The Soviets have two ground-based lasers that could be used against satellites. The Soviets also have the technological capability to conduct electronic warfare against space systems.[238]

In addition the report devotes a full page to directed-energy weapons:

> The Soviets are also pursuing technologies to support laser weapons development. This includes R&D on efficient electrical power sources and on the development of high-quality optical components. The USSR has developed a rocket-driven magnetohydrodynamic generator ... The Soviets also continue an intensive effort aimed at the development of high-power microwave and millimeter-wave sources for radio frequency weapons ... Many Western weapons systems would be vulnerable to such a weapon, which not only could damage critical electronic components but also inflict disorientation or physical injury on personnel. Finally, there is considerable research on the development of destructive particle-beam weapons. Such weapons could deliver intense energy particles at the speed of light, capable of penetrating the exterior of a target, destroying key internal components or igniting fuels and munitions. While much of the Soviet R&D effort in this field is on par with that in the West, there are difficult technological problems to be solved. Technology to support development of such weapons is not expected to be available before the mid-1990s.[239]

In the fourth Pentagon report of April 1985, the description of Soviet activities in the field of 'Laser and Energy Weapons' is further upgraded:

> Soviet directed-energy development programs involve BMD as well as ASAT and air-defence concepts.

However, most of the details deal with speculations about possible future developments and applications:

> By the late 1980s, the Soviets could have prototypes for ground-based lasers for ballistic missile defence. Testing of the components for a large-scale deployment system could begin in the early 1990s ... The Soviets could ... be ready to deploy a ground-based laser BMD by the early-to-mid-1990s ... : The Soviets have a vigorous

program underway for particle beam development and could have a prototype space-based system ready for testing in the late 1990s.

The Soviets have begun to develop at least three types of high-energy laser weapons for air defence. These include lasers intended for defence of high-value strategic targets in the USSR, for point defence of ships at sea, and for air defence of theater forces. Following past practice, they are likely to deploy air defence lasers to complement, rather than replace, interceptors and surface-to-air missiles (SAMs). The strategic defence laser is probably in at least prototype stage of development and could be operational by the late 1980s. It most likely will be deployed in conjunction with SAMs in a point defence role ... The shipborne lasers probably will not be operational until after the end of the decade. The theater force lasers may be operational sometime sooner and are likly to be capable of structurally damaging aircraft at close ranges and producing electro-optical and eye damage at greater distances. The Soviets are also developing an airborne laser. Assuming a successful development effort, limited initial deployment could begin in the early 1990s. Such a laser platform could have missions including anti-satellite operations, protection of high-value airborne assets, and cruise missile defence.

The Soviets are working technologies ... [which] include space-based kinetic energy, ground- and space-based laser, particle beam, and radiofrequency weapons.

In comparison with the United States directed-energy programme it is claimed that:

[the Soviet programme] is much larger than the US effort. They have built over half a dozen major R&D facilities and test ranges, and they have over 10,000 scientists and engineers associated with laser development ... The Soviets have now progressed beyond technology research, in some cases to the development of prototype laser weapons. They already have ground-based lasers that could be used to interfere with US satellites. In the late 1980s, they could have prototype space-based laser weapons for use against satellites ... Ground-based laser anti-satellite (ASAT) facilities ... could be available at the end of the 1980s and would greatly increase the Soviets' laser ASAT capability beyond that currently at their test site at Sary Shagan. They may deploy operational systems of space-based lasers for anti-satellite purposes in the 1990s, if their technology developments prove successful, and they can be expected to pursue development of space-based laser systems for ballistic missile defense for possible deployment after the year 2000.

Since the early 1970s, the Soviets have had a research program to explore the technical feasibility of a particle beam weapon in space. A prototype space-based particle beam weapon intended only to disrupt satellite electronic equipment could be tested in the early 1990s. One designed to destroy satellites could be tested in space in the mid-1990s.

The Soviets have conducted research in the use of strong radio frequency (RF) signals that have the potential to interfere with or destroy components of missiles, satellites, and re-entry vehicles. In the 1990s, the Soviets could test a ground-based RF weapon capable of damaging satellites.

In the area of kinetic energy weapons . . . the Soviets have a variety of research programs underway. These systems could result in a near-term, short-range, space-based system useful for satellite or space station defence or for close-in attack by a maneuvering satellite. Longer-range, space-based systems probably could not be developed until the mid-1990s or even later.[240]

All these claims were repeated in the October 1985 joint report of the United States Department of Defense and the State Department, 'Soviet Strategic Defense Programs'.[241] However, these worst case projections are in conflict with other reports of the United States department of Defense and with the September 1985 Office of Technology Assessment of the US Congress, which concluded:

However, the quality of that work [in advanced technologies] is difficult to determine, and its significance is therefore highly controversial. In large part, we are limited to observing what goes into their efforts (e.g., the amount of floor space at various research laboratories, the observable activity at test sites) and what does not come out (e.g., absence or cessation of publication on topics known to be under investigation, indicating that the activity has been classified.

In terms of basic technological capabilities, however, the United States remains ahead of the Soviet Union in key areas required for advanced BMD systems (my emphasis), including sensors, signal processing, optics, microelectronics, computers, and software. The United States is roughly equivalent to the Soviets in other relevant areas such as directed energy and power sources. According to the

Under Secretary of Defense for Research and Engineering, the Soviet Union does not surpass the United States in any of the 20 'basic technologies that have the greatest potential for significantly improving military capabilities in the next 10 to 20 years'.[242]

These internal assessments of Soviet technological capabilities appear to be more realistic than the projections of what the Soviets might be able to do which have been published for public consumption.

(b) Soviet beam weapons as a topic of press reporting

Public concern about Soviet beam weapons and speculations in the Western press were aroused in early 1977 when *Aviation Week & Space Technology* published an article based on interviews with the retired former assistant director of the Air Force intelligence, General Keegan, who charged that the 'Soviet Union is developing a charged particle beam device designed to destroy US intercontinental and submarine-launched ballistic nuclear warheads'.[243] In July 1980 Senators Wallop and Garn 'cited a National Intelligence Estimate which expects the Soviet Union to test a space-based laser weapon in the mid-1980s'[244] and *Aviation Week* warned that a

> directed-energy weapon that could be the first step in a revolutionary concept of warfare is being constructed by the Soviet Union at Saryshagan . . . Many US intelligence analysts believe the weapon is an early prototype of a new-design charge-particle beam device, and that it may be used within a year or so in tests against ballistic missile targets.[245]

However, as of August, 1986 no such tests have taken place. In 1981 James P. Wade, the Principal Deputy Undersecretary for Research and Engineering, remained rather doubtful: 'I would believe if the Soviets would deploy a space-based system in 1985 and 1986, it most likely would be ineffective, as far as the ballistic RVs go'.[246] On 21 January 1982 *Defence Daily*, referring to Department of Defense reports, claimed that a Soviet space-based laser 'has a near-term potential application for anti-satellite and anti-aircraft applications, i.e. in the 1990s, with the BMD potential more probable in the next century. However, it emphasises that 'the program faces serious technical problems, extremely high costs and vulnerability to attack'.[247]

In March 1982 leading United States Air Force generals doubted reports 'about Soviet space-based lasers posing a threat to United States satellites any time soon'. The Air Force research director, Lt General Kelly H. Burke, opposed a United States 'crash effort to develop space-based lasers' arguing 'I personally am not encouraged by what I've seen in the application of space-based lasers up to this point'.[248]

On 23 March 1984 John Gardner, one of the chief SDIO officials rejected the alarmism in some press reports by stating:

> We do not believe that the Soviets could deploy a laser weapon capable of effectively defending against ballistic missiles, by the end of this decade, although they could be in a position to exploit such technology as an ASAT weapon.[249]

Many of the projections on behalf of the United States alarmists who went public to obtain additional funds for their pet R&D programmes were not true as of August 1986. Nevertheless, the alarmists provided sufficient 'evidence' for hawkish interpretors of the Soviet threat.

(c) Soviet laser and particle beam weapons – a controversial issue within the strategic community in the West

Depending on their general interpretation of Soviet military programmes, United States strategic analysts have differed in their assessment of Soviet directed-energy weapons programmes. Payne has more or less accepted the alarmist reports,[250] Yost has relied primarily on the Department of Defense assessments,[251] and Meyer and Hecht have remained rather sceptical. Meyer mentioned several of the technical constraints for directed-energy weapons in general and for ground-based lasers in particular:

> The biggest problem [of ground-based lasers] is that they have limited attack windows against enemy satellites – prospective targets must pass overhead. Moreover, ground-based lasers are 'fair weather' weapons – unlike the US Postal Service they will not work in rain, sleet, snow, or cloud cover ...
> While it is difficult to assess from open source information the state of Soviet technology pertinent to a ground-based weapon, some observations can be made about the prospects for a Soviet space-based laser system in the near future. The operational problems of Soviet space platforms are well documented ... In essence, the Soviets have exceptional trouble keeping their automated space

probes functioning for more than a few months. The combination of complexity, longevity, and reliability seems to have been beyond Soviet technical capabilities. The technical demands of a laser ASAT system will be far greater.[252]

For Meyer, who has confronted official claims with available information on Soviet space performances:

> there can be no doubt that the Soviet military is active in space. And, in the future, the Soviet Union will certainly increase its military investment in space ... Soviet military systems in space whether serving in support missions or as weapons, are only one component of a large and diverse military force posture. The question should not be how much is the Soviet Union putting into space, but rather how much is it getting out of space?[253]

Jeff Hecht, an independent laser expert, based his opinion on the following three arguments

> First, the Soviet Union historically tends toward making risky demonstrations early in a development program ...
> The second reason is what I call the 'Soviet submarine effect', the selective leaking of confidential information to the press ...
> [For example *Aviation Week*'s] coverage of X-ray lasers started with an unnamed source evidently unhappy because the Pentagon was unwilling to support the program.[254]
> The third problem is the 'threat inflation' that tends to occur during intelligence gathering. The information gathered during intelligence operations inherently tends to be imprecise. Technical details may be misunderstood, conversations may be only partly overheard, future goals may be confused with the present realities, and the most interesting details may not be resolved by a spy satellite image. Intelligence analysts interpreting the information have to estimate a range of possibilities. Military planners often assume the 'worst' case, which seems prudent when planning how to deal with possible threats. However, all too often worst-case estimates are transformed into seemingly authoritative reports of new equipment about to roll off Soviet assembly lines, when in reality the Soviet Union may have built only a single experimental model that doesn't work very well.[255]

Selective leaking, worst-case analyses and the overall threat inflation was easily demonstrated based on the authoritative Pentagon reports on *Soviet Military Power* before and after President Reagan's 'Star

Wars' speech. Given the Soviet overemphasis on secrecy in all military-related activities, the analyst who does not want to be misused as a mere sophisticated argument producer for military programmes must caution against taking contradictory public assessments at face value. He must reveal the tactics of 'overselling the threat' and 'overselling the remedy'.

IX CONCLUSIONS – SOVIET MILITARY SPACE WEAPONS: DETERMINANTS OF AND IMPLICATIONS FOR EUROPEAN SECURITY

Given the lack of Soviet sources concerning its military space potentials and its research and development efforts on the one hand and the bias of the *Soviet Military Power* official reports and of many reports in the United States trade press on the other, it is impossible to present an accurate description of Soviet military space weapons based on open sources.

The independent analyst may reach different conclusions based on his or her overall assumptions and the evaluation of the available sources. The detailed analysis of the four official reports on *Soviet Military Power* before and after President Reagan's 'Star Wars' address indicate a drastic shift in the concept of the 'Soviet space threat'. Obviously the imagined Soviet space threat has grown since 1984 after the credibility of President Reagan's strategic vision had increasingly come under attack in the United States. In order to discuss the initial question as to whether SDI has been a *unilateral decision*, an *unavoidable reaction* or a *hedge* against future Soviet breakout of the ABM Treaty a brief survey of the different interpretative models to explain United States and Soviet military space efforts may be appropriate. Therefore, the implications of SDI for United States – Soviet relations and of Soviet and American BMD activities in Europe will be discussed.

(a) Models relating to BMD weapons systems for explaining United States and Soviet arms competition in outer space

Five different conceptual-type models for explaining the United States – Soviet competition in outer space may be distinguished:

1. the military activities of both influence each other directly (*action–reaction model* of an arms race);[256]

2. both pursue their military programmes independent of each other (*model of Eigendynamik*);[257]
3. the Soviet Union has been the pace-setter to which the United States have to respond (*Soviet military buildup thesis*);[258]
4. the United States by initiating SDI has become a pace-setter in the militarisation of outer space (*model of United States initiative*);[259]
5. the United States has to take the lead in the R&D effort in order to deter a potential future Soviet treaty violation (*anticipatory reaction* based on worst-case assumptions).[260]

None of these pure models appear to be sufficient to explain the United States–Soviet military space competition since the Sputnik shock of 1957.

The drastic increases in United States expenditures for military space activities from FY80 to FY86[261] cannot be explained simply as a *direct reaction* to similar Soviet programmes,[262] but neither can it be explained as an initiative completely unrelated to the Soviet military armament processes. Colin S. Gray has distinguished between six patterns of United States–Soviet interactions: (1) *Eigendynamik*, (2) random response, (3) macro response, (4) limited response, (5) differential response, and (6) mechanistic response. A combination of two of these models may contribute to an explanation of the United States–Soviet military space competition: the model of *Eigendynamik* and the model of a *macro response*. According to Gray the latter model 'is an attempt to politicise what is all too often a technical exercise'; and often 'broad responses to broad perceived challenges may have to await favourable electoral political circumstances, and may not be reflected in weapons developed and deployed for many years to come. Nevertheless, as a signal of political intent... declared shifts in the arms race strategy should be understood as having an immediate political effect'.[263]

Since the mid-1950s three macro responses may be distinguished which had a direct impact on the overall military effort in general and on the military space programme of either superpower in particular:

1. the *Sputnik shock* and the assumed missile gap;[264]
2. the *Cuban missile crisis* and the *Soviet humiliation*;[265]
3. the *Soviet intervention in Afghanistan* and the Iranian hostage crisis.[266]

The Sputnik shock provided the political climate in the United States for a major increase in overall military expenditures, both for strategic nuclear and conventional forces and for a major civilian and increasingly military space effort within the first year of the Kennedy Administration (*Kennedy impulse*).

The Soviet retreat during the Cuban missile crisis became instrumental for the replacement of Khrushchev by Leonid Brezhnev in 1964 and for the initiation of a major and persistent Soviet strategic buildup during the 1960s and its continuation in the 1970s (*Brezhnev impulse*).

The Iranian hostage crisis and the Soviet intervention in Afghanistan provoked a major shift in the United States official perception of the Soviet threat – a shift that had started in influential private groups as early as the Yom Kippur war of 1973. After the defeat of President Carter, President Reagan implemented many of the policy suggestions of the coalition of détente critics which had emerged in 1976 as the Committee on the Present Danger. In current United States dollars, United States defence expenditures doubled between FY80 and FY85, as did funds for the military space effort between 1980 to 1984 (*Reagan impulse*).

Three major statements indicated a drastic shift in the United States attitude towards the military use of outer space. On 4 July 1982, President Reagan in his directive on space policy emphasised the priority of security concerns. With his so-called 'Star Wars' speech of 23 March 1985, he initiated the SDI and in his State of the Union speech of 25 January 1984, he proposed the deployment of a manned space station in orbit by 1992. Have these three *initiatives* been in *response to specific Soviet challenges* in the areas of ASAT, BMD and manned space stations, or have they rather been part of the third macro response: the Reagan impulse?

(b) From a 'strategic revolution' to a reaction to Soviet BMD activities

In order to interpret the American SDI, two levels of analysis may be distinguished:

1. the *political or declaratory level*, and
2. the *specific planning or programme level*.

This chapter has focused primarily on the political level of declaratory politics, where new weapons systems are usually justified either as *direct reactions* to present Soviet programmes or as *anticipatory reactions* to potential future programmes of the main opponent. However, on the specific planning or programme level, the analysis offered by Herbert York in 1970 in his *Race to Oblivion* may also apply to SDI;

> I have found in the majority of those cases that the rate and scale of the individual steps has, in the final analysis, been determined by unilateral actions of the United States ... Why have we led the entire world in this mad rush toward ultimate absurdity?
> The reason is not that our leaders have been less sensitive to the dangers of the arms race ... Rather the reasons are that we are richer and more powerful, that our science and technology are more dynamic, that we generate more ideas of all kinds.[267]

The consequence of the continuing competition between those specialising in offensive systems and others emphasising defensive systems has been – in the words of the first of the first director of the Lawrence Livermore Weapons Laboratory – 'an arms race with oneself.'[268]

Within the United States – and presumably also within the Soviet Defence Department – there has been a continuing battle between those supporting offensive nuclear forces (the United States Air Force in the United States and the Strategic Rocket Forces in the USSR) and those calling for strategic defence systems (ABM, or BMD respectively). Within the United States Department of Defence, BMD has been the primary responsibility of the United States Army, and more recently SDI has become the primary task of a special organisation (SDIO). In the Soviet Union, BMD has been the exclusive task of the Air Defence Forces. Representatives of these commands or units in their internal budgetary requests (and, to some extent, in their published writings) have always stressed the need for a continued R&D effort for BMD. Between 1959 and 1983 the United States spent on the average between 1–2,000 million dollars annually for R&D programmes, which can be seen as predecessors to SDI.[269] And there is no doubt that the Soviet Union may have spent at least as much – or five times as much[270] as the United States.

President Reagan's three unilateral decisions to go ahead with ASAT and to initiate long-term research and potential deployment effort for a multilayered BMD as well as for a manned space station may hardly be interpreted as a direct (micro) response to equivalent Soviet programmes. They were rather part of a macro reaction which was provoked by the United States electoral response to the perceived Soviet military buildup, to the Soviet intervention in Afghanistan and possibly even to the American humiliation in the Iranian hostage crisis.

On the level of declaratory politics, a gradual shift has taken place in the effort to justify SDI. On 23 March 1983, President Reagan characterised his 'Star Wars' vision as one 'which holds the promise of changing the course of history' (overselling the remedy), but the pendulum swung to the other side in the 1984 and 1985 editions of the Pentagon's publication on *Soviet Military Power*, in which a major upgrade of the emphasis on Soviet military space activities and programmes can be detected in comparison with the two previous editions. The shift has become especially obvious with respect to the two Soviet air defence missiles SA-10 and SA-X-12: in the first edition the Pentagon claimed that the SA-10 'may have some capability against cruise missiles'. Only some thirty months later, the Department of Defense had recast its assessment: 'Both the SA-10 and SA-X-12 may have the potential to intercept some types of US strategic ballistic missiles as well'.[271] This shift can hardly be a consequence of dramatic new evidence but is rather one of many examples of the traditional tactics of 'overselling the threat'.

While it is true that the Soviet Union has the only operational ASAT and BMD weapons and has a major quantitative superiority over the United States and NATO as far as SAMs for air defence are concerned, nevertheless according to Pentagon assessments (Table 3.2) the United States is superior in fifteen of twenty 'basic important technology areas' and in the additional five the United States and the USSR are equal, and only one of these areas (laser) is a potential SDI component. Given these official technology assessments, the rather alarmist references to Soviet future potentials, lacking any reference to the performance of Soviet BMD and ASAT systems, appear to be doubtful to any independent observer (see also Table 3.3).

(c) Implications of SDI for United States–Soviet relations and for the defence of Western Europe

Given the technological inferiority of the Soviet Union – especially in those technologies that would be vital for any space-based multilayered BMD such as computers, electro-optical sensors, guidance and navigation, materials, microelectronic materials, production, radar sensors, robotics, machine intelligence and telecommunications

Table 3.2 Relative United States/USSR standing in the 20 most important basic technology areas*[272]

Basic Technologies	US Superior	US/USSR Equal	USSR Superior
1. Aerodynamics/Fluid Dynamics		×	
2. Computers and Software	←×		
3. Conventional Warhead (including all chemical explosives)		×	
4. Directed Energy (laser)		×	
5. Electro-optical sensor	×→		
6. Guidance and Navigation	×→		
7. Life Sciences (Human factors/Genetic engineering)	×		
8. Materials (Lightweight, high-strength, high temperature)	×→		
9. Micro-electronic materials and integrated circuit manufacturing	×→		
10. Nuclear Warhead		×	
11. Optics	×→		
12. Power Sources (Mobile) (Includes energy storage)		×	
13. Production/Manufacturing (Includes automated control)	×		
14. Propulsion (Aerospace and ground vehicles)	×→		
15. Radar Sensor	×→		
16. Robotics and machine Intelligence	×		
17. Signal Processing	×		
18. Signature Reduction (stealth)	×		
19. Submarine Detection	×		
20. Telecommunications (includes fibre optics)	×		

Notes:

*1. The list is limited to 20 technologies, which in aggregate were selected within the objective of providing a valid base for comparing overall United States and USSR basic technology. The list is in alphabetical order. These technologies are 'on the shelf' and available for application. (The technologies are not intended to compare technology level in currently *deployed* military systems.)

2. The technologies selected have the potential for significantly *changing* the military capability in the next 10–20 years. The technologies are not static; they are improving or have the potential for significant improvements; new technologies may appear on future lists.

3. The arrows denote that the relative technology level is changing significantly in the direction indicated.

4. The judgements represent consensus within each basic technology area.

Table 3.3 Relative United States/USSR technology level in deployed military systems*[273]

Deployed System	US Superior	US/USSR Equal	USSR Superior
Strategic			
ICBM		x→	
SSBN	x		
SLBM	x→		
BOMBER	x		
SAMs			x
BMD			x
ASAT			x
Cruise Missile	x		
Tactical			
Land forces			
SAMs (including naval)		x	
Tanks		x	
Artillery		x	
Infantry Combat Vehicles		x	
Anti-tank Guided Missiles		x	
Attack Helicopters (VTOL)		x	
Chemical Warfare			x
Ballistic Missiles		x	
Air Forces			
Fighter-attack Aircraft	x→		
Air-to-air Missiles	x		
PGM	x→		
Air Lift	x		
Naval Forces			
SSNs	x→		
Anti-submarine Warfare	x→		
Sea-based Air	x		
Surface Combatants	x→		
Naval Cruise Missile		x→	
Mine Warfare		x	
Amphibious Warfare	x→		
Command, Control, Communications and Intelligence			
Communications	x→		
Electronic Countermeasures/ECCM	x		
Early Warning (Includes surveillance & reconnaissance)	x		
Training Simulators	x		

Notes:
*1. These are comparisons of system technology level only, and are not necessarily a measure of effectiveness. The comparisons are not dependent on scenario, tactics, quantity, training or other operational factors. Systems farther than 1 year from IOC are not considered.
2. The arrows denote that the relative technology level is changing significantly in the direction indicated.
3. Relative comparisons of technology levels shown depict gross standing only; countries may be superior, equal or inferior in subcategories of a given technology in a deployed military system.

– it seems extremely unlikely that the Soviet Union could take the lead from the United States in SDI technology in the near future. If the United States government should reject limitation of its SDI programme in future arms control efforts, it would mean major political and military–technological consequences:

1. It is doubtful, and highly unlikely, that the Soviet Union would accept a position of technological inferiority with major strategic and political implications.
2. If the Soviet Union should decide to counter the United States SDI effort, either by cheaper countermeasures (see IV above) or by upgrading its rudimentary BMD to an area defence system, an intensified arms race can be predicted similar to the missile race of the 1960s, but many times more expensive.
3. Such an arms race would not be conducive to a policy of détente in Europe. As a consequence, given the different assessments in the United States and in Western Europe of the potential contribution of détente, transatlantic relations may further erode.

David Holloway has listed the following potential Soviet military (technical) responses to the SDI:

> Three broad options are open to the Soviet Union, either separately or in combination: it can upgrade its retaliatory forces, it can develop weapons that could destroy the space-based BMD system, or it can deploy its own BMD system. Each of these responses needs to be assessed in terms of its technical feasibility, economic cost, and contribution to Soviet military policy.[274]

According to Holloway:

> the most obvious Soviet response to SDI would be to upgrade their offensive forces in order to ensure that they can penetrate, evade, or overwhelm the defence. They could increase the number of their offensive ballistic missiles, thereby complicating the task of the defence, and diversify their forces by deploying nuclear weapons in space or by making greater use of bombers and cruise missiles, which would not be vulnerable to BMD. In order to complicate boost-phase interception ... the Soviet Union could shield its launchers with reflective or ablative materials, thus raising the power requirement for the defence's kill mechanisms. It could also

shorten the boost phase by developing fast-burn ICBMs. A payload penalty of about 10 to 30 per cent might have to be paid in order to reduce the boost phase from 300 to 60 seconds, but the gain would be an enormous increase in the difficulty of boost-phase interception.[275]

In order to overload and to confuse the mid-course and the terminal defence level of a multilayered BMD, the Soviet Union could deploy:

> more RVs on its launchers. The Soviet ICBM force, with its large throw-weight, is particularly well-suited to countermeasures of this kind ... Decoys could also be deployed, with possibly even hundreds of decoys to an RV. This would present the defence with the formidable problem of identifying the real RVs in mid-course where there is no atmospheric drag to sort them out from the decoys.[276]

In Holloway's assessment, it appears highly unlikely that the Soviet Union would 'abandon its ICBM force, or that it will necessarily eliminate the Soviet pre-emptive option'. The Soviets could also develop space mines, ASAT weapons and ground-based lasers which could attack a space-based United States BMD. And finally, the Soviet Union could decide to deploy its own BMD as a deliberate attempt 'to enhance the survivability of its offensive missiles. So far, it has tried to cope with ICBM vulnerability by developing mobile ICBMs, by diversifying its strategic forces, and perhaps by adopting a launch-under-attack policy, but it might deploy BMD if the problem became serious enough'.[277]

Holloway has also pointed to potential Soviet reactions if the United States should proceed to BMD deployment:

> They might decide to make a 'pre-emptive breakout' from the ABM treaty by building their own conventional BMD system on a nationwide scale. They might do this to show that the Soviet Union would not be overtaken in a BMD race, to complicate United States military plans, or in the hope of gaining some advantage in arms-control negotiations.[278]

After discussing the possible implications of a United States and a Soviet space-based BMD on Soviet–American relations and on arms control, Holloway concluded:

> If defensive systems are to contribute to a safer and more stable strategic relationship between the United States and the Soviet

Union, they will have to be embedded in a strict arms-control regime that limits offensive systems. In the current political and technological circumstances, however, the attempt to build defences may well push the other side into expanding and upgrading its offensive forces. It is thus a paradox of the present superpower rivalry that the effort to build BMD can, and very well will, undermine the very condition that is needed to ensure that BMD contributes to a safer world.[279]

(d) Possible implications of potential Soviet responses to SDI for the security of Western Europe

If the Soviet Union should decide to enhance its retaliatory forces by shifting from ballistic missiles to Cruise missiles and bombers Western Europe would be affected first; a threefold defence system might be a direct consequence: against short- and medium-range Soviet ballistic missiles, against Cruise missiles and against bombers and fighter aircraft.

If the Soviets should decide to focus on space attack systems – for example, space mines, ASAT and ground-based lasers – both the civilian and the military C^3I satellites would become vulnerable. Given the greater reliance of NATO on space-based C^3I systems in relation to the Soviet military, such a development would be to the clear disadvantage of NATO.

If the Soviet leadership should decide to develop and deploy its own area defence BMD system, according to Yost, at least four major consequences could be predicted:

1. The credibility of NATO's flexible response strategy could be reduced and the United States guarantee of an extended deterrence could be degraded.
2. The Soviet prospects to win a conventional war would be increased, especially if the Soviets used its BMD to reduce the utility of NATO's emergent technologies, its FOFA concept and its planned JTACMs.
3. The potential control of the Soviet Union over the process of escalation could be enlarged, and the Soviet Union could feel confident to limit future conflicts to Europe in order to avoid an intercontinental nuclear war.
4. The credibility of the British and the French nuclear deterrence could also be reduced.[280]

If the implications of a Soviet BMD effort in response to the SDI should be as severe for Western Europe, why should the West Europeans encourage the United States SDI programme or an EDI which would provoke such severe consequences for their own security?

(e) Implications of SDI, EDI and of a Soviet BMD for the security of Western Europe

At least two broad scenarios are foreseeable: a *competition scenario*, which is bound to lead to a multifaceted arms race, and a *co-operation scenario*, which could open new possibilities for confidence building, arms reduction and détente in Europe.

In a *competition scenario* SDI would remain unrestrained by any presently existing or future arms control agreement; the ABM Treaty, the SALT II Treaty, the Outer Space Treaty and possibly also the Limited Test Ban Treaty would be casualties. Both the United States and the Soviet Union could be expected, in a political worst case, to enlarge their defensive and offensive forces as far as budgetary constraints allow. The consequences for Western Europe would be severe – it could either join the United States in SDI, at the same time building up its own EDI, or it might be tempted to reach a special deal with the Soviet Union by opting for a neutral position between superpowers as a consequence of major Soviet offers and concessions both on the conventional and the nuclear level.

If Western Europe should opt for a *co-operative approach* – as has been indicated by the support of all major West European leaders for the ABM Treaty[281] – then Western Europe has to be extremely careful not to contribute to the erosion of the Treaty, which does not permit the deployment of space-based or area defence systems. The West Europeans should become active in tightening the ABM Treaty and closing its major loopholes and grey areas: the ABM/ASAT loophole, the uncertainty about the disputed Soviet radars, and the grey area between SAMs for air defence and an ABM system against strategic ballistic missiles. This last grey area especially affects the security of Western Europe. For this reason, the Europeans should press the United States to put both the Soviet SA-10 and SA-X-12 and the planned United States ATBM systems on the conference table, either in the context of the Standing Consultative Commission or of the Geneva Talks.[282]

Notes

1. President Reagan's speech of 23 March 1983, *Weekly Compilation of Presidential Documents*, 12, 28 March 1983 (Washington, DC: United States Government Printing Office, 1983) pp. 423–66.
2. Theodore J. Lowi, 'Making Democracy Safe for the World: National Politics and Foreign Policy', in James N. Rosenau (ed.), *Domestic Sources of Foreign Policy*, (New York: The Free Press, 1967) ch. 11, p. 315.
3. Michael J. Deane, 'Soviet Military Doctrine and Defensive Deployment Concepts: Implications for Soviet Ballistic Missile Defense', in Jacquelyn K. Davis *et al.* (eds), *The Soviet Union and Ballistic Missile Defense* (Cambridge, Mass.: Institute for Foreign Policy Analysis, 1980) pp. 42, 50.
4. Jonathan B. Stein, *From H-Bomb to Star Wars. The Politics of Strategic Decision Making* (Lexington, Mass.; Toronto: Lexington Books, 1984). p. 84.
5. David S. Yost, 'Soviet Ballistic Missile Defense and European Security' (paper prepared for presentation at the Netherlands Institute of International Relations conference on 'The American Defense Initiative: Implications for West European Security', The Hague, 26–7 April 1985) p. 5.
6. David Holloway, 'Doctrine and Technology in Soviet Armaments Policy', in Derek Leebaert (ed.), *Soviet Military Thinking* (London, Boston, Sydney: George Allen & Unwin, 1981) ch. 9, pp. 260–1.
7. Holloway, 'Doctrine and Technology', p. 266.
8. Holloway, 'Doctrine and Technology', p. 276.
9. See the major political statements in the Soviet press: *Pravda, Izvestia et al.*
10. See, for example, the classified journal *Voennaya Mysl*, which has been translated by the CIA and made available through the Library of Congress in Washington, DC.
11. See for example, Raymond L. Garthoff, *Détente and Confrontation: American–Soviet Relations from Nixon to Reagan* (Washington, DC: Brookings Institution, 1985).
12. See, for example, Rose Gottemoeller, 'Decisionmaking for Arms Limitation in the Soviet Union', in Hans Günter Brauch and Duncan L. Clarke (eds), *Decisionmaking for Arms Limitation – Assessments and Prospects* (Cambridge, Mass.: Ballinger, 1983) pp. 53–80.
13. See, for example, the Galosh paraded in 1964.
14. These data are of sufficient quality in order to recognise SS-20 bases and other major military installations in the Soviet Union.
15. Raymond L. Garthoff, 'BMD and East–West Relations', in Ashton B. Carter and David N. Schwartz (eds), *Ballistic Missile Defense* (Washington, DC: Brookings Institution, 1984) ch. 8, pp. 275–329.
16. IISS, *The Military Balance 1985–1986* (London: IISS, 1985).
17. SIPRI, *World Armaments and Disarmament – SIPRI Yearbook 1985* (London, Philadelphia: Taylor & Francis, 1985).

18. US Department of Defense, *Report of the Secretary of Defense Caspar W. Weinberger to the Congress on the FY 1986 Budget, FY 1987 Authorization Request and FY 1986-90 Defense Programs, 4 February 1985* (Washington, DC: United States Government Printing Office, 1985).
19. See for details in the research bibliography Hans Günter Brauch, Rainer Fischbach and Thomas Bast, 'Eine Forschungsbibliographie zur Militarisierung des Weltraums', *Die Friedens-Warte* 65 (1985) pp. 216–81; Hans Günter Brauch and Rainer Fischbach, *Military Use of Outer Space – A Research Bibliography*, AFES Papier (Stuttgart: AG Friedensforschung und Europäische Sicherheitspolitik, Institut für Politikwissenschaft, February 1986); Hans Günter Brauch and Rainer Fischbach, *Die Militärische Nutzung des Weltraums – Eine Forschungsbibliographie*, Militärpolitik und Rüstungsbegrenzung, vol. 8 (Berlin: Berlin-Verlag Arno Spitz, 1986).
20. US Congress, Office of Technology Assessment, *Ballistic Missile Defense Technologies*, OTA-ISC-254 (Washington, DC: United States Government Printing Office, September 1985).
21. US Department of Defense, *Soviet Military Power 1985* (Washington, DC: United States Government Printing Office, 1985).
22. US Department of Defense and Department of State, *Soviet Strategic Defence Programs* (Washington, DC: United States Government Printing Office, October 1985); reprinted in: *National Defense*, November 1985.
23. Klaus von Beyme, *Die Sowjetunion in der Weltpolitik* (München–Zürich: R. Piper, 1983) p. 12.
24. Carl J. Friedrich (ed.), *Totalitarianism* (New York: Grosset & Dunlap, 1964); R. C. Tucker, *The Soviet Political Mind: Studies in Stalinism and Post-Stalin Change* (New York: Praeger, 1963); Marshall D. Shulman, *Stalin's Foreign Policy Reappraised* (Cambridge, Mass.: Harvard University Press, 1963) p. 3.
25. Hannes Adomeit and Robert Boardman, 'The comparative study of communist foreign policy', in: Hannes Adomeit and Robert Boardman (eds), *Foreign Policy Making in Communist Countries. A Comparative Approach* (New York, London, Sydney, Toronto: Praeger, 1979) p. 2.
26. Hannes Adomeit, 'Soviet foreign policy making: The internal mechanism of global commitment', in Adomeit and Boardman, *Foreign Policy Making*, p. 17.
27. Adomeit, 'Soviet foreign policy making', p. 18.
28. Raymond L. Garthoff, *Soviet Strategy in the Nuclear Age* (New York: Praeger, 1958); Herbert S. Dinerstein, *War and the Soviet Union* (New York: Prager, 1962); Stephen M. Meyer, 'Soviet Perspectives on the Paths to Nuclear War', in Graham T. Allison, Albert Carnesale, Joseph S. Nye, Jr (eds), *Hawks, Doves & Owls. An Agenda for Avoiding Nuclear War* (New York, London: W. W. Norton, 1985) ch. 7, pp. 167–205.
29. A. A. Sidorenko, *The Offensive (A Soviet View) Moscow 1970* (Washington, DC: United States Government Printing Office, 1976).

30. Richard Pipes, *US–Soviet Relations in the Era of Détente* (Boulder, Co. Westview Press, 1981); Joseph D. Douglass, Jr, *The Soviet Theater Nuclear Offensive* (Washington, DC: United States Government Printing Office, 1976).
31. Garthoff, *Détente and Confrontation*; 'BMD and East–West Relations; *Soviet Strategy*.
32. Fred Charles Iklé, 'Nuclear Strategy: Can There Be a Happy Ending?, *Foreign Affairs* 63 (Spring 1985): pp. 810–26; Yost, *Soviet Ballistic Missile Defense*.
33. Raymond L. Garthoff, 'Mutual Deterrence, Parity and Strategic Arms Limitation in Soviet Policy', in Derek Leebaert (ed.), *Soviet Military Thinking* (London, Boston, Sydney: George Allen & Unwin, 1981) ch. 5, pp. 92–124; David Holloway, *The Soviet Union and the Arms Race* (New Haven, London: Yale University Press, 1983); Douglas M. Hart, 'The Hermeneutics of Soviet Military Doctrine', *Washington Quarterly* (Spring 1984) p. 78.
34. Paul C. Warnke, Testimony before the United States Congress, Senate Committee on Banking, Housing, and Urban Affairs, *Civil Defense*, 8 January 1979 (Washington, DC: United States Government Printing Office, 1979); Roman Kolkowicz, 'Soviet–American Strategic Relations: Implications for Arms Control', in Roman Kolkowicz, Matthew D. Gallagher, and Benjamin S. Lambeth, with Walter C. Clemens, Jr and Peter W. Colm, *The Soviet Union and Arms Control: A Superpower Dilemma* (Baltimore: Johns Hopkins University Press, 1970); Donald W. Hanson, 'Is Soviet Strategic Doctrine Superior?', *International Security* 7 (Winter 1982/83).
35. Raymond L. Garthoff, 'Mutual Deterrence and Strategic Arms Limitation in Soviet Policy', *International Security* 3 (Summer 1978).
36. Richard Pipes, 'Why the Soviet Union Thinks it Could Fight and Win a Nuclear War', *Commentary* 64 (July 1977).
37. James McConnell, 'The Interacting Evolution of Soviet and American Military Doctrines', *Center for Naval Analysis Memorandum 80-1313.00*, 17 September 1980.
38. Christopher Jones, 'Soviet Military Doctrine and Warsaw Pact Exercises', in Derek Leebaert (ed.), *Soviet Military Thinking* (London et al.: George Allen & Unwin, 1981); Christopher Jones, 'Soviet Military Doctrine: The Political Dimension', in William Kincade and Jeffrey Porro (eds), *Negotiating Security* (Washington, DC: Carnegie Endowment for International Peace, 1979); Chrisopher D. Jones, *Soviet Influence in Eastern Europe–Political Authority and the Warsaw Pact* (New York: Praeger, 1981).
39. David Holloway, 'Military Power and Political Purpose in Soviet Policy', *Daedalus* (Fall 1980).
40. Hart, 'The Hermeneutics of Soviet Military Doctrine', p. 86.
41. Hart, 'The Hermeneutics of Soviet Military Doctrine, p. 86.
42. Pipes, 'Why the Soviet Union Thinks it Could Fight and Win'; Douglass, *The Soviet Theater*; Joseph D. Douglass, Jr and Amoretta M. Hoeber, *Soviet Strategy for Nuclear War* (Stanford: Hoover Institution, 1979).

43. William E. Odom, 'Trends in the Balance of Military Power Between East and West', in *The Conduct of East–West Relations in the 1980s, Part III*, Adelphi Paper 191 (London: International Institute for Strategic Studies, 1984); Peter H. Vigor, *The Soviet View of War, Peace and Neutrality* (London, Boston: Routledge and Kegan Paul, 1975); Benjamin S. Lambeth, *Risk and Uncertainty in Soviet Deliberations About War*, R-2687-AF (Santa Monica, California: Rand Corporation, October 1981); Jack L. Snyder, *The Soviet Strategic Culture: Implications for Limited Nuclear Options*, R-2154-AF (Santa Monica, California: Rand Corporation, September 1977); Thomas W. Wolfe, *The SALT Experience* (Cambridge, Mass.: Ballinger, 1979).
44. Garthoff, 'BMD and East–West Relations'; 'Mutual Deterrence'; for a critical view see in Benjamin S. Lambeth, *The State of Western Research on Soviet Military Strategy and Policy*, N-2230-AF (Santa Monica, California: Rand Corporation, October 1984).
45. See for details II(e) especially notes 109 and 110.
46. David Holloway, *The Soviet Union and the Arms Race*, p. 29.
47. *Dictionary of Basic Military Terms, A Soviet View* (Washington, DC: United States Government Printing Office, 1977) p. 37.
48. Holloway, *The Soviet Union and the Arms Race*, p. 29.
49. *Dictionary of Basic Military Terms*, p. 38.
50. *Dictionary of Basic Military Terms*, p. 39.
51. Holloway, *The Soviet Union and the Arms Race*, p. 55.
52. V. D. Sokolovskiy (ed.), *Military Strategy*, 3rd edn (New York: Crane, Russak and Co., 1975).
53. Holloway, *The Soviet Union and the Arms Race*, p. 57.
54. Holloway, p. 57.
55. Michail A Milstein, 'Strategic Arms Limitation and Military Strategic Concepts', in David Carlton and Carlo Schaerf (eds), *Arms Control and Technological Innovation* (London: Croom Helm, 1977) p. 201.
56. Garthoff, 'BMD and East–West Relations', p. 289.
57. N. Talensky, 'Anti-Missile Systems and Disarmament', in *International Affairs* (Moscow) 10 (October 1964) p. 18.
58. I. Anureyev, 'Determining the Correlation of Forces in Terms of Nuclear Weapons', *Voyenna mysl (Military Thought)* 6 (June 1967) translated in Foreign Press Digest 0112/68, 11 July 1968.
59. Anureyev, p. 38.
60. V. D. Sokolovskiy (ed.), *Voennaya Strategiya* (Moscow: Voenizdat, 1968).
61. Garthoff, 'BMD and East–West Relations', p. 312.
62. Garthoff, p. 313.
63. Yost, 'Soviet Ballistic Missile Defense', p. 39.
64. Garthoff, 'BMD and East–West Relations', p. 313.
65. US Department of Defense, *Soviet Military Power 1981* (Washington DC: United States Government Printing Office, September 1981, hereafter *SMP 81*); *Soviet Military Power 1983* (Washington, DC: United States Government Printing Office, April 1983, hereafter *SMP 83*); *Soviet Military Power 1984* (Washington, DC: United States Government Printing Office, April 1984, hereafter *SMP 84*);

Soviet Military Power 1985 (Washington, DC: United States Government Printing Office, April 1985, hereafter *SMP 85*).
66. *SMP 83*, pp. 14–15.
67. *SMP 84*, p. 32.
68. See, for example, the many publications of the Committee on the Present Danger in Washington, DC ever since 1976.
69. Pipes, 'Why the Soviet Union Thinks it Could Fight and Win'; *US–Soviet Relations*; Douglass, *The Soviet Theater*.
70. Fritz W. Ermath, 'Contrasts in American and Soviet Strategic Thought', in Derek Leebaert (ed.), *Soviet Military Thinking* (London *et al.*: George Allen & Unwin, 1981) ch. 3, pp. 50–69.
71. Yost, 'Soviet Ballistic Missile Defense', see there for additional literature.
72. Pipes, 'Why the Soviet Union Thinks it Could Fight and Win'.
73. Dennis G. Hall, 'Space & the Soviet View of War', in Peter A. Swan (compiler), *Military Space Doctrine The Great Frontier, vol. 3, A Book of Readings for the USAFA Military Space Doctrine Symposium, 1–3 April 1981* (Colorado Springs; United States Air Force Academy, 1981) pp. 849–57.
74. 'The Real Paul Warnke', *The New Republic* (26 March 1977) p. 23, as quoted in Pipes 'Why the Soviet Union Thinks it Could Fight and Win'.
75. Richard Pipes, 'Why the Soviet Union Thinks It Could Fight and Win a Nuclear War', in Pipes, *US–Soviet Relations*, p. 155.
76. Pipes, *US–Soviet Relations*, p. 168.
77. William R. van Cleave, 'Soviet Doctrine and Strategy: A Developing American View', in Lawrence L. Whetten (ed.), *The Future of Soviet Military Power* (New York: Crane, Russack, 1976) p. 50ff.
78. Douglass, *The Soviet Theater*; Douglass, Hoeber, *Soviet Strategy for Nuclear War*; Lewis Allen Frank, *Soviet Nuclear Planning: A Point of View on SALT* (Washington, DC: American Enterprise Institute for Public Policy Research, 1977).
79. Michael J. Deane, *The Role of Strategic Defense in Soviet Strategy* (Miami: Advanced International Studies Institute, 1980); Leon Gouré, William G. Hyland, Colin S. Gray, *The Emerging Strategic Environment: Implications for Ballistic Missile Defense* (Cambridge, Mass.: Institute for Foreign Policy Analysis, 1979); William Schneider, Jr, Donald G. Brennan, William A. Davis, Jr and Hans Rühle, *US Strategic–Nuclear Policy and Ballistic Missile Defense: The 1980s and Beyond* (Cambridge, Mass.: Institute for Foreign Policy Analysis, 1980).
80. Deane, *The Role of Strategic Defense*, p. 13.
81. Jacquelyn K. Davis, Uri Ra'anan, Robert L. Pfaltzgraff, Jr, Michael J. Deane and John M. Collins, *The Soviet Union and Ballistic Missile Defense* (Cambridge, Mass.: Institute for Foreign Policy Analysis, 1980) p. 44.
82. Deane *The Role of Strategic Defense*, p. 60.
83. Yost, *Soviet Ballistic Missile Defense*, p. 32 and especially note 118.
84. Howard M. Hensel, 'Soviet Media Images of Superpower Space Policy', (paper presented to a conference organised by the Georgia

Institute of Technology on 'International Space Policy: Options for the Twentieth Century and Beyond', Atlanta, 16-17 May 1985) p. 5.
85. Keith Payne and Dan Stroude, 'Arms Control: The Soviet Approach and its Implications', in *Soviet Union* 10, pts 2-3 (1983) pp. 229-32, 234-6, 243, quoted from Hensel, 'Soviet Media Images', p. 5.
86. Hensel, p. 5.
87. James A. Williams, 'The Ambitious Soviet Space Program', in *Defense* (February 1985) p. 11.
88. Williams, p. 16.
89. Uri Ra'anan, 'The Soviet Approach to Space: Personalities and Military Doctrine', in Uri Ra'anan and Robert L. Pfaltzgraff' Jr (eds), *International Security Dimensions of Space* (Hamden: Archon Books, 1984 p. 55.
90. See Garthoff, *Soviet Strategy in the Nuclear Age; Détente and Confrontation*; 'BMD and East-West Relations'; 'Mutual Deterrence'.
91. See for details also section II(e).
92. Christoph Bertram, *Arms Control and Technological Change – Elements of a New Approach*, Adelphi Paper 146 (London: International Institute for Strategic Studies, 1978).
93. See Hensel, 'Soviet Media Images'; Garthoff, 'BMD and East-West Relations'; Sidney Drell, Philip Farley, and David Holloway, 'Preserving the ABM Treaty', *International Security* 9 (Fall 1984); Drell, Farley and Holloway, *The Reagan Strategic Defense Initiative: A Technical, Political and Arms Control Assessment* (Stanford: Center for International Security and Arms Control, 1984).
94. Raymond L. Garthoff, 'The Soviet Military and SALT', in Jiri Valenta and William Potter (eds), *Soviet Decisionmaking for National Security* (London, Boston, Sydney: George Allen & Unwin, 1984).
95. Garthoff, 'Mutual Deterrence', p. 94.
96. Garthoff, p. 97.
97. Garthoff, p. 105.
98. Garthoff, p. 107; see note 40.
99. Garthoff, p. 119; see note 85.
100. Garthoff, p. 120.
101. Garthoff, 'BMD and East-West Relations', p. 294.
102. Garthoff, p. 302.
103. Garthoff, p. 307; see note 70.
104. Garthoff, p. 314.
105. Garthoff, p. 327; see note 121.
106. Drell, Farley and Holloway, *The Reagan Strategic Defense Initiative*, p. 18.
107. Drell, Farley and Holloway, p. 27.
108. Robert P. Berman and John C. Baker, *Soviet Strategic Forces – Requirements and Responses* (Washington, DC: Brookings Institution, 1982) p. 32.
109. Stanley Sienkiewicz, 'Soviet Nuclear Doctrine and the Prospects for Strategic Arms Control', in Derek Leebaert (ed.), *Soviet Military Thinking* (London *et al.*: George Allen & Unwin, 1981) ch. 4, pp. 73-91.

110. Ermath, 'Contrasts in American and Soviet Strategic Thought'; Fritz Ermath, 'The Evolution of Soviet Doctrine' (paper presented to the 27th Annual Conference of the International Institute for Strategic Studies, Berlin, 12–15 September 1985).
111. Scienkiewicz, 'Soviet Nuclear Doctrine', pp. 86–8.
112. Ermath, 'Contrasts in American and Soviet Strategic Thought', p. 51.
113. Ermath, p. 58.
114. Ermath, p. 67.
115. Sayre Stevens, 'The Soviet BMD Program', in Ashton B. Carter and David N. Schwartz (eds), *Ballistic Missile Defense* (Washington, DC: Brookings Institution, 1984) ch. 5, pp. 185–6.
116. Sayre Stevens, p. 188.
117. Hensel, 'Soviet Media Images', pp. 19–20.
118. Hensel, pp. 23–4.
119. Fred Charles Iklé, 'Nuclear Strategy: Can There Be a Happy Ending?', *Foreign Affairs* 63 (Spring 1985) pp. 810–26 (820).
120. Vernon V. Asperturian, 'The Stalinist Legacy in Soviet National Security Decisionmaking', in Jiri Valenta and William Potter (eds), *Soviet Decisionmaking for National Security* (London et al.: George Allen & Unwin, 1984) pp. 23–73.
121. Holloway, *The Soviet Union and the Arms Race*, p. 160.
122. Valenta and Potter, *Soviet Decisionmaking*; Gottemoeller, 'Decisionmaking for Arms Limitation'; Holloway, *The Soviet Union and the Arms Race*; Wolfe, *The SALT Experience*; Karl F. Spielmann, Analyzing Soviet Strategic Arms Decisions (Boulder, Co.: Westview Press, 1978); Arthur J. Alexander, *Decisionmaking in Soviet Weapons Procurement, Adelphi Paper* 147/48, 1978–79 (London: International Institute for Strategic Studies, 1978/79); David Holloway, 'Military Technology', in Ronald Amann, Julian Cooper and R. W. Davies (eds), *The Technological Level of Soviet Industry* (New Haven, London: Yale University Press, 1977).
123. Stephen M. Meyer, 'Space and Soviet Military Planning', in William J. Durch (ed.), *National Interests and the Military Use of Space* (Cambridge, Mass.: Ballinger, 1984) ch. 3, pp. 61–88 (64–5).
124. Berman and Baker, *Soviet Strategic Forces*, p. 11.
125. Berman and Baker, *Soviet Strategic Forces*, p. 19.
126. Meyer, 'Space and Soviet Military Planning', pp. 65–6.
127. See chapter 14 in this volume.
128. John Tirman (ed.), *The Fallacy of Star Wars* (New York: Vintage Books, 1984).
129. R. Z. Sagdeyev, A. A. Kokoshin for the Committee of Soviet Scientists for Peace Against Nuclear Threat: *Strategic and International Political Consequences of Creating a Space-Based Anti-Missile System Using Directed Energy Weapons* (Moscow: Institute of Space Research, USSR Academy of Sciences, 1984).
130. Sagdeyev and Kokoshkin, pp. 22–23.
131. See chapter 14 in this volume.
132. Paul H. Nitze, 'SDI: The Soviet Program', *Current Policy* 717 (Washington, DC: United States Department of State, July 1985).

133. 'Soviets Seen Spending 2% of GNP for Space', *Defense Daily*, 30 April, 1982, p. 357; 'Soviets Seen Outspending US On Space By $ 3 Billion', *Defense Daily*, 23 September 1982, p. 94; 'Congressman Says Soviets Spending $45 Billion A Year on Space', *Defense Daily*, 1 June 1983, p. 171.
134. George C. Wilson, 'US Sees Wider Gap In Space Weaponry', *International Herald Tribune*, 4 March 1982.
135. Walter Andrews, 'Soviet Military Space Threat Bared', *Army Times*, 8 March 1982.
136. Nitze, 'SDI: The Soviet Program'.
137. Bhupendra Jasani (SIPRI) (ed.), *Outer Space – A New Dimension of the Arms Race* (London: Taylor & Francis 1982) p. 27.
138. Meyer, 'Space and Soviet Military Planning'; Smith (see below note 146).
139. *SMP 81*, p. 79.
140. *SMP 83*, pp. 65–69.
141. *SMP 84*, p. 32.
142. *SMP 84*, p. 47.
143. *SMP 85*, p. 55.
144. *SMP 85*, p. 58.
145. Nicholas L. Johnson, *The Soviet Year in Space: 1982* (Colorado Springs: Teledyne Brown Engineering, 1983); Johnson, *The Soviet Year in Space: 1983* (1984); Johnson, *The Soviet Year in Space: 1984* (1985).
146. Marcia S. Smith, *Space Activities of the United States, Soviet Union and other Launching Countries/Organizations: 1957–1983* (Washington, DC: Library of Congress, Congressional Research Service, 15 January 1984); Marcia S. Smith, 'Evolution of the Soviet Space Program from Sputnik to Salyut and Beyond', in Uri Ra'anan & Robert L. Pfaltzgraff, Jr (eds), *International Security Dimensions of Space* (Hamden: Archon Books, 1984) pp. 285–304.
147. Bhupendra Jasani (SIPRI), *Outer Space – Battlefield of the Future?* (London: Taylor & Francis, 1978); Bhupendra Jasani (ed.), *Outer Space – A New Dimension of the Arms Race* (London: Taylor & Francis, 1982); Bhupendra Jasani (ed.), *Space Weapons – The Arms Control Dilemma* (London; Philadelphia: Taylor & Francis, 1984); see also his contributions to the *SIPRI Yearbooks World Armaments and Disarmament* from 1972 to the present.
148. Meyer, 'Space and Soviet Military Planning', p. 67.
149. Meyer, pp. 68–9.
150. Meyer, p. 76.
151. Meyer, pp. 77–8.
152. *SMP 81*, p. 68.
153. *SMP 83*, pp. 67–8.
154. *SMP 84*, pp. 34–5.
155. *SMP 85*, p. 44.
156. Nitze, 'SDI: The Soviet Program'.
157. '2 Magazines Say Soviet Lasers Destroyed a US Space Satellite', *New York Times*, 23 November 1976.

158. Richard Burt, 'US Says Russians Develop Satellite-Killing Laser', *New York Times*, 22 May 1980.
159. 'Pentagon Denies Report of Soviet Killer Satellite', *International Herald Tribune*, 29 October 1981.
160. 'Washington Roundup: Cosmos Threat', *Aviation Week & Space Technology*, 30 November 1981, pp. 17.
161. Charles W. Corddry, 'Soviet said to improve anti-satellite arms', *The Sun*, 10 February 1982.
162. 'General Sees Russia Close to Laser Arm', *International Herald Tribune*, 24 April 1982.
163. *Defense Daily*, 21 March 1983; *Defense Daily*, 23 May 1983.
164. Donald Kerr, 'Implications of anti-satellite weapons for ABM issues', in Bhupendra Jasani (ed.), *Space Weapons – The Arms Control Dimension*, p. 117.
165. Clarence A. Robinson, Jr, 'Antisatellite Weaponry and Possible Defense Technologies against Killer Satellites', in Uri Ra'anan and Robert L. Pfaltzgraff, Jr (eds), *International Security Dimensions of Space* p. 71.
166. M. Smith, in Ra'anan and Pfaltzgraff (eds), p. 297.
167. Marcia Smith, 'Satellite and missile ASAT systems and potential verification problems associated with the existing Soviet systems', in Jasani (ed.), *Space Weapons – The Arms Control Dimension*, p. 85.
168. Johnson, *The Soviet Year in Space*: 1982, p. 26; Johnson, *The Soviet Year in Space*: 1983, p. 39.
169. Paul B. Stares, *Space Weapons & US Strategy: Origins and Development* (London; Sydney: Croom Helm, 1984), p. 262.
170. Meyer, 'Space and Soviet Military Planning' p. 78.
177. Meyer, p. 80.
172. Stares, *Space Weapons*, p. 261.
173. David N. Schwartz, 'Past and Present: The Historical Legacy', in Ashton B. Carter and David N. Schwartz (eds), *Ballistic Missile Defense* (Washington, DC: Brookings Institution, 1984), p. 331; Stevens, 'The Soviet BMD Program', p. 189f.
174. For the early history see: Stevens 'The Soviet BMD Program', p. 191ff; Lawrence Freedman, *US Intelligence and the Soviet Strategic Threat* (Boulder, Co.: Westview Press, 1978) p. 87; John Prados, *The Soviet Estimate: US Intelligence Analysis and Russian Military Strength* (New York: Dial, 1982) p. 155.
175. Stevens, 'The Soviet BMD Program', p. 265.
176. Stevens, p. 197.
177. Yost, 'Soviet Ballistic Missile Defense', p. 5; Mark E. Miller, *Soviet Strategic Power and Doctrine: The Quest for Superiority* (Bethesda, Md: Advanced International Studies Institute, 1982) p. 173.
178. Yost, 'Soviet Ballistic Missile Defense', p. 5.
179. Garthoff, 'BMD and East–West Relations', p. 30; Garthoff, 'The Soviet Military and SALT', p. 147.
180. Freedman, *US Intelligence*, p. 89; Paul H. Nitze, 'Comments', *Foreign Policy* 16 (Fall 1974) p. 82.

181. About the United States debate on ABM in the 1960s and early 1970s see Garthoff, 'BMD and East-West Relations', p. 300ff; Schwartz, 'Past and Present'; Ernest J. Yanarella, *The Missile Defense Controversy: Strategy, Technology, and Politics, 1955-1972* (Lexington; Kentucky: University Press of Kentucky, 1977); Fred Kaplan, *The Wizards of Armageddon* (New York: Simon & Schuster, 1983).
182. See Garthoff, 'BMD and East-West Relations', p. 292 and notes 29 and 30.
183. See Garthoff, pp. 292-3.
184. Garthoff, p. 295.
185. Garthoff, pp. 298-9.
186. Garthoff, p. 302.
187. Garthoff, p. 306.
188. Garthoff, p. 306.
189. Garthoff, p. 307.
190. Garthoff, p. 312.
191. Garthoff, p. 313.
192. Yost, 'Soviet Ballistic Missile Defense', p. 8 and notes 25-9.
193. Yost, p. 8 and note 26.
194. Yost, p. 9.
195. *SMP 81*, p. 68.
196. *SMP 83*, p. 28.
197. *SMP 83*, p. 68.
198. *SMP 84*, p. 34.
199. *SMP 85*, pp. 47-8.
200. 'Soviet Antimissile Lead Is Feared', *New York Times*, 3 December 1983, p. 4.
201. 'Washington Roundup: Soviet BMD', *Aviation Week & Space Technology*, 5 December 1983, p. 15.
202. Clarence A. Robinson, Jr, 'Soviets Accelerate Missile Defense', *Aviation Week & Space Technology*, 16 January 1984, pp. 14-16.
203. Robinson, Jr., pp. 14-16.
204. 'Soviet Union is Rated Decade Ahead of US on Missile Defenses', *Washington Post*, 9 March 1984.
205. Wayne Biddle, 'Pentagon Scientists Backtracked At Secret Hearing on Space Arms', *New York Times*, 22 December 1984.
206. Bill Gertz, 'CIA: Soviets are developing their own "star wars" system', *Washington Times*, 10 May 1985, p. 1.
207. Walter Andrews, 'US, Soviets "star wars" race even', *Washington Times*, 21 June 1985, p. 1.
208. Charles W. Corddry, 'CIA study sees Soviet near to missile defense', *Baltimore Sun*, 25 June 1985, p. 1.
209. See Office of Technology Assessment, OTA-ISC-254, p. 243; Testimony of National Intelligence Officer Lawrence K. Gershwin before a joint session of the Subcommittee on Strategic and Theater Nuclear Forces of the Senate Armed Services Committee and the Defense Subcommittee of the Senate Committee on Appropriations, 26 June 1985.
210. Rebecca V. Strode, 'Space-Based Lasers for Ballistic Missile Defense:

Soviet Policy Options', in Keith B. Payne (ed.), *Laser Weapons in Space, Policy and Doctrine* (Boulder Co.: Westview Press 1983) p. 137.
211. Strode, p. 139.
212. Strode, p. 149.
213. Strode, p. 150.
214. Stevens, 'The Soviet BMD Program', p. 213.
215. Stevens, p. 214.
216. Stevens, pp. 216–7.
217. Stevens, p. 218.
218. See chapter 15 in this volume.
219. Gerard Smith, *Doubletalk: The Story of SALT I* (Garden City, NY: Doubleday, 1980); John Newhouse, *Cold Dawn. The Story of SALT* (New York: Holt, Rinehart & Winston, 1973); Mason Willrich and John B. Rhinelander (eds), *SALT, The Moscow Agreements and Beyond* (London: Collier Macmillan, 1974).
220. US Congress, Senate, Armed Services Committee, Hearing, *Military Implications of the Treaty on Anti-Ballistic Missile Systems and the Interim Agreement on Limitation of Strategic Offensive Arms* (Washington, DC: US Government Printing Office 1972) pp. 257–60.
221. Jack P. Ruina, 'US and Soviet Strategic Arsenals', in M. Willrich and Rhinelander (eds), *SALT*, pp. 34–65.
222. *SMP 81*, p. 67.
223. *SMP 83*, p. 29.
224. US Congress, House, Committee on Armed Services, Hearings Authorization for Appropriations for Fiscal Year 1984, part 5, p 247.
225. *SMP 84*, p. 38.
226. *SMP 85*, pp. 48–50.
227. *SMP 85*, p. 48.
228. *SMP 84*, p. 34.
229. Jack Anderson, 'Soviet Missile May Be Peril to US Weapon', *Washington Post*, April 5 1983, p. C15.
230. 'New Soviet Missile Defences', *Foreign Report*, 14 April 1983.
231. 'Soviets Deploying ABM-Capable Air Defense System', *Defense Daily*, 9 March 1984, p. 50; Walter Andrews, 'Pentagon aide says Soviets deploying weapon convertible to ABM defense', *Washington Times*, 9 March 1985, p. 3.
232. Clarence A. Robinson, Jr, 'Soviets Making Gains in Air Defense', *Aviation Week & Space Technology*, 2 April 1984, pp. 22–3.
233. Stevens, 'The Soviet BMD Program, pp. 215–16.
234. US Department of Defense, *Report to the Congress on the Strategic Defense Initiative* (Washington, DC: Department of Defense, April 1985).
235. *SMP 81*, pp. 74–6.
236. *SMP 83*, pp. 67–75.
237. *SMP 84*, p. 34.
238. *SMP 84*, p. 38.
239. *SMP 84*, p. 106.
240. *SMP 85*, pp. 43–5.

241. See note 22.
242. See Office of Technology Assessment, OTA-ISC-254, p. 244; *The FY 1986 Department of Defense Program for Research, Development and Acquisition*, Statement by the Undersecretary of Defense, Research and Engineering, 99th Congress, 1st session, 1985, pp. II-3 and II-4.
243. Clarence A. Robinson, Jr, 'Soviets push for beam weapon', *Aviation Week & Space Technology*, 2 May 1977, pp. 16–22; 'Soviets build directed-energy weapon', *Aviation Week & Space Technology*, 28 July 1980, pp. 47–50.
244. 'Soviet Tests in Mid-1980s', *Soviet Aerospace*, 7 July 1980, p. 79.
245. Jeff Hecht, *Beam Weapons. The Next Arms Race* (New York, London: Plenum Press, 1984).
246. 'Soviets Seen Readying Laser Satellite Weapon for Mid-'80s Launch', *Aerospace Daily*, 12 January 1982, p. 51.
247. 'Possibility of Soviet Space Laser in "Few Years" Not Ruled Out', *Defense Daily*, 21 January 1982, p. 96.
248. George C. Wilson, 'US Aides Differ on Soviet Lasers', *International Herald Tribune*, 11 March 1982.
249. Answer by Mr Gardner, SDIO in US Congress, Senate, Committee on Armed Services, *Hearings, Department of Defense, Authorization for Appropriations, Fiscal Year 1985*, vol. 6, pp. 2990–1.
250. Payne (ed.), *Laser Weapons*.
251. Yost, 'Soviet Ballistic Missile Defense', pp. 21–3.
252. Meyer, 'Space and Soviet Military Planning', p. 81–2.
253. Meyer, pp. 82–85.
254. Hecht, *Beam Weapons*, p. 345.
255. Hecht, pp. 345–6.
256. See Lewis Fry Richardson, *Arms and Insecurity: A Mathematical Study of the Causes and Origins of War* (London: Pittsburgh: Boxwood Press, 1960); Hans Günter Brauch, *Struktureller Wandel und Rüstungspolitik der USA (1940–1950), Zur Weltführungsrolle und ihren innenpolitischen Bedingungen* (London: Ann Arbor: University Microfilms, 1977).
257. Dieter Senghaas, *Rüstung und Militarismus* (Frankfurt: Suhrkamp, 1972).
258. See the papers of the Committee on the Present Danger.
259. Brauch, *Struktureller Wandel*; Hans Günter Brauch, 'Allgemeine Entwicklungslinien der Waffentechnik–Weltraumrüstung der USA', *Atempto* 70/71 (1984) pp. 57–63; Herbert York, *Race to Oblivion* (New York: Simon and Schuster, 1970).
260. Bernard T. Feld, T. Greenwood, G. W. Rathjens and S. Weinberg (eds), *Impact of New Weapons Technologies on the Arms Race* (Cambridge, Mass.: MIT Press, 1971).
261. Brauch, *Struktureller Wandel*; John Pike. *The Strategic Defense Initiative – Budget and Program – A Staff Study* (Washington, DC: Federation of American Scientists, 10 February 1985).
262. Hans Günter Brauch, Rustungsdynamik und Waffentechnik', in Beate Kohler-Koch (ed.), *Technik und Internationale Politik* (Baden-Baden: Nomos, 1986) pp. 411–48.

263. Colin S. Gray, *The Soviet–American Arms Race* (Farnborough: Saxon House, 1976) pp. 104–5.
264. Walter A. McDougall, *The Heavens and the Earth. A Political History of the Space Age* (New York: Basic Books, 1985).
265. Graham T. Allison, *Essence of Decision. Explaining the Cuban Missile Crisis* (Boston: Little, Brown, 1971); Dan Caldwell, *American–Soviet Relations – From 1947 to the Nixon–Kissinger Grand Design* (Westport; London: Greenwood, 1981); Robin Edmonds, *Soviet Foreign Policy. The Brezhnev Years* (Oxford; New York: Oxford University Press, 1983).
266. See the extensive literature in journals such as *Foreign Affairs, Foreign Policy, Washington Quarterly*.
267. Herbert York, *Race to Oblivion*, pp. 238–9.
268. York, pp. 228–39.
269. Pike, *The Strategic Defense Initiative*, p. 28.
270. David Yost, 'Sowjetische Bemühungen um Raketenabwehr und ihre möglichen Auswirkungen auf die Sicherheit der NATO-Länder', *Europa Archiv* 50, no. 18 (25 September 1985) pp. 541–50; E. C. Aldridge, Jr and Robert L. Maus, Jr, 'SALT Implications of BMD Options', *US Arms Control Objectives and the Implications for Ballistic Missile Defense*, Proceedings of a Symposium held at the Center for Science and International Affairs, Harvard University, 1–2 November 1979 (Cambridge, Mass.: Center for Science and International Affairs, 1980).
271. See above, section VII.
272. US Congress, Senate, Committee on Appropriations, *Hearings, Department of Defense Appropriations Fiscal Year 1985*, part 1 (Washington DC: United States Government Printing Office, 1984) p. 578.
273. Committee on Appropriations, part 1, p. 579.
274. David Holloway, 'The Strategic Defense Initiative and the Soviet Union', *Daedalus* 114 (Summer 1985), Weapons in Space, vol. 2, p. 268.
275. Holloway, p. 269.
276. Holloway, p. 270.
277. Holloway, p. 272.
278. Holloway, p. 272.
279. Holloway, p. 272.
280. Yost, 'Sowjetische Bemühungen', pp. 545–7.
281. See chapters 13 and 15 in this volume and Hans Günter Brauch, *Antitactical Missile Defence. Will the European Version of SDI Undermine the ABM Treaty*, AFES–Papier 1 (Stuttgart: AG Friedensforschung und Europäische Sicherheitspolitik, Institut für Politikwissenschaft, Universität Stuttgart, July 1985).
282. See chapter 14 in this volume. For a recent official United States government interpretation of Soviet statements with regard to SDI, see *The Soviet propaganda against the US Strategic Defense Initiative* (Washington, DC: US Arms Control and Disarmament Agency, August 1986).

Part II
European Perceptions of and Reactions to SDI

Part II

4 SDI – The British Response
Trevor Taylor

The Strategic Defense Initiative (SDI) combined momentous long-term strategic, arms control, political and economic implications with uncertainty about its technological feasibility, purposes and future political appeal within the United States. Scarcely surprisingly there were a whole range of British responses – few of which, it must be said, involved a wholehearted welcome for President Reagan's vision and policy.

Defence policy in the United Kingdom is normally made, discussed and criticised by a comparatively small group consisting of what loosely may be called 'professionals' in defence analysis. These consist of politicians, Ministry of Defence (and to a degree Foreign and Commonwealth Office) officials, members of research institutes, writers for the serious press, academics and some pressure group leaders. The first thing to note, perhaps, is that these almost unanimously viewed the SDI as a most significant development, a major challenge for Europe, the North Atlantic Alliance and the peace movement. No one doubted its significance, although some felt there was no great urgency in the need for a response to it.

Yet despite its perceived importance among defence elites, the SDI has had little popular impact. One commentator could properly observe in May 1985 that 'The British public debate on SDI and the possible effects of the abrogation of the ABM Treaty has not yet really begun'.[1] Unlike Cruise missiles, Trident or even United Kingdom membership of the European Community, the SDI was not an issue to have caught the British public's imagination. Only speculation is possible on why this was so. The implications of the SDI were so complex that it was not easy for people to feel they had any sense of understanding of the issues. Public mystification – and consequent lack of interest – was all the more understandable given that some of those elites who had traditionally argued that nuclear deterrence was effective, necessary and not undesirable suddenly changed to seeing it as an evil to be removed. On the other hand, those who had opposed deterrence as dangerous and immoral were pressed into recognising that it could be lived with for a little while longer. In general the SDI had few unambiguous effects. While some experts credited it with helping to get the USSR back to strategic arms control talks, others saw it as the major reason why such talks could not progress. Since arms control was scarcely advancing quickly before SDI, it was hard to perceive the Reagan initiative as the major arms control obstacle.

As well as being a complex issue for the British public to handle, the SDI did not require the immediate commitment of British territory or money. This could change, as George Keyworth has indicated,[2] especially if a BMD system is accepted for Western Europe, but for the moment the lack of a direct impact on Britain of the SDI is perhaps another reason for public lack of interest in it.

Whatever the causes, the SDI did not mobilise British public opinion and this reinforced the tendency of many members of parliament not to get involved. Few MPs are drawn anyway to in-depth study of defence and high-technology issues and there had been no major debate in the House of Commons on the SDI by the summer of 1985. The House of Commons Defence Committee planned an investigation for the autumn of 1985. Interest was greater in the House of Lords, where a short debate on SDI was held early in 1985.

Should the public get involved with the SDI, it is likely to oppose it – partly because of its likely consequences for a US–Soviet arms race[3] and partly because it might come to threaten the viability of a British deterrent.

It is in this context that the nature and development of the United Kingdom government's formal position can be considered.

I EVOLUTION OF THE UNITED KINGDOM GOVERNMENT POSITION

From Reagan's 23 March 1983 speech until late 1984, United Kingdom policy had a strong element of 'wait and see', as Paul Stares also points out elsewhere in this volume. Many officials and non-governmental analysts were unimpressed (or stunned) with what Reagan had to say, but the Prime Minister was an almost instinctive supporter of Reagan's ambition and determination. The easiest course was to set discussions under way within government but to say little in public until it became clearer what the United States was actually going to do.

When Mrs Thatcher went to Washington in December 1984, much

had clarified. The SDIO had been set up with the distinguished General Abrahamson as its head, and with a substantive planned budget. President Reagan had been re-elected after a campaign in which his commitment to a comprehensive BMD system had been reiterated. The SDI was clearly going to be around for at least the whole of the second Reagan term, and perhaps longer. In the United Kingdom, meanwhile, fears had grown that the SDI could lead to an uncontrolled arms race in space, a prospect which Mrs Thatcher viewed with distaste. At the same time there was recognition of the futility of outright opposition to the United States line, requiring British adoption of the argument that the United States should not try to defend its population. Such outright opposition – which could have damaged Alliance relations – could also have been interpreted as going along with Soviet efforts to get the SDI to be abandoned.

The four points agreed by Mrs Thatcher and President Reagan at the end of her December visit were a compromise reflecting these various pressures. These points, as reported by the Prime Minister to the House of Commons, were:

1. the United States (and Western) aim is not to achieve superiority, but to maintain balance, taking account of Soviet developments;
2. SDI-related developments would, in view of Treaty obligations, have to be a matter for negotiation;
3. the overall aim is to enhance, not undercut, deterrence;
4. East–West negotiation should aim to achieve security with reduced levels of offensive systems on both sides.

The Prime Minister welcomed the resumption of strategic arms control talks and insisted that the USSR would not succeed in splitting the United States and the United Kingdom over this matter.[4]

Points 1 and 3 reflected a British concern that Western security could not be assured by Western action alone, and that Soviet perceptions and responses needed to be taken into account. The insistence that deterrence was to be maintained not abandoned reflected two British concerns. The first was that the concept of an impenetrable shield, behind which a superpower could have complete confidence that it could not be damaged, was quite unrealistic. As analysts have noted in and beyond the United States, no-one could be sure that 'on the day' such a system would work perfectly, especially as it would not be possible to test a defensive system beforehand. Moreover both superpowers – even assuming they each possessed a defensive system perceived as effective – would continue to work on offensive systems in case the other should make a major technological breakthrough.

The second British source of attachment to nuclear deterrence was the conventional balance in Europe. The then Defence Minister Heseltine argued that the Western Alliance would never be able to rely fully on conventional forces to deter even a conventional attack from the Warsaw Pact. The reasoning here was that, since the USSR's geographical position means that it will always want to have greater conventional strength than the West can accept, the West must always be able to fall back upon the nuclear escalation threat.[5]

A variant on this argument was heard often from defence analysts in the United Kingdom when they considered a situation in which both superpowers were protected by an impermeable shield. This would mean that the United States could no longer effectively threaten the use of nuclear weapons to defend Western Europe and would leave Western Europe 'safe for conventional war'.[6] Those who rejected this view – and who tended to feel that the United States nuclear guarantee to Europe would be strengthened should the United States not face the certainty of nuclear destruction if it used its nuclear weapons – presumably assumed that only the United States rather than both superpowers would have the benefit of a defensive shield. British reservations about a BMD position which would prevent the successful delivery by either side of nuclear weapons rested on the same grounds as its opposition to any NATO 'no first use' commitment.

The second point agreed in Washington, relating to negotiations before deployment, left open the question of with whom negotiations were to be held. There is no doubt that Mrs Thatcher meant that both America's allies who had not been consulted before the 23 March speech and the USSR should be involved. However, there were some in the United States Administration who felt that negotiations should be confined to the United States and its allies, and that the second point could be interpreted to mean this.

The fourth point referring to arms control talks reflected the British concern that an uncontrolled arms race in space should be avoided. This was a central point made by the United Kingdom government on numerous occasions. The apparent preference of the Thatcher government was that the SDI should be negotiated away as part of a package of arms control/disarmament measures reducing the number of offensive missiles.

The four points at the heart of Britain's SDI policy were reiterated by Foreign Secretary Geoffrey Howe in his major policy statement at the Royal United Services Institute in March 1985.[7] The speech sought to achieve balance. It emphasised the value of deterrence and of the 'historic' ABM Treaty. It referred to the Soviet buildup of offensive strategic nuclear forces, and of its anti-satellite (ASAT) and strategic defence programmes. Reagan's 23 March speech was also described as 'historic'. The speech surveyed the existing military uses of space, noting that there was nothing new about the militarisation of space. Yet it pointed to the destabilising effects of ASAT warfare and to the links between an ABM and an ASAT capability.

The speech also raised a series of questions about the SDI, its cost, its impact upon the credibility of the United States commitment to Europe, the countermeasures it could provoke (including the advance of non-ballistic weapons), and its impact upon the thinking of the USSR. 'If the technology does work, what will be its psychological impact on the other side? President Reagan has repeatedly made it clear that he does not seek superiority. But we would have to ensure that the perceptions of others were not different'.[8] The speech concluded, not with SDI *per se*, but with a plea for major arms control progress – 'radical cuts in offensive missiles might make the need for active defences superfluous'[9] – and a gentle reminder was given that the United States too was committed to avoiding an arms race in space.

> We warmly welcome the renewal of US–Soviet talks. The joint communiqué agreed between them as the basis for the talks describes their aim as the prevention of an arms race in space. We applaud and endorse that aim.[10]

Despite the criticism from some members of the Reagan team which followed, the Howe speech stood as the major British formal policy statement on the SDI. Britain insisted on the United States right – even duty – to match Soviet research in the BMD field, but preferred that arms control agreements should lead to the avoidance of a new dimension to the East–West arms race. By emphasising the need for continued allied consultation on SDI, to which the United States was anyway committed, the United Kingdom hoped to have an opportunity to influence and moderate the further development of SDI. The United Kingdom government did recognise that the SDI had been instrumental in getting the Soviet Union back to strategic arms control talks in Geneva.

In public, the United Kingdom Ministry of Defence was less forthright on SDI than the Prime Minister or the Foreign Office. The starting point for the Ministry of Defence was that, on its assessment, the USSR would not be able to do anything in even 30 years which could compromise the effectiveness of Trident as a deterrent system. The Ministry could thus argue with some justification that, whatever the long-term significance of the theoretical debates on the impact of BMD, the SDI was not affecting current decisions in the Ministry of Defence and would not do so for some years. In one respect, the emphasis on Soviet BMD capabilities served to emphasise the wisdom of the choice of Trident as opposed to some simpler system which might become vulnerable earlier to Soviet defences. Paul Stares makes similar observations in his contribution to this volume. The Defence Ministry line of recognising the importance of the SDI while playing down its urgency stood in something of a contrast with that of Foreign Minister Howe, who said that doctrine should lead technological developments in defence. But it kept the Ministry of Defence out of the limelight on an issue where no major gains could be expected from the adoption of a prominent position. Moreover, it gave scope for hoping that, over time, technological advances might be so limited as to bring home the cost and difficulties of a comprehensive BMD system to the Reagan Administration and its successor. It was also compatible with the view that the United States Congress, by establishing its intention to monitor SDI progress and to trim SDI expenditure, would do an adequate job of restricting the Administration without United States allies having to get too involved. But overall the SDI had few friends in Whitehall.

In short the determinants of United Kingdom formal policy on SDI, which was one of qualified support, were of the opinion that:

1. the United States should match Soviet research efforts;
2. an extension of uncontrolled arms competition into space was unwelcome;
3. such an arms race did not, however, threaten the viability of the British deterrent; and
4. outright opposition to United States aspirations was likely only to provoke resentment and to have little positive effect.

On the specific issue of participation in SDI work, by the summer of 1985 the government had not developed a full public position although it was expected to do so in the autumn and Mrs Thatcher made clear at the start of 1985 that she wanted British firms to work on SDI projects – the United Kingdom government felt that the United States would make technological breakthroughs through the SDI programme which would have applications in many conventional military and civil products. Consequently, the government wanted Britain to be involved and it was also quick to appreciate that the EUREKA concept of President Mitterrand could also contribute to preventing a widening of the transatlantic technology gap.

At least two other considerations were also at work. One was that it would have been politically difficult to stop British companies bidding for SDI work once the United States government had invited them to contribute. In 1985 there was a stream of American SDI advocates and 'talent scouts' passing through Britain to show that the United States really was interested in British involvement. The second was the real fear that, if research of the most advanced kind in SDI areas was not available in Britain, the best of British electronics, artificial intelligence and data processing minds would be tempted away to the United States. Some reports suggested that a brain drain had already begun.[11]

Nevertheless, consideration of how precisely the United Kingdom government should participate took some time, longer than some in industry would have preferred. This was because the government was unsure just what British research had to offer the SDIO and because the United States had not make clear whether it would place major contracts overseas and whether it would expect allied governments to contribute to funding. Could United Kingdom companies hope at most to be subcontractors to American firms? Given the history of technology transfer problems between Europe and the United States under Reagan, important questions arose related to the technology transfer arrangements to which Washington would agree. Many in Britain felt that Washington would not allow sensitive technology to leave the United States. What restrictions would be placed on the commercial application of SDI-related advances? In the first part of 1985 different members of the United States Administration tended to give different messages on these and other issues. Nevertheless, things became clearer in the summer: rail gun technology, battery power systems, software and battle management systems, optical computers and conventional missiles had emerged as potential areas for British contributions and the United Kingdom had sent evaluation personnel (including Defence Minister Heseltine himself) to Washington to discuss American needs and constraints at first hand.

Participation related also to the issue of how closely Britain wanted to be associated with the SDI. Three broad modes of association could be distinguished, although it was not clear which would prevail:

1. full United Kingdom official participation in the SDI through the use of United Kingdom funds, and through government supervision, on behalf of the United States government, of British companies working on SDI projects. Whitehall would also assist with United States evaluation of British firms bidding for contracts.
2. no British funds being committed, but with United Kingdom firms being encouraged and helped by the government to bid for contracts. United Kingdom government research establishments could be allowed to undertake SDI work using American money.
3. British firms being allowed but not encouraged to work on the SDI. No government support, advice or protection would be forthcoming and the government could also make clear a preference that companies work on projects related to EUREKA if possible. Government research establishments would not work on SDI projects. This latter point was of some significance given the range of work in such institutions: British rail gun research, for instance, was being carried out at the Royal Armament R&D Establishment.[12]

In the autumn of 1985 Britain was apparently ready to adopt a posture along the lines of option 2 with elements of option 1 if the United States government would reciprocate. In Washington in July Defence Minister Heseltine reportedly sought an understanding that Britain would be awarded SDI contracts between 1–2,000 million dollars and have satisfactory assurances on technology transfer. The United States Administration was somewhat taken aback and reluctant to enter into any such commitments. Certainly the positive British response on the United States offer of participation tested the water to see how serious the United States was on this issue. By the autumn of 1985, it seemed likely that the United States–British participation agreement would be reached by the end of the year, not least because the United States wanted to be able to show that the SDI had support within the Alliance. It seemed likely, however, that any agreement would not earmark any work for the United Kingdom, although it could well highlight areas of United Kingdom technological expertise. Nor would any work arising from the agreement give Britain a firm technological base from which Britain could make judgements on the overall feasibility of BMD. Despite its enthusiasm for United States contracts, it seemed unlikely that the British government would change its research priorities by allocating its own funds to the SDI.[13]

Interestingly, Mr Weinberger's original formal letter of invitation to participate in SDI work, issued on 26 March 1985,[14] was not taken seriously with its sixty-day deadline for replies. If anything the deadline, which was later abandoned, served to strengthen the views of those who, like the French government, believed that the United States was offering participation as a crude means of getting backing for SDI in general. In December 1985 a United States–United Kingdom agreement on participation was finally concluded. Although its terms were kept secret, it was understood to identify eighteen areas of technology where British experience was recognised as on a par with that of the United States. It had provisions for technology transfer and exchanges of information perceived in the Whitehall as a very good deal for the United Kingdom. Secretary of Defense Weinberger said that the deal meant Britain should get a significant amount of work from the SDI programme[15] although experience of decades of efforts to sell military equipment to the United States left many in the United Kingdom government and industry with minimal expectation that British industry would be allowed to contribute much to the SDI.

There were some other issues where the government line was understandably blurred. One was the extent to which Britain wanted a coherent European line on the SDI. Here the United Kingdom government did not emphasise the importance of coherence on strategic defence but it took part in the European discussions on the topic, particularly in the WEU. Moreover it quickly came to support the French EUREKA initiative which had some links with the SDI. Both were originally concerned with similar technologies, both were dramatic concepts appealing to scientists keen to work in the most advanced technologies, and France was aware that Germany in particular might endorse the SDI to get access to new technologies if no European alternative was available.

There were two problems for Britain within the European dimension. The first was that there were still many in Whitehall worried about the development of a European caucus in NATO, feeling that it would permanently divide the Alliance and that the United States reacted better to bilateral contacts. Such a caucus had, however, been largely accepted on equipment and on the industrial side of defence, with the Independent European Programme Group (IEPG) having a recognised role. Britain was thus seemingly ready to countenance a European position on participation but was reluctant to move towards a co-ordinated European line on the SDI in general.[16]

The second problem was the overt hostility of France to the SDI. Britain supported a United States research programme while France, although its hostility to the SDI had been tempered somewhat, did not. This difference disguised the many common reservations which Britain and France held. Overall, the impression in mid-1985 was that the SDI represented another challenge to Britain's claim of being a good European and Britain looked as if it were failing the test, at least in French eyes.

Another feature of United Kingdom government policy was a reluctance to get deeply involved in BMD feasibility issues. Publicly the government pointed out that a perfect defence was impossible but left open how near to the perfect state might be achieved. Privately United Kingdom officials were aware of the rising optimism in 1984 and 1985 of United States scientists associated with the SDI and recognised that a point defence system using ground-based missiles could be developed fairly rapidly. Beyond that, uncertainty had to be accepted although scientific opinion outside government often tended to be doubtful about BMD possibilities and critical of the SDI in general.[17] United Kingdom official reserve on feasibility disguised a point of potential difference with the United States. Whereas Britain viewed the SDI programme as research to help decide *whether* to proceed with BMD, the Reagan Administration tended to see it as research on *how* to proceed.

II THE POLITICAL PARTIES

In the Labour Party, with defence a sensitive and internally divisive issue, there was little public discussion of the SDI, but the broad lines of Labour's opposition to it were set by Neil Kinnock before Mrs Thatcher left for the United States in December 1984. He argued that the SDI was a threat to the cohesion of NATO, massively expensive and risky, an impediment to arms control and disadvantageous to Europe. 'We in Europe would be caught in an alley between intercontinental defensive and offensive weapons'.[18] Denis Healey, a more authoritative Labour voice on strategic matters, condemned BMD as undermining the basis of post-1945 stability. Moreover, should the United States give its BMD knowledge to the USSR, 'it would make a farce of the present government's policy of maintaining a national nuclear deterrent'.[19] Given Labour's current opposition to a national United Kingdom deterrent, this latter point seemed to be an argument which the government could use in Washington rather than a particular Labour concern.[20]

A more formal statement of Labour's policy on the SDI was not produced until 11 July 1985 when the Party's National Executive Committee adopted a three-page fourteen-point resolution condemning the SDI as threatening second-strike capabilities, damaging the prospects for arms control and representing an attempted technological solution to a political problem. It listed the actions the USSR might take to offset SDI effects and called for a non-provocative, conventional defence. It supported those in Europe who had opposed participation in SDI work and called for a co-ordinated European rejection of the SDI, a notable point given Labour's traditional antipathy towards European co-operation. Finally, it called specifically for the strengthening of the arms control agreements restricting BMD and ASAT development.

The Liberal–Social Democratic Alliance shared many of the views outlined above, having presented a detailed analysis somewhat earlier. Many Alliance arguments were outlined by MP John Cartwright in the spring of 1984,[21] and then on the eve of Mrs Thatcher's departure for Washington, SDP leader David Owen wrote an open letter to her summarising the findings of an SDP working party on defence and disarmament. The Owen letter listed a wide range of objections to the SDI:

1. a totally invulnerable defence was impossible, in space as elsewhere;
2. the SDI would undermine deterrence and therefore be destabilising;
3. the arms race would be accelerated because of the responses forthcoming from the USSR;
4. East–West relations would be damaged and arms control prospects hit by the ending of the ABM Treaty;
5. NATO would be strained as it would make allies fear a United States decoupling from Europe;

6. nuclear proliferation would become more likely as Third World countries would recognise that the United States had really no interest in meeting its disarmament commitments under the Non-Proliferation Treaty;
7. the SDI would cause a wasteful diversion of resources away from the conventional defence sector where they were most needed;
8. the cost of BMD development and deployment would damage the United States economy and hence that of the rest of the world;
9. the technological feasibility of BMD was as yet unknown;
10. there would be no defence against non-ballistic or short-range ballistic means of nuclear weapon delivery; and
11. BMD would require dangerous automated responses.

The SDI, concluded Owen, was an ill thought out, unrealistic and dangerous programme.[22] Other Alliance views were that Britain should not accept SDI-related contracts and should work for the primacy of arms control.[23] Alliance representatives were also among those who felt that the designation of the SDI as purely a research programme was meaningless, since the United States political, military and bureaucratic pressures for further effort would be irresistible once $26 thousand million had been spent. The Alliance peer Lord Kennet went further and suggested that development could well be included in the SDI programme itself.[24]

Those opposing the SDI also included some Tory MPs, with former Prime Minister Edward Heath and a former Foreign and Defence Minister, Francis Pym being their more notable spokesmen. In an address at Chatham House, Heath called the SDI 'decoupling, destabilising, and a diversion of resources'.[25]

In contrast there were a few within the Conservative Party who held the SDI in more affection than did their Prime Minister. Among backbench MPs the more prominent of these attended the Assembly of the WEU where they helped to block anti-SDI resolutions. They have written letters to the press spelling out their position[26] and, interestingly, members of this group did not just back the SDI on grounds of its asserted feasibility and possibilities for reducing the importance of nuclear weapons,[27] they also built on it to justify a major programme of European co-operative activity in space. Two of the most pro-SDI European documents were the WEU Assembly Reports from its Committee on Scientific, Technological and Aerospace Questions on *The Military Use of Space*.[28] The Rapporteur for these reports was the backbench Tory MP John Wilkinson, who had considerable experience in defence and aerospace. Their argument was supportive of the SDI, of European space co-operation and of the idea that the WEU should be the focus of European SDI-related co-operation.

Given fears in the United Kingdom government of the arms race implications of the SDI, it is notable that Mr Wilkinson felt that arms control should be strengthened in the long term. 'Advanced defences', he wrote, 'should reduce the potential of ballistic missiles thus providing an incentive for negotiated reductions of them'.[29] He also cited Mr Kissinger extensively on this issue, whose argument was for the value of limited rather than would-be comprehensive defences, and he reproduced the argument of Mr Kissinger and Mr Brzezinski that, even before SDI, arms control had reached a dead end. All this was then used to advocate European participation in the United States space station project. Early in 1985 European governments in fact decided on a range of co-operative steps in the European Space Agency (ESA) to continue the buildup of European space capabilities.

For Mr Wilkinson and other SDI advocates, two central points were that the USSR was anyway travelling down the BMD road[30] and that the effects of a limited BMD deployment would be to complicate the plans and lower the uncertainty of a potential aggressor contemplating a first strike.

The political forces favouring the SDI had Lord Chalfont as their chief spokesman, ironically once a Labour Disarmament Minister. Chalfont asserted that BMD offered possibilities for reducing nuclear weapon stocks, strengthening stability, decreasing reliance on the threat of United States suicidal retaliation and reducing the chances of a successful conventional attack on Western Europe, all points firmly contested by SDI opponents. In his speech in the House of Lords debate, however, Chalfont stressed that the SDI was only a research programme, not a violation of the ABM Treaty, and that Reagan was committed to consulting Allies before proceeding to any development phase. He cautioned against seeing BMD as vastly expensive and emphasised that the SDI concerned a layered but limited defence 'whatever President Reagan may have said in his original speech, or what he may have said since'.[31] In his judgement, the USSR was likely to respond by building up its defensive rather than its offensive forces, a prospect which did not alarm him.

To summarise, of the three main political groupings in the United Kingdom, two were openly hostile to the SDI while the Conservative Party was somewhat divided on the issue. The government itself was supportive of the ABM Treaty and keen to avoid an arms race in space, yet it was loath to offer open opposition to the SDI, especially as it did not want to see United Kingdom industry miss out on the technological advances involved. The basic position was that those who were positively enthusiastic about the SDI in British politics, although vocal, were few in number.

On the fringe of formal politics in the United Kingdom was the peace movement, centred on the Campaign for Nuclear Disarmament (CND). Insofar as the SDI was meant to win over anti-nuclear opinion by presenting an alternative route to a world without the danger of nuclear destruction, it simply failed to erode CND support. In the United Kingdom as a whole there was little acceptance of the claim that effective defence against nuclear delivery systems was feasible and the British peace movement viewed the SDI primarily in terms of its negative arms race implications.

In terms of support for the peace movement, although committed members of the peace movement saw the SDI as further evidence of the alarming nature of the Reagan Administration, the failure of the SDI to make a great impact on British public feeling meant that the peace movement gained no meaningful increase in backing or exposure as a consequence. Cruise missile deployment and the Trident programme were more central issues for the peace movement than the SDI until the end of 1985.

III ACADEMICS AND DEFENCE ANALYSTS

Much of the public debate about the SDI was conducted among journalists, policy analysts and academics. Of the three most serious national newspapers, only *The Times* came out clearly in support of the SDI, although its initial position was otherwise: it had responded to the 23 March speech by arguing that BMD deployment would be 'destabilising, contributing to uncertainty and suspicion'. It felt that the speech could well hurt arms control talks and alarm America's allies. Only later did it turn to backing the SDI and then it did so with a vengeance, with vitriolic criticism of the Howe speech at the RUSI, accusing him of supporting Soviet views. Basically *The Times* was impressed by the feasibility and morality of SDI technology and was also concerned about Soviet efforts. The conservative weekly *The Economist* has argued that the SDI, so long as it does not result in a complete umbrella for the United States, would be essentially defensive and coupling the United States more closely to Europe. But on the other hand it has also supported arms control, and would be prepared to slow down BMD research to get it. Additionally it has argued against the development of high-altitude ASAT weapons, so its overall position offers the SDI far from unconditional support. Somewhat surprisingly the pro-government *Daily Telegraph* has spoken out against the SDI while the *Observer* has not committed itself editorially. Feeling on the paper, however, is felt to be opposed to the SDI. The somewhat left of centre newspaper the *Guardian* steadily opposed the SDI while the *Financial Times* provided the most thorough scientific, military and political analyses of the SDI. On balance, the *Financial Times*' scientific writers were fully aware of the technological problems of BMD and its political writers tended to view the SDI's wider consequences as negative.[32]

The SDI gained little support from Britain's research institute/academic community, having to rely mainly on the small Institute for European Defence and Strategic Studies (IEDSS). The IISS pronounced the SDI to be 'the security policy centrepiece of the second Reagan Administration', reviewed the claims made for it, and argued that 'even if strategic defences were to prove feasible, they could damage stability rather than strengthen it'. The utility of the nuclear threat for Western defence was recognised: 'it is not at all certain that Western interests as a whole would be served by having to rely exclusively on conventional forces to protect those interests.[33] The Deputy Director of the IISS, Mr Jonathan Alford, regularly expressed his misgivings about the SDI, as did Professor Lawrence Freedman, perhaps the British strategist best known in the United States. Freedman's paper to the IISS 1984 conference was a thorough British critical analysis of the SDI, concluding that disarmament was a cheaper and more effective way of reducing the offensive threat. It explored in detail the relationship with the USSR on which a meaningful United States BMD system would have to rest.

> Without limitations on offensive arms, this Administration may be reduced to arguing, as the Nixon Administration was when promoting the Safeguard ABM system fifteen years ago, that the Soviet Union will maintain sufficient offensive forces to warrant making

the effort but not build them up to a level that would overwhelm the defence.[34]

At the Royal Institute of International Affairs (RIIA), the Director, Admiral Sir James Eberle, saw the SDI as divisive within NATO, damaging to arms control and unnecessary, since the research needed to keep abreast of the USSR and evaluate BMD possibilities could have been conducted without the political drama and organisational commitment of the SDI. The early years of the SDI involved only modest increases in BMD spending and these could have been introduced quietly, without the disruption which the SDI caused. One of the RIIA research staff, Phil Williams, pointed to the uncertainty about the purposes and funding of the SDI but identified one element of certainty as the need for Western Europe to have a nuclear dimension to its defence, a dimension which the SDI sought to erode. Moreover he saw the arms race implications of the SDI as leading the USSR to build up its offensive forces. He also recognised, as do many others, that a state with a limited defensive capability could give the impression that it was seriously considering a first strike, leaving its defensive forces to deal with any retaliatory forces. Recognising that unequivocal opposition to the SDI would be politically counter-productive, he came out in favour of European support for research but for little else. This was essentially the United Kingdom government position. Williams also noted that SDI advocates in the United States would try to use any European support for the SDI to their advantage in internal American debates.[35]

British scientists, scarcely an organised or coherent group, made little visible response to the SDI. Certainly many of them had no objection to getting SDI money if possible for their research. Individuals among them wrote letters to the press both for and against SDI with the most impressive contribution coming from eighty artificial intelligence scientists who wrote a collective letter to the *Guardian* (published on 5 July 1985) which attacked the SDI at a time when Vice-President Bush was in Europe trying to get support for it. In their view, BMD could not rid the world of the threat of nuclear war and would increase the chances of such a war. The pro-disarmament British Pugwash Group had not taken a formal stance on the SDI by October 1985 although it was holding a meeting on the subject in December and its opposition could be anticipated. The leading British scientific magazine, *New Scientist*, has published reports sceptical of the technological possibilities for BMD and which have pointed out that much nuclear-powered X-ray laser work for BMD would be illegal under current treaties. It has noted the view of Farooq Hussein of RUSI that Soviet research work does not in itself justify the SDI, but editorially it has kept a low key, arguing that opponents of the SDI should be more subtle in their opposition and regretting that BMD should be taken so seriously by governments when so little is known about it. Civil projects need much more homework to be done on them before they can attract substantive resources.[36]

All in all, among those who might be tarred with the brush of being intellectuals, the SDI had minimal support in the United Kingdom, but there was recognition of the disadvantages of opposing it too openly. An important consideration was the view held of the SDI's political future in the United States, and here opinion varied. there were those who believed it would be neglected or abandoned by Reagan's successor as technological shortcomings became clear and resource shortages felt. On the other hand, there were those who believed that the American interests in favour of BMD would have unstoppable momentum by the time Reagan left office. Some in the United Kingdom perceived that the BMD advocates in the United States would try to get a decision to proceed to development as soon as 1988 so as to limit the choices of Reagan's successor. The idea that an objective, considered decision would be taken on the development phase, along the lines suggested by Paul Nitze in his New Strategic Concept (NEC),[37] commanded little respect, although the view of the SDI as a research-only programme was a cornerstone of formal United Kingdom government policy. For many British analysts, the entire process by which the SDI emerged, starting with a sudden presidential exposition of an idea for which the back-up work had not been done, highlighted the limitations of United States decision-making and the United States political system.

IV CONCLUSION

Official United Kingdom policy on the SDI was the product of the determination of the government that no wedges were going to be driven between Britain and the United States on this issue, recognition that Reagan was committed to the SDI, acceptance that it had helped to get strategic arms talks restarted, concern about the industrial and technological advances involved, and dismay about the political and strategic consequences which could follow. At most, Britain wanted the United States to move very cautiously on BMD and wanted Alliance consultations so that it could inject fact and argument into United States thinking.

British wariness about upsetting the United States was reflected in the absence of any British push for formal consideration of where research (allowed under the ABM Treaty) ended and where development began. This sensitive but central issue was largely ignored in public.

Pertinent to SDI was the absence of British faith in technology alone to solve political problems. The notion of an impermeable umbrella covering the United States or the United States and Europe, which chronologically, emotionally and politically was the basis for the SDI, found little endorsement in the United Kingdom. The determined enemy could always find a way round, underneath or through, and the destructiveness of nuclear weapons meant that only a few need penetrate for the defence to be valueless. Moreover, many government scientists, while recognising that some BMD was possible, believed that it would be very expensive and that effective offensive countermeasures would always be cheaper. Given the cost even of Patriot, few in the United Kingdom were impressed with the idea that a European effort at BMD/improved air defence might be worthwhile. For the British, the main route forward lay in improving relations with the enemy while maintaining deterrence. However, the political attraction of a 'European SDI' could not be easily dismissed and it remained possible that particular European needs could form a basic guide for the kinds of SDI research which Britain might take on.

Basically, there were comparatively few in the United Kingdom who shared the background beliefs making a major Western BMD effort appealing. These background beliefs may be said to include a view of the USSR as untrustworthy, deceitful and aggressive while having limited technological competence. Such views lead to the conclusion that effective arms control agreements with the USSR will be nearly impossible to agree on while offering the prospect that the West can establish a politically meaningful technological lead.

One fundamental uncertainty about the SDI is whether it will eventually be oriented towards a partial or a comprehensive defence, but in Britain the comprehensive concept was widely seen as being for United States domestic consumption only and a real issue was the kind of partial defence envisaged. There is a central distinction between a partial defence which is partial because it defends only a limited number of points (using ground-based ABMs and perhaps space-based surveillance and tracking systems) and a partial defence which is 'leaky' while covering a wide area. The former was seen in Britain as feasible in the relatively short term, but the SDI was deliberately meant to consider a layered defence including space-based systems of area defence. It was the 'leaky' area defence prospect particularly which led to some British fears of a Soviet perception that the SDI indicated a United States move towards a first strike capability. Even a partially layered system might be able to deal comprehensively with a limited number of Soviet retaliatory missiles.

A public debate about how space- and ground-based BMD systems would affect ideas about the concept of limited nuclear warfare as a means of strengthening deterrence had not really begun in Britain, although Dr Freedman has argued that superpower BMD systems, by reducing the options for the limited use of nuclear weapons, would damage NATO rather than Warsaw Pact doctrines.[38] It would probably be easier for Britain to accept the deployment of United States hard-point, ground-based BMD systems rather than a layered space-based effort, especially were the USSR to increase the vulnerability of SLBMs by making marked advances in anti-submarine warfare.

Although NATO survived the initial years of the SDI, it was striking how many in the United Kingdom felt that Reagan's initiative, taken without Allied consultation, was divisive within NATO. Certainly the SDI placed many British supporters of NATO in the same camp as the peace movement on this issue.

Perhaps of most concern in the final analysis will be the resources aspects of the SDI, with their link to the burden-sharing debate in the Alliance. Advocates of the SDI and BMD, while tending to play down the likely costs, will have to explain where any extra money will come from and/or what will be given up instead. British responses to the SDI should gain increased sharpness should it become clear that efforts in the BMD area are hitting other aspects of Allied defence.

By November 1986, after the Reykjavik summit, the SDI acquired the positive advantage for Britain and other West European states of holding up a Soviet–US arms control agreement which would have much weakened the nuclear element in Europe's defence. Britain's position on the SDI, as laid out in the December 1984 'Four Points', had received implicit support from much of Western Europe, in particular in the Council of the Western European Union. In terms of participation, Britain had gained only modest contracts worth between $25 and $30 million, although the government was making real efforts in Washington to assist British industry. With the US Congress voting a substantial sum ($3.5 billion) for the SDI, it was clear that the US BMD effort would continue but, nevertheless, funding was not at the level Reagan had requested, and no major technological breakthroughs were anticipated. With a new President to be elected in 1988, most of whatever urgency the SDI had possessed for the UK government had disappeared. The ATBM question, however, was firmly on the long-term agenda.

Notes

1. John Roper, 'The British nuclear deterrent and new developments in ballistic missile defence', *The World Today* (May 1985) p. 95.
2. G. Keyworth, US Information Service (USIS) London, 10 July 1984.
3. P. E. Gallis, M. M. Lowenthal and M. S. Smith, 'The Strategic Defense Initiative and United States Alliance Strategy' (Washington, DC: Library of Congress Congressional Research Service, 1 February 1985) p. 39.
4. *Hansard*, Issue 1331, 9 January 1985, Written Answers Col. 441.
5. 'The Week in Politics', Channel 4 Television (UK), 14 June 1985.
6. See, for instance, T. Garden, 'Space and Strategic Defense', *The Hawk* (March 1984) p. 75.
7. Text in the *Journal of the RUSI* 130/2 (March 1985) pp. 3–8.
8. Journal of the RUSI 130/2 p. 6.
9. Journal of the RUSI 130/2 p. 8.
10. Journal of the RUSI 130/2 p. 8.
11. Report of the Advisory Board of the Research Councils, *Guardian*, 14 June 1985.
12. *Observer*, 9 June 1985 and *Financial Times*, 13 July 1985.
13. *Financial Times*, 31 July 1985, *The Times*, 16 May 1985 and *Sunday Times*, 13 October 1985.
14. *The Times*, 18 June 1985.
15. See T. Taylor, 'Britain's Response to the Strategic Defense Initiative', *International Affairs*, vol. 62, no. 2 (Spring 1986) pp. 217–30.
16. *Financial Times* and *The Times*, 19 July 1985.
17. See for example Gallis *et al.* 'The Strategic Defense Initiative', p. 24, letter in *Guardian*, 3 June 1985, and Lord Zuckerman in the House of Lords, *Hansard* 30 January 1985, Cols 694ff.
18. *Guardian*, 20 December 1984.
19. House of Commons, reported in *The Times*, 13 December 1984.
20. *Guardian*, 23 May 1985 and 26 April 1985.
21. *Guardian*, 15 June 1984.
22. *Guardian*, 21 December 1984.
23. *Guardian*, 21 December 1984, 1 March 1985 and a letter from Lord Kennet in *The Times*, 19 December 1984.

24. *Hansard*, House of Lords, 30 January 1985, Col. 700.
25. *Financial Times*, 13 March 1985, see also *The Times*, 9 November 1984, *Guardian*, 20 December 1984.
26. *Guardian*, 21 April 1985, *Daily Telegraph*, 10 July 1985 and 5 March 1985.
27. *The Times*, 19 December 1984.
28. Assembly of the WEU, Doc. 876, 15 May 1985, and Doc. 993, 8 November 1984.
29. Assembly of the WEU, Doc. 993, p. 10.
30. Doc. 993, p. 21.
31. *Hansard*, House of Lords, 25 January 1985, Col. 704. See also Lord Chalfont, 'SDI: The Case For The Defence', Occasional Paper no. 12, (London: Institute for European Defence and Strategic Studies, 1985).
32. *The Times*, editorials, 25 March 1983 and 25 March 1984. See also *The Economist*, 28 September 1985, p. 19, 14 September 1985, p. 17, and 5 October 1985, p. 17; and *Daily Telegraph*, 21 June 1984.
33. *Strategic Survey 1984–5*, (London: International Institute for Strategic Studies, 1985), p. 4. The IEDSS has published two pro-SDI occasional papers, the first (no. 10) by W. Kaltefleiter, 'The Strategic Defence Initiative: Some Implications for Europe' (1984), and the second (no. 12) by Lord Chalfont, 'SDI: The Case for the Defence'.
34. L. Freedman, 'The Star Wars Debate: The Western Alliance and Ballistic Missile Defence, Part II' in *Adelphi Paper* 199, New Technology and Western Security Policy, Part III (London: International Institute for Strategic Studies, 1985).
35. 'Western European Security and the SDI' (paper for the Workshop on the SDI, Netherlands Institute of International Relations, The Hague, 26–7 April 1985).
36. See *New Scientist*, 10 January 1985; 28 February 1985; 21 March 1985 and 5 October 1985.
37. *Guardian*, 22 February 1985; *Sunday Times*, 24 February 1985.
38. Freedman, 'The Star Wars Debate'.

5 French Political Reaction to SDI – The Debate on the Nature of Deterrence
Alain Carton

I INTRODUCTION

France, as an independent nuclear power with small but militarily significant deterrent forces, has a special perspective in the debate on ballistic missile defence (BMD). The ABM Treaty both validated and simplified the French position. The planned modernisation of the French nuclear potential, with a marked increase in the number and penetration ability of its warheads, might also contribute to maintain a minimum deterrent against a forseeable Soviet ABM.[1]

These deployments were originally planned during the former military planning law 1976–82, before the current debate on SDI and BMD systems had started. However, the emergence of a new Soviet ABM programme and of a strategic defence programme in the United States has reopened the debate among defence experts as well as politicians, and it might complicate the task of defining a plausible nuclear strategy and of rallying public support behind it.

At present it has to be emphasised that the French reaction to SDI cannot be isolated from other European assessments and simply considered a petty and selfish withdrawal behind its own national deterrent. The effectiveness of the French strategic deterrent forces is not only in its own interest. This was explicitly recognised in the 'Declaration on Atlantic Relations', which was signed by the heads of government of the NATO countries on 26 June 1974.[2]

The French view is based on the desire to maintain the credibility of its deterrent and to avoid the dangers, in the European context, of a new East–West arms race.

The question has been how to react against the combined effects of political indifference of the European partners and of the lack of collective will in the field of European co-operation in science and technology, the consequence of which has allowed the United States and the Soviet Union to play the dominant role in all matters concerned with outer space.

II THE REACTION AT GOVERNMENT LEVEL

The reactions of the French government to President Reagan's 'Star Wars' speech and to the renewed emphasis on BMD have been cautious and sceptical in view of the French evaluation of nuclear balance and its definition of nuclear stability.

French officials have expressed concern about the political aspects of SDI. There has also been scepticism as to whether it is possible from a technical point of view to shift drastically the advantage from offence to defence. The American ASAT and ABM programmes are partly responsible for the current questioning of strategic concepts in NATO, in particular the concept of structural deterrence with the ability to threaten an agressor with second-strike retaliation. This was put into question by the Reagan Administration's belief that the concept of deterrence can be revolutionised in the fields of both conventional and nuclear warfare. In this respect the new ABM programmes can be linked to military plans for a conventional deterrent based on the new weapon technologies. The ultimate objective, as defined by President Reagan, is to reduce or even eliminate the dependence on nuclear arms.

French officials remain doubtful as to whether it is possible to rely on modern technology to achieve this purpose. This seems however to be the idea behind both the SDI programme as well as the FOFA concept for Central Europe – increased conventional military preparedness combined with the hope that further technological progress can overcome the threat of nuclear war.[3]

(a) The first French reaction

Two phases may be distinguished in the official French reaction to SDI. The first is from Winter 1984 – that is, from Mitterand's speech in The Hague and the first reaction of the Ministry of Foreign affairs, to the intensification of public criticism and political debate about the American initiative during Winter 1985. The second stage is from Winter 1985 to the present.

During the first stage two major objectives were pursued:

1. convincing the superpowers to return to the negotiating table in order to prevent the militarisation of space and to maintain the ABM treaty;
2. launching the concept of a 'European Space Community' based on an independent means of space reconnaissance as was announced by President Mitterrand in The Hague in February 1984.[4]

As far as the first objective is concerned persons within the French government and other major officials have stressed that research and projects relating to the military use of outer space give rise to a number of serious questions:

1. they involve the risk of stimulating an arms race in weapons which do not enter space (non-ballistic nuclear weapons, cruise missiles and highly sophisticated weapons).
2. they may lead to a new division between countries which possess ABM and ASAT weapons in outer space and those countries which do not, but are nevertheless threatened by ballistic missiles.
3. last but not least, the principle of nuclear deterrence may be weakened.

While acknowledging that the complete demilitarisation of space was not longer a realistic aim, the French government, in a paper presented to the Conference on Disarmament in Geneva by the French representative, Mr François de la Gorce, on 12 June 1984, stressed the need to reach negotiated agreement on the following points:[5]

1. strict limitations on ASAT systems, especially those operating in high orbit.
2. a ban for a five-year period on the deployment of directed energy weapons systems, on the ground, in the atmosphere or in space, which are capable of destroying ballistic missiles and the prohibition of any related testing;
3. the reinforcement of the existing system of notification of the characteristics and tasks of objects launched into space as established by the Registration Convention on 14 June 1975;

4. a commitment by the United States and the USSR to extend the non-interference provisions of the SALT treaties to cover the satellites of third parties.

The French position is, however, based on the concept that the passive use of outer space – satellite networks for verification of treaties and for observation – is inescapable in order to maintain the credibility of independence and of the national deterrent.[6] Therefore the negotiations proposed by the French delegation in Geneva should focus on the active military use of space and they should lead to commitments limiting ASAT systems and strive for a ban on ASAT testing. The commitments should enable verification, and they should improve the existing systems of notification of the launching of objects into space.

Finally, the special meaning of the ABM Treaty for France must be discussed. France has supported the ABM Treaty and stressed its positive effects on global strategic balance. It is one of the few real arms limitation agreements and it supports both the American strategy and the French interpretation of deterrence.

As far as the American research programme (SDI) and the existing Soviet systems are concerned, the French government would prefer to maintain the ABM limitations or to supplement them in accordance with technological developments and thus avoid 'unfair deployments'. In the French view, it is unrealistic to prohibit research in the field of space technology and it cannot be *a priori* condemned. However, further developments and testing of components in the absence of negotiations would be a violation of the treaties and must therefore be condemned.

In his address to the UNO General Assembly on 11 June 1982, Mr Claude Cheysson, the then Minister of Foreign Affairs in the Mauroy government, expressed France's willingness to sign a global ABM Treaty which would prohibit ABM systems on an international level and which would provide for a system of verification.

As early as 1972 France announced in its Defence White Book that it would not test or deploy any ABM systems. The reasons for this decision were the high costs and the poor performance of such systems, especially in view of the very short flight times of missiles between the USSR and France. This position has remained unchanged.[7]

As far as the negotiating stances of both superpowers relating to ABM and ASAT are concerned, the official French view stresses the subtleties. The Soviet offer to negotiate a moratorium on the testing of ASAT systems and ABM research is being evaluated as a diplomatic failure. French officials have expressed the opinion that one of the Kremlin's objectives has been politically to divide the United States from its European allies. On the other hand, the United States position that future discussions on space matters should be linked with the START and the INF negotiations has also been received with doubt: 'The United States have to adopt a stance towards the Soviet Union in which the reaffirmation of American power is accompanied by negotiations on space matters'.[8]

In this context the French position concerning the interaction between the deterrent aspects of nuclear weapons and the protection of the civilian population from the damages of nuclear warfare should be mentioned. The same concept of nuclear weapons can be observed as the basis for both the ABM debate as well as the civil defence issue. If the Soviet deployment of INF in Europe is considered a threat to the balance between the French nuclear deterrent and Soviet military posture, the result could be a stronger effort to protect the civilian population. This would mean that the nuclear forces would be analysed from a military perspective and not purely as a deterrent. However, if the credibility of the French nuclear doctrine rests on the capability to respond to any threat at a level chosen by the political power, civil defence can have only the role of organising and mobilising the population during a crisis. If civil defence is a part of the system of global security and thus defined and limited by law, then the assumption is wrong that it should be used to intercept tactical measures which are part of a battle concept and not of a deterrent.[9]

France has also taken the initiative in the realisation of the second objective: the realisation of a European presence in space. It has called for the deployment of a military reconnaissance satellite which would be administered and controlled by the European countries and would thus provide additional protection in the event of aggression. This initiative is based on the principle that Europe will only be able to sustain an adequate level of technology if it acquires a system of space transport and an independent means of reconnaissance in space. A joint effort in this matter seems essential. However the overall cost would be so great that such a project would be impossible before a stronger, more unified European will developed. This is what President Mitterrand meant in his appeal for a 'European Space Community':

There is still a broad area in which we can organise our security however. Not only by conventional armaments but also by the new means about to erupt on the world scene. We must already look beyond the nuclear era if we are not to fall behind in a future that is closer than one might think. I will quote only one example, the conquest of space. If Europe were able to launch its own manned space station allowing it to observe, transmit and consequently avert all possible threats, it would have taken a big step towards its own defence. Nor should one forget advances in electronic calculation and artificial memory, or the known capability to fire projectiles at the speed of light. To my mind, a European space community would be the response best adapted to the military realities of tomorrow.[10]

The vision of a European role in the military use of space, as expressed by President Mitterrand in this statement, has provoked a great deal of controversy in both the French political establishments as well as those of the European Allies. It is true that in some areas Europe has acquired enough expertise to be able to engineer certain components of such a space station. However, the mandate of the European Space Agency, the only European institution which would come into question, specifies that only civil projects can be undertaken.

In 1984 the French National Space Agency (*Centre National d'Etudes Spatiales*, (*CNES*) submitted to the government a complete programme for the future including:

1. the development and deployment of civilian and military observation satellites;
2. the gradual realisation of European autonomy in space station technology, including the development of the ARIANE V launcher and the HERMES space shuttle.[11]

France has been interested in space research since the 1960s, especially in its civil use. Military aspects have been considered but budgetary constraints have not allowed any specific programmes to be initiated. The heavy launchers required can be developed only with the assistance and close co-operation of the European countries.

The withdrawal in 1973 from the European observation and civil technology rocket project (the so-called EUROPA project) had the result that the European countries were dependent on the United States for the chance to participate in space programmes. At present it appears that only the ARIANE rocket could provide the launch capability required for both military and civil applications.

It is very important for France, given the special relations it maintains with a number of overseas countries, to have at its disposal an independent communications system.[12] A Defence Ministry programme initiated in 1979 is at present in progress: the satellite radio-communications system (*Système de Radiocommunication Utilisant un Satellite, SYRACUSE*) will improve the armed forces' long distance communication.

Several medium-term projects have been initiated by the military, including satellite communication systems to replace SYRACUSE in the early 1990s:

1. FY83 and FY84 were devoted to testing prototypes.
2. FY85 has allowed the installation of the first stations and has guaranteed continuation until 1987.

This is the first example of co-operation between the Defence Ministry and the Post and Telecommunications Ministry. In political terms the realisation of such a system has been given high priority, irrespective of whether it is a purely national project or whether it is realised in co-operation with other European countries.

Mr Hubert Curien, the then Minister of Research and Technology and the former director of the National Space Agency (*Centre National d'Etudes Spatiales, CNES*), has declared that 'the combining of civilian and military activities is essential given the great costs. It would enable the military to benefit from the existing results of the civilian programme. A programme such as the Military Optical Reconnaissance Satellite [*Satellite Militaire de Reconnaissance Optique, SAMRO*] could not otherwise be envisaged.'[13]

The Franco–German Reconnaissance Satellite has not yet been approved by the Germans. There are several reasons for this, especially the financial aspect and concerns about the military significance for the Federal Republic. In reply to a French parliamentarian, Mr Curien added that 'France could build a reconnaissance satellite by itself for a cost of about 10 thousand million francs. It would be more difficult to build an early warning satellite. Even though civilian programmes such as ARIANE or HERMES could be realised by France alone, international co-operation is necessary in order to assure sufficient outlets'. Mr Curien has repeatedly stressed that the European partners can be mobilised only by a high technology programme.

(b) The development of the French response

In Winter 1985 the second phase of French reaction began with an intensification of public criticism and the formulation of a counter strategy on the political level.

Initially French officials believed that the development of the components needed for SDI would fundamentally alter the present foundation of deterrence.

The French nuclear doctrine is based on the close linking of the American strategic forces protecting the American sanctuary and the 'deterrence defence linkage' which protects Europe and extends across the Atlantic Alliance. According to French doctrine the French threat to use nuclear weapons has to be 'meaningful'. It has to signify to the aggressor that France's vital interests could always be protected by the threat of nuclear retaliation.[14]

Deterrent capability implies that a significant amount of the retaliatory capacity would survive a first strike. A stable mutual deterrence relationship is said to exist when both sides have such a survivable second-strike capability. If either or both sides could hope to defend both the delivery systems and the civilian population, deterrence would no longer rest on the ability to inflict unacceptable damage in retaliation to aggression.

However, a limited defence of hardened targets of the retaliatory forces would not necessarily alter the character of deterrence any more than other efforts to reduce the vulnerability of the second-strike capability. This position was made clear by Mr Hubert Vedrine, an advisor to President Mitterand, in Winter 1985.[15] He expressed concern about uncontrolled developments which could lead to destabilisation, uncertainties and new developments in the arms race. Even the perfect shield would not be possible. Military spending would drastically increase, and the only result would be the introduction of new systems. The reasons for the government's position were not the fear that France's nuclear deterrent would suddenly become obsolete. On the contrary their position is based, according to Mr Vedrine, rather on the requirements of international balance than on the particular interests of France. Under these circumstances no country could rationally expect to be able to intercept the French deterrent if it should be used.[16]

A month later at an international conference attended by experts and politicians from both sides of the Atlantic Charles Hernu expressed the same position:

> If such systems should be deployed it is to be assumed that the old dialectic between bullet and shield would also apply to nuclear weapons as well. For the very reason that no defence against nuclear weapons was conceivable, these weapons had been exempted thereof.

And he added that

> it is to be assumed from history that the deployment of defensive systems would lead to a renewed competition in the realm of offensive armament.[17]

Following the questions concerning the deterrent doctrine, the second major cause of criticism of SDI has been the feasibility of the project. Under no circumstances would it be possible to suggest that the offensive nuclear system will disappear. It would be very dangerous to paint such a utopian picture in an attempt to influence public opinion. This could only be achieved by means of disinformation concerning the role played by nuclear weapons in deterring other types of aggresssion – conventional or chemical – and by means of disinformation concerning the fact that nuclear weapons cannot be 'disinvented' and that any research into a shield system will be coupled with research in systems to penetrate such a shield.[18]

With or without SDI, space seems to be one of the major challenges for the European countries.[19] At stake is the status of the European countries in the fields of science, technology and industry. However, France maintains the position that the peaceful use of space is a prerequisite for stability at the lowest possible level of armament.

In the French view the first challenge the European countries will have to face concerns technology. In reaction to the American offer to participate in SDI and to the ultimatum given to European leaders by Mr Weinberger, France has decided to accept this challenge both diplomatically and industrially. In 1985 the Foreign Ministry took over the initiative from the Defence Ministry in the development of a 'counter strategy' to the pressure being exercised by the Americans. The first step in response to SDI was the initiation of a project for a European organisation for joint research in the field of advanced technology – the EUREKA project: 'The first challenge for Europe is technological. The military challenge will come much later and maybe it is not possible to agree unilaterally on its form in advance. It is the philosophy of the EUREKA project that the strengthening of the technological capabilities in Europe will make co-ordinated reaction easier'.[20]

During the Bonn Summit in May 1985 President Mitterrand and the French delegation refused to bargain concerning the EUREKA project – it was not a European summit. They stressed the position that France is currently not interested in participation in the SDI research programme.[21] In his traditional opening speech at the annual space show in Le Bourget, President Mitterrand developed his conception of the role of Europe in space in view of the SDI programme: 'The two projects [SDI and EUREKA] are not incompatible, but competitive, especially in financial terms. The costs for one and the costs for the other must be met out of the same budget ... The question is: will the sharing be favourable to the one or to the other? At present it is impossible to answer this question'.[22]

There are three major points in the French position on SDI:[23]

1. with or without SDI, the 'European Community of Technology' is an imperative. The research programmes will provide civil and military results in the fields of observation and communication and will contribute to the strategic balance and to the verification of arms control agreements.
2. SDI is a research programme of which the consequences at present remain unknown, especially from a military point of view. For this reason the Americans have decided to maintain its deterrence capability in the future. The Trident and the MX programmes are the best examples of this.
3. Maintaining deterrence is necessary for maintaining peace and stability in Europe. This position was underlined by the ministerial release at the conclusion of the WEU meeting in Bonn in April 1985. Deterrence contributes to the strategic balance in the area of 'offensive' forces. For this reason, France is engaged in the modernisation of its nuclear forces at the so-called 'minimal deterrence' level.[24]

'European technology', 'rationality of nuclear deterrence' (the so-called 'non-warfighting strategy') and the 'modernisation of the nuclear forces' are the conceptual pillars of the French reaction to SDI. This has become even clearer with the establishment of a 'Space Agency' within the Defence Ministry in June 1985. Some Defence Ministry officials have argued that it would permit the gain of another dimension for defence and above all it would, by strengthening the deterrent, noticeably extend France's diplomatic freedom.[25]

Finally, France has been able to strengthen its position within the Atlantic Alliance. At the Spring Session of the Foreign Ministers of the Atlantic Alliance in June 1985, the French Foreign Minister Mr Roland Dumas refused to give his support for SDI. 'We will not accept in Lisbon what we refused in Bonn'.[26] The French effort to build a European community of technology was supported by the German delegation and by the German Foreign Minister, Hans-Dietrich Genscher. The diplomatic victory was not absolute, but debate about the technological challenge and about the EUREKA project will provide the political basis for a common reaction of the European countries to the pressure being exercised by the Americans.

III THE POLITICAL DEBATE

The reaction of the French political parties to the SDI programme are an indicator of the respective standpoints in view of the French nuclear deterrent.

As far as the former ruling party, the Socialists, is concerned, the debate centred on the transatlantic relations within the Alliance and on the government's security policy. The French socialist delegations to the WEU and to the North Atlantic Assembly have supported the official French view in order to strengthen the European socialist front against the American venture.

Jacques Huntzinger, the then secretary for foreign affairs of the Socialist Party, has expressed the close links with European political developments. He has stated that the American disengagement illustrated by SDI, the German interest in European co-operation and the evolution of the French nuclear deterrent forces are all indications that the horizon of French defence capabilities are becoming more and more European.[27] Moreover, the French Socialist Party invited members of the Socialist International to come to Paris and this led to a common communiqué on SDI and space technology. The meeting was prepared by the French and German delegations. Both parties expressed 'the same concern and the same doubts about SDI and share the intention to reach the same position concerning space matters'.

The general secretary of the Socialist Party, Mr Lionel Jospin, was even more specific in saying that 'behind the strategic pretext the American administration is aiming to obtain the financial resources and the concentrated efforts needed for a fast scientific and technological advance in order to react to the European endeavours, which, in some fields, are more advanced'.[28]

The Communist Party has adopted a position that may be described as being both nationalist- and Soviet-orientated. The major enemies, as seen by the Communist Party, are both the Socialist government and the Reagan Administration, and thus current domestic decisions are being made in accordance with international developments. The Communist newspaper, *L'Humanité*, for example, has denounced the creation of a space agency within the Defence Ministry as a French contribution to the arms race in space.[29] In the same vein, EUREKA has been presented as a supplement to SDI, which also has the objective of consolidating European support behind the American project. For the Communist Party EUREKA is an example of the abandoning of the doctrine of French nuclear independence.

Mr René le Guen, a member of the party's central committee, has written in *L'Humanité* that militarisation of French research may very well contribute to the indirect approval of the SDI programme: 'Multinational companies are using the spin-off from the SDI and EUREKA projects to develop their own strategies . . . In addition, France makes its own bullets, but the technological advances are being sold by misleading the Europeans in the name of Atlantic integration'.[30]

In the other former opposition parties, in the centre and on the right, different positions have gradually evolved. The Gaullist *RPR (Rassemblement pour la République)*, led by Jacques Chirac, has shown an increasingly 'revisionist' position on military and foreign policy matters. Despite the convictions of some of the more orthodox Gaullists such as the former Foreign Minister, Mr Couve de Murville, and the former Prime Minister, Mr Michel Debrè, several RPR security experts including François Fillon, Michel Noir and Jacques Baumel, deputies of the National Assembly, have assessed SDI rather more positively. In his opening remarks to a conference on space matters, Mr Baumel, who is also the chairman of *Fondation du Futur*, claimed that only a European solution based on the co-operation between France and the German Federal Republic could be the response to the American challenge introduced by SDI. 'In the first stage, Europe can and must build an observation satellite for information transmission . . . in outer space, the real battlefield of the future. And afterwards, why should we not devise a manned space station?'[31]

Jacques Chirac has however said: 'Europe must join the United States in the American SDI project'. The Gaullist leader has charged that President Mitterand's position at the Bonn summit was unnecessarily aggressive. According to Chirac, 'the solidarity of the Western nations in the face of the Soviet Union should be reinforced'.[32] For the same reasons the Gaullists have received the French EUREKA project with scepticism. Because of the lack of sufficient funding and the mixed civilian and military nature of the project it is in their view an inadequate response to the SDI.[33]

The other former opposition party, the liberal *UDF* (*Union pour la Démocratie Française*) also holds the opinion that France should exercise more solidarity with the Atlantic Alliance. For a long time they have called for a deeper integration into the Alliance and closer military co-operation with the United States. EUREKA is considered as only a supplement to possible participation in the American SDI research programme.

The 'consensus' as shown by this general survey of the French official and unofficial reaction is, however, becoming more and more awkward in view of the technological progress and the intensive technological competition within the Alliance.

IV THE DEBATE BETWEEN THE GOVERNMENT AND OTHER OPINION LEADERS – THE INDUSTRIAL CHALLENGE

In addition to the political debate on EUREKA and SDI there has also been an industrial debate among the French companies with a stake in advanced technology. Matra and Aerospatiale, for example – the two main French companies in the fields of electronics and aeronautics – have presented different points of view on the matter. Mr Lagardère, the director of Matra, has declared that defence policies lead to an enormous concentration benefitting technological innovation and that the SDI research programme could enhance this innovation. It would be impossible to say in advance that Matra would not participate in the programme. He said he was interested in both the SDI and the EUREKA project, and it would be inappropriate to consider EUREKA a competitor to SDI. In his view, the projects should be complementary in order to maximise the efficiency of efforts on both continents.[34]

Aerospatiale, the main producer of French nuclear defence technology, has taken the opposite position, clearly favouring the EUREKA project. This may be described as the more 'nationalist' position. A statement by the director of Aerospatiale, Mr Matre, serves as a good example of this position: 'There is no similarity between EUREKA and the American project. The objectives are radically different. SDI is an American programme which concerns American defence. Europe cannot stay behind in the development of its own technologies and, without declaring war on the United States, we have to show that Europe will control the basic technologies necessary to maintain its position in the world'.[34]

These two extremes are good examples of the schools of thought comprising the current phase of development planning of future military research within the French government.

On the one hand, French officials have emphasised the strategic risks inherent in SDI: French political decisions must take into account that SDI will continue to have a destabilising effect on a deterrent system based on mutual vulnerability. Consequently, all policy decisions must be in favour of strengthening France's offensive nuclear capacity in order to maintain credibility. On the other hand, civilian spin-off of the SDI investments is expected to be significant and French companies are interested in participating in the American research programme in order not to lose touch technologically. The current situation is similar to that of the 1960s: immense funds are being principally invested in basic research such as physics, computer science, electronics and hardware systems. The technologies being developed for conventional defence and space weapons will also have important consequences for civilian production technology.

V CONCLUSION

As has been shown, the American SDI plans have set off strong and varying reactions in France. This chapter has focused on the strategic and technological aspects and it must be recognised that these are becoming more and more intertwined. Neither the French government nor its leading officials have ever rejected the fact that research in the new technologies of outer space has become a top priority. However, France has been obliged to restate its concept of deterrence and its assessment of strategic balance.

France will have to face two major issues in the coming years:

1. would a major BMD initiative provide a widespread Soviet strategic defence system which in turn would weaken the credibility of the French nuclear missile force?
2. would the SDI programme alter the framework of the French contribution to the strategic balance in Europe and of its contribution to solidarity with the Allies in terms of conventional warfare in defence of the so-called vital interests?

At stake is the status of the European countries in the fields of science, technology and industry. France maintains that the peaceful use of space is a prerequisite for stability at the lowest possible level of armament.[35]

Notes

1. Charles Hernu, Minister of Defence, *Rapport au Parlement sur l'exécution et la réévalution de la loi no. 83–606 du 8 juillet portant approbation de la programmation militaire 1984–1985* (Paris: Ministry of Defence, June 1985).
2. NATO, *NATO Handbook* (Brussels: North Atlantic Treaty Organisation Information Service, 1985).
3. The initial reaction from the Ministry of Foreign Affairs: 'Audition de M. Claude Cheysson devant la Commission des Affaires Etrangères de l'Assemblée Nationale, 5 juillet 1984' *Negociations stratégiques et militarisation de l'espace* (Paris: Document du Quai d'Orsay, 4 July 1984).
4. For the text of the speech by the President of the Republic in The Hague on 7 February 1984, see 'Pas d'Europe libre sans Européens dans l'espace', *Revue de Défense Nationale*, August–September (1984) pp. 156ff.
5. 'French Statement to the Conference on Disarmament', *Survival* 26 (1984) pp. 235–7.
6. M. Pichard, 'Dossier: "Espace"', *Journées nationales Sciences et Défense*, Ecole Polytechnique (Paris: Ministry of Defence, April 1983).
7. *Livre Blanc de la Défense Nationale* (Paris: Ministry of Defence, June 1972) p. 18.
8. See note 3.
9. 'Speech of M. Laurent Fabius, then Prime Minister of France at the Institute of Higher National Defence Studies (excerpts) *Survival*, 26 (1984) pp. 280–2.
10. 'Extract from the speech by President Mitterrand in the Second Chamber of the States-General, The Hague, 7 February 1984', WEU, Doc. 976, 15 May 1984; *The military use of space, Report submitted on behalf of the Committee on Scientific, Technological and Aerospace Questions by Mr Wilkinson, Rapporteur* (Paris: Western European Union, 1984).
11. 'Le futur de L'Europe spatiale', *Air et Cosmos*, 1000, 5 May 1984.

12. Jeanou Lacaze, 'La menace militaire a l'an 2000', *Revue des Sciences Morales et Politiques* (Winter 1985).
13. Statement by Mr Hubert Curien, Minister of Research and Technology, *Hearings before the Committee of Defence and Armed Forces*, (Paris: National Assembly, 25 April 1985).
14. Ministry of Defence, *Loi de programmation militaire 1984–1988, exposé des motifs* (Paris: Ministry of Defence, 1984).
15. Interview with Mr Hubert Vedrine, Adviser on Foreign Affairs in the Cabinet of the President, Radio France Internationale, 30 January 1985.
16. Vedrine, 30 January 1985.
17. 'Allocution de M. Charles Hernu, le 9 Fevrier 1985 devant la "Wehrkunde"', translated from the German by the editor.
18. Ministry of Foreign Affairs, *Document du Quai d'Orsay, SDI – position française* (Paris: Ministry of Foreign Affairs, 8 May 1985).
19. See chapter 9 in this volume.
20. Roland Dumas, Address at the Meeting of Foreign Ministers of the WEU in Bonn, 23 April 1985.
21. François Mitterrand, Press Conference after the summit of the highly industrialised countries in Bonn, May 1985.
22. François Mitterrand, Press Conference for the inauguration of the 36th Aeronautics and Space Show, Paris Le Bourget, 31 May 1985.
23. See answer of Charles Hernu, Minister of Defence, *Journal Officiel* (Paris: National Assembly, 25 June 1984) for the session of 24 June 1984 under the heading 'Initiative de Défense stratégique'.
24. See chapter 11 in this volume.
25. Critias (a pseudonym for a group of high-level civil servants), 'Stratégie de l'espace, espace d'une strategie', *Le Monde*, 7 June 1985.
26. See the statement by Roland Dumas, the then French Minister of Foreign Affairs at the Meeting of the North Atlantic Council in Lisbon, 7 June 1985.
27. Jacques Huntzinger at a press conference in Paris, 4 June 1985.
28. Lionel Jospin at a press conference with Willy Brandt in Paris, 21 May 1985.
29. Yves Moreau, 'Projets insensés', *l'Humanité*, 5 June 1985.
30. René le Guen, 'Eurêka ou comment ne pas s'y perdre', *l'Humanité*, 28 July 1985.
31. 'La nouvelle stratégie de l'espace', Colloque a l'Assemblée Nationale, *Les cahiers de la Fondation du Futur* (Paris: Fondation du Futur, September 1984) p. 15.
32. Press conference with Jacques Chirac for American journalists in Paris, 23 July 1985.
33. See Michel Noir and F. Fillon, 'Il ne suffit pas de dire: Eurêka!', *Le Monde*, 1 August 1985.
34. 'Des industriels partagés entre Eurêka et IDS', *Le Monde*, 9–10 June 1985.
35. For the French debate until Spring 1987, see Alain Carton, 'French Position on the Strategic Defense and on Defensive Systems', Paper to be presented to the 28th Annual ISA Convention in Washington D.C., 15–18 April, 1987.

6 The Political Debate in the Federal Republic of Germany
Hans Günter Brauch

I FROM INF TO SDI – THE SECOND DIVISIVE SECURITY DEBATE OF THE 1980s

No defence issues since the adoption of NATO's strategy of flexible response and of the political goals of the 1967 'Harmel Report' have been as divisive politically within governments, coalitions, parties, churches and trade unions in Western Europe. While the 1979 INF decision caused the formation of a new independent peace movement and posed a basic challenge to NATO's nuclear posture and its first use option as well as to the search for alternative political and military security and arms control policies,[1] President Reagan's strategic vision of overcoming deterrence based on offensive nuclear weapons neither halted the disintegration of public nuclear policy acceptance and the erosion of the common foreign and security policy consensus within the Alliance nor persuaded the West European governments (who had implemented the INF decision at a high political price) and the European anti-nuclear movement to jump on the SDI bandwagon. Instead President Reagan's 'Star Wars' vision and the subsequent SDI programme became even more divisive within coalition governments – in West Germany in particular – between France and the Federal Republic of Germany, within the Western European Union, and within the European Communities and the North Atlantic Alliance.[2] SDI has become an effective political tool in neutralising the intra-European debate on a Europeanisation of European affairs and on a stronger and more independent European pillar within the Alliance, in pushing the Soviet Union on to the economic defensive and in challenging the West European Allies in the area of technology. America's position of leadership which during the era of détente had been challenged by friends and foes alike, has thus been reinforced.[3]

In the first half of his Administration, President Carter failed in his attempts to create a new foreign policy consensus within the United States based on Wilsonian concepts. In the second half of his Administration, his attempts to push through the INF decision were not successful in regaining the leadership position within NATO.[4] President Reagan, however, seems to have been more successful in both respects – in recreating both a conservative cold war consensus at home and a dominant political position abroad *vis-à-vis* both the European Allies and the Soviet Union. However, this short-term gain may become a longer-term liability. Nevertheless, the approach pursued by Carter in the second half of his Administration was similar to that subsequently pursued by his successor: both searched for the solutions to political problems in the areas of technology and military hardware. In the view of one American analyst, the implementation of the INF decision was 'becoming a bellweather in the United States as to European willingness to continue to support the NATO alliance'.[5] Political support for President Reagan's SDI has become another prerequisite for Alliance cohesion, and independent conceptual thinking and opposition to SDI has been denounced as wedge-driving[6] which could undercut the political support of the American people for the defence of Europe.

In the early 1980s the INF decision was legitimised by United States officials as an indispensable tool in regaining 'escalation dominance'.[7] By spring 1985 'escalation dominance' had, in the words of Fred Iklé, 'been overtaken by massive changes in the nuclear arsenals'.[8] The INF decision has been implemented since 1983 despite massive opposition within the European public. In early 1985 Iklé made the following comment: 'In the long run, reliance on a balance of mutual vulnerability would favour totalitarian regimes, with a demoralising effect on democracies'.[9] Western European public support was more or less irrelevant for the Reagan Administration during the INF debate, but American public support has suddenly become a major ingredient of the SDI debate.[10] The 'class differences' within NATO could hardly have been demonstrated more clearly.

Both the INF and the SDI debate have so far demonstrated that the alliance consensus of defence readiness and détente adopted in 1967 has disappeared. In the early 1980s the basic components of the Harmel Report – defence and détente – were questioned from two opposing angles: Parts of the European and the American peace and freeze movements attacked the nuclear element of NATO's military strategy and force posture, and détente was already declared dead by the Carter Administration after the Soviet intervention in Afghanistan.[11]

However, changes in public opinion and the different threat perceptions – a major factor behind the structural crisis within NATO – cannot be overcome by technical solutions, be they INF or SDI. Any defence policy must be based on three scarce resources: *public support* both for NATO's strategy and for the defence burden;[12] *manpower*, for example, the number of soldiers, the quality of their training and their motivation to defend the values and the democratic institutions;[13] and the *available financial resources* to equip the armed forces with the means to perform their defensive missions.[14]

In the Federal Republic of Germany the INF debate caused the erosion of the broad consensus in security affairs which had emerged in 1961 after the Social Democratic Party accepted West German integration into NATO and NATO strategy.[15] On 6 March 1983, the Kohl government won the federal elections after a campaign in which both the Christian Democrats and the Liberals supported both the INF decision and NATO's nuclear strategy of mutual vulnerability.[16] President Reagan's strategic heresy of 23 March 1983 was – not surprisingly – harshly criticised by many INF supporters and NATO-minded journalists. Christoph Bertram, the former director of the IISS argued in the 1 April 1983 edition of the liberal weekly *Die Zeit* that the United States president had damaged 'the sensitive linkage of security by deterrence and negotiations. He has undermined the precarious domestic consensus on the NATO double decision in the deployment countries while he should have strengthened it by new signs of compromise in Geneva'.[17]

In Spring 1983 both the Kohl government and the parliamentary opposition interpreted President Reagan's speech as a tactical move to undercut the American freeze movement and the antinuclear criticisms of the American Catholic bishops.[18] The INF debate was in full swing and absorbed all the energies of the government, the opposition parties and the peace movement alike. When the West German officials were finally officially informed by a United States briefing team in February 1984, most experts were either confused or irritated.[19]

While the peace movement in early 1984 was concentrating its energies on the major challenge of the new American deep-strike concepts, AirLand Battle, AirLand Battle 2000 and the new NATO long-term defence guidance on FOFA,[20] it was the West German Defence Minister, Manfred Wörner, who openly criticised the United

States position following a briefing on the SDI at the NPG meeting in Cesme in early April 1984: 'I can't see that it would provide greater protection or stability'.[21] Walther Stützle, the former head of the Defence Planning Staff, wrote on 11 April 1984 about the rare event 'that all five parties represented in the Bundestag unanimously oppose Washington's plans'.[22] 'What brings these otherwise divergent politicians together,' Elizabeth Pond writes in the *Christian Science Monitor* (referring to Karsten Voigt, Manfred Wörner and Franz Josef Strauss), 'is the conviction that the Reagan administration's plans for space would harm the West more than the Soviet Union – and fatally decouple West European from American defence'.[23]

After Manfred Wörner changed his position in July 1984 and came out in support for the SDI research programme,[24] SDI replaced INF as the most controversial security issue in 1985. SDI has increasingly polarised the domestic security debate within the present coalition government and especially between the Christian Social Union (CSU) of Franz Josef Strauss and the Liberal Party (FDP) of Hans-Dietrich Genscher. Major casualties of the shift in the West German position towards a conditional support for the SDI research programme have been the Franco-German effort to revive and space co-operation and the joint Franco–German effort to revive and strengthen the WEU after the Rome meeting in October 1984.[25]

The SDI controversy within the Kohl coalition government, between Genscher on the one hand and Strauss and Wörner on the other, has reflected two different tendencies in West German foreign policy since the 1950s – between the Atlanticists and the Europeanists.[26] France tried to influence this intra-coalition squabble with European counteroffers to American proposals: less than three weeks after United States Defence Secretary Weinberger's invitation of 27 March 1985 to join the SDI research programme, France tabled a civilian alternative – EUREKA.[27] When in early July 1985 the parliamentary leader of the ruling Christian Democrats, Alfred Dregger, suggested a 'second SDI for Europe together with the Americans',[28] the then French Defence Minister Hernu responded with the idea of a joint European missile defence effort.[29] In the second part of 1985 the public debate stimulated by the challenge posed by the SDI research programme focused on the linkage between SDI, EUREKA and EDI.

The emerging SDI debate has been taking place on five levels – the political, the military–strategic, the arms control, the economic and technological as well as the ideological. SDI has not contributed to a new domestic consensus on security and foreign policy but has caused disunity on the intra-European level and has been successful in polarising the German security debate.

In this chapter the following questions will be addressed:

1. How has the position of the Kohl government on SDI developed between March 1983 and the signing of the SDI agreement in April 1986?
2. What have been the reactions of the five parties in the Bundestag and what have been the major parliamentary SDI-related activites?
3. How have the business world, the trade unions and the scientific community responded to the American challenge posed by SDI?
4. How have the media and the public reacted to President Reagan's 'Star Wars' vision and to possible West German participation in the SDI programme?
5. What effects has the SDI debate already had in the Federal Republic and what consequences may be expected if SDI should actually emerge from the military drawing boards in the 1990s?

II FROM BENIGN NEGLECT TO CONDITIONAL SUPPORT OF SDI – THE EMERGING POSITION OF THE FEDERAL GOVERNMENT ON SDI, EUREKA AND EDI

The political controversy within the conservative/liberal coalition government of Chancellor Helmut Kohl on the terms of German participation in the SDI research programme evolved during 1985 from two divergent political tendencies – the Atlanticist *v.* the European perspective – and from two political poles – the conservative CSU and the liberal FDP.

The former chairman of the liberal FDP, Hans-Dietrich Genscher, Vice Chancellor and Foreign Minister in both the Schmidt and Kohl governments, had engineered the change of government from left-liberal to right-liberal after withdrawing the liberal ministers from the Schmidt government in September 1982. Genscher considered himself as one of the major guardians of the Harmel consensus of 1967: maintaining a credible *defence* and pursuing an active *Ostpolitik*. For Genscher, arguing primarily on political and arms control grounds, the American SDI appeared to undercut both a policy of closer co-operation among the West European countries in foreign and security policy both in the context of the *European Political Co-operation* (e.g., the Genscher–Colombo Initiative that had led to the Declaration of European Unity adopted at Stuttgart in June 1983) and of the *WEU* (e.g., the Rome Declaration of the WEU Foreign and Defence Ministers of October 1984) and the policy of détente between the governments of Western and Eastern Europe.[30]

Genscher's primary opponent was the Christian Social Union (CSU) and its chairman, Franz Josef Strauss, the Minister President of Bavaria, who had been defeated in 1980 as the candidate for Chancellor of the CDU/CSU in the federal election. Strauss, a long-time Gaullist or Europeanist, said in April 1984 that 'the US space weapons programme is evidence that the US-European partnership in nuclear security "is no longer functioning" '.[31] However, on 30 March 1985 Strauss gave his full support for German participation in the SDI research programme. Strauss argued primarily in technological, economic and military–strategic terms in support of SDI and simultaneously for EDI against Soviet SRBMs.[32] The ideological argument of replacing the present nuclear doctrine of mutual assured destruction (MAD) by a doctrine of mutual assured survival (MAS) was primarily stressed by the Secretary General of the Christian Democratic Union, Mr Geissler, who condemned the critics of German participation in the SDI as being 'immoral'.[33] Finally, tactical considerations and coalition politicking influenced the position of Chancellor Kohl on SDI. Eight stages in the political reaction of the federal government may be distinguished:

1. From March 1983 to March 1984 the official reaction was dominated by benign neglect, and an official silence on both President Reagan's vision and the SDI research programme.[34]
2. Between April and July 1984 Defence Minister Wörner shifted his position from public criticism of the negative impact of SDI for stability and arms control at the NPG meeting in Cesme[35] to open endorsement three months later in Washington if the United States' nuclear umbrella continued to extend over Western Europe.[36]
3. From June to November 1984 both leading spokesmen of the Christian Democrats and the Liberals called for arms control solutions to prevent a militarisation of space, and

they urged the Reagan Administration to resume arms control negotiations with the Soviet Union including space- and SDI-related issues.[37]

4. After President Reagan's re-election the West German government, although sharing many of the concerns which had been expressed by security experts,[38] moved to a double strategy: political endorsement of the SDI research programme combined with a set of political conditions and the pursuit of a common West European technological offensive as a direct reaction to the technological challenge posed by the United States-controlled SDI research programme.[39]

5. After Chancellor Kohl's Wehrkunde speech on 9 February 1985 the public debate focused on the formulation of a set of conditions – in close consultation with several other West European governments – which would determine the general West German position on the SDI programme and on German participation in the research phase[40] (from February 1985–December 1985).

6. As a reaction to the technological challenge posed by the SDI research programme, Genscher supported the French initiative for a civilian alternative EUREKA,[41] and called for a European Community of Technology.[42]

7. Since June 1985 a third object of public debate has emerged in addition to German participation in both the American SDI and the European EUREKA research programmes: the proposal by both United States government officials and European conservative and liberal politicians for a EDI.[43]

8. Some three years after President Reagan initiated the transatlantic debate on strategic defence and two years after the SDIO was established, a memorandum of understanding was signed in April 1986 which will govern the terms of the German participation in the research programme.[44]

Having described the early stages of the West German official reaction to the SDI programme elsewhere[45] this chapter will focus only on the latter four stages from February 1985 to April 1986.

(a) Formulating the conditions for political support of the SDI research programme

After President Reagan's re-election in November 1984 the West German Chancellor, Helmut Kohl, gradually moved to a policy of conditional support for the SDI research programme, as can be seen from seven major statements and position papers of his government during 1985:

1. in a speech to the Wehrkunde Conference in Munich on 9 February 1985;[46]
2. in a statement of the Federal Security Council on 27 March 1985;[47]
3. in a declaration of governmental policy to the Bundestag by Chancellor Kohl on 18 April 1985;[48]
4. in an address to the North Atlantic Assembly in Stuttgart on 20 May 1985;[49]
5. in the Defence White Paper of June 1985;[50]
6. in a report by the Federal Minister of Defence on the SDI from 20 August 1985;[51]
7. in the decision of the federal government of 18 December 1985 to initiate specific negotiations with the United States government on the terms of German participation in the SDI research.[52]

These seven statements and papers provide the basis of the ambivalent official West German attitude towards the SDI. In his Munich speech, Chancellor Kohl although supporting the SDI research programme in general terms nevertheless avoided a final assessment of its potential implications for European security. Instead he offered the following criteria for an evaluation both of the military and of the arms control aspects of President Reagan's SDI:

1. The current defence strategy must continue to be valid for the Alliance as long as more effective alternatives for the prevention of war do not exist.
2. The political and strategic unity of the Alliance must be preserved. There must be no zones of differing security in the Alliance and no decoupling of European security from that of North America.
3. The development of strategic defensive systems must take place on a co-operative basis, that is on the basis of arrangements between the superpowers. The objective must be to strengthen strategic stability.
4. Strategic instabilities must be avoided, particularly in a possible transition phase. A clear renunciation of superiority is necessary.

5. German and European interests must be taken into account on the basis of recurrent and intensive consultations.
Since the SDI programme will result in spin-offs for civilian life and industry, the government must observe and assess work on the programme from this standpoint.
6. If, as announced, the United States carries out the SDI research programme in the planned manner, and if the European and German interests mentioned are taken into account, the following criteria will also have to be considered:
    (a) The Soviet Union is also carrying out research in the area of strategic defensive weapons systems. Presumably the Soviets are farther along in their programme than is generally assumed.
    (b) The American space programme is a strong incentive for Soviet willingness to engage in comprehensive negotiations.[53]

In his Munich speech Chancellor Kohl referred to the following elements of strategic stability:

1. maintenance of a high and unbearable risk against a first strike;
2. search for improved war avoidance capabilities including the defensive elements of deterrence strategy;
3. definition of a new balance between offensive and defensive weapons on the basis of the available new technologies;
4. renunciation of superiority and of destabilising effects on the superpower relationship;
5. effective crisis management in order to avoid the outbreak of war as a consequence of technical or human failure.[54]

On 27 March 1985 the West German government, in a statement on SDI, repeated the conditions for its support of the American SDI programme by stressing the need for a more stable relationship between the United States of America and the USSR in the context of NATO's Harmel report of 1967. An arms race in outer space was to be prevented and the ABM Treaty should be strengthened. In accordance with the other members of the Alliance, the German government stressed that NATO's strategy of flexible response was to be maintained as well as the political and the strategic unity of risk within the Alliance. It stressed the need for common European position in the consultation with the United States on the SDI.[55]

On 18 April 1985, during a parliamentary debate on the SDI, Kohl repeated his previous conditions for a German support for the SDI research programme and stressed simultaneously the need for a closer European co-operation in the high-technology area as has been suggested by the French Foreign Minister in his proposal for a European institution to co-ordinate research in the high-technology field.[56]

One common feature in all official statements of the Federal government – as well as in the major statements of the coalition parties (CDU/CSU and FDP) – was the continued support for the ABM Treaty. In a joint resolution on 28 February 1985 the coalition parties emphasised that the 'ABM Treaty should be fully maintained'.[57] On 27 March the federal government stressed 'that the ABM Treaty should be strengthened as long as no other common agreement has been reached'. And on 18 April Kohl declared 'in the short and medium term the observance of the ABM Treaty should have priority'. Any decision beyond research should be the result of a co-operative solution reflecting the interest in strategic stability and a drastic reduction of the offensive potential.

Between March and July 1985 different nuances of the conditioned support have been emphasised by Kohl, Genscher and Wörner in an effort not to offend either the Americans or the French. The technological appetite as a major motivating force in support of the participation in the SDI research programme has gradually been replaced by the fear of a 'brain drain' and 'technological exploitation'. EUREKA provided an alternative for those interested in funds for high technology projects. On 20 March 1985 Chancellor Kohl stressed that bilateral arms control talks could make the 'deployment of space-based systems increasingly superfluous'. At the same time he advocated 'that the Europeans develop a joint position and that they bring this to bear with our American allies'.[58]

In early May 1985, after the Bonn economic summit, Chancellor Kohl had manoeuvered himself into a difficult foreign policy position 'having damaged his standing with the White House and alienated his relations with Paris'.[59] With his strong verbal support for research into SDI Kohl temporarily distanced himself from Mitterrand's call for a common European technological offensive. In his Stuttgart speech on 20 May Kohl toned down his earlier endorsement for Reagan's SDI and moved closer to the French project EUREKA. Chancellor Kohl portrayed the SDI of President Reagan both as 'an opportunity and as a risk' for the North Atlantic alliance. At the same time, he stressed that 'Europe will have to deal with the

American technological and political–strategic challenge', emphasising the need for a closer West European co-operation on security affairs.⁶⁰ His comments reflected disparate comments within his own government. Foreign Minister Genscher, in close co-operation with his French counterpart, Roland Dumas, had quietly rallied support behind EUREKA among his European colleagues, while stressing publicly that space defence plans were 'incompatible' with détente and blocking all hope for progress at the Geneva arms talk. Genscher's cautious approach was fully supported by General Wolfgang Altenburg, the highest ranking military officer of the Bundeswehr, who had stated at Stuttgart that SDI raises more questions than answers and that its enormous cost could divert United States resources from conventional defence, thus weakening the Alliance's overall deterrence.⁶¹

In late May Kohl's chief foreign policy advisor Horst Teltschick was quoted in *TIME* to the effect that 'the Chancellor will only endorse the programme when we are sure there will be a give and take in the exchange of technology on both sides ... If there is no chance for a two-way street, then I think the Chancellor isn't interested. Of course, individual German companies could still be involved, but without participation by the government ... What does not interest West Germany is selling components to a project without knowing what is going on in the project as a whole and without bringing technical knowhow back to Germany'. Teltschik stressed the need for European consensus on SDI because SDI would affect all West Europeans: 'it is decisive for our security, NATO strategy, East–West relations and technological and economic development. This agreement with our European partners is in some respects essential'.⁶²

At the Spring meeting of NATO's foreign ministers in Estoril, Portugal, German Foreign Minister Genscher insisted in talks with George Shultz that the meeting should emphasise that NATO's present strategy of flexible response remains valid, thereby rejecting the strategic considerations of United States Deputy Secretary of Defence Fred Iklé,⁶³ while his French colleague successfully objected to any official endorsement of SDI in the NATO communiqué.⁶⁴

The German Defence White Paper of June 1985 described the SDI research programme as justified, politically necessary and as being in the security interest of the West as a whole. According to this White Paper both NATO's strategy of flexible response would remain valid and the aim of the federal government would remain unchanged to create peace with fewer weapons and to establish a higher degree of stability between West and East. However, after repeating the criteria of the first three statements, the White Paper emphasised: 'In the short and medium term the observance of the ABM Treaty would take precedence'. It would be indispensable, in the view of the federal government, that any decision beyond research would have to be taken in a co-operative framework which would guarantee an improvement in strategic stability, a drastic reduction in offensive systems and a negotiated balance of offensive and defensive systems.[65]

In a report by Secretary of Defence Wörner on the SDI which was supplied to the Bundestag Defence Committee in preparation of a public hearing on SDI, it was stated that in two of the five major SDI research programmes – in optics and in the field of subsystems of space technology – German industry has a competitive edge. As to the options for Western Europe and for the Federal Republic of Germany, the report admitted that a space-based BMD system would provide no protection against Soviet SRBMs, cruise missiles and aircraft. 'Given the limitations of the American SDI approach and the specific threats of NATO Europe, a supplementary research programme will be in the European interest'.[66] Of the three available options the federal government faces 1) no participation, 2) a participation in the SDI research programme, and 3) initiation of a European research programme; the report opted for the second alternative on the basis of a bilateral treaty or memorandum of understanding:

> A European SDI research programme with the participation of the Federal Republic of Germany is only useful in co-operation with the United States. For reasons of efficiency and cost effectiveness, a European defence system would only be suitable as part of a comprehensive defensive system. The United States are considering an extension of their research programme in this respect.
> For security, military-strategic, technological and economic reasons German participation in the SDI research programme appears to be imperative. In this respect, a combination of the options 'participation in the American SDI research programme' and 'initiation of a European research programme' in close co-operation with the United States appears to provide the most promising solution.[67]

On 18 December 1985 the cabinet rejected Wörner's preferred option. While reiterating its political support for Reagan's SDI research programme, Chancellor Kohl's cabinet also declared that it would not pursue any direct government role or provide any public funding for the research effort. *Washington Post* correspondent William Drozdiak commented on this decision:

> By stressing the business aspects and muting the security repercussions of SDI, Bonn clearly hoped to stifle a protracted feud between Mr Kohl's Christian Democrats and their junior partner, the Free Democrats, over the wisdom of embracing the controversial project....
>
> Mr Kohl and other Christian Democrats have advocated a staunch political endorsement of the programme to demonstrate allied support for Mr Reagan and to give him a stronger hand in the Geneva arms talks. But Foreign Minister Hans-Dietrich Genscher and other members of the Free Democratic Party have expressed fears that a close role in SDI could damage Bonn's relations with Eastern Europe.
>
> Mr Genscher is known to be wary of tempering with the North Atlantic Treaty Organization policy of nuclear deterrence.[68]

The cabinet voted unanimously to send Economics Minister Martin Bangemann to Washington in January 1986 to seek conditions for the exchange of scientific research and technology between the two countries. The *International Herald Tribune* correctly commented on 'Bonn's Ambiguity on SDI':

> The declaration could hardly be more laboured and anticlimactic. Although it contained a passing reference to a more positive declaration made in April, it bears little resemblance to the ringing pledges of political support for Mr Reagan's initiative that Chancellor Helmut Kohl has made in the past. ... It is difficult not to suspect that Mr Bangemann's mission is a face-saving device intended to permit Mr Kohl to continue in his role of Mr Reagan's trusted friend while at the same time seeking to reduce the danger of provoking the Russians and prevent the dispute within his coalition from getting out of hand. Horst Teltschick, the Chancellor's foreign policy advisor, had argued that it was the German companies that are interested in SDI that had requested the government to become their emissary and protector on such issues as pricing, technology transfers and property rights on which they may disagree with the American authorities.[69]

This ambiguous cabinet decision had been the lowest common denominator within the coalition government after nine months of open dispute between the coalition parties in general and between Strauss and Genscher in particular. It contained no reference to Wörner's proposal that an EDI against Soviet SRBMs and cruise missiles should be created. The decision was based on the recommendations of two high-level delegations of German government officials and industrialists, both headed by Horst Teltschik, which visited the United States in June and again in September 1985. According to press accounts, the first Teltschik delegation visit concluded that only a very narrow framework existed for German and European participation in SDI research, given the limited American willingness to permit their allies a voice in the programme.[70] After the second Teltschik delegation visit in September 1985 the expectations of United States officials that an agreement or an exchange of letters of understanding were pending were premature.[71]

What have been the essential political, economic and technological as well as military and strategic issues which have dominated the dispute within the Kohl coalition government, and what major changes in the assessment of SDI by Defence Minister Wörner and his close advisors, Hans Rühle and Lothar Rühl, could be observed, from the early rejection in 1983 and 1984 to the endorsement in 1985–6?

(b) The political and arms control aspects of the German debate

The West German perception of the SDI and the intensive political debate touched upon all major elements of West German foreign policy: (a) on East–West relations; (b) on arms control in general and on the ABM Treaty in particular; (c) on the relationship with the United States and (d) on that with its European partners, especially (e) on the French–German relations and (f) on the political balance within the conservative–liberal coalition government. SDI polarised the two existing tendencies of West German foreign policy – the Atlanticist v. the Europeanist orientation. Even though Chancellor Kohl claimed 'The oft-stated dilemma of choosing between Paris and Washington does not affect us',[72] the conflicting offers from Washington and Paris nevertheless directly fuelled the intra-coalition quarrelling between Genscher and Strauss. While leading FDP spokesmen called for the resignation of Kohl's national security advisor Horst Teltschik for his public endorsement of a German participation in the

SDI,[73] Strauss and CSU members of the Bundestag publicly attacked Genscher for his scepticism about the SDI and opposed any extension of Genscher's term after the next election in 1987.[74]

Foreign Minister Genscher (FDP)[75] and his two State Ministers, Alois Mertes (CDU)[76] and Jürgen Möllemann (FDP)[77] endorsed Sir Geoffrey Howe's political and strategic misgivings about the SDI. On 18 March 1985 Genscher expressed his severe concerns in an article, 'a new chapter in the West–East relations', saying that the Europeans should not limit their role in the superpower dialogue to one of bystanders. Genscher explicitly called for a 'strengthening of the European pillar in the Atlantic Alliance by reviving the Western European Union.'[78] Genscher, who had appeared a determined Atlanticist during the INF controversy had gradually shifted towards a Europeanist perspective. Within the cabinet Genscher became the most outspoken promoter of the French EUREKA and the European technology proposals as the joint response to the technological challenge posed by the SDI. It was the Foreign Office which had contributed the many conditions appended to the general German support for President Reagan's 'long-term research effort'. Instead of challenging the United States Administration openly, West German officials favoured an outcome of the resumed superpower dialogue that would 'avoid an arms race in space'.[79] Chancellor Kohl, addressing the CDU party convention in Essen in March 1985, suggested that would 'avoid an arms race in space'.[79] Chancellor Kohl, addressing the CDU party convention in Essen in March 1985, suggested Konstantin Chernenko the German Chancellor expressed the hope: 'The more successful one can be in Geneva in reducing offensive nuclear weapons the more superfluous it could be to deploy space-based weapons'.[80] In mid-October 1985 Chancellor Kohl quoted American assurances about a possible 'renunciation of the deployment of such a system if it were an obvious inducement for the other side to increase drastically its offensive weapons'.[81] A few days before the New York summit of October 1985 Genscher warned the United States against believing it could influence the Soviets at the arms control talks by the arms buildup and expressed the fear that the SDI might become a schizomycete within the alliance.[82]

One assumption shared by the various West German officials about the SDI research programme was that SDI research would take place only within the framework allowed by the ABM Treaty, and that no decision would be made about the development or deployment of new defence systems without consulting the Allies and negotiating with the Soviet Union in order to avoid an expansion of the Soviet offensive potential.[83] During 1985 the federal government stressed in all relevant statements its support for the ABM Treaty. At the NATO meeting in Lisbon in June 1985, and in Brussels in October, Mr Genscher, in close co-operation with Mr Roland Dumas, his French counterpart, strongly defended the 'firewall' of the ABM treaty which separates the research from the development phase in order to counter American efforts to reinterpret the ABM Treaty.[84]

For *Washington Post* correspondent William Drozdiak, the dilemma SDI had posed for Chancellor Kohl at the economic summit in Bonn was reminiscent of the debate in the past 'when Bonn was torn between choosing Gaullist or Atlanticist options in foreign policy. Conscious of their basic security interests, successive West German governments have bent to Washington's line in such disputes'.[85] In an effort not to alienate either the United States or France, the federal government backed both SDI and EUREKA, with Mr Kohl favouring an Atlanticist orientation and Mr Genscher a Europeanist perspective. These different nuances in an attempt to accommodate both perspectives are the reason for the ambiguity in the official West German position on the SDI, apparent since Spring 1985. A major casualty of the official West German position on SDI have been the efforts to strengthen the European pillar of NATO and to develop better German–French relations in the field of security.

In March 1985 Genscher warned NATO against abandoning the strategy of flexible response and against splitting the strategic unity of the alliance, while Kohl advocated 'that the Europeans develop a joint position and that they bring this to bear with [the] American allies. President von Weizsäcker was quoted as having reservations about European and West German participation in the SDI.[86] In early May Chancellor Kohl appeared to have shifted completely, supporting President Reagan's favoured defence programme, but his reference of 20 May that SDI constitutes 'both an opportunity and a gamble' – was interpreted as an effort to shift away from his endorsement of the SDI during the Bonn summit. In an effort to please both United States and French expectations on the SDI issue, the Kohl government appeared in both Allied capitals as speaking with two voices, thereby damaging its credibility and limiting its influence in both Washington and Paris.

With the German option in favour of a participation in the American space station and with the rejection of the French shuttle programme HERMES the French expectations for a longer-term and closer co-operation with West Germany in security and technological matters experienced a major setback. Mr Genscher countered the increasing French criticism about the 'contradictions' and the diffuse nature of the West German non-policy on the SDI by becoming a major backer of EUREKA, while the director of the Foreign Policy Planning staff Konrad Seitz launched the idea of an EDI during a talk in London in April 1985 as part of the European technological response to the SDI programme.[87]

While some American journalists speculated in October 1985 that Bonn was bidding for time with regard to its participation in SDI while 'French and German experts have been working quietly and intensely to put several key space technology research projects on track',[88] French–German relations had seriously deteriorated by December 1985 in the view of all major French papers.[89] One month after the decision of the West German government to enter into official negotiations with the United States on participation in the SDI research programme, Chanellor Kohl discussed the possibilities of a closer co-operation in the field of air defence during a one-day meeting with French President Mitterrand in Baden-Baden on 16 January 1986.[90]

The West German reaction to the American SDI challenge has demonstrated the dispute between the Atlanticists and the Europeanists within the cabinet and between the 'steel helmet' faction and the 'Genscherites' within the ruling Christian Democratic Union. Instead of formulating a consistent foreign and security policy and trying to use the bargaining potential Chancellor Schmidt had developed during the 1970s, the SDI dispute within the government coalition was a major reason for the decreasing influence of West German foreign policy both in relation with its American and French allies and with the Soviet Union. Not surprisingly Mr Gorbachev by-passed Bonn during his first visits to London and Paris in 1984 and 1985. Increasing suspicion by Allies and opponents alike about the future course of West German foreign policy was the price to be paid for the effort to please both Washington and Paris at the same time. In reality, both were alienated. Genscher's Europeanist option in support of EUREKA was blocked by both the Atlanticists faction within the CDU/CSU, by the fiscal conservative Finance Minister Mr Stoltenberg, and by the technological appetite of the minister presidents of Bavaria and Baden-Württemberg, Mr Strauss and Mr Späth, who expected contracts, jobs and a technological impulse from a participation of the West German industry in the SDI research programme.

(c) The economic and technological aspects of the West German debate

Economic interest and technological appetite emerged as a major aspect of the West German debate on the terms of the participation of West German firms in the SDI research programme. Chancellor Kohl, in his Wehrkunde speech in February 1985, indicated that the SDI would produce major technological innovation in the United States and that a technological decoupling of the Federal Republic of Germany and of the other European allies must be avoided.[91] In his 18 April 1985 declaration of governmental policy to the Bundestag, Chancellor Kohl emphasised the interest 'in applying research results which will have revolutionary implications for civilian applications for our economy'. However, the Chancellor added 'this economic and technological interest alone will not determine our decision on eventual participation in the research programme'. The following are criteria for a possible future co-operation in the SDI programme:

1. fair partnership and a free exchange of research results must be guaranteed;
2. a technological one-way-street must be avoided;
3. one complete research area must be secured for the Europeans;
4. direct influence on the whole project must be allowed.[92]

Chancellor Kohl expressed the hope of keeping and even extending influence on any eventual reformulation of Alliance strategy by participating in technology research. In his May 1985 address to the North Atlantic Assembly, Chancellor Kohl stated:

> The federal government will take up the American suggestion and examine thoroughly the possibilities of co-operation on the research programme. Our guiding principles will be German and European interests and the interests of the Atlantic Alliance ... The oft-asserted dilemma of choosing between Paris and Washington does not affect us ... We accept and support the basic idea of an amalgamation and pooling of European high-technology resources and capabilities ... Europe will have to respond to this technological, political and strategic challenge from the United States.[93]

In its decision of 18 December 1985, the federal cabinet clearly emphasised the economic and technological aspects related to

German participation in the SDI. It was the Economic Minister who was asked to initiate negotiations with the United States government on an improvement of the framework for a mutual transfer of the scientific research results in order to strengthen the legal position of the research institutes and companies participating in the SDI. The federal government rejected both direct participation and the appropriation of funds for co-operative projects.[94]

In the debate on the economic and technological aspects of German participation in the SDI research programme, two schools of thinking emerged within the cabinet and within the coalition parties:

1. the *technology enthusiasts*, represented by Strauss and Späth, stressed the need for official participation in the SDI programme if only for technological reasons;
2. the *sceptics*, represented by the Minister of Science and Technology, Mr Heinz Riesenhuber, countered that German participation in the SDI research programme on technological reasons only could not be legitimised.

After intensive consultations in Washington in February 1985, the Minister President of Baden Württemberg, Lothar Späth,[95] called for German participation in the SDI research programme not for military but for purely economic reasons to avoid technological decoupling. Späth interpreted the SDI as an American answer to the Japanese technological offensive in the competition for power and for the markets of tomorrow. In his view, SDI represented the West's most comprehensive reaction in the field of research to the realisation that modern civilian and military products rely on identical basic technologies. SDI should strengthen the dominant position of the United States both in economic and scientific terms. If Western Europe rejected participation in the SDI research programme or if the United States insisted on discriminating terms based on an overemphasis on security considerations, in Späth's view:

> Western Europe would irreversibly fall back into the second and third rank in the technological area within ten or fifteen years at the latest. After a period of presumably great and in the last analysis fruitless efforts a destabilised economic structure would be a consequence that would lead to both political and social concussions. Domestic instability would be supplemented by instability in foreign affairs: if technological competence is intended to be more important as a power factor than the number of deployed weapons, incompetence would imply imponderable risks.[96]

After a second visit to Washington in May 1985 Späth's technology enthusiasm was replaced by the concern 'that the Americans will buy into our industrial capacity, block it for other research projects and that in addition to a large capital and material drain, they will secure for themselves a major proportion of our research capacity without direct European participation in the benefits'.[97] Späth had also become rather pessimistic as far as the possibility of a technology transfer to Europe and a participation of German small business in the SDI research project was concerned. Instead of his former enthusiasm for the SDI he pleaded for a common European effort in the area of high technology in the framework of EUREKA.

From the very beginning Science and Technology Minister Riesenhuber interpreted the SDI as a purely military programme that could not be justified by economic and technological considerations alone.[98] In May 1985 Riesenhuber defined 'fair partnership' for possible German participation in the SDI as 'the free exchange of knowhow and technology among equals'.[99] Given the limited number of technical experts in West Germany, Riesenhuber asserted that a simultaneous participation in both SDI and EUREKA would be unrealistic. In the meantime Genscher pleaded for German participation in EUREKA and in a European technology to avoid a direct brain drain into the United States.[100] 'Technological co-operation should become a part of a co-operative European order of peace and security', Mr Genscher told a group of young businessmen in May 1985.[101]

In early September 1985 Mr Riesenhuber argued that SDI could not be legitimised by the civilian spin-off, but purely on military and strategic terms and he was also doubtful about the civilian applications of several SDI projects.[102] A major issue of the debate within the cabinet during 1985 was the contradictory evaluations of the potential civilian spin-off for West German industry in the area of high technology. In two studies for the Ministry of Science and Technology the claim that SDI would produce major spin-off for West Germany's high-technology industry was disputed, and the past problems in technology transfer between the United States and West Germany were analysed rather critically.[103] However, these scientific studies did not dissuade the technological enthusiasts within the government, and the spokesmen of those German high-tech firms which hoped for SDI contracts.

The terms for technology transfer between the United States and West Germany became evident during the two Teltschik missions to Washington in June and September 1985: a true 'two-way-street' based on the principle of equal partnership would be unacceptable for the United States Department of Defence. By December 1985 the criteria for technological co-operation which Chancellor Kohl had formulated in April 1985 were no longer realisable.[104] In January 1986, after completing his talks with United States officials in Washington, Martin Bangemann, the Minister of Economic Affairs, admitted that without a financial contribution, influence on the architecture would be impossible. He also said that the West German government does not strive for influence on the SDI programme, thus back-tracking from the goal stated by Chancellor Kohl in April 1985. In its decision of 18 December 1985, at the insistence of Genscher and Bangemann, the federal cabinet rejected any financial contribution to the SDI effort.[105]

However, the technological challenge posed by the SDI was not without consequences for closer high-tech co-operation among the West European countries. The concept of a European technological community had been discussed within the European community since the Goodge Report. In early April 1985 the French government (in close co-operation with the German Foreign Minister Genscher) launched the idea of a European Research Co-ordination Agency (EUREKA) with the goal of enhancing the economic competitiveness of West European industry *vis-à-vis* the United States and Japan.[106] What was originally conceived by France as a diplomatic card to stop the German drift towards SDI and to prevent an isolation of France developed a rare momentum of its own. Within three months the member nations of the European Community and EFTA met in Paris to develop a framework of EUREKA which was further specified at the meeting in Hanover in November 1985.[107]

Within the federal cabinet the support for EUREKA was not unanimous. While Genscher has been the primary force behind closer French–German and broader European co-operation in the field of high technology,[108] the Minister of Science and Technology Riesenhuber objected to the creation of new research bureaucracy and to the initiation of large European research projects with a rather limited civilian utility. On 2 July 1985 Riesenhuber stated the following four criteria for EUREKA projects:

1. consequence of EUREKA's common research programmes should be an impetus to civilian technology.
2. EUREKA should comprise ambitious programmes (e.g., development of computers with artificial intelligence) which

could be conducted more efficiently in an international framework
3. EUREKA should avoid clumsy bureaucratic organisation;
4. industrial projects in the context of EUREKA should include an appropriate contribution from private industry.[109]

Between August and November 1985 a new dispute emerged between Genscher, Riesenhuber and the Minister of Finance Stoltenberg on the German financial contribution to the EUREKA project. Although Genscher announced a German contribution of 330 million Marks for FY86 in order to complement President Mitterrand's July 1985 pledge made during the first EUREKA conference in Paris, Stoltenberg objected to appropriating any additional funds for EUREKA, and Riesenhuber refused to cut existing projects within his Science budget in favour of EUREKA. In October it was reported that West Germany would not enter any financial commitment for EUREKA projects until the third EUREKA meeting in London in May 1986.[110] While the federal government did not specify its likely contribution for EUREKA projects in FY86, the Budget Committee of the Bundestag appropriated up to 40 million Marks for EUREKA.[111]

The two Minister Presidents (Strauss and Späth) disagreed in their evaluation of EUREKA. While Späth called upon the European governments and the European Parliament to initiate joint European research efforts, Strauss was rather critical of the results of the Hanover EUREKA meeting, and preferred German participation in the SDI project instead.[112]

While the governments were still consulting and developing schemes for co-ordinating joint research projects, the four largest electronics companies were reported to be co-operating in developing new electronics ventures in both civilian and military sectors under the umbrella of EUREKA. These companies (including Siemens AG of West Germany, General Electric Co. of Britain, N. V. Philips of the Netherlands and Thomson of France) agreed to study developing joint research projects in advanced sectors of electronics, new generations of microcomputers, gallium arsenide integrated circuits, microwave components, high-density memories, flat screens and a wide range of sensors, in an effort that could well require $1 billion in initial funding.[113]

In late June 1985 a French–German forum in Munich had proposed six projects under EUREKA in which EC and other West

European nations would be welcome to participate'. Based on an article in the West German financial paper *Handelsblatt* Elizabeth Pond reported in the *Christian Science Monitor*: 'Research fields would include supercomputers (with suggested funding of 800 million marks, or $270 million over two years); integrated circuits "with the least structure" (3.5 billion marks, or $1.2 billion over ten years); integrated circuits with switching times of less than a picosecond (500 million marks, or $170 million over ten years); flexible automatic montage systems in an integrated factory (500 million marks, or $170 million, over ten years); software tools (700 million marks, or $233 million over ten years); and smart robots capable of learning (300 million marks, or $100 million, over seven years)'.[114] In early July 1985 two other EUREKA-related agreements were reported between the French companies Aerospatiale and Matra with West Germany's Messerschmitt-Boelkow-Blohm (MBB) on computer-aided manufacturing techniques and software for use with new, complex development programmes and between Matra and MBB on feasibility studies on a laser data transmission platform equipped with an active reflector mirror.[115] These examples indicate that SDI functioned as a catalyst for closer inter-governmental co-operation and between the high technology companies of different West European countries.

Despite these first EUREKA-related transnational projects, a major disparity could be observed between the declared intention of closer European technological co-operation and the absence of financial commitment to implement the EUREKA project in order to strengthen the relative position of Western Europe in the global competition for future high-technology markets.

In mid-July 1985, after a detailed discussion of the technological challenge SDI presented for Europe, Konrad Seitz (the head of the planning staff in the Foreign Ministry and one of Genscher's close advisors) called for a broader European initiative to create a common sphere for research and technology which combined both civilian and military projects:

> EUREKA is directed at civil projects. At the same time there should be European projects in defence co-operation. The European members of NATO spend £6–7 thousand million each year on military research and development. Here, too, it is a question of finally overcoming national egoism and – within the framework of

NATO strategy – agreeing upon a European defence initiative which would use the new technologies for three major objectives:

1. The construction of an integrated air defence system (against aircraft, short-range ballistic missiles, cruise missiles) together with the Americans.
2. The development of 'intelligent' weapons which seek their own targets and development of information and delivery systems for NATO's Follow-on-Forces-Attack (FOFA) concept for long-range defence against enemy second-echelon attack.
3. The development of a European multisensor surveillance satellite, which is at present under discussion by France and West Germany.[116]

In Mr Seitz's view 'a combination of EUREKA and a European defence initiative ... should enable Europe to make the same advances in technology which can be expected for America as a result of SDI and NASA's project for a manned space station. They would at the same time make Europe a more equal partner for the United States and open the way to large-scale technology exchange'. In addition, Mr Seitz claimed: 'Combined civil and military projects are at the same time the most rapid means of creating a European common market in the new technologies'.[117] While Mr Seitz tried to legitimise EDI as part of a European reaction to the technological offensive posed by the American SDI programme, the West German Defence Ministry has fallen back on the more traditional arguments that SDI and EDI are legitimate Western responses to similar Soviet research programmes and military systems.

(d) The military and strategic aspects of the West German debate

The initial reactions of Defence Minister Wörner and the head of the Defence Planning Staff Hans Rühle to both President Reagan's strategic vision and the subsequent SDI research programme were highly critical. Already in 1980 Hans Rühle had a view of 'profound scepticism' regarding a missile defence option for Europe: 'this option appears neither technically practicable within acceptable financial limits at present, nor is there any prospect of an optimised combination of strategic defence systems in the foreseeable future'.[118] On 1 April 1983 Rühle was still sceptical that the American BMD system could protect the United States against Soviet ICBMs and he doubted whether such a defence system would be viable against Soviet IRBMs and SRBMs targeted against Western Europe:

> Should, as may be assumed, both superpowers follow the road charted by Reagan, their territories would become invulnerable sanctuaries, while Europe, even if it deployed a corresponding defence system, would be rid of few of its security policy concerns. Although in such a case protection from the Soviet ballistic missiles, e.g., the SS-20 were guaranteed, Soviet cruise missiles, short-range missiles and low-flying bombers could not be prevented from penetrating Western Europe.
>
> What is worse, all conventional arms systems would increase in importance, recalling prenuclear times – not a particularly pleasant perspective in view of the existing conventional imbalance in favour of the Soviet Union.[119]

The German Defence White Paper of October 1983 praised the ABM Treaty as a major achievement of the arms control efforts of the 1970s, one which had prevented an uncontrolled deployment of BMD systems. Rühle's early scepticism was echoed a year later by Minister of Defence Manfred Wörner who said in an interview after a meeting of the NPG in Cesme: 'My impression is the Europeans were broadly united in their critical questions. I can't see that [SDI] would provide greater protection or stability. I can only hope it would give an incentive for arms control'.[120] A few days after the NPG meeting in Cesme Wörner told the *Hannoversche Allgemeine Zeitung* that a United States antimissile system could destabilise the East–West balance, 'decouple' the United States and Western Europe, and even lead to a splitting apart of the Western Alliance.[121]

After July 1984, Mr Wörner gradually changed from an outspoken critic of the SDI project to a major supporter of both the American ground- and space-based BMD system as well as a supplementing European defence initiative against Soviet short-range systems.[122]

In January 1985 Hans Rühle argued that the United States SDI research project was legitimate given the Soviet advantage in the area of BMD systems,[123] and in November 1985 in the news magazine *Der Spiegel* he countered the arguments of the scientific critics of the SDI project.[124]

The SDI report of the Minister of Defence of 20 August 1985 provided a detailed analysis of the military strategic consequences of the SDI for NATO, and was the result of a major turnabout during late 1984 and early 1985. It was now argued – more or less in agreement with the rationale presented by the United States Department of Defense to the Europeans – that the deployment of an American BMD would improve the second-strike capability of the ICBMs, and would enhance the credibility of nuclear deterrence. Defensive systems would, according to the new rationale, prevent a destabilisation of the strategic balance and renew the security the United States had tried to obtain in vain by a mutual limitation of the strategic offensive capabilities. In such a case, the Soviets would have to retarget their ICBMs against cities, something which would make little sense militarily.

On the other hand, the Soviet deployment of a strategic BMD would endanger the ability to make selective nuclear strikes as a means of controlled escalation with the goal of war termination. Both United States and NATO options for nuclear escalation would be seriously hampered. The official West German SDI report argued that the following consequences of strategic defence systems would in the long run, have major political and military implications:

1. The capability of the American and the Soviet strategic nuclear forces would be limited to the function of counter-deterrence and to nuclear reaction in response to a massive attack.
2. The capability of both the United States strategic systems and the MRBMs for a selective strike against targets on the territory of the aggressor would be impeded.
3. The credibility of the British and the French nuclear forces would be severely challenged.
4. The superior medium- and short-range potential of the Soviet Union would be increased if Western Europe were not included in a defensive system.

In addition, the future conventional employment means of the Soviet medium-and short-range nuclear potentials would increase the conventional superiority of the Warsaw Pact in Europe.

The official SDI report of the Defence Minister drew two conclusions:

1. the efforts for an improvement of conventional defences should be continued;
2. a defence system against conventional and nuclear attack

options of Soviet medium- and short-range ballistic missiles, cruise missiles and aircraft should be deployed.[125]

The call for an EDI emerged from the prior criticism that a division of the NATO territory should not be permitted.

Franz Josef Strauss,[126] an initial critic of the SDI, was among the first conservative West German politicians who, along with the leader of the parliamentary CDU Alfred Dregger,[127] and former Defence Minister von Hassel,[128] called for an EDI in co-operation with the United States.

In June 1985 some unidentified German experts were publicly quoted after the return of a delegation to the United States headed by Horst Teltschik as saying that a European BMD against Soviet TNF should fulfill four conditions. It should be developed independently from the SDI, it should be cost-effective, a close co-ordination and mutual exchange of results with the United States should be guaranteed and such a programme should not violate the ABM Treaty.[129]

In early July 1985 Mr Dregger reported on American plans to launch a second research programme, ('a smaller SDI for Europe') in co-operation with the Europeans for defence against SRBMs. Such a second project should be planned, financed and developed together to avoid costly parallel developments. Such a second-defence research programme should apply also against aircraft and cruise missiles.[130]

In September 1985 Mr Wörner publicly endorsed the idea of an EDI in close co-operation with the United States, and in October 1985 he rejected all proposals for an autonomous West European defence policy independent of the United States.[131] In early December 1985 Mr Wörner discussed the EDI concept informally with other NATO Defence Ministers in a series of bilateral talks 'on the fringe' of a meeting of the Defence Planning Committee in Brussels.[132] However, by the end of February 1986 no details about a preferred European Defence Architecture had been published by the German Defence Department, nor have any cost estimates been made available.

In several speeches and interviews Inspector General Altenburg expressed his concerns about the fiscal and strategic consequences for West Germany's defence planning. If both the United States and the Soviet Union go ahead with their BMD programmes, additional European efforts for an extended air defence (including ballistic and Cruise missiles) will be required.[133]

On 17 January 1986 Lothar Rühl, State Secretary in the Defence Department, discussed the conditions and the consequences of a missile defence for Europe on NATO's strategy and arms control policy in the East–West context.[134] The deployment of a regional Soviet ATM system and the increasing threat of Soviet short- and medium-range ballistic and cruise missiles would require an extension of existing air defence capabilities in Europe, independent of the United States decisions on SDI, as part of the overall NATO air defence effort.

To what extent have these political and arms control, economic and technological as well as military and strategic considerations been reflected in the 1986 German–American memorandum of understanding?

(e) The terms of the German–American agreement on participation of German companies in the SDI research programme

After more than a year of an agonising public debate within the coalition government and after months of tough bilateral negotiations two documents were finally signed on 27 March 1986 in Washington by United States Defence Secretary Caspar Weinberger and Economics Minister Martin Bangemann: a memorandum of understanding (MOU) that would allow the direct participation by West German companies in research on SDI projects and a general agreement under which the two countries would share technology in commercial and other fields that might develop from the research.[135] Both agreements were classified at the insistence of the United States government. However, on 18 April and on 20 April 1986 both agreements were published in full by the popular newspaper *Express* in Cologne.[136]

Both the initial presentation of these two agreements to the public by government officials and the political reaction and press coverage remained highly controversial. While Mr Bangemann and Mr Genscher stressed the civilian nature of both agreements and their full agreement with the ABM Treaty, Mr Wörner and Mr Strauss interpreted the MOU as a military agreement. This dispute on the nature of the two SDI-related agreements was reflected in a special reference to West German companies based in West Berlin that were not allowed to participate in military projects due to the demilitarised status of Berlin.[137]

The initial editorial reaction has ranged from muted to negative:

while the conservative *Frankfurter Allgemeine Zeitung* pointed to the ambiguity and the secrecy of both agreements, the *Frankfurter Rundschau* referred to a partial success of the German delegation that the ABM Treaty was to be maintained while the editor-in-chief of the liberal weekly *Die Zeit*, Mr Theo Sommer, openly criticised both agreements as 'A step in the wrong direction. The SDI agreement will not be beneficial, it undermines our security'. The economic impact of the SDI accord for German industry would be irrelevant, while the government had undercut its own interests in the area of defence, arms control and disarmament and with respect to a closer West European co-operation.[138] While Rolf Zundel interpreted the United States insistence on classification of both agreements as an expression of a policy of 'divide et impera', both the opposition parties and leading spokesmen of the coalition parties, Mr Strauss and Mr Baum[139] criticised the secrecy and called for the publication of both agreements. While Chancellor Kohl justified both agreements in Parliament on 17 April 1986 as an effort to 'prevent a technological and strategic decoupling' and Mr Bangemann emphasised their civilian nature and the achievement for an improved information exchange and technology transfer,[140] the West German business community remained rather sceptical as to the scope and the potential impact of SDI for European and German companies.[141]

The Joint Understanding of Principles – as published one day after the controversial SDI debate in Parliament – after referring to the existing bilateral agreements on trade, mutual assistance of customs administrations, and on the handling of secret documents – discussed in most general terms – the new challenges for technological co-operation and the need for an increased co-operation in order to prevent the transfer of sensitive technologies that would affect the mutual security. This Joint Understanding would also apply for Berlin within the rights and responsibilities of the three allied powers: France, the United Kingdom and United States.

The preamble of the bilateral SDI accord – noting the previous West German declarations on SDI of 27 March, 18 April and 18 December 1985 and Minister Weinberger's invitation of March 1985 for other NATO countries to join the American SDI research programme – stressed the common interest of both parties 'in the creation of a broad foundation for a comprehensive participation of German companies and research institutions and of other agencies interested to participate in the SDI research programme.' In detail, the MOU focused on the following aspects:

2. *Implementation* (for the individual SDI projects individual agreements with specific terms were to be negotiated).
3. *Existing accord* (which stressed the existing national and international obligations of both parties and on behalf of the United States 'the adherence to the American–Soviet ABM Treaty of 1972' and which mentioned all relevant bilateral agreements on dealing with secret documents, mutual co-operation on research and development, exchange of patents and price–cost evaluations for defence contracts.
4. *Definition of terms* ('Classification', 'Technical data', 'computer software', 'background information', 'foreground information', 'protected information by property rights', 'German participation').
5. Mechanisms for *co-operation and acquisition in SDI research*.
6. *Exchange of information and rights of intellectual property*.
7. *Protection of information*.
8. Rights with respect to *intellectual property and the use of information*.
9. *Additional information*.
10. *Final clauses* (e.g., withdrawal clause after three months).[142]

This MOU – without the general reference in the preamble – did not meet the set of conditions outlined in the declaration of the West German Security Council of 27 March 1985 and in the declaration of governmental policy of Chancellor Kohl of 18 April 1985.[143] The West German government was not offered any information with respect to the overall SDI architecture and no active role in the strategic decision making process.

Even more controversial have been the letters exchanged between United States Assistant Secretary of Defense, Mr Richard Perle, and Mr Lorenz Schomerus (a high official in the West German Economics Ministry) in which the United States government called upon the Federal Republic of Germany to tighten its export controls for sensitive technologies in the framework of the COCOM embargo list. In his letter, Mr Perle interpreted the Joint Declaration of Principles and the MOU in terms of international law rather as a 'declaration of political intent than a legally valid document'. According to Perle's letter para. 3 of the MOU would be facilitated by bilateral talks on the COCOM list prior to the discussion in the COCOM framework. According to the United States view both sides have agreed in urgent cases not to grant export permits for embargoed goods. On behalf of the West German government, Mr Schomerus guaranteed an improved and tightened implementation of the COCOM embargo list: 'We agree with you that additional joint efforts will be necessary in order to make COCOM a more powerful instrument and to increase the daily co-operation among COCOM-members... I agree with your interpretation of [para] 3 of the MOU provided that it will be in agreement with the laws and regulations of the Federal Republic of Germany'.[144]

'A needless shame – the Federal Government has been outfoxed by the Pentagon', stated Kurt Becker's editorial on the SDI accord in the liberal weekly *Die Zeit* on 25 April 1985. Becker confronted the meagre result with the high expectations the Kohl government had raised as the price for its political support for the SDI and the participation of German companies in its research phase. Given the many disadvantages, the result has been 'miserable'. In matters of technology transfer – especially in the trade with the East – the federal government has bowed to the philosophy of the Pentagon in using technology as an instrument for weakening the Soviet Union. 'The Chancellor has made an error with respect to SDI both in terms of domestic and foreign policy. He has unnecessarily made a fool of himself'. Chancellor Schmidt's former government spokesman wrote.[145]

After the publication of both agreements and the accompanying letters the official position of the Kohl government celebrating the SDI accord as a major achievement was ridiculed by the press and harshly attacked by the opposition parties and even by leading members of Mr Bangemann's liberal party (FDP).

Mr Roth, the leading economics spokesman of the SPD, attacked the Kohl government for offering the Pentagon a direct say in the German trade towards Eastern Europe – amounting to 30,000 million DM per year – for a few million dollars it may expect from the SDI budget in return.[146] On 23 April 1986, Mr Vogel, the leader of the SPD in Parliament, castigated the government for their political support for the next round of the arms race. The SDI accord would be detrimental to German business because it allowed the Pentagon a power of veto in German exports. Furthermore it would isolate West Germany in Europe and within NATO. While the Greens interpreted the SDI accord as a 'document of capitulation', the SPD announced that it would withdraw from the SDI accord within three months of returning to power.[147]

The SDI debate ended in an anticlimax. Hardly any of the conditions raised by the federal government in 1985 had been included in the text of the published accords and exchanges of letters and many of the concerns of the critics in the Parliament, within business and within the scientific community appeared in retrospect to be justified.

How did the position of the political parties on the SDI in general and on the terms of a German participation in its research phase in particular gradually evolve from March 1983 to late February 1986?

III THE POSITIONS OF THE POLITICAL PARTIES ON SDI

On 11 April 1984 Walther Stützle, head of the Defence Planning Staff until September 1982, could still write 'that all five parties represented in the Bundestag unanimously oppose Washington's plans'.[148] In the meantime the negative consensus towards SDI which prevailed till Summer 1984 has disappeared. SDI (and increasingly EUREKA and EDI or extended air defence) have polarised the political debate within and between these five parties – the ruling Christian Democratic Union (CDU), the Christian Social Union (CSU), the Free Democratic Party (FDP) and the two opposition parties: the Social Democratic Party (SPD) and the Greens. After the deployment controversy about Pershing II and cruise missiles was over SDI dominated the security debate within and outside Parliament during 1985. By early 1986 the attitudes set out in Table 6.1 had emerged on these three related issues and on an alternative conventional defence posture. The following brief survey will focus on the shifts which have occurred in the position of the five parties on SDI from March 1983 to February 1986.

(a) The evolution of the CDU response to SDI, EUREKA and EDI

The initial reaction of senior and influential members of the CDU to President Reagan's 'Star Wars' vision and to the SDI programme was dominated by restrained scepticism. In 1983 and 1984 all major CDU spokesmen commenting on the SDI stressed a preference for arms control talks between the superpowers to avoid an arms race in space.[149]

The authoritative spokesmen articulating the position of the CDU

Table 6.1 Attitudes of 5 West German parties on 4 defence issues

Issues	Parties				
	Christian Democratic Union (CDU)	Christian Social Union (CSU)	Free Democratic Party (FDP)	Social Democratic Party (SPD)	Greens
Strategic Defence Initiative (SDI)	+	+	(−) (+)	−	−
EUREKA	+	(+)	+	+	−
European Defence Initiative (EDI)	+	+	+	(−)	−
Alternative conventional defence	−	−	(−)	+	(+)

Notes: + = In support − = Opposed (−) = Leaning towards opposition

parliamentary party have been Alfred Dregger, the chairman of the CDU/CSU faction in the Bundestag; Volker Rühe, his deputy and the foreign policy spokesmen; Willy Wimmer, the defence spokesman of the CDU/CSU faction; Jürgen Todenhöfer, the spokesman of the CDU/CSU MPs in the subcommittee on arms control and disarmament; Christian Lenzer, the spokesman for research and technology matters of the CDU/CSU faction and Heiner Geissler, the secretary general of the CDU.

While Jürgen Todenhöfer had been an early supporter of the SDI programme and of an active participation of German scientists, Willy Wimmer and Volker Rühe gradually shifted from a highly sceptical position to a restrained support for the SDI research programme following the old adage 'If you can't beat 'em, join 'em'. After President Reagan's re-election in November 1984, the CDU/CSU faction in the Bundestag gradually moved to a position of 'jump onto the bandwaggon or die'.

Representative of the rather sceptical attitude towards the SDI were the first statements of Volker Rühe and Willy Wimmer in April 1984. In his first comment on the SDI on 4 April 1984 Mr Rühe called for comprehensive briefings of the Allies during the research phase. An equal protection of the European allies was to be guaranteed from the beginning, and two classes of NATO countries were to be avoided. The research efforts should not undermine disarmament negotiations, and the Soviet Union should be invited to take part in arms control talks on such systems in order to avoid an arms race in space.[150]

In a joint position paper for the internal debate within the CDU/CSU faction on the SDI of 9 October 1984 Mr Rühe and Wimmer concluded:

1. The SDI research programme is necessary for security reasons in order to evaluate the technical possibilities of the Soviet Union and to have a hedge against a breakthrough in BMD technology.
2. The effective deterrence strategy of flexible response is to be maintained.
3. The CDU/CSU emphatically supports the prevention of an arms race in space and welcomes the American readiness for the immediate initiation of negotiations on space weapons.
4. The research effort should not interfere with the arms control efforts regarding the existing nuclear forces.

5. As a consequence of the introduction of a defensive system no zones of graduated vulnerability within the Alliance may be permitted.
6. The CDU/CSU welcomes the United States readiness to inform and to consult with its Allies and to permit their participation in the research programme.[151]

Two weeks after Chancellor Kohl's Munich speech in support of the SDI Rühe remained sceptical whether European participation would be possible for Alliance and economic reasons. Rühe repeatedly stressed that a co-operative approach on the part of both superpowers in the Geneva talks would be needed.[152] On 25 February 1985 Mr Wimmer was quoted in the conservative daily *Die Welt* to the effect that the arms trade between the United States and West Germany had shifted to a relationship of ten to one in favour of the United States and that the United States was increasingly withholding modern technologies from West Germany.[153] On 24 May 1985 Rühe stated that past experience with the technology transfer from the United States had not been very encouraging.[154] And in an interview with *Der Spiegel*, Mr Rühe admitted on 23 December 1985 that in the past bilateral co-operation – for example, in the negotiations for the Space Lab and for DI – had been unsatisfactory. In June 1985 Rühe already supported EUREKA, and in December he came out in support of an extended air defence.[155]

Alfred Dregger, the parliamentary leader of the CDU/CSU in the Bundestag, gradually emerged as the major proponent of German participation in the SDI research programme and of the initiation of a 'second SDI for Europe' or an 'EDI'. Nevertheless, he stressed that a one-sided technology transfer from Europe to the United States would be unacceptable.[156]

In April 1985 Mr Dregger began calling for an EDI to counter the Soviet SRBMs and cruise missiles in order to avoid the most unfavourable situation of both superpowers being protected and Europe remaining the unprotected battlefield.[157] During his visit to Washington in early June 1985 Mr Dregger suggested the establishment of a consultative council between the United States government and representatives from those countries who were interested in joining the SDI research programme.[158] In his view, the Europeans should concentrate on research projects to counter the Soviet ballistic and cruise missile threat confronting Western Europe.[159]

Mr Lenzer, the CDU/CSU spokesman for science and research in the Bundestag, welcomed the French EUREKA programme and suggested that the different European space projects – ARIANE, COLUMBUS, HERMES and the planned German–French surveillance satellite – should be integrated into a common concept in order to strengthen 'the autonomy and independence in unmanned and possibly even manned spaceflight by the turn of the century'. In addition, Mr Lenzer called for 'full systems responsibility and a free technology transfer between American and European partners': 'Only in this context could SDI and EUREKA be useful for the European and German economy'.[160]

In August 1984 Mr Todenhöfer, irrespective of the then prevailing cautious scepticism, called for 'direct participation of our scientists' in the SDI programme and in October 1984 he became the first proponent within the parliamentary CDU/CSU of a complementary EDI.[161] On 21 February 1985 Mr Todenhöfer published ten points for support of German participation in the SDI research programme, in which he echoed both President Reagan's long-term vision and the shorter-term goals of the SDI programme.[162] Only one major CDU spokesman (Mr Geissler) supported the SDI on moral grounds as an attempt to overcome the present dilemmas of NATO's nuclear doctrine.[163].

(b) The CSU – the major supporter of SDI and EDI

After initial doubts about the implications of President Reagan's strategic vision for Europe Mr Franz Josef Strauss – the Minister President of Bavaria and the Chairman of the CDU's Bavarian sister party, the Christian Social Union (CSU) – has become the major proponent of German participation in the SDI. Mr Strauss argued for participation in the SDI research phase on the basis of a formal government agreement (Memorandum of Understanding) which would require parliamentary ratification. SDI, in Mr Strauss's view, should be supplemented by research against short- and medium-range missiles. In Mr Strauss's opinion German participation in the SDI would offer an opportunity for testing the United States willingness to maintain a co-operative partnership based on equal rights, while non-participation would lead to 'creeping estrangement' among the allies. SDI would become the drive behind the third innovation thrust of American industry after the war. Non-participation would mean being left behind technologically at the level of second-rate industrial states.[164] However, if the United States insisted on a technological one-way-street, Mr Strauss was quoted as saying he would prefer a joint European effort to counter the Soviet threat.[165] In early November 1985[166] the CSU supported unanimously a German–American agreement similar to the British–American MOU. After the federal cabinet decided on 18 December to enter into direct negotiations with the United States on German participation, Mr Strauss publicly disagreed with the mandate for Mr Bangemann, which emphasised economic details over security aspects.[167]

(c) The FDP – scepticism about SDI, full support for EUREKA and restrained support for EDI

It was the liberal party – and in particular Foreign Minister Genscher – supported by his State Minister, Mr Möllemann – who prevented an early German–American agreement on official participation of the Federal Republic in the SDI programme and who were instrumental in watering down the mandate for the German–American SDI talks to be headed by its Party Chairman and Economics Minister Bangemann.

Since April 1984 Mr Olaf Feldmann[168] – a disarmament expert and member of the Defence Committee in Parliament – has been the most persistent critic of the SDI programme, primarily for arms control and fiscal reasons. The chairman of the FDP faction in Parliament, Mr Wolfgang Mischnick, fully supported the foreign policy and arms control considerations of Mr Genscher. In April 1985 he called for the formulation of a common European position within the framework of the WEU.[169]

In early June 1985 at a small-scale party conference in Neuss, the 125 party delegates unanimously supported a resolution that just stopped short of an outright rejection of the SDI: 'The West German Government must on no account consider participation in SDI alone, but must seek a common West European response'. Otherwise the relations with Eastern Europe could be harmed. Prior to the vote, 'Mr Bangemann had laid down four conditions the FDP were attaching to Germany's possible participation in SDI research. They were: full equality for all participants, a unified European stance, an unrestricted mutual flow of technological information, and no automatic realisation of the military plans after the research phase'.[170] In the same party resolution, German participation in the EUREKA project was supported, with the goal of improving French–German co-operation.

Mr Genscher's scepticism about German participation in the SDI research programme was shared by the party leadership, in the federal states who had directly attacked Mr Teltschik after his public endorsement of a German–American agreement on SDI in early October.[171] On 13 December 1985, after an intensive public and internal debate, the FDP board discussed the conditions for the cabinet decision to be adopted on 18 December 1985.[172] Although Count Lambsdorff, the former Economics Minister, supported a governmental agreement, Mr Genscher pushed through his criticism of any scaling down of German participation for research institutes and industry.[173] On 18 December 1985 Chancellor Kohl had to give in to many of the conditions made by his junior partner and Foreign Minister, Mr Genscher, much to the chagrin of the SDI supporters within the cabinet (Mr Wörner, and especially Mr Strauss, who subsequently harshly criticised the cabinet decision).[174]

While the press focused primarily on the clashes within the cabinet on the non-decision over German participation in the SDI, the full support of the FDP for EUREKA and the plea by its foreign policy spokesman, Mr Schäfer[175] and by Mr Genscher's close advisor Mr Seitz,[176] in support of an EDI hardly provoked any controversy. With its criticism of the SDI, the FDP had all the headlines it needed to win back a part of its former supporters in two state elections while its indirect support for an EDI contributed to the stability of the Kohl government in the pre-election year 1986.

(d) The SPD – opposition to SDI and full support for EUREKA and for a technological Europe

The SPD was consistent in its criticisms of the two military programmes SDI and EDI, and in its full support for the French EUREKA proposal within the broader framework of a Europeanist orientation and a closer French–German co-operation. Differences emerged only about the participation of the EC in the EUREKA project. While the members of the European Parliament called for some role for the EC and the Parliament in EUREKA, a majority of the members of the Bundestag objected to any increase in the influence of the European bureaucracy.[177]

The major SPD spokesmen in the SDI debate have been its Party Chairman, Mr Willy Brandt, the chairman of the parliamentary SPD, Mr Hans-Jochen Vogel, and his two deputies Mr Horst Ehmke and Mr Wolfgang Roth, the major foreign and economic spokesman of the parliamentary party, as well as Mr Egon Bahr and the SPD spokesman in the foreign affairs committee (Mr Karsten Voigt) and in the subcommittee for arms control and disarmament affairs (Mr Hermann Scheer).

In September 1984 the SPD tabled a Draft Treaty on the limitation of the military use of outer space which relied on a proposal formulated by a group of natural scientists and lawyers which had been presented to an International Congress on the Militarisation of Space in Göttingen in July 1984.[178]

In the first reaction to President Reagan's 'Star Wars' speech, Mr Bahr, one of the architects of the West German Ostpolitik, had applauded President Reagan's criticism of the deterrence doctrine but he had differed with his vision to deploy a space-based BMD system.[179] Arms control concerns were in the forefront of criticism by Mr Scheer and Mr Voigt during 1984. On 24 September 1984 Mr Scheer put his full support behind the French space initiative in Geneva on the occasion of the introduction of the space Draft Treaty into the Bundestag. Mr Scheer pointed to six major aims of this proposed treaty:

1. The obligation that no space projects of other states should be destroyed or damaged or disabled and that no such systems should be developed.
2. The destruction of all existing ASAT systems.
3. The renunciation of the deployment of space-based weapons irrespective of whether they would be targeted against objects in space, in the air or on the ground.
4. The obligation not to use any systems in space for the direct guidance of nuclear weapons and to deploy any command centres in space.
5. The obligation not to interfere with the technical means of verification.
6. The establishment of a Standing Consultative Commission of all parties to the treaty.[180]

On 26 February 1985 the presidency of the SPD objected to any German participation in the SDI programme for political, arms control, economic and strategic reasons. In its statement, the SPD pointed to a series of major dangers:

1. The precarious stability of the assured second-strike capability would be removed.

2. It would be highly unlikely that offensive weapons would be neutralised by defensive weapons.
3. SDI could remove the strategic unity of the Alliance by creating zones of different security.
4. The United States refusal to introduce the SDI into the Geneva talks would reduce the changes for concrete results at the negotiating table.

Instead of participating in the SDI for technological reasons, the SPD suggested an active participation in European space programmes such as a European reconaissance satellite and a space shuttle.[181]

On 9 April 1985 Mr Ehmke, in presenting a paper on space co-operation, opted for joint European programmes such as the French proposal for HERMES and he described a decision on German participation in the United States space station as premature.[182] A few days later the chairman of the Socialist International Mr Brandt, called upon the member parties to initiate a peaceful European space initiative.[183]

On 23 May 1985, Helmut Schmidt advised his successor in the Chancellory in an open letter not to endanger French–German co-operation with SDI. For strategic reasons, and given the political interests of France and Britain, the ABM Treaty should be maintained and strengthened. 'For these reasons, from an Alliance perspective participation of the federal government in the American developments would not be advisable. Isolated participation of the Federal Republic of Germany would endanger the internal cohesion among the European partners of the alliance'.[184] As to the expected economic spin-off, Mr Schmidt reminded Chancellor Kohl that the United States government could be tempted to reduce the economic freedom of action of European companies, either through COCOM or other new technology transfer rules. For economic and political reasons, former Chancellor Schmidt supported closer co-operation of the European countries in the high-technology field including EUREKA.[185]

On 21 May 1985 Egon Bahr argued in *Der Spiegel* that Europe should emphasise the development of its future economic strength, by closer co-operation to its mutual advantage in order to make détente irreversible. If Western Europe is to be more independent in the future, Bahr argued, it will have to guarantee a conventional defence capability which will deter purely conventional aggression. A new Ostpolitik would have to be based on European co-operation in the area of security and economy. 'This would be a way that would provide Europe with a future global role in accordance with its strength and its capabilities.'[186]

For the SPD, a critical attitude towards SDI and closer German–French co-operation in the framework of a Europeanist orientation are elements of a new foreign policy consensus which has emerged after the divisive internal INF debate.[187]

Horst Ehmke, the major foreign policy spokesman of the parliamentary party, argued that conflicts of interests between East and West could not be solved technically but only politically. SDI might not only result in a tightening of the United States technology transfer rules in relation with its Allies, it would also contribute to a militarisation of basic research, thereby reducing the innovative potential of society and restraining the freedom of science. In the technological competition with the United States and Japan, the West European countries should not become an appendix of the military–industrial complex of the United States. In order to enhance the technological competitiveness of Western Europe, Mr Ehmke fully supported the French EUREKA project.[188]

On 18 November 1985 the SPD party board called upon the federal government not to participate in the development of an EDI because it might be interpreted as political and military support for the American plans to introduce SDI. Such a system would be in conflict with the ABM Treaty, it would destabilise East–West relations and it would initiate an arms race.

Instead the SPD supported the search for new technological posssibilities for an adequate non-nuclear air defence in close co-operation with the other NATO countries.[189] On 4 December 1985 it introduced two draft resolutions into the parliamentary debate which called upon the government neither to take part in the SDI nor to support EDI.[190]

(e) The Greens – opposition to SDI, EDI and EUREKA

In 1984 detailed comments on behalf of the Greens regarding President Reagan's SDI plans were scarce. The Greens opposed both the American SDI programme and the French aspirations of Europe as a third space power. During the parliamentary debate on the SPD draft treaty Ms Kelly rejected the German scientists' initiative 'as being insufficient'. She objected also to American plans for a European defence system in the context of Counterair '90 and of a modified Patriot-ATM.[191]

In several draft resolutions they introduced into the Bundestag, the Greens opposed any participation in the American SDI programme and called for a prohibition of the space-based BMD, ASAT and NAVSTAR systems.[192] The Greens also rejected the EUREKA programme as – according to their interpretation – a military project which would offer no useful civil applications.[193] In a draft resolution of 23 October 1985 the Greens called upon the federal government not to permit the deployment of missile defence systems on the territory of the Federal Republic of Germany or to provide any fiscal resources for this purpose.[194]

What was the reaction within the Bundestag to these numerous resolutions and parliamentary questions to the federal government?

IV THE BUNDESTAG AS A FORUM FOR THE DEBATE ON SDI, EDI AND EUREKA

Since November 1984 both arms control and the question of German participation in the SDI research programme have become a major issue of the debate in the Bundestag. On four occasions – 8 November 1984, 18 April 1984, 18 October 1985 and 13 December 1985 – SDI was the major topic of parliamentary controversy, and on 8 November EUREKA became a topic of concern immediately after the second EUREKA conference.[195] In addition the Committee on Science and Technology of the Bundestag organised a two-day hearing on 'space research–space technology' on 11–12 November 1985;[196] and the Defence and the Foreign Affairs Committees of the Bundestag held another two-day public hearing on 9–10 December 1985 on SDI.[197] Numerous motions were tabled – especially by the Greens – on the many facets of the SDI, EDI and EUREKA issues.[198]

On the occasion of the introduction of the 'Treaty on the Limitation of the Military Use of Outer Space', the SPD MP Mr Scheer supported the project of a European observation satellite for verification purposes. His CDU colleague Mr Wimmer offered four criteria for the evaluation of the SDI project:

1. SDI research should not automatically lead to a change of NATO's deterrence strategy.

2. The deployment of defensive systems should not lead to zones of graduated vulnerability.
3. The danger of regional nuclear and conventional conflicts has to be reduced.
4. A missile defence system should not lead to intensified armament efforts.

Ms Kelly on behalf of the Greens attacked the French–German project of an observation satellite as an element of a united European nuclear force while Mr Ronneburger (FDP) rejected the verification provisions of the SPD draft treaty as being insufficient.[199]

On 18 April 1985 Chancellor Kohl opened the first SDI debate with a declaration of his policy towards the United States President's initiative, specifying the criteria for evaluation and possible participation in the research phase. On behalf of the SPD, Mr Ehmke responded by saying that SDI would not protect Western Europe but would instead increase the danger of a war limited to Europe. The opposition spokesman called for a European civilian programme in the area of basic technologies in order to enhance the competitiveness of European companies in the global competition. The CDU spokesman, Mr Dregger, fully supported a shift in the deterrence theory to defensive systems, and the spokesman of the Greens backed the proposal for an International Satellite Monitoring Agency (ISMA) to verify arms control compliance. Foreign Minister Genscher focused on the aim of enhancing strategic stability as the major goal of the resumed arms control negotiations between both superpowers and on the firewall separating research and development in the ABM Treaty, which should not be torn down by the SDI. Mr Schäfer, the major foreign policy spokesman of the FDP, pointed to various negative implications of the SDI. It could not protect Western Europe and it would indirectly involve tremendous fiscal resources. The defence spokesman of the parliamentary CDU, Mr Wimmer, focused primarily on the Soviet space projects and interpreted the American SDI as a legitimate response. His colleague, Mr Rühe, emphasised that SDI would have to take place within the constraints of the ABM Treaty.[200]

On 18 October 1985 the Bundestag used the opportunity of a vote on the SPD space Draft Treaty for another controversial debate on German participation in the American SDI research programme. While Mr Scheer on behalf of the SPD called for a prohibition of both ASAT and the space-based BMD effort, Ms Geiger (CDU/CSU) spoke for a supplementary EDI against Soviet SRBMs and cruise missiles within the framework of the ABM Treaty. Mr Feldmann (FDP) criticised the efforts of the Reagan Administration to reinterpret the ABM Treaty. He also rejected the argument, pointing to the Japanese example, that SDI would produce a major civilian spin-off. Mr Schierholz (Greens), revising the negative attitude Ms Kelly had expressed in November 1984, fully supported the German scientists' Draft Treaty tabled by the SPD. In his detailed contribution Mr Genscher (FDP) re-emphasised the position of the federal government that the restrictive interpretation of the ABM Treaty should be adhered to and the Treaty itself be strengthened.[201]

The parliamentary debate on German participation in the SDI and on a possible second SDI for Europe (EDI) climaxed on 13 December 1985 in the discussion and vote of several motions and draft resolutions on behalf of the SPD and the Greens which called upon the government not to participate in both SDI and EDI. Not surprisingly all draft resolutions by the parliamentary opposition were rejected by the solid majority of the coalition parties.[202]

Two days after the second EUREKA conference in Hanover on 5 and 6 November 1985, both Foreign Minister Genscher and the Minister for Science and Technology, Mr Riesenhuber, informed the Bundestag about the role EUREKA will play in the technological renaissance of Europe.[203] Both the coalition parties and the SPD welcomed the new initiative for a European Community of Technology while Ms Kelly of the Greens rejected the EUREKA initiative because it might be misused for military purposes. On behalf of the SPD, Mr Ehmke criticised the federal government for its reluctance to provide sufficient public funds to implement the EUREKA initiative, while the spokesman for science policy, Mr Vosen (SPD), complained about the exclusion of the European Community from the EUREKA effort. Speaking for the coalition, Mr Riesenhuber defended the official line that private companies should take the initiative and that public funds should be used only to supplement – but not to replace – private funding. Speaking for the CSU, Mr Klein objected to the creation of a new EUREKA bureaucracy and to an interpretation of EUREKA as an alternative to the American SDI.[204]

From 11–12 November the Committee on Science and Technology of the Bundestag held a two-day hearing on space research and space technology with twenty scientists and experts from industry testifying. All scientific and industrial experts pleaded for a European autonomy in the peaceful research and use of outer space. In order to strengthen the position of the Federal Republic a majority of the experts supported German participation in the French shuttle project HERMES which had been postponed by the federal government for financial reasons.[205]

A week before the decision of the federal cabinet, the Foreign Affairs and the Defence Committees of the Bundestag invited fourteen experts from industry, research institutes and universities to testify on the technological dimensions of the SDI and on the implications of BMDs in East and West on disarmament, arms control, on East–West relations and on the strategic implications for both alliances.[206] Among the experts, scepticism and well founded rejection was the prevailing attitude. Only three of the fourteen experts supported German participation on political and military grounds: Uwe Nerlich, General Franz-Joseph Schulze (retired) and the close advisor to the Chancellor Kohl, Professor Michael Stürmer, while the two representatives from industry – Dr Lohr (for the electrical industry) and Dr von Freyend (Federation of German Industries) – and Dr Hoff from the University of Tübingen pointed to potential spin-offs.[207]

Both the parliamentary debates on SDI and EUREKA and the two hearings had only a limited impact on the decision of the federal government. However, the prevailing scepticism in the second SDI hearing reinforced the critics within the junior coalition party and it provided additional arguments in support of the reluctant position of the Deputy Chancellor and Foreign Minister Genscher. The second hearing also documented the divergent views within business and the scientific community.

V SDI AND EUREKA: THE REACTION OF WEST GERMAN BUSINESS AND THE TRADE UNIONS (DGB)

While the German industry was divided as far as German participation in the SDI and in the EUREKA programmes was concerned, the West German Trade Union Congress (DGB) supported EUREKA but opposed any German participation in the American SDI. In a decision of the DGB Board of 2 July 1985, SDI was rejected because it would undermine both the strategic balance and the concept of a common security and because it would stimulate an arms competition between defensive and the offensive systems. The DGB called upon the federal government not to support the SDI programme politically or financially.[208]

After Chancellor Kohl's declaration of government policy on 18

April 1985 two high-level meetings between government officials and representatives of major industrial concerns and research institutes took place in the Chancellor's office in May and again in August 1985 in order to prepare the two information missions to Washington headed by Mr Teltschik, Mr Kohl's foreign policy advisor.[209] Some of the largest West German companies – for example Siemens – made their decision on participation in the SDI dependent on a clear governmental position. Siemens welcomed EUREKA.[210] The representatives of at least four German companies spoke out in support of German participation in the SDI: those from SEL, a former German subsidiary of ITT, AEG, and the two aerospace companies Dornier and MBB. Heinz Dürr, the chief executive of AEG, expected a tremendous impetus for new technologies from German participation in the SDI. Others, like Roland Mecklinger, a member of the board of SEL, interpreted SDI as a pure research project in the most advanced technologies which would strengthen the United States economy in the global competition. In his opinion, West German firms would get a fair chance only if their industrial interests were protected by a governmental agreement.[211] Given the high degree of similarity in the SDI and in EUREKA research projects, Mecklinger felt it lacked candour to support EUREKA and at the same time discredit the SDI.[212]

At least two West German companies were interested in cooperation with the SDIO even prior to a German–American MOU, the aerospace companies Messerschmitt–Bölkow–Blohm (MBB) and Dornier Aerospace. SDIO had indicated its special interest in the MBB's SPAS satellite and in the Instrument Pointing System of Dornier.[213] In late October 1985 it was reported that Dornier, in co-operation with Sperry, had applied for a SDI contract in the field of optical pointing systems.[214]

Many firms were wary that direct participation could lead to a one-way-street: with them contributing knowhow which they would be prevented from using afterwards for civilian projects. Mr Weisweiler, the director of Mannesman, told *Der Spiegel* in May 1985: 'In my opinion SDI is not important for German industry. If a political decision should be made in favour of SDI, then guarantees must be given that the federal government can benefit from the use of the results. Participation should take place only under the condition that we will not become a subcontractor. There must be no technological one-way-street'.[215] Marcus Bierich, the President of the Bosch Company was equally sceptical: 'We can do without SDI and Eureka'. A similar opinion was expressed by the head of Zeiss, Mr Horst

Skoludek: 'If we turn to a certain area then we shall not do this because of EUREKA or SDI but only for the simple reason that we consider the project promising in the long run.'[216] The president of the German Industrial and Trade Council (DIHT), Mr Otto Wolff von Amerongen, called upon the federal government to conduct a foreign policy cost-benefit analysis before joining the SDI. His general scepticism towards the SDI included the EUREKA project as well.[217] The general manager of the DIHT, Mr Franz Schoser, reminded the federal government in January 1986 that it should keep in mind the German trade interests in Eastern Europe which would be much larger than the potential benefits from German participation in the SDI project.[218] Representatives of the Federal Association of German Industries (BDI) expressed their interest in a government agreement which would protect their interests in the following fields:

1. technology transfer, including determination of the rights for private use of transferred knowhow;
2. questions of secrecy;
3. questions of ownership and of transfer of users' rights;
4. questions of price setting and price control.[219]

Representatives of the BDI supported EUREKA as well, especially in areas that would require joint European standards – for example in modern train systems, communication systems and environment technologies. EUREKA should contribute to the creation of a 'common market' in the high-technology area.[220]

Among industrialists, politicians and economists the potential spin-off from German participation in the American SDI programme remained controversial. While the supporters pointed to the consequences of the Manhattan and the Apollo Projects for the industrial development of the United States, the critics referred to Japan and to the European countries, who had achieved a high degree of modernisation independent of the defence budget.[221]

VI THE REACTION OF THE WEST GERMAN SCIENTIFIC COMMUNITY TO THE SCIENTIFIC AND MILITARY CHALLENGE POSED BY THE UNITED STATES SDI PROGRAMME

A major element of the German SDI controversy has been debated within the German scientific community: among the natural scientists at the universities, at publicly-funded research institutes (for example the Max Planck Society) and to a lesser degree within industrial companies. The SDI critics were more numerous and outspoken while the SDI supporters remained relatively silent. The debate within the sixty institutes of the Max Planck Society was reflected in a statement by its president, Professor Heinz A. Staab, who indicated an interest in the European research programme, but 'secret research' within the Max Planck Society was inconceivable. He also repeated the objections the Max Planck Society, the Deutsche Forschungsgemeinschaft (DFG) and the Conference of West German University Presidents (WRK) had formulated against German participation in the space station COLUMBUS.[222]

The critical debate among the natural scientists received a major impulse from a meeting of some 2,500 scientists on 'Responsibility for Peace – Scientists warn against Militarisation of Space' in Göttingen in July 1984.[223] The most vocal critic was Professor Dr Hans Peter Dürr, a student of both Heisenberg and Teller and now a director at the highly-regarded Max Planck Institute on Physics and Astrophysics in Munich.[224] His criticism of the technical aspects of the SDI were published in two major news magazines: *Der Spiegel* and *Stern*. 'Instead of contributing to a stable and secure peace, the SDI will most likely lead to additional destabilisation', Dürr concluded.[225]

In early July 1985 350 natural scientists (among them many leading scientists of the Max Planck Society) announced in an open letter to Chancellor Kohl that they would not participate in any research which would contribute to the SDI.[226]

On the 40th anniversary of Hiroshima leading natural scientists (among them the 1984 Nobel prize-winner Professor Georges Köhler, Professor Werner Buckel, now the President of the European Association of Physicists, Professor Dürr and others) appealed to the federal government 'not to participate in the research, development and deployment of space weapons and instead to work for strict compliance with existing arms control treaties and to use its influence for a limitation of the military use of space by negotiated treaties.'[227]

In November 1985, 380 scientists and technicians from Berlin announced that they would not participate in the SDI, as did 315 of 1,000 scientists of the German Synchroton (DESY) in Hamburg. Even nineteen professors of the University of the Armed Forces in Hamburg joined this pledge.[228] In the same month peace weeks were conducted at about fifty West German universities with panel discussions, lectures and meetings about the SDI. Professor Buckel estimated that of the 15,000–18,000 physicists in the Federal Republic of Germany approximately 2,000–3,000 were opposed while some 500 were assumed to be supportive with the rest being undecided.[229] Only a few SDI supporters expressed their support in letters to the editor and no organisation of natural scientists went public in support of the SDI.[230]

What impact did the political, economic and strategic debates among scientists on SDI in general and on German participation in particular have on the media and on public opinion?

VII THE REACTION OF THE MEDIA TO THE SDI DEBATE

The initial reaction in the West German press to President Reagan's speech of 23 March 1983 was highly critical. 'Ronald Reagan's horror show' stated Dieter Schröder's editorial in the *Süddeutsche Zeitung*:

> To make the wellbeing and the survival of mankind dependent on progress in weapons technologies neither reflects experience nor logic, even though deterrence has provided the human race with a long period of peace . . . The search for new, presumably invulnerable weapons indicates the helplessness technological progress has created for the politicians. Reagan's effort to persuade Congress with futuristic horror visions to support his enormous defence budget does not provide a way out.[231]

Walther Stützle, the former head of the Defence Planning Staff, asked whether the forces unleashed by President Reagan's vision would do even more damage to NATO and East–West relations.[232] Ulrich Mackensen, in a commentary in the liberal daily *Frankfurter Rundschau* saw only negative results: a new arms race, a further increase in the influence of the United States defence industry on foreign policy and a weakening of the prospects for peace and co-operation.[233] Christoph Bertram, the former director of the International Institute for Strategic Studies, wrote in an editorial in the weekly *Die Zeit*: 'Reagan's vision is not only unrealistic it is also detrimental to Western interests. Neither technically nor financially can it be realised in the near future'. In Bertram's assessment, the President had damaged the 'sensitive linkage of security by deterrence and negotiations. He has undermined the precarious domestic consensus on the NATO double decision in the deployment countries while he should have strengthened it by new signs of compromise in Geneva'.²³⁴

Immediately after Secretary of Defence Wörner's frank criticism of the SDI project at Cesme, Fritz Ullrich Fack's editorial in the conservative *Frankfurter Allgemeine Zeitung* stated:

> Those short-range systems which would be sufficient to destroy Europe if fired from the Western rim of the Soviet empire cannot be countered, even by the most perfect laser weapons in space. The envisaged development in the field of cruise missiles, as far as Europe will be concerned, will not permit any support from 'Star Wars' concepts.

Fack repeated Mr Wörner's concern that Reagan's SDI project would lead to zones of different security, thereby furthering the tendency towards transatlantic decoupling.²³⁵

With the change in the official policy towards the SDI, a gradual shift in editorial policy could be observed. The larger newspapers close to the CDU/CSU – *Die Welt, Frankfurter Allgemeine Zeitung, Rheinischer Merkur* – tended to favour SDI during 1985 and stated that Germany could not afford to be politically decoupled from the United States. Others, like *Frankfurter Rundschau, Süddeutsche Zeitung* and the weeklies *Die Zeit, Der Spiegel* and *Stern* remained sceptical. During 1984 and 1985 several books provided detailed arguments for both the pro and the contra stances in this second intensive security debate of the 1980s.²³⁶

VIII THE IMPACT OF SDI ON PUBLIC OPINION

According to a public opinion poll taken in March 1985, some 60 per cent of the West German population opposed SDI, 13 per cent were in favour and 23 per cent remained undecided. Even among the CDU/CSU and FDP supporters, according to this poll, 47 per cent were opposed and only 28 per cent were in favour, with 25 per cent remaining undecided. Among the supporters of the opposition parties, the SPD and the Greens, the percentage of those opposed to the SDI was 78 per cent.²³⁷

According to another opinion poll of a Mannheim-based research group, 46 per cent of the West German population were opposed to participation in the SDI research programme, while 31 per cent were in favour and 22 per cent remained undecided.²³⁸

IX CONCLUSION

The SDI debate, the second major security debate in West Germany in the 1980s – in particular the debate about German participation – has polarised the cabinet, the parliament, the parties, the business world and the scientific community and further fragmented the precarious concensus on national security matters. Many former INF supporters joined the SDI critics but only a few INF critics joined forces with the SDI supporters on ethical or moral grounds.

Until early 1986 the debate took place nearly exclusively on the declaratory level. The direct or indirect costs of potential participation of German companies and research institutes in the SDI programme remained uncertain. No cost estimates have been published about the fiscal burden of an EDI or about a comprehensive air defence system as has been called for by both Mr Wörner and a close advisor to Mr Genscher.[239] How both a continued modernisation of NATO's conventional forces and SDI and EDI are to be financed, given the tremendous manpower challenge that will confront the Bundeswehr in the mid-1990s,[240] remains unclear. The debate on the longer-term strategic and geopolitical implications of SDI and EDI is only just beginning.[241]

X FROM EDI TO EXTENDED AIR DEFENCE AND REACTIONS TO REYKJAVIK

In early 1986, the official approach regarding the defence against tactical ballistic missiles shifted from a *top-down approach* (application of SDI-related technologies for the defence of Western Europe) to a *bottom-up approach* (modernisation of traditional air defence systems, e.g. of the Patriot for AT(B)M functions). Simultaneously, the official threat analysis of the Federal Government towards the Soviet TBMs changed drastically. After the Reykjavik summit, the different tendencies within the coalition government that had dominated the whole SDI debate re-emerged: while Foreign Minister Genscher supported the zero-zero option for INF, members of the majority parties CDU/CSU indicated their criticism by calling on the United States government to make any INF agreement dependent on major reductions in the Soviet SRINF and on parity of the conventional forces in Europe. With the shift in the threat perception with respect to Soviet SRINF between October 1985 and early 1986, these systems have acquired three different political functions:

(a) to legitimise an extended air defense (M. Wörner since early 1986);
(b) to call for new preconditions before an INF agreement could be signed (CDU/CSU members in the Bundestag);
(c) to legitimise the call for parity in SRINF systems or for a levelling-up by introducing new Western tactical ballistic missiles (F. J. Strauss).

The public debate on these three aspects became one defence-related issue during the Federal election campaign from October 1986 to January 1987 that was won by the present government.

Notes

1. Hans Günter Brauch, 'INF and the Current NATO discussion on Alliance Strategy', in Hans-Henrik Holm and Nikolaj Petersen (eds), *The European Missiles Crisis: Nuclear Weapons and Security Policy* (London: Pinter, 1983) pp. 156–202; Hans Günter Brauch, *Die Raketen kommen! Vom NATO-Doppelbeschluss bis zur Stationierung* (Köln: Bund, 1983).
2. Hans Günter Brauch, 'From Strategic to Tactical Defense? European Reactions to the "Star Wars" Vision', in John McIntyre (ed.), *International Space Policy* (New York: Greenwood, 1987).
3. See Holm and Petersen (eds), *The European Missile Crisis*.
4. See Raymond L. Garthoff, *Detente and Confrontation: American–Soviet Relations–From Nixon to Reagan* (Washington: Brookings Institution, 1985); Ernst-Otto Czempiel (ed.), *Amerikanische Aussenpolitik im Wandel, Von der Entspannungspolitik Nixons zur Konfrontation unter Reagan* (Stuttgart, Berlin, Köln, Mainz: Kohlhammer, 1982).
5. Lynn E. Davis, 'The INF policy of the United States', in Holm and Petersen (eds), *The European Missile Crisis*, pp. 67ff.
6. See the public warning by Richard Burt, then Assistant Secretary of State for European Affairs in February 1985.
7. See Marsha McGraw and Jeffrey D. Porro (eds), *Nuclear Weapons in Europe: Modernization and Limitation* (Toronto, Lexington: Lexington Books, 1983).
8. Fred Charles Iklé, 'Nuclear Strategy: Can There Be a Happy Ending?', *Foreign Affairs*, 63/4 (Spring 1985) pp. 800–26, especially p. 820.
9. Iklé, p. 824.
10. Iklé, p. 821ff.
11. Hans Günter Brauch, 'What resources for defense–Rejoinder', Netherlands Institute of International Relations, Clingendael (ed.), *Conventional Balance in Europe: Problems, Strategies and Technologies*, Zoetermeer, 11–13 May 1984, pp. 43–9.

12. Wolfgang R. Vogt, 'The Acceptance Question and the Legitimacy of NATO's Nuclear Defence Posture in the FRG', in Hans Günter Brauch and Robert Kennedy (eds), *Alternative Conventional Defense Postures in the European Theater–The Future of the Military Balance and Domestic Constraints* (forthcoming in 1987).
13. Bernd Grass, 'The Manpower Shortage of the Bundeswehr till the Year 2000', in Brauch and Kennedy (eds), *Alternative Conventional Defense*.
14. Hartmut Bebermeyer, 'The Fiscal Crisis of the Bundeswehr', in Brauch and Kennedy (eds), *Alternative Conventional Defense*.
15. Lothar Wilker, *Die Sicherheitspolitik der SPD 1955–1966. Zwischen Wiedervereinigung und Bündnisorientierung* (Bonn-Bad Godesberg: Verlag Neue Gesellschaft, 1977); Hans Günter Brauch, 'Arms Control and Disarmament Decisionmaking in the Federal Republic of Germany', in Hans Günter Brauch and Duncan L. Clarke (eds), *Decisionmaking for Arms Limitation–Assessments and Prospects* (Cambridge, Mass.: Ballinger, 1983) pp. 131–74.
16. See note 1.
17. Christoph Bertram, 'Ein Schritt vor, ein Schritt zurück–Reagans Pläne zur Raketenabwehr unterhöhlen die Chancen der Genfer Verhandlungen', *Die Zeit*, 1 April 1983.
18. Dieter Schröder, 'Ronald Reagans Horror-Vision', *Süddeutsche Zeitung* 25 March 1983, p. 4.
19. Reuter, 'US-Vertreter verwirrten Europäer', *Frankfurter Rundschau* 29 February 1984.
20. Hermann Scheer, 'Der Ost–West-Konflikt und die Antwort der NATO–Eine kritische Auseinandersetzung mit dem AirLand Battle-Konzept', in Hans Günter Brauch (ed.), *Sicherheitspolitik am Ende? Eine Bestandsaufnahme, Perspektiven und neue Ansätze* (Gerlingen: Bleicher Verlag, 1984) pp. 204–13.
21. NAA, *Military Committee, General Report on Alliance Security, Mr Michael Forrestal (Canada), General Rapporteur* (Brussels: North Atlantic Assembly, November 1984) p. 23.
22. Walther Stützle, 'Weltraum-Gefahr', *Stuttgarter Zeitung*, 11 April 1984.
23. Elizabeth Pond, 'Europe fears "star wars" may destroy, not defend West', *Christian Science Monitor*, 12 April 1984, p. 1.
24. Robert Ruby, 'Possible US space-based defense causes some discomfort in Europe', *Baltimore Sun*, 11 July 1984, p. 2.
25. See reference to Lellouche and Moisi, chapter 14 in this volume, notes 157 and 158.
26. See Brauch, 'Arms Control'.
27. See chapter 9 by Carton in this volume.
28. 'Ein "kleines" SDI-Programm gegen Kurzstreckenraketen? Dregger berichtet von amerikanischen Überlegungen', *Frankfurter Allgemeine Zeitung*, 2 July 1985.
29. 'Paris lädt zu Weltraumprojekt ein – Hernu schlägt enge militärische Zusammenarbeit in Europa vor', *Frankfurter Rundschau*, 22 July 1985.
30. See chapter 8 in this volume.

31. William Drozdiak, 'Bonn Worried by US Plans for Space Weaponry', *Washington Post*, 11 April 1984, p. 29.
32. Fritz Wirth, 'Bonn sucht Zweifel in USA zu zerstreuen', *Die Welt*, 30 March 1985; 'Strauss fordert europäische SDI', *Frankfurter Allgemeine Zeitung*, 23 April 1985; 'Strauss sprach mit Reagan auch über Schutz für Europa', *Süddeutsche Zeitung*, 8 May 1985.
33. 'Geissler: SDI-Gegner vertreten unmoralische Position', *Neue Osnabrücker Zeitung*, 20 April 1985; 'Geissler nennt Weltraumrüstungs-Gegner unmoralisch', *Süddeutsche Zeitung*, 22 April 1985; 'SDI gibt uns eine grosse Chance', *Union in Deutschland*, 25 April 1985.
34. For details see Hans Günter Brauch, *Antitactical Missile Defense-Will the European Version of SDI Undermine the ABM-Treaty, AFES Papier 1* (Stuttgart: AG Friedensforschung und Europäische Sicherheitspolitik (AFES), Institut für Politikwissenschaft, Universität Stuttgart, July 1985); see also Brauch, in McIntyre (ed.), *International Space Policy*.
35. '"Star Wars": Misgivings in Bonn', *Los Angeles Times*, 15 April 1984; William Drozdiak, 'Bonn Expresses Concern Over US Space Strategy', *International Herald Tribune*, 12 April 1984; James Markham, 'Bonn Is Worried by US Arms Research', *New York Times*, 14 April 1984; Elizabeth Pond, 'West Germany reacts to "star wars" defense – Critics say system would harm Alliance not Soviets', *Christian Science Monitor*, 14–20 April 1984; ' "Star Wars", plan hit by W. German', *Washington Times*, 10 April 1984, p. 6.
36. Charles W. Corddry, 'Bonn endorses "star wars" effort', *Baltimore Sun*, 13 July 1984, p. 2; Hans Günter Brauch, 'Sicherheitspolitische Wende–Wie Manfred Wörner vom Kritiker zu einem "Star Wars" –Befürworter wurde', *Sozialdemokratischer Pressedienst*, 3 August 1984.
37. See Brauch, in McIntyre (ed.), *International Space Policy*; Lutz Krusche, 'Genscher für Signal an Moskau', *Frankfurter Rundschau*, 9 May 1984; ' "Vertrauensvolle" Gespräche in Washington', *Frankfurter Allgemeine Zeitung*, 10 May 1984; Amy Stromberg, 'Bonn's foreign minister wants Star Wars ban', *Washington Times*, 22 May 1984, p. 6; Alexander Szandar, 'Der "Krieg der Sterne" aus Bonner Sicht–Ein Konzept bereitet Kopfzerbrechen', *Süddeutsche Zeitung*, 6 September 1984.
38. Christoph Bertram, 'Kein Frieden in den Sternen – Bonn sollte sich dem Wettrüsten im All entgegenstellen', *Die Zeit*, 21 September 1984.
39. See for details Brauch in McIntyre (ed.), *International Space Policy*.
40. See below in detail for sources.
41. See Chapter 9 by Carton in this volume.
42. See chapter 8 by Brauch in this volume.
43. See chapter 14 by Brauch in this volume.
44. The 'secret' text has been published in: *Frankfurter Rundschau*, 19 April 1986, p. 10.
45. See Brauch, *Antitactical Missile Defense*; Brauch, in McIntyre (ed.), *International Space Policy*.
46. Helmut Kohl, 'Die Bundesrepublik Deutschland und Europa im Nord-

atlantischen Bündnis', *Europäische Wehrkunde* 34 (March 1985) pp. 133–40.
47. 'Stellungnahme der Bundesregierung zur Strategischen Verteidigungsinitiative (SDI) des Präsidenten der Vereinigten Staaten von Amerika', in Bundespresse- und Informationsamt der Bundesregierung (ed.), *SDI-Dokumentation* (Bonn: BPIA, April 1985) pp. 64–65.
48. 'Erklärung der Bundesregierung zur Strategischen Verteidigungsinitiative des Präsidenten der Vereinigten Staaten von Amerika', in *Deutscher Bundestag, 10. Wahlperiode, 132. Sitzung 18 April 1985*, pp. 9715ff.
49. Helmut Kohl, Address to the North Atlantic Assembly, 20 May 1985, in NAA: *Official Record of the Spring Plenary Sitting, held in the Landtag of Baden-Württemberg, Stuttgart, Monday 20 May 1985*, AC 121, SA/CR1(85)3 (Brussels: North Atlantic Assembly, June 1985) pp. 3–8.
50. Der Bundesminister der Verteidigung, *Weissbuch 1985, Zur Lage und Entwicklung der Bundeswehr* (Bonn: BMVg, June 1985) pp. 30–3.
51. Der Bundesminister der Verteidigung, *Bericht zur Strategischen Verteidigungsinitiative (SDI) der USA* (Bonn, BMVg, 20 August 1985).
52. William Drozdiak, 'Bonn Seeks Role in SDI for Firms, Rejects Funding by Government', *International Herald Tribune*, 19 December 1985; Hans-Herbert Gaebel, 'Alles bleibt offen', *Frankfurter Rundschau*, 19 December 1985; 'West German Cabinet Approves Role in SDI; Negotiations On Formal Agreement Set for January', *Aviation Week & Space Technology*, 23 December 1985.
53. 'US Strategic Defense Initiative – Government Press Spokesman Peter Boenisch at a Press Conference, 13 February 1985', broadcast by Informationsfunk der Bundesregierung.
54. See note 46.
55. See note 48.
56. Christoph Bertram, 'Lockrufe ausÜbersee. Was bringt die Beteiligung am amerikanischen Programm?', *Die Zeit*, 26 April 1985.
57. 'Entschliessungsantrag der Fraktion der CDU/CSU und der FDP im Unterausschuss Abrüstung und Rüstungskontrolle des Deutschen Bundestages' approved 28 February 1985.
58. William Drozdiak, 'Kohl: Europeans Need Joint "Star Wars" Stand. Unity Seen Boosting Influence on Prospect', *Washington Post*, 21 March 1985, p. 16.
59. Jerôme Dumoulin, 'Le prix des étoiles', *L'Express*, 31 May 1985; Angela Nacken, 'Zwischen Eureka und SDI', *Frankfurter Allgemeine Zeitung*, 10 May 1985, p. 1.
60. See note 49.
61. James M. Markham, 'Kohl, in Speech to NATO Aides, Seems Cooler to "Star Wars"', *New York Times*, 21 May 1985; William Drozdiak, 'Kohl Hedges Backing for "Star Wars" Risk as Well as Opportunity Seen for Western Alliance', *Washington Post*, 21 May 1985, p. 21.
62. '"We Need a Lot More Answers" A top aide to Chancellor Kohl speaks out', *Time*, 26 May 1985.
63. 'Aussenpolitik: Viele Hühner', *Der Spiegel*, 24/1984, 10 June 1985, p. 29.

64. See chapter 8 by Brauch in this volume.
65. *Weissbuch 1985*, pp. 30–3.
66. *Weissbuch 1985*, p. 30–1.
67. *Weissbuch 1985*, p. 33.
68. See note 52 above.
69. Editorial by *International Herald Tribune*, 'Bonn's Ambiguity on SDI', *International Herald Tribune*, 21 December 1985.
70. Rüdiger Moniac, 'SDI: Bonn sondiert in Washington. Kohls Berater Teltschik soll Expertengespräche über Forschungsbeteiligung vorbereiten', *Die Welt*, 8 June 1985; Fritz Wirth, 'Bonn sondiert Bedingungen für SDI-Teinahme', *Die Welt*, 13 June 1985; 'Ernüchterung in Bonn über Beteiligung an SDI-Bonn erwägt europäisches Raketenabwehrsystem–wenig Interesse in USA an Mitsprache', *General-Anzeiger*, 21 June 1985.
71. Hedrick Smith, 'Bonn Pact Is Seen Over Star Wars', *New York Times*, 18 September 1985, p. 7; William Drozdiak, 'Bonn Is Likely To Join In SDI Research Effort', *International Herald Tribune*, 16 September 1985; W. Drozdiak, 'Bonn Group Comes to US for SDI Talks', *Washington Post*, 5 September 1985, p. 1; David B. Ottaway, 'Agreements Near with Bonn and London on SDI Research', *Washington Post*, 17 September 1985, p. 1.
72. See note 49 above.
73. 'FDP reagiert empört', *Frankfurter Rundschau*, 3 October 1985.
74. 'Strauss und Waigel mahnen mit Blick auf ein SDI-Rahmenabkommen zur Eile', *Frankfurter Allgemeine Zeitung*, 14 October 1985; 'Strauss preist Londons Kurs', *Frankfurter Rundschau*, 5 November 1985; 'Strauss will von Kohl eine Antwort', *Süddeutsche Zeitung*, 14 November 1985; Norbert A. Skorz, 'Warnschuss aus Bayern', *Bonner Rundschau*, 14 November 1985.
75. 'Bonner Vorbehalte gegen Reagans Weltraumpläne, *Neue Zürcher Zeitung*, 21 March 1985.
76. 'Bonn endorses Howe's "Star Wars" misgivings', *The Times*, 26 March 1985.
77. 'Möllemanns "Klarstellungen" von Dallas in Bonn zurechtgerückt', *Frankfurter Allgemeine Zeitung*, 2 April 1985.
78. Hans-Dietrich Genscher, 'Ein neues Kapitel in den West–Ost-Beziehungen', *Der Bundesminister des Auswärtigen informiert*, 18 March 1985.
79. James M. Markham, 'West Germany Cooler to Star Wars', *New York Times*, 28 March 1985.
80. 'Genscher und Kohl halten Distanz zu den Weltraumwaffenplänen Amerikas', *Süddeutsche Zeitung*, 25 March 1985.
81. 'US views to halt Star Wars if it spurs Soviet arms buildup', *Atlanta Constitution*, 15 October 1985, p. 4; Eghard Mörbitz, 'Bundeskanzler hält Verzicht der USA auf Weltraumrüstung für möglich', *Frankfurter Rundschau*, 14 October 1985.
82. 'Genscher warnt USA vor Überrüstung', *Rhein-Neckar-Zeitung*, 23 October 1985.
83. 'Bonn bittet Washington um Aufklärung – Bundesregierung über Auslegung des ABM-Vertrages durch die USA beunruhigt', *Süddeutsche Zeitung*, 12 October 1985.

84. 'Wir werden SDI nicht wegverhandeln', *Der Spiegel*, 43/1985, pp. 19–22. 'Die Frage der Verlässlichkeit"', *Der Spiegel*, 44/1985, pp. 17–20.
85. William Drozdiak, 'Kohl Caught in Dilemma', *Washington Post*, 7 May 1985, p. 22; Elizabeth Pond, 'W. Germany tries to balance European, US space research', *Christian Science Monitor*, 3 May 1985; p. 9.
86. Anna Tomforde, 'Genscher joins critics of Star Wars', *Guardian*, 19 March 1985; William Drozdiak, 'Kohl: Europeans Need Joint "Star Wars" Stand', *Washington Post*, 21 March 1985; Frank Johnson, 'Bonn President has reservations on SDI', *The Times*, 22 March 1985.
87. Konrad Seitz, 'SDI: the technological challenge for Europe', *World Today*, September 1985, pp. 154–7; 'Für Genscher ist ein Einvernehmen mit Paris wichtiger als SDI-Beteiligung', *Frankfurter Allgemeine Zeitung*, 8 May 1985; Rudolph Chimelli, 'Weltraumwaffen entzweien Bonn und Paris', *Süddeutsche Zeitung*, 1 March 1985; 'Kohl ruiniert sein Ansehen', *Der Spiegel*, 22/1985, 27 May 1985, pp. 19–21.
88. Louis Wiznitzer, 'Star wars brings West Germany and France together', *Christian Science Monitor*, 1 October 1985, p. 10.
89. Hans Klein, 'Die "Achse" verliert sich in den Sternen', *Rhein-Neckar-Zeitung*, 19 December 1985.
90. 'Mitterrands Fragen an Kohl werden drängender', *Frankfurter Allgemeine Zeitung*, 16 January 1986; 'Bonn und Paris prüfen "erweiterte Luftverteidigung"', *Rhein-Neckar-Zeitung*, 16 January 1986; Hans-Hagen Bremer, 'Gemeinsame Abwehr als Thema', *Frankfurter Rundschau*, 16 January 1986.
91. See note 46.
92. See note 48.
93. See note 49.
94. 'Der SDI-Beschluss im Wortlaut', *Rhein-Neckar-Zeitung*, 19 December 1985; 'Bangemann erweiterter Verhandlungsauftrag kompliziert die Gespräche in Washington', *Frankfurter Allgemeine Zeitung*, 17 January 1986; 'Genscher will den Staat aus der SDI-Forschung heraushalten–Der Bundesaussenminister schiebt die Meinung eines FDP-Landesverbandes in den Vordergrund', *Frankfurter Allgemeine Zeitung*, 10 October 1985.
95. Dietrich Möller, 'Späth wirbt für Zusammenarbeit bei SDI', *Badische Neueste Nachrichten*, 16 February 1985; 'Späth nimmt das amerikanische Weltraumprojekt in Schutz', *Frankfurter Allgemeine Zeitung*, 28 February 1985.
96. 'Wissen die Europäer, was sie riskieren? Baden-Württembergs Ministerpräsident Lothar Späth über SDI und die europäische Technologie-Lücke', *Der Spiegel* 39/1985, 11 March 1985; Wolfgang Borgmann, 'In Moskau gilt Späth als interessanter Gesprächspartner', *Stuttgarter Zeitung*, 20 March 1985.
97. '"Meine SDI-Pläne sind nicht mehr aktuell", Interview der "Stuttgarter Zeitung" mit Ministerpräsident Späth', *Stuttgarter Zeitung*, 22 June 1985; 'Soll Bonn sich am Krieg der Sterne beteiligen–Streitge-

spräch zwischen Oskar Lafontaine und Lothar Späth', *STERN*, 23/1985, 30 May 1985.
98. 'Rüstung-Was weiss ich denn', *Der Spiegel*, 15/1985, 8 April 1985, pp. 21–3; 'Bonn: Eigener SDI-Beitrag', *Der Spiegel*, 18/1985, 29 April 1985.
99. 'Minister Riesenhuber: Meine Bedingungen für SDI', *Quick*, 23 May 1985.
100. 'Regierung geht auf Abstand zu SDI–Genscher und Riesenhuber verschärfen ihre Kritik am Star-Wars-Programm', *Frankfurter Rundschau*, 28 May 1985.
101. Wolf J. Bell, 'Genscher mahnt zu behutsamer Debatte über SDI', *General Anzeiger*, 11 May 1985.
102. 'Spiegel Gespräch: "Mit zivilem Nutzen nicht zu rechtfertigen" Forschungsminister Heinz Riesenhuber über die technologische Zusammenarbeit bei SDI und Eureka', *Der Spiegel*, 37/1985, 9 September 1985; 'Zweifel Riesenhubers an SDI', *Süddeutsche Zeitung*, 9 September 1985; 'Beifall in Bonn für Riesenhuber – Rückt die Bundesregierung vorsichtig von SDI ab?', *Hannoversche Allgemeine*, 10 September 1985; Udo Bergdoll, 'Kontrastierende Begleitmusik zu einer Bonner Mission', *Süddeutsche Zeitung*, 12 September 1985; 'Gefahr erkannt', *Die Zeit*, 13 September 1985.
103. Wolfgang Hofmann, 'Der Krieg ist nicht der Vater. Zum Thema SDI: Militärische Forschung bringt nur geringen zivilen Nutzen', *Die Zeit*, 10 January 1986, pp. 17–18.
104. 'Germans Insist on Technology Gains as Part of SDI Cooperation', *Aviation Week & Space Technology*, 8 April 1985, p. 21; 'Weltraum-Rüstung: Auf dem Tablett', *Der Spiegel*, 28/1985, 8 July 1985, p. 76.
105. Günther Bading, 'SDI-Forschung kostet die Bonner Kasse keinen Pfennig – WELT-Gespräch mit Minister Schäuble/Konkrete Ergebnisse im Herbst', *Die Welt*, 12 July 1985.
106. For details see Brauch in McIntyre (ed.), *International Space Policy*; 'Französische Initiative für eine europäische "Star Wars"-Version', *Neue Zürcher Zeitung*, 20 April 1985.
107. See chapter 9 by Carton in this volume.
108. 'Genscher spricht über Eureka-Finanzierung – "Technologiegemeinschaft" in Europa?', *Frankfurter Allgemeine Zeitung*, 22 May 1985; Axel Krause, 'Bonn Backs European Technology Push', *International Herald Tribune*, 23 May 1985; 'Für Bonn hat Paris Vorrang', *Süddeutsche Zeitung*, 25 May 1985; 'Warnung vor einer technologischen Provinz', *Frankfurter Allgemeine Zeitung*, 28 May 1985; Horst Schreitter-Schwarzenfeld, 'Ein europäisches Luftschloss gegen US-Science-fiction', *Frankfurter Rundschau*, 7 June 1985; 'Genscher: Durch Eureka "neue Perspektiven"', *General-Anzeiger*, 19 July 1985.
109. 'Im Gespräch: Europäischer Kraftakt', *Frankfurter Rundschau*, 2 July 1985; 'Forschungspolitik: Riesenhuber: Eureka wird kein neuer Subventionstopf', *Die Welt*, 3 August 1985; 'Riesenhuber will keinen neuen staatlichen Subventionstopf schaffen', *Handelsblatt*, 5 August 1985.

110. 'Streit um Eureka', *Der Spiegel*, 33/1985, 12 August 1985; Gerda Struck, 'Milliardenbetrag für Eureka-Projekt–Bonner Minister einigen sich über Finanzierung', *Frankfurter Rundschau*, 6 September 1985; 'Teures Eureka', *Der Spiegel*, 41/1985, 7 October 1985; Gerda Struck, 'Bonn lehnt zusätzliche Mittel für Eureka ab', *Frankfurter Rundschau*, 18 October 1985; 'Kohl will keine finanziellen Zusagen geben – Über erste Eureka-projekte soll erst 1986 entschieden werden', *Frankfurter Rundschau*, 12 October 1985; 'Eureka: Finanzielles Signal', *Der Spiegel*, 44/1985, 28 October 1985; Horst Schreitter-Schwarzenfeld, 'Kohl verweist auf Wirtschaft – Kanzler empfiehlt Eureka-Projekt der privaten Initiative', *Frankfurter Rundschau*, 6 November 1985.
111. 'Bis zu 40 Millionen Mark für Eureka', *Süddeutsche Zeitung*, 9 November 1985.
112. 'Späth: Mit Eureka nicht warten', *Frankfurter Allgemeine Zeitung*, 1 November 1985; 'Strauss kritisiert Eureka-Initiative', *Frankfurter Allgemeine Zeitung*, 30 November 1985; 'Strauss hält Eureka für überflüssig', *Neue Osnabrücker Zeitung*, 29 November 1985.
113. Axel Krause, 'Electronic Plans Set in Europe, 4 Firms Draft Eureka Project for $1 Billion', *International Herald Tribune*, 27 June 1985; Axel Krause, 'Consortium Proposed to Coordinate European Work on Eureka, SDI', *International Herald Tribune*, 25 June 1985, p. 3; 'Forschung/Deutsche Unternehmen bisher ohne konkretes SDI-Angebot–Europäische Elektronikunternehmen kooperieren bei Eureka-Projekten', *Handelsblatt*, 27 June 1985; 'Eureka-Forschungsprojek–Vier Elektronikkonzerne sagen ihre Unterstützung zu', *Die Welt*, 27 June 1985; 'Accord avant le Sommet des dix: Les quatre grands de l'électronique européenne s'engagent dans Eurêka', *Le Monde*, 27 June 1985.
114. Elizabeth Pond, 'W. Europe decides to pull on defense research. "Eureka" would keep some "star wars" type research in Europe', *Christian Science Monitor*, 28 June 1985, p. 8.
115. Jeffrey M. Lenorowitz, 'European Ministers Meet in Paris to Discuss Eureka Prospects', *Aviation Week & Space Technology*, 15 July 1985, pp. 21–2.
116. Konrad Seitz, 'SDI: the technological challenge for Europe', *The World Today*, September 1985, pp. 154–7.
117. Seitz, pp. 154–7.
118. Hans Rühle, 'A European Perspective on the US–Soviet Strategic–Military Relationship', in William Schneider, Jr et al., *US Strategic–Nuclear Policy and Ballistic Missile Defense: The 1980s and Beyond* (Cambridge: Institute for Foreign Policy Analysis, 1980), p. 51.
119. Hans Rühle, 'Reagans Verteidigungsvision: Löcher im Drahtverhau der Sicherheitsdoktrin – Das Defensivkonzept der Zukunft löst Europas Probleme kaum', *Rheinischer Merkur, Christ und Welt*, 1 April 1985. The English translation has been quoted from: Hubertus G. Hoffman, 'A Missile Defense for Europe?', *Strategic Review* (Summer 1985) p. 47.

120. Fred Hiatt, 'US Antisatellite Plan Draws Fire – NATO Ministers Reportedly Express Skepticism, Anxiety', *Washington Post*, k4 April 1984, p. 18.
121. Walter Pincus, '"Star Wars" Defense Plan Broadened, with Protection for European Allies', *Washington Post*, 25 April 1984, p. 1.
122. See notes above, especially notes 2, 34, 35.
123. Hans Rühle, 'Tschernenkos "Krieg der Sterne" Ein geheimes Kapitel der sowjetischen Raketen-Rüstung', *Frankfurter Allgemeine Zeitung*, 22 January 1985, p. 9.
124. Hans Rühle, '"An die Grenzen der Technologie". Der Planungschef des Verteidigungsministeriums, Hans Rühle, zur wissenschaftlichen Kritik am SDI-Projekt', *Der Spiegel*, 48/1985, 25 November 1985, pp. 155–9.
125. See *Bericht zur Strategischen Verteidigungs Initiative*, pp. 17–20.
126. 'Strauss fordert europäisches SDI – Kein "Konkurrenzunternehmen", sondern "Ergänzung"', *Frankfurter Allgemeine Zeitung*, 23 April 1985.
127. 'Abwehrsysteme gegen Kurzstreckenraketen und Marschflugkörper gefordert–Dregger befürchtet Abwanderung deutscher Wissenschaftler', *Frankfurter Allgemeine Zeitung*, 23 April 1985; 'Ein "kleines" SDI-Programm gegen Kurzstreckenraketen? Dregger berichtet von amerikanischen Überlegungen/Mitwirkung europäischer Länder und Firmen möglich', *Frankfurter Allgemeine Zeitung*, 2 July 1985; Günther Bading, 'Ein "kleines SDI" für Europa als "integrierte Gemeinschaftsaufgabe" – Welt-Gespräch mit dem CDU/CSU-Fraktionsvorsitzenden Alfred Dregger', *Die Welt*, 4 July 1985.
128. Kai-Uwe von Hassel, '"SDI hat einen wichtigen Impuls für den Rüstungskontrolldialog gegeben". Von Hassel bringt die "Europäische Verteidigungsinitiative" auf den Weg', *Die Welt*, 22 June 1985.
129. 'Ernüchterung in Bonn über Beteiligung an SDI – Bonn erwägt europäisches Raketenabwehrsystem. Wenig Interesse in USA an Mitsprache', *General-Anzeiger*, 21 June 1985.
130. 'Bonn erwägt europäische Raketenabwehr. Staatlich finanzierte Beteiligung an SDI offenbar nicht mehr im Gespräch', *Süddeutsche Zeitung*, 22 June 1985.
131. 'Wörner will Europäische Verteidigungsinitiative', *Süddeutsche Zeitung*, 5 September 1985; 'Wörner: Das SDI-Vorhaben ist nicht nur legitim, sondern notwendig', *Frankfurter Allgemeine Zeitung*, 25 October 1985.
132. Michael Freazel, 'German Minister Proposes Initiative to Improve European Defenses', *Aviation Week & Space Technology*, 9 December 1985; 'Abwehrpläne der Verteidigungsminister', *Rhein-Neckar-Zeitung*, 4 December 1985; 'Bemühungen Bonns um verbesserten Schutz gegen Angriffe aus der Luft-Trainingszentrum für Tiefflüge', *Frankfurter Allgemeine Zeitung*, 4 December 1985; 'Wintertagung der NATO in Brüssel', *Neue Zürcher Zeitung*, 5 December 1985.
133. 'Altenburg befürwortet Teilnahme am SDI-Programm', *Hamburger*

Abendblatt, 19 April 1985; Karl Feldmeyer, 'Die Risikogemeinschaft Gespräch mit Generalinspekteur Altenburg', *Frankfurter Allgemeine Zeitung*, 29 November 1985.
134. Lothar Ruehl, 'Eine Raketenabwehr auch für Europa. Voraussetzungen und Folgen für Strategie und Rüstungskontrolle im Ost–West-Verhältnis', *Frankfurter Allgemeine Zeitung*, 17 January 1985, p. 10.
135. 'West Germany Agrees to Join in Research on Space-Based Defense', *International Herald Tribune*, 28 March 1986; Horst Schreitter-Schwarzenfeld, 'Abkommen über SDI wird geheimgehalten – Bangemann über Abschluss mit USA zufrieden', *Frankfurter Rundschau*, 29 March 1986; 'Nach Bangemanns Washingtoner Unterschriften liegt der Rahmen der deutschen Beteiligung fest', *Frankfurter Allgemeine Zeitung*, 29 March 1986; 'Rau sieht die Wirtschaftsbeziehungen "unter einen militärischen Imperativ gestellt"', *Frankfurter Allgemeine Zeitung*, 29 March 1986.
136. 'Der geheime SDI-Vertrag', *Express*, 18 April 1986, p. 6; '"Strategisch sensitive Technologien müssen geschützt werden". Das geheime Rahmenabkommen über eine deutsche Beteiligung an der SDI-Forschung/Ein Abdruck aus dem Kölner "Express"', *Frankfurter Rundschau*, 19 April 1986, p. 10; '"Vor unbefugter Offenlegung schützen". Die SDI-Vereinbarungen sollen vor allem den Osthandel reglementieren', *Der Spiegel*, 17, 1986, 21 April 1986, pp. 27–8.
137. 'Amerikanisch-deutsches SDI-Abkommen', *Neue Zürcher Zeitung*, 31 March 1986; Claus Genrich, 'Genscher bürgt für Bangemann – Nach der Unterzeichnung der SDI-Texte muss die FDP-"Basis" überzeugt werden', *Frankfurter Allgemeine Zeitung*, 29 March 1986.
138. Theo Sommer, 'Ein Schritt in die falsche Richtung. Das SDI-Abkommen bringt nichts, gefährdet aber unsere Sicherheit', *Die Zeit*, 28 March 1986, p. 1; Horst Schreitter-Schwarzenfeld, 'Übertriebene Anstrengung', *Frankfurter Rundschau*, 29 March 1986; 'Streitstoff in Fülle', *Frankfurter Allgemeine Zeitung*, 29 March 1986.
139. Rolf Zundel, 'Deutsch-amerikanische Beziehungen–Der Regenschirm wird eingerollt. Bonn und das SDI-Abkommen: Wie sich die Bundesregierung in unnötigen Problemen verhedderte", *Die Zeit*, 28 March 1986; 'Strauss kritisiert Geheimhaltung der SDI-Abkommen', *Frankfurter Allgemeine Zeitung*, 8 April 1986; "Baum: SDI-Abkommen veröffentlichen', *Frankfurter Rundschau*, 12 April 1986.
140. 'Bangemann hebt den zivilen Charakter der SDI-Verträge hervor', *Frankfurter Allgemeine Zeitung*, 18 April 1986; 'SDI-Abkommen verteidigt – Kohl: Nicht von der Entwicklung abkoppeln lassen', *Frankfurter Rundschau*, 18 April 1986.
141. 'Auch die Wirtschaft rätselt über SDI', *Frankfurter Allgemeine Zeitung*, 3 April 1986; 'Rodenstock skeptisch über Nutzen der SDI-Forschung', *Frankfurter Allgemeine Zeitung*, 15 April 1986.
142. See note 136.
143. See section IIa in this chapter.
144. 'SDI: Die geheimen Briefe', *Express*, 20 April 1986.
145. Kurt Becker, 'SDI-Abkommen – Eine unnötige Blamage – Die Bun-

desregierung hat sich vom Pentagon übervorteilen lassen', *Die Zeit*, 25 April 1986, p. 7.
146. Horst Schreitter-Schwarzenfeld, 'Vetorecht beim Osthandel', *Frankfurter Rundschau*, 19 April 1986.
147. 'Die Opposition bekräftigt ihre Kritik an den SDI-Abkommen', *Frankfurter Allgemeine Zeitung*, 24 April 1986.
148. Walther Stützle, 'Weltraum-Gefahr', *Stuttgarter Zeitung*, 11 April 1984.
149. 'Würzbach: Andropows Angebot annehmen. Für Verhandlungen über Weltraumwaffen-Moskau will Moratorium', *Neue Osnabrücker Zeitung*, 19 August 1983.
150. Volker Rühe and Willy Wimmer, 'Über den derzeitigen Stand der fraktionsinternen Beratungen über Weltraumsysteme berichten', *CDU/CSU Fraktion, Pressereferat*, 9 October 1984; Alexander Szandar, 'Perspective: Bonn changes stance over Star wars concept', *The German Tribune*, 1140, 23 September 1984, p. 5.
151. Volker Rühe, 'Zu den amerikanischen Überlegungen über ein weltraumgestütztes Raketenabwehrsystem erklärt', *Pressedienst CDU/CSU-Fraktion*, 4 April 1984.
152. 'Skepsis in der CDU/CSU: Bonn soll Weltraumwaffenpläne besonders gründlich prüfen–Vizefraktionschef Rühe: Frage einer europäischen Beteiligung noch offen/Plädoyer für vorbeugende Rüstungskontrolle', *Süddeutsche Zeitung*, 22 February 1985; Thomas Meyer, 'Rühe gab Haltung Kohls zu Weltraumplan wieder', *Rhein-Sieg-Anzeiger*, 23 February 1985.
153. Fritz Wirth, 'SDI-Kooperation noch nicht konkret', *Die Welt*, 25 February 1985.
154. Martin S. Lambeck, 'Interview mit dem CDU-Politiker Volker Rühe – Nur wenig Geld für SDI im Bundes-Etat', *Hamburger Abendblatt*, 24 May 1985.
155. '"Sicherheit der USA und Europas ist unteilbar"-Rühe: Sorgfältige Prüfung des SDI-Projekts", *Westfälische Nachrichten*, 27 June 1985. '"SDI ist komplizierter als die Mondlandung", Spiegel-Interview mit dem CDU/CSU-Fraktionsvize Volker Rühe über die Rüstungs- und Aussenpolitik der Union', *Der Spiegel*, 52/1985, 23 December 1985, pp. 21–4.
156. Manfred Schell, 'Dregger für "Partnerschaft der Gleichheit"', *Die Welt*, 1 April 1985.
157. 'Abwehrsystem gegen Kurzstreckenraketen und Marschflugkörper gefordert–Dregger befürwortet Abwanderung deutscher Wissenschaftler', *Frankfurter Allgemeine Zeitung*, 23 April 1985.
158. 'Dregger schlägt "SD-Konsultativrat" vor. Deutsche Sicherheitsinteressen', *Frankfurter Allgemeine Zeitung*, 7 June 1985.
159. Karl Feldmeyer, 'Westeuropa darf nicht zu einer Zone minderer Sicherheit werden/Dregger: Amerikanische und sowjetische Raketenabwehrsysteme sind nicht zu verhindern', *Frankfurter Allgemeine Zeitung*, 12 June 1985.
160. Christian Lenzer, 'SDI and EUREKA – Beide Weltraumprojekte schliessen einander nicht aus', *Deutschland-Union-Dienst*, 29 May 1985.

161. 'Todenhöfer: Bonn an Forschung für Weltraumwaffen beteiligen. Appell an die USA blosse Information reicht nicht aus', *Neue Osnabrücker Zeitung*, 4 August 1984; 'Todenhöfer für Beteiligung in "Sternenkriegsforschung"', *Süddeutsche Zeitung*, 6 August 1984; 'Plädoyer für eine aktive Beteiligung der Europäer', *Welt am Sonntag*, 14 October 1984; 'Todenhöfer: Europa muss mitreden können. Stärkere Beteiligung an amerikanisch-sowjetischen Dialog verlangt', *Süddeutsche Zeitung*, 17 December 1984.
162. Jürgen Todenhöfer, 'Zehn Thesen zur Beteiligung an SDI', *Die Welt*, 21 February 1985; 'Union dringt auf SDI-Abkommen mit den USA – Meinungsverschiedenheit mit den Freien Demokraten erneut hervorgetreten', *Süddeutsche Zeitung*, 21 October 1985.
163. 'Geissler sorgt sich um SDI-Diskussion', *Süddeutsche Zeitung*, 18 April 1985.
164. 'Strauss fordert europäisches SDI', *Frankfurter Allgemeine Zeitung* 26 April 1985; 'Strauss legt sich für SDI ins Zeug', *Rhein-Neckar-Zeitung*, 6 July 1985; 'Strauss: SDI ist die Nagelprobe', *Rhein-Neckar-Zeitung*, 21 November 1985.
165. 'Schwenk – Strauss zu SDI', *Nürnberger Nachrichten*, 19 September 1985.
166. Peter Schmalz, 'CSU-Vorstand verlangt einstimmig ein Rahmenabkommen über SDI – Skepsis zu Genf/Bayern wird rheinland-pfälzische 218-Initiative unterstützen', *Die Welt*, 5 November 1985; Max-Hermann Bloch, ' "Sternenkrieg hat längst begonnen" – Strauss befürchtet bei Zögern mit Beitritt zu SDI unaufholbaren Rückstand für Europa', *Augsburger Allgemeine*, 5 November 1985.
167. 'Strauss sieht die CSU getäuscht', *Rhein-Neckar-Zeitung*, 13 January 1986; Carl-Christian Kaiser, 'Trommelschlag aus Bayern', *Die Zeit*, 17 January 1986.
168. 'FDP-Abrüstungsexperte gegen Beteiligung an Weltraumplänen', *Süddeutsche Zeitung*, 22 October 1984; 'Abrüstungsexperte der FDP widerspricht Waigels SDI-These', *Süddeutsche Zeitung*, 11 April 1985.
169. 'Mischnick bei Fraktionsvorsitzendenkonferenz in Düsseldorf: SDI Forschung im Rahmen geltender Verträge', *fdk Tagesdienst*, 20 April 1985.
170. Anna Tomforde, 'FDP warns Kohl against Star Wars', *Guardian*, 3 June 1985; 'Die Liberalen drängen auf eine europäische Technologiegemeinschaft', *Handelsblatt*, 3 June 1985; 'Die FDP kritisiert SDI-Pläne', *General-Anzeiger*, 3 June 1985.
171. Stephan A. Casdorff, 'Koalitionspartner sind über SDI zerstritten', *Bonner Rundschau*, 7 October 1985; Günther Bading, 'In der FDP Vorbehalte gegen SDI-Vereinbarung', *Die Welt*, 11 November 1985.
172. Claus Genrich, 'Bangemanns Blumen blühen – Nach Genschers SDI-Querelen rückt die FDP ihren Vorsitzenden ins Blickfeld', *Frankfurter Allgemeine Zeitung*, 28 November 1985; 'Die FDP tut sich schwer mit Washingtons Weltraumprogramm', *Frankfurter Allgemeine Zeitung*, 29 November 1985; Claus Genrich, 'Haussmann warnt die FDP vor SDI-Streit', *Frankfurter Allgemeine Zeitung*, 2 December 1985; 'Die Minister der FDP signieren die SDI-

Vereinbarungen nicht', *Frankfurter Allgemeine Zeitung*, 5 December 1985; 'FDP sucht Ausweg aus ihrem SDI-Dilemma', *Süddeutsche Zeitung*, 12 December 1985.
173. 'SDI-Effektives Netz', *Der Spiegel*, 51/1985, 16 December 1985, pp. 31–2.
174. See notes above and detailed article, *Der Spiegel* 22/1985, 23 December 1985, pp. 21–24; Claus Genrich, 'Bonn stellt sich auf Bangemanns Verhandlungsführung ein', *Frankfurter Allgemeine Zeitung*, 18 December 1985.
175. 'Die SPD zwischen Eureka und SDI', *Frankfurter Allgemeine Zeitung*, 28 November 1985.
176. Reiner Labusch, Eckart Maus und Wolfgang Send (eds), *Weltraum ohne Waffen. Naturwissenschaftler warnen vor der Militarisierung des Weltraums* (München: C. Bertelsmann, 1984).
177. See also my chapter 8 in this volume.
178. See Labusch/Maus/Send (eds), for the results of this Congress.
179. Egon Bahr, 'Partnerschaft wäre besser. Neue Instabilitäten durch Reagans Strategie', *Vorwärts*, 31 March 1983.
180. Deutscher Bundestag, 10. Wahlperiode, *Drucksache 10/2040, 26 September 1984, Antrag der Fraktion der SPD: Vertrag zur Begrenzung der militärischen Nutzung des Weltraums* (Bonn: Verlag Heger, 26 September 1984); see also the statement by Dr Scheer, in Deutscher Bundestag, 10. Wahlperiode, *98. Sitzung, 8 November 1984, Bundestags-Protokolle*, pp. 7054–7056.
181. 'SPD-Präsidium: Amerikanische Weltraumrüstung gefährdet die strategische Einheit der NATO: Eindeutiges Nein um "Krieg der Sterne"', *Vorwärts*, 2 March 1985.
182. 'Forschungspolitik/SPD zur Weltraum-Kooperation – Die Europäer sollen ihre Interessen bündeln', *Handelsblatt*, 10 April 1985.
183. 'Brandt für europäische Weltraum-Initiative', *VWD Europa*, 19 April 1985.
184. 'Brief des ehemaligen Bundeskanzlers Helmut Schmidt an seinen Nachfolger: "Enger Schulterschluss mit Paris" – Vor isolierter SDI-Beteiligung gewarnt – Sorge um Europa', *Vorwärts*, 1 June 1985, p. 19.
185. 'Der frühere Bundeskanzler Schmidt warnt vor deutscher SDI-Beteiligung', *Frankfurter Allgemeine Zeitung*, 30 May 1985.
186. Egon Bahr, 'Spiegel Essay: Europas Antwort auf SDI', *Der Spiegel* 20/1985, 21 May 1985, pp. 126–7; Kurt Becker, 'Militarisierung des Weltraums – Es geht um Grundfragen der Allianz – Die Sozialdemokraten setzen auf die Zusammenarbeit mit Frankreich', *Die Zeit*, 7 June 1985; Rüdiger Moniac, 'Meinungen von Sozialdemokraten und Amerikanern prallten hart aufeinander', *Die Welt*, 1 June 1985; Sten Martenson, 'Alle Fragen kreisen um den "Krieg der Sterne"', *Stuttgarter Zeitung*, 31 May 1985.
187. Hans Günter Brauch, *Die Raketen kommen! Vom NATO-Doppelbeschluss bis zur Stationierung* (Köln: Bund, 1983).
188. 'Gegen Weltraumwaffen warum – Die zwölf entscheidenden Argumente gegan das SDI-Programm', *Sozialdemokrat-Magazin*, no. 7 (July 1985).

189. 'SPD:EVI darf nicht entwickelt werden', *Parlamentarisch-Politischer Pressedienst*, 19 November 1985.
190. 'SPD-Anträge gegen SDI und EVI', *Süddeutsche Zeitung*, 5 December 1985.
191. Petra Kelly, 'Wettlauf im Weltraum', *Abendzeitung*, 27 September 1984.
192. 'Entschliessungsantrag der Abgeordneten Frau Borgmann u.a.', *Deutscher Bundestag, 10. Wahlperiode, Drucksache 10/3388* (Bonn: Verlag Heger 23 May 1985); 'Entschliessungsantrag der Abgeordneten Frau Borgmann u.a. zur Grossen Anfrage der Fraktion Die Grünen' – *Drucksache 10/2378; Militarisierung des Weltraums, Deutscher Bundestag, 10. Wahlperiode, Drucksache 10/3396* (Bonn: Verlag Heger, 23 May 1985).
193. 'Grüne bestreiten zivilen Nutzen von SDI – Studie des Bremer Wirtschaftsinstituts/Bedenken auch gegenüber Eureka', *Frankfurter Allgemeine Zeitung*, 1 June 1985.
194. 'Antrag der Abgeordneten Lange u.a. und der Fraktion Die Grünen: Westeuropäische Raketenabwehr und Europäische Verteidigungsinitiative (EVI)', *Deutscher Bundestag, 10. Wahlperiode, Drucksache 10/4073* (Bonn: Verlag Heger, 23 October 1985); 'Kleine Anfrage der Abgeordneten Frau Borgmann u.a.: Zulässige "SDI-Forschungsarbeiten" nach dem ABM-Vertrag', *Deutscher Bundestag, 10. Wahlperiode, Drucksache 10/4403* (Bonn: Verlag Heger, 29 November 1985).
195. See *Deutscher Bundestag, 10. Wahlperiode, 98. Sitzung, 8 November 1984, Bundestagsprotokoll*, pp. 7054ff.
196. The official minutes of this hearing may be obtained from the Committee on Research and Technology of the Deutsche Bundestag.
197. 'Stenographisches Protokoll der Sitzungen des Auswärtigen Ausschusses und des Verteidigungsausschusses – Öffentliche Informationssitzungen – am Montag, dem 9 Dezember, und Dienstag, den 10 Dezember 1985: Tagesordnung: Offentliche Anhörung zum Thema "Strategische Verteidigungsinitiative (SDI)', *Deutscher Bundestag, 10. Wahlperiode, 712–2450 and 715–2450* (Bonn: Verlag Heger, 1986).
198. See, for example note 194 above.
199. See note 180 above.
200. *Deutscher Bundestag, 10. Wahlperiode, 132. Sitzung, Protokoll, 18 April 1985* (Bonn: Verlag Heger, 18 April 1985).
201. *Deutscher Bundestag, Plenarprotokoll 10/166, Stenographischer Bericht, 166. Sitzung, Freitag, den 18 Oktober 1985* (Bonn: Verlag Heger, 18 October 1985) pp. 12439–57.
202. *Deutscher Bundestag, Plenarprotokoll 10/185, Stenographischer Bericht, 185. Sitzung, Freitag, den 13 1985* (Bonn: Verlag Heger 13 December 1986).
203. 'Erklärung der Bundesregierung zur Zweiten EUREKA-Ministerkonferenz', *Bulletin, Presse- und Informationsamt der Bundesregierung*, 123, p. 1069, 9 November 1985.
204. 'Aus Mitteln des Bundeshaushaltes – Bis zu 40 Millionen Mark zu Eureka – Subventionen sollen anderen Titeln des Forschungsetats

entnommen werden', *Süddeutsche Zeitung*, 9 November 1985; 'Parteienstreit im Bundestag: Was darf Eureka den Staat kosten? Die SPD verlangt massive Zuschüsse, Stoltenberg lehnt neue Subventionen ab/ Debatte im Parlament', *Frankfurter Allgemeine Zeitung*, 9 November 1985.
205. 'SDI-Streit im Forschungsausschuss – Mögliche deutsche Beteiligung Hauptthema bei Anhörung über Weltraumnutzung', *Süddeutsche Zeitung* 12 November 1985.
206. The following experts were invited to testify: Hans Günter Brauch, Hans-Peter Dürr, Horst Fischer, Günter Hoff, Eckhard John von Freyend, Rainer Labusch, Helmut Lohr, Eckhard Lübkemeier, Uwe Nerlich, Jürgen Schneider, Franz-Joseph Schulze, Dieter Senghaas, Karl-Peter Stratmann, Michael Stürmer.
207. 'Bei der Bonner SDI-Anhörung überwiegt die differenzierte Ablehnung – Die Folgen für Europa/"Abschreckungspartner Sowjetunion"', *Frankfurter Allgemeine Zeitung*, 11 December 1985, p. 4; 'Ergebnis der SDI-Anhörung für Bonn ohne Belang', *Süddeutsche Zeitung*, 10 December 1985. Wolfgang Hoffman, 'Der Krieg ist nicht der Vater – Zum Thema SDI: Militärische Forschung bringt nur geringen zivilen Nutzen', *Die Zeit*, 10 January 1986, pp. 17–18.
208. 'SPD und DGB bekräftigen Absage an SDI', *Süddeutsche Zeitung*, 3 July 1985; 'Erklärung zur strategischen Verteidigungsinitiative (SDI) der USA', *DGB Informations Dienst* ID 18, 5 July 1985.
209. 'Die Industrie drängt sich nicht nach SDI – Wirtschaftsdelegation soll Informationen einholen', *Frankfurter Allgemeine Zeitung*, 25 April 1985; 'Entscheidung über SDI noch nicht abzusehen – Gespräch mit Vertretern von Wirtschaft und Industrie im Kanzleramt', *Frankfurter Allgemeine Zeitung*, 14 May 1985; 'Noch viele Fragen zum SDI-Projekt', *Süddeutsche Zeitung*, 15 May 1985; 'Bonn auf SDI-Kurs', *Hannoversche Allgemeine*, 10 August 1985; Eduard Neumaier, 'Die Bundesregierung muss bald über SDI entscheiden', *Stuttgarter Zeitung*, 10 August 1985.
210. 'SDI/Siemens: US-Töchter können nicht gehindert werden – Kaske: Bonn soll über Teilnahme entscheiden', *Handelsblatt*, 9 July 1985.
211. 'SDI-Forschungsprogramm/Abkommen muss den Austausch und die spätere Verwendung von Forschungsergebnissen sicherstellen. Der Schub für die zivile Technologie wird heute vielfach eher unterschätzt', *Handelsblatt*, 21 August 1985.
212. '"Wer kuscht, hat keine Chance" – SEL-Vorstandsvorsitzender Roland Mecklinger zur deutschen Beteiligung an der Weltraum-Raketenabwehr', *Der Spiegel*, 47/1985, 18 November 1985.
213. Hans Jörg Sottorf, "Satellit und Radar sollen im All als Abfangjäger eingesetzt werden', *Handelsblatt*, 10 September 1985; 'Teilnahme des deutschen Konzerns MBB an SDI', *Neue Zürcher Zeitung*, 29 July 1985; 'MBB will SDI–"Leitfirma" sein', *Frankfurter Rundschau*, 20 December 1985.
214. 'Dornier mit Partner auf SDI-Kurs – Zusammenarbeit mit Sperry–Gehen mit IPS ins Rennen', *Rhein-Neckar-Zeitung*, 26 October 1985.
215. '"In der Mikroelektronik sind wir Spitze", Mannesmann-Chef Weis-

weiler über Osthandel, SDI und Wettbewerbsfähigkeit der Deutschen', *Der Spiegel*, 13 May 1985.
216. 'Eureka contra SDI: Hohes Mass an Konfusion', *Wirtschaftswoche*, 6 December 1985; 'Interview: Ambitionen als Vorgeiger', *Wirtschaftswoche*, 17 January 1986.
217. 'Wolff skeptisch gegenüber SDI', *Süddeutsche Zeitung*, 4 December 1985; 'Wolff von Amerongen äussert Skepsis gegenüber SDI und Eureka', *Frankfurter Rundschau*, 4 December 1985; 'Distanzierte Äusserungen auch zu der europäischen Forschungsinitiative', *Handelsblatt*, 4 December 1985.
218. 'Angst um den Osthandel–DIHT sieht Gefährdung durch Bonner Beteiligung an SDI', *Frankfurter Rundschau*, 11 January 1986; 'Osthandel und SDI', *Frankfurter Allgemeine Zeitung*, 11 January 1986.
219. Siegfried Mann, 'SDI und Eureka – Die ideologische Aufgeregtheit übersieht den zu erwartenden US-Terraingewinn am Markt', *Handelsblatt*, 18 September 1985; 'Rüstungsindustrie – Angst vor SDI', *Wirtschaftswoche*, 31 May 1985.
220. 'Unterstützung für Eureka durch die deutsche Industrie', *Frankfurter Allgemeine Zeitung*, 26 July 1985.
221. SDI/Entscheidungen erst nach Rückkehr der Expertendelegation aus USA – Gewerkschaftsinstitute bestreitet einen Schub für die zivile Technologie', *Handelsblatt*, 3 September 1985; Ronald Bock, 'SDI volkswirtschaftlich schädlich', *Vorwärts*, 1 June 1985; Klaus Broichhausen, 'Sterntaler aus dem Weltraum?', *Frankfurter Allgemeine Zeitung*, 11 June 1985; 'Pro und contra SDI', *Wirtschaftswoche*, 20 September 1985, pp. 104–20.
222. '"Geheime Forschung an Instituten nicht vorstellbar" Max-Planck-Gesellschaft ist an Eureka interessiert', *General-Anzeiger*, 28 November 1985; 'Geheimniskrämerei über SDI verstimmt Max-Planck-Gesellschaft", *Frankfurter Rundschau*, 29 November 1985.
223. See Labusch/Maus/Send (eds), *Weltraum ohne Waffen – Naturwissenschaftler*.
224. Karl Stankiewitz, 'Wissenschaftler: SDI ist "technologischer Unsinn", Der Münchner Teilchenphysiker Dürr hat Weltraumwaffen-Modelle durchgerechnet', *Frankfurter Neue Presse*, 17 May 1985.
225. Hans Peter Dürr, 'Der Himmel wird zum Vorhof der Hölle: Professor Dürr über den Wahnwitz der "Strategischen Verteidigungsinitiative" (SDI)', *Der Spiegel* 29/1985, 15 July 1985, pp. 28–42; Hans-Peter Dürr, 'Could "Star Wars" Work? A leading European scientist's caveats', *World Press Review*, September 1985.
226. '"SDI ist keine Verteidigungsinitiative"–350 deutsche Wissenschaftler lehnen Mitarbeit daran ab. Offener Brief an Bundeskanzler Kohl', *Süddeutsche Zeitung*, 4 July 1985; 'Dokumentation: Offener Brief von 350 Wissenschaftlern: "Wir lehnen das SDI-Projekt ab"', *Die Tageszeitung*, 4 July 1985; '40 Wissenschaftler begründen "Wir machen nicht mit"', *STERN*, 18 July 1985.
227. 'Im Wortlaut: "Wettrüsten wird angeheizt"', *Frankfurter Rundschau*, 7 August 1985; 'Naturwissenschaftler warnen Bonn vor SDI', *Süd-

deutsche Zeitung, 7 August 1985; 'West German scientists urge "star wars" boycott', Baltimore Sun, 16 August 1985, p. 17.
228. 'Physik ersetzt Politik nicht', Frankfurter Rundschau, 5 November 1985; Karsten Plog, 'Physiker lehnen SDI ab – Mitarbeiter einer Grossforschungsanlage schrieben an Kohl', Frankfurter Rundschau, 12 November 1985; Karsten Plog, 'Warnung vor Rüstung im All – Hochschullehrer der Bundeswehr appellieren an Politiker', Frankfurter Rundschau, 13 November 1985.
229. 'Wissenschaftler warnen vor SDI-Programm', Frankfurter Allgemeine Zeitung, 8 August 1984.
230. See, for example, Dr J. Häger, Süddeutsche Zeitung, 20 July 1985.
231. Dieter Schröder, 'Ronald Reagans Horror-Vision', Süddeutsche Zeitung, 25 March 1983, p. 4.
232. Walther Stützle, 'Reagans Vision', Stuttgarter Zeitung, 25 March 1983.
233. Ulrich Mackensen, 'Krieg der Sterne?', Frankfurter Rundschau, 28 March 1983.
234. Christoph Bertram, 'Ein Schritt vor, ein Schritt zurück – Reagans Pläne zur Raketenabwehr unterhöhlen die Chancen der Genfer Verhandlungen', Die Zeit, 1 April 1983.
235. Fritz Ulrich Fack, 'Eine amerikanische Trumpfkarte?', Frankfurter Allgemeine Zeitung, 26 April 1984.
236. Among the publications critical of the SDI: see Labusch, Maus and Send (eds), Weltraum ohne Waffen-Naturwissenschaftler. Hans Günter Brauch, Angriff aus dem All – Der Rüstungswettlauf im Weltraum (Berlin-Bonn: Dietz, 1984); in favour of the SDI: Wolfgang Schreiber, Die Strategische Verteidigungsinitiative (Melle: Verlag Ernst Knoth, 1985).
237. 'Mehrheit gegen SDI', Frankfurter Rundschau, 6 April 1985.
238. 'Gegen SDI-Beteiligung', Parl. Polit.-Pressedienst, 5 November 1985.
239. Hans Günter Brauch, 30 Thesen und 10 Bewertungen zur Strategischen Verteidigungsinitiative (SDI) und zur Europäischen Verteidigungsinitiative (EVI), AFES Papier 2 (Stuttgart: AG Friedensforschung und Europäische Sicherheitspolitik (AFES), Institut für Politikwissenschaft, Universität Stuttgart, January 1986); Hans Günter Brauch, Militärische Komponenten einer Europäischen Verteidigungsinitiative Amerikanische militärische Planungen zur Abwehr sowjetischer ballistischer Raketen in Europa, AFES Papier 3 (Stuttgart: AFES, February 1986); Hans Günter Brauch and Rainer Fischbach, Military Use of Outer Space – A Research Bibliography, AFES Papier 4 (Stuttgart: AFES, February 1986).
240. Hartmut Bebermeyer and Bernd Grass, 'Unsere Streitkräfte auf dem Wege in die Ressourcenkrise', in Hans Günter Brauch (ed.), Sicherheitspolitik am Ende? Eine Bestandsaufnahme, Perspektiven und neue Ansätze (Gerlingen: Bleicher, 1984).
241. See the chapter by Peter Wilson in this volume.
242. All the references on the most recent German debate until December 1986 have been included in the Bibliography in the section on 'The European Debate'.

7 Perceptions and Reactions to SDI in the Benelux Countries
Robert J. Berloznik

In the literature on European defence policy, the position of the smaller countries, such as the Benelux Countries, has generally been overlooked. The reasons for this are found not only in the relative 'weight' other countries have in the existing military and political alliances, but also in the fact that their positions on specific defence topics do not differ substantially from that of the 'bigger' nations. This was true until recently. The position of Belgium and the Netherlands regarding the NATO dual-track decision indicates that their attitude can indeed differ. The internal political and public pressure on the governments in both countries forced them to postpone the deployment decision. This relatively new situation of a strong internal discussion on defence-related matters also influences the way SDI is being perceived and reacted to. The Dutch government did not take its final decision on cruise missile deployment until 7 November 1985. Therefore the discussion on SDI has been influenced by this pre-deployment atmosphere: the two questions constantly appear related to each other. The Belgian SDI debate has a post-deployment character. Because a part of the cruise missiles have already been deployed in late March 1985, SDI is generally being perceived as a new controversial defence issue following on the cruise missile debate. The reactions to SDI in both Belgium and the Netherlands should be seen in these specific contexts.

A further introductory remark must be made about the difference in the political culture between Belgium and Luxembourg on one hand and the Netherlands on the other. According to the Dutch law policy-making is a public affair; the administration on all political levels has to be open to the public. Consequently defence policy issues are discussed openly in public and in parliament. However, in Belgium and Luxembourg no similar law exists, and therefore all defence-related issues and all foreign policy decisions are made by the government, and they are subsequently either approved or disapproved by the parliament. For these reasons parliamentary debates on defence-related matters are limited in Belgium and Luxembourg while in the Netherlands such debates are the rule rather than the exception.

I THE REACTION TO SDI IN BELGIUM

(a) The official government reaction

The first Belgian government official who commented on SDI was the Prime Minister himself, Mr Wilfried Martens, during an interview on Luxembourg television on 21 March 1985. The immediate reason for the interview was the political trouble over possible deployment of cruise missiles in Belgium. At the end the Prime Minister was asked to comment on SDI, which had become one of the main topics in the political debate. In response to the question whether he was a proponent of SDI, the Prime Minister answered that solidarity between the United States and Europe is essential, and that Europe should co-operate in SDI if everything is clear to all parties involved. This opinion of the country's leading politician provoked questions from the opposition whether the Prime Minister coupled the question of co-operating in SDI to the question of the deployment of cruise missiles. He answered: 'solidarity ... must be expressed in ... the deployment of the middle-range nuclear missiles. On the other hand, this solidarity supposes, if SDI can be realised, that Europe would be involved and protected by the SDI'.[1]

The view of the Minister of Foreign Relations, Mr Leo Tindemans, was expressed in an interview for the French Belgian radio on 12 March 1985. On his departure to Moscow to attend the funeral of Soviet leader Chernenko he said that it would be useless to oppose the American space defence plans, because it is impossible to forbid scientists to use their brains.[2] In the same interview Mr Tindemans declared that research into space weapons did not necessarily mean that they would be developed and deployed.

The next important and meaningful step was taken by the Minister of Defence, Mr A. Vreven. On 26–7 March 1985 he attended the Spring session of the NATO NPG in Luxembourg. During this meeting United States Defense Secretary Caspar Weinberger detailed his government's plans for a layered space defence. The official communiqué published after the meeting claimed a unanimous support for these plans by the fourteen countries involved. This reference and the fact that at this NATO meeting the official American invitation to participate in the research program was handed over to the Allies, provoked questions in the Belgian parliament about the involvement and the position of the government. On 28 March 1985, in response to a parliamentary question in the Chamber of Representatives by the leading MP of the Flemish Socialist Party, Mr Tobback, the Prime Minister admitted that there had been a unanimous decision in the (NPG) about SDI, and he pointed out that there had not yet been consultation on it inside the cabinet, 'but there have been declarations by Mr Tindemans and myself which show that we are proponents of a participation in the research programme.'[3]

In this statement of 28 March 1985, Prime Minister Martens implicitly admitted that his Minister of Defence had agreed with the United States plans without preliminary deliberation within the government. This provoked various responses at different levels – and not just within the political opposition – to the Euro-missile debate under way since 1979. At the NATO meeting in December 1979 where the famous dual track decision was adopted, the then Defence Minister, Mr José Demarets, took a position which later appeared to be not exactly identical with that of the government. Later, in December 1981 (in the infrastructure committee of NATO) the then Minister of Defence, Mr Swaelen, entered a commitment on a deployment time schedule for cruise missiles, which was not preceeded by governmental decision. These political precedents gave rise in an early phase to questions as to how government policy concerning participation in SDI was being made. A few weeks later this still unclear and ambiguous position of the Belgian government was aggravated by a statement by the United States Department spokesmen, Mr Burch, who declared that Mr Vreven had admitted in a private conversation with Defense Secretary Weinberger that Belgium would positively engage itself in the SDI.[4] Mr Vreven denied this emphatically and said that Mr Burch must have made a slip of the tongue to the question why a high-ranking official like Mr Burch would do something like this, he answered: 'Some take their wishes for real'. In the same interview, Defence Minister Vreven stressed that no principal decision had been taken by the Belgian government. But later on, he admitted that bilateral contacts had already been established between Belgian commercial firms and the American government, and that, as a matter of principle, the Belgian government could not and would not intervene in such contacts, even if the decision were made not to participate officially in SDI.[5]

We can conclude that in this first phase of decision-making several leading governmental officials expressed themselves in favour of SDI, without consulting either the cabinet or the parliament.

Meanwhile, the French government launched its own alternative project to SDI: the European Research Coordination Agency (EUREKA).[6] With the introduction of EUREKA, a second phase in the SDI decision-making process began. While at first the government took a rather positive attitude towards Belgian participation in SDI, it was only after the introduction of EUREKA that internal governmental opposition against SDI grew in favour of the French alternative. Initially EUREKA was supported primarily by some French-speaking Ministers. The Minister of Scientific Policy, the French-speaking Christian Democrat Mr Philippe Maystadt, was known to be among the first in the government to support a participation in EUREKA. However, this would not exclude any participation in SDI.[7] Later EUREKA gained general support from political leaders on both sides of the language border.

On 9 April 1985, Mr Tindemans, the Minister of Foreign Relations sent a letter to Premier Martens proposing the establishment of a think tank, or a 'Task Force', to study mainly the technological aspects of SDI. The first time SDI was raised in the cabinet was on 17 May 1985. The result was a communiqué that referred to the WEU text of 13 April 1985 in which the WEU countries had agreed 'to proceed with talks to produce a co-ordinated reaction of the governments to the American initiative'. The cabinet also officially approved the establishment of the Task Force. Ambassador Baekelandt was appointed as its chairman. The Task Force is composed of representatives from nine departmental administrations. The description of its tasks is vague. Its objective is to collect information about EUREKA as well as SDI. At a later stage, commercial firms and university research institutions will be involved in the Task Force.

According to the Flemish Catholic newspaper *De Standaard*, the government also agreed to the point of the NATO communiqué of Luxemburg that stated that: 'This research . . . is in NATO's security interests and should continue'. But in the published governmental decision of 17 May 1985, this was not mentioned.[8] Such a decision would be important, because it could lead to a possible agreement on a future change of NATO doctrine which could be interpreted as an agreement to the eventual deployment of SDI components in Belgium. Whatever it was that was agreed upon on 17 May, it opened the way for different interpretations and it caused some confusion about the government's policy concerning SDI.

In May 1985 the Foreign Affairs Department published a document on its foreign policy.[9] It stated that Belgium had not yet reached a decision on SDI, because the concept was still insufficiently defined and its strategic implications remained unclear. This document distinguished two separate aspects: the politico–military or strategic aspect and the scientific–technological aspect. This second aspect is the topic to be studied by the special Task Force. As to the first aspect it was stated that 'because no deployment will take place within the next 15 years, we still have plenty of time to consider these questions before we can clearly express ourselves on this part'.

(b) The political debate and the views of the opposition, the ruling parties and the peace movement

The present Belgian governing majority consists of Christian Democrats and Liberals. Due to the specific linguistic situation each political tendency exists in the form of a political party in both the Flemish and the Walloon parts of the country (with the exception of some small linguistic parties). Not every political party did express an opinion on SDI until the end of 1985. Nor has there been a profound political debate in the parliament as in most of the other NATO countries. But there have been some parliamentary questions and motions from ministers. An attempt to reconstruct some of the political arguments will be made below.

The most extensive response by a member of the government took place on 22 May, 1985 in the House of Representatives at the initiative of the *Front Démocratique des Francophones* (FDF), a linguistic opposition party.[10] The party had been motivated by its experience during the discussion about the installation of cruise missiles, which was introduced into the Belgian parliament only at the very end of the decision-making process. Besides problems raised by SDI itself – such as the time it would take to be implemented, the huge costs involved or the uncertainty as to whether it would be both feasible and effective – the FDF criticised the establishment of the Task Force: given its unspecified purpose it would permit many different interpretations. The FDF advised the government to be careful in its decision concerning SDI and to give priority to EUREKA.

This argument was also put forward by the Parti Socialiste (PS). Although the PS did not intervene directly in the interpellation, its executive committee rejected on 29 April 1985 any SDI participation. According to the Parti Socialiste, SDI would provide new momentum to the arms race, it would mean an enormous waste of means and the results would be uncertain. The PS stressed that the technological problems should be resolved first among the Europeans themselves.[11] In defence matters, the Flemish Socialist Party (SP) is sometimes referred to as the most radical party in the Belgian political spectrum. Its parliamentary spokesman, Mr Tobback, has called SDI an absurdity'.[12] The SDI, in the view of the SP, would bring the deterrence strategy to an end and provide protection for the superpowers, while the rest of the world would remain a battlefield. It would violate both the Atlantic Treaty and the ABM Treaty. Participation in an SDI research programme, as proposed by the United States, would reduce Europe to the status of a subcontractor in the area of new technologies. The only goal the United States would achieve with this tactic would be a 'brain drain'. The huge costs would affect possible Belgian initiatives concerning EUREKA. SDI would thus harm European interests.[13] According to the SP the main fact concerning the government's policy on Belgian participation in SDI is that, even if the decision has not yet been officially made, the government's enthusiasm for the American project has hardly been hidden. Although the realisation of the SDI would imply sending weapons into space, the government has claimed that the SDI would enhance the demilitarisation of outer space.[14]

The argument that the development of new military technology would have a civilian spin-off, has been rejected. Japan's example of technological innovation without relying on military research proved the contrary.[15] The Flemish Socialist Party even organised a forum on the SDI in Brussels on 6 June 1985. Besides background papers on the SDI there was a panel discussion where the SDI was described as a problem of technological blackmail. At the same Forum the President of the SP, Mr Karel van Miert, pleaded for support of EUREKA, and stated that if certain conditions were met, the SP, would consider participation. Within the party not everyone agreed with this position. Several Flemish socialists defended the idea of a third possibility, because EUREKA is not an alternative to SDI, although some 90 per cent of its projects would be compatible with SDI.[16]

The ecologist party AGALEV's spokesman, Mr Dierickx, introduced the first motion on SDI in parliament.[17] Belgium should refuse participation in the SDI on the same grounds as the Danish Parliament had done. He called on the Belgian government to wait until the parliament had discussed the matter in detail (which had not yet occurred) before making any decisions. AGALEV also rejected the assumption that the military industry should be the driving force behind a revitalisation of the civilian economy.[18]

The Communist Party (KP/PC) stated that the way the SDI issue had been handled was strongly reminiscent of the situation before the installation of the cruise missiles, with its contradictory governmental declarations and ambiguous decisions. The SDI according to the KP/PC would mean a significant new step in the arms race. Although it had been proposed as a defensive strategy, it would perform offensive functions that would provoke Soviet countermeasures. Whatever the scenario for its future deployment would be, SDI would be unacceptable for the Soviet Union. Participation in the SDI research programme should be rejected and the funds better spent on peaceful technologies.[19]

The ruling Christian Democratic and Liberal parties have not formulated any position on SDI opposing the governmental position. However, a declaration of the chairman of the Flemish Christian Democratic Party (CVP), the biggest political party in the country is worth mentioning. At the conference of the Union of European Christian Democrats in Barcelona in early June 1985, he suggested that a more prudent approach to SDI had emerged in the governmental position after the proposal of EUREKA. The CVP chairman, Mr Swaelen, said that clear answers would be needed on the fundamental questions of the transition from offensive to defensive weapons, and whether a transition from nuclear to non-nuclear armament would be possible and desirable. Questions should also be answered about the consequences of the SDI for peace in the world and for the security of Europe.[20] There were no obvious differences on the SDI between the ruling parties, but only a difference in approach. While the Christian Democrats were prudent in raising strategic questions, the liberal coalition partner was rather pragmatic. In the liberal press, comments and articles have mainly indicated the opportunities for the Belgian industry if Belgium participated in the SDI.[21] Although the Liberal parties have not played a very active role in the political debate about SDI, their relation with the influential Atlantic lobby has been obvious. It may be said that at least some important members of the Liberal ruling parties have played a very active role in lobbying behind the screens for the SDI. The influential general, Robert Close, a member of the World Anti-Communist League, as well as of the Advisory Board of High Frontier Europe and the chairman of an

Atlantic Peace and Security Research Institute in Brussels, is also a senator for the French-speaking Liberal Party (*Parti Réformateur Libéral*). The institute published an affirmative paper on SDI as early as March 1984.

The Belgian peace movement unsurprisingly has been opposed to the SDI, which they consistently refer to as 'Star Wars'. The first official public reaction was made at the end of May 1985. Two Flemish organisations '*Overlegcentrum voor de Vrede*' (*OCV*) and the '*Vlaams Aktiecomité tegen Atoomwapens*' (VAKA) which represent the great majority of Flemish peace organisations, offered a joint public statement on the SDI.[22] They described the SDI as a new threat for humanity and as a new phase in the arms race. They requested the government not to co-operate with the American initiative. A participation in the SDI based on eventual technological and economical advantages should not be accepted by the government. The refusal of Denmark and Norway to join the SDI research project had shown that a NATO member can assume an independent attitude that is responsive to the will of its citizens.

A rather interesting and original point of view has been that of *Pax Christi*, the main Catholic peace organisation. It considered SDI together with other recently formulated military policies such as the AirLand Battle doctrine and the FOFA concept. The overall change would result from a search for ways out of the nuclear blind alley in general, and out of the deterrence doctrine in particular. SDI could lead to the shift from a deterrence strategy to a defensive strategy, while the Airland Battle and FOFA would indicate a shift from nuclear to conventional arms. Both developments would result in a strategic shift from nuclear deterrence to a conventional defence. According to Pax Christi this strategic shift and its related questions would constitute the greatest challenge for the future strategy of the peace movement.[23]

(c) **The scientific debate in Belgium**

There has not yet been a debate among scientists on the SDI and its implications. This is mainly due to the fact that in Belgium there is no adequate forum that is suited for this purpose – a periodical or study group, for example. But there are some publications that reflect the existing points of view.

First of all there is the above-mentioned study published by the institute of General Robert Close.[24] This paper is important because, although it was published as early as June 1984, it largely reflects ideas which were later adopted by NATO as part of its official policy. The study argues that SDI could be very valuable for American defence strategy, but for Europe the implementation of strategic defence systems may have negative consequences for its defence interests. SDI would divert immense budgetary resources which are now being used for conventional defence. The mutual deployment of strategic defence systems by the two superpowers would neutralise their nuclear forces, making it unlikely that either side would use atomic weapons and thereby increasing the possibility of a conventional war in Europe.

A more critical view of SDI has been expressed in a series of documents published by the *Groupe de Recherche et d'information sur lá Paix* (GRIP).[25] The main aspects of SDI have been treated in separate publications. A physicist has assessed the technical feasibility of SDI systems, and a defence expert has analysed the possible impact of SDI on the arms race. A further study deals with the history of ABM defence systems and the ABM Treaty. This series is published in French and distributed mainly in the French-speaking part of Belgium, but it has become widely used by SDI critics as a source of information.

A third publication has been prepared by the International Association for University Peace days.[26] It tries to present an objective and non-expert view on all aspects of BMD with special focus on SDI. It is available in both French and Dutch and has been presented to the heads of all the Belgian universities as a source of information for the academic community in Belgium.

(d) The response of the industrial sector to SDI

Due to the official government policy of allocating defence contracts equally between the two language communities, there exists in Belgium an official list of forty defence companies. The companies that could participate in SDI research are on this list. The companies primarily referred to are aerospace companies such as Sabca and Sonaca, and in the area of laser technology CBC Electrics (*Compagnie Belge des Lasers*). Representatives of all possible contractors have met twice in March and June 1985 at the office of the National Employers Organisation. Although a strong interest exists within these companies, they are aware that Belgian industry can offer little as a candidate for a participation in SDI research.[27] No information was available about companies sending delegations to Washington to explore such possibilities until the end of 1985.

II THE REACTION TO SDI IN THE NETHERLANDS

(a) The official government reaction

As in most West European countries, the reaction to the American SDI programme in the Netherlands was not immediate.

According to high-ranking officials in the Foreign Ministry among the reasons for this late reaction was a lack of information and not just a misperception or misjudgement, as it has been generally seen. They wonder whether this has been just the result of carelessness or part of a well-considered tactic. In 1983, an official Dutch delegation which went to the United States to study the anti-satellite weapons programme was told that they should not pay too much attention to President Reagan's dream because there would not be sufficient funds to realise it. And when the Dutch Defence Minister, Mr Jan Van Houwelingen, visited the United States in 1984 to study 'emerging technologies', he was carefully kept away from all SDI-related research. However, in 1985 everything was turned around. Information about SDI started to flood into the departments. In March 1985 in Luxembourg, when United States Defence Secretary Weinberger urged the allies to respond to the American invitation to participate in the SDI research within sixty days, the Dutch Defence Department officials felt, 'as if a gun was pointed at their heads.'[29] Subsequently SDI was received with some reservation in the relevant departments.

An impression as to how SDI has been assessed by the Department of Defence may be gained from a public statement made on 5 March 1985 by H. P. M. Kreemers, a member of the Department's general policy directory. He referred to one fundamental condition: the new defensive strategy should be stabilising.[30] Given the fundamental distrust between the superpowers, Kreemers argued, it would be better to shift to a strategy aimed at preventing war rather than to believe in a strategy that is based on the principle that in case of war the enemy's weapons would be destroyed. Mr Kreemers also expressed his fear of a possible decoupling of the security interests of the United States and Europe as a result of the deployment of a strategic defence system in the United States. He concluded that a lot of questions still remain unanswered, but the SDI research programme should not *a priori* be rejected.

Obviously this has also been the position of the Defence Minister, Mr De Ruiter. He has always been very careful with comments on the concept of strategic defence itself. Nevertheless, he did not hide his reservation about SDI. He has publicly stated that he would support the SDI only if he were certain that it would not endanger the balance of power and if the Russians would also see advantages in the SDI programme.[31] But, because it was still too early to assess the SDI properly, his advisors have been instructed to study the problem as thoroughly as possible.

In early March 1985, the Dutch Parliament debated the SDI for the first time. The debate revealed that the government did not yet have a clear position of its own. Consequently, there was neither a rejection nor an approval. A common position on space weapons should be achieved in a West European context. For this purpose, the Foreign Affairs Minister, Mr Van Den Broek, prepared a memorandum entitled 'Common Principles' that was circulated in the WEU in early 1985. During the parliamentary debate Mr van den Broek had to admit that the answers to a lot of questions did not yet exist. This applied to such problems as the financial consequences of European participation and to the questions of technology transfer.[32] Although no decision was reached during the first debate, Mr van den Broek stressed that a distinction should be made between the concept of space-based missile defence and the concept of the SDI research programme. He concluded that if the Netherlands or Western Europe were to support the research, this would not mean that they would be in accord with the realisation of the SDI.

Mr van den Broek's ideas were presented in a more structured form on 10 May 1985 when he devoted an entire speech to the SDI in The Hague. In the meantime the idea of EUREKA had been launched by the French, and this had already influenced his views on the SDI. The European countries should not reject the SDI in principle, nor should they accept it unconditionally. Their cooperation had to be made dependent on a step-by-step evaluation of the SDI programme. During such an evaluation the European countries[33] could use political and strategic criteria which are of importance to the security in Europe'.[33] He made two additional conditions. Technological transfer should be guaranteed in both directions, and it should be made clear that eventual Dutch participation in the SDI research would not anticipate a final position on the SDI in general.

Mr van den Broek stressed that such a final assessment on the concept itself would be developed gradually as the consequences of a critical examination. However, he rejected the reasoning that once the research had started, the rest could not be stopped any more, because financial constraints would impose themselves as an imperative. The EUREKA idea should be worked out within the EEC 'but it cannot be used to avoid the discussion on SDI'.

However, what was new and significant about his speech was that it opened a Dutch option on what has since been called a Tactical Defense Initiative (TDI). 'The Dutch government attaches a "great value" to the American "intention" to find defensive systems against short-range missiles, cruise missiles and bombers, that "can not be diverted" as the Soviet SS-20 by strategic defence'. This policy declaration by the Foreign Minister was made official in the Dutch *Staatscourant* on 14 May 1985.

During the next few months nothing essential changed in this governmental position. During the second parliamentary debate on 20 June 1985, the two responsible Ministers aired and defended these policy options. The only new move by the government was that the decision on the research participation was postponed until after the summer recess.

Meanwhile, the government had formalised its policy in setting up a study group on the SDI. Although the 'Coördinatiegroep SDI' was composed only of representatives from the two relevant departments, its structure was complex. It consisted of two independent subgroups. One was occupied with political matters of SDI and was to make a political assessment of the programme. The second had to study the technological questions and the opportunities the SDI might offer for the Dutch industry.

(b) The political debate and the views of the ruling and opposition parties and the peace movement

As has already been pointed out the different positions of the Dutch political parties have become very clear during the parliamentary debates. Because the Dutch political spectrum consists of numerous small parties with only one or two MPs, only the ruling parties and the most important opposition parties will be included in the survey. As in Belgium, the governmental coalition consists of the Christian Democratic and Liberal Party. The position of the liberal VVD (Volkspartij voor Vrijheid en Demokratie) comes closest to the governmental policy. In early April 1985, the VVD spokesman on Defence matters, Mr J. C. C. Voorhoeve, outlined his party's point of view on SDI. Five principal questions would have to be answered: Would SDI work? Would it be affordable? Would it affect stability? Would it enhance the security of Europe? Would SDI reduce the nuclear armament on both sides?[34] As long as these questions remain open, the VVD objects to deployment of new weapons. But, in their view, in order to answer these questions a research program would be needed first. Later in Parliament the VVD elaborated its view.[35] They agreed to the participation in the SDI research under the condition that an evaluation process took place, based on the five criteria mentioned above. A sixth was also added: How survivable would such a system be? Mr Voorheove further pointed out that the potential Dutch share in the SDI research would be very modest. The VVD defended a West European participation in the SDI in the context of a TDI. EUREKA was interpreted as a positive idea, but still too vague for a firm commitment.

During the first parliamentary debate in March 1985, signs of a split within the CDA (*Christen Demokratisch Appel*) on the question of SDI had become obvious. It did not come as a surprise, because during the discussion on the cruise missile deployment the same internal division on defence matters had appeared. This situation is embodied in the differences between the CDA chairman, Mr Bukman, and the CDA spokesman and defence specialist, Mr de Boer. While Mr de Boer is known within the majority coalition for his critical attitude on defence, Mr Bukman represents the conservative, pro-Atlantic wing of the CDA. When asked whether he shared the enthusiastic reaction of his party leader to SDI, Mr de Boer answered: 'It is very clear that I am not inspired with the same enthusiasm'.[36] Already during that first debate, Mr de Boer argued that non-military technological co-operation in Europe would be preferable to a participation in the SDI.[37] However, this was not yet official policy. The fact that the Foreign Affairs Minister, Mr Van Den Broek, committed himself during the debate to stressing the necessity of a common European position on SDI can be explained also as a reaction to the pressure that the CDA faction exercised on its minister.

While the position of the CDA MPs was not the official one in March, it became the party's policy view during the parliamentary debate in June. Mr de Boer argued that the West European countries would be better off if they could manage their own technological development. In a later stage it would then be easier to exchange the results of the research on a basis of equality.[38] He based his rejection to participation in the SDI research programme on his 35 years of experience within NATO, during which he learned that all attempts at mutual co-operation with the United States had always resulted in Europe becoming a mere subcontractor with the patents, technology and information disappearing to the United States. 'There is no reason to assume' he declared, 'that suddenly there is an absolute reciprocal exchange of research results'.[39] Mr De Boer's remarks on an eventual TDI for Europe are also noteworthy. He calls it 'a blow in the air', because nobody – neither the Americans nor the West Europeans – had seriously thought about it or had proposed the necessary technical means.[40]

On different items related to SDI, the CDA MPs in the parliament were thus in disagreement with the government. In fact, a majority in parliament opposed participation in the SDI research. Until now no vote of confidence has been provoked, and it would be doubtful that the government would fall over these questions. The experience of the decision-making process during the cruise missile debate has shown that finding acceptable compromises to satisfy the ruling majority, is one of the strong features of the Dutch government.

In the opposition, the socialist party PVDA (*Partij van de Arbeid*) and the small leftist parties are opposed to SDI. We will consider only the positions of the PVDA, and the Communist Party (CPN).

The major opposition party, the PVDA, has criticised the concept of the SDI as well as Dutch government policy with regard to the SDI.[41] The cornerstone of the PVDA's criticism of SDI has been that ever since its announcement the concept itself has quietly been changed several times. For example, the idea that the system could be leakproof, non-nuclear and exclusively defensive has been abandoned and nothing has been heard recently about a transfer of the results of the SDI research to the Soviet Union.

The PVDA has raised important objections to the official Dutch policy on SDI. A participation in the SDI research would not be neutral in the present East–West relationship. On the contrary, it would have immediate consequences, and be a major destabilising influence. Even if the present international situation were not so tense as it is, the SDI research project would not be neutral: according to the PVDA the question must be asked and answered whether all these means – financial as well as human – would not be better invested in other priorities. It would be an illusion to think that the results of such an expensive and extensive research project could be easily discussed or rejected. The PVDA raised other objections, too: SDI would negatively influence the negotiations between the Soviet Union and the United States in Geneva. Only in theory would SDI research not be in violation of the ABM Treaty. The PVDA has opted for EUREKA because they consider it doubtful whether the technology transfer in the SDI context would be reciprocal.

The Dutch Communist Party (CPN) has interpreted the SDI as a deliberate blow against the peace movement from the United States. It would also constitute a blow against a potential technological and economic bridge between the Soviet Union and the West. The CPN has also criticised the huge economic and social costs involved. Its destabilising consequences could be dangerous for all mankind. In a technological and an economic sense, SDI would imply a brain drain for Europe. EUREKA has been rejected by the CPN because it was considered to be too vague, its content in fact being similar to the SDI projects.[42]

The opposition against SDI within the peace movement is slowly gaining ground in the Netherlands. It must be kept in mind, however, that the Dutch peace movement was still very active against the deployment of cruise missiles, and it seems to have been taken by surprise by the sudden appearance of the SDI.

The first peace organisation that presented clearly formulated statements on SDI was Pax Christi. Pax Christi was asked to formulate some advice for the Dutch Bishops' Conference on SDI. According to Pax Christi, in a document of 17 April 1985,[43] the American plan should be rejected for several reasons. SDI was primarily seen as a new phase in the arms race. It would not be enough to reject SDI on political, military, strategic, technological and economic grounds; the peace movement would have to go deeper. Its proponents tend to justify SDI by using concepts and slogans from the peace movement itself, Pax Christi sees the present challenge to the peace movement in searching for answers and formulating alternatives.

On 15 June 1985 the IKV (*Interkerkelijk Vredesberaad*) published a note by its Daily Board in which they rejected SDI for many reasons. They referred to the SDI as 'an expression of the American striving for complete independence from other countries in the world' which is 'completely contrary to the conception of the peace movement, where solidarity and co-operation is a condition for the future of Europe and the entire world community'.[44] The IKV has also tried to stimulate a discussion within the political parties. A discussion paper about peace policy was prepared for this purpose. Mr Mient Jan Faber, the secretary of IKV and the virtual spokesman of the Dutch peace movement, expressed his ideas on 'Star Wars' in a newspaper forum on 29 June 1985.[45] He developed two theses: SDI affects NATO's deterrence philosophy, and it leads to a decoupling of the security interests of the United States and Western Europe. The solution he offered is simple but it is closely related to the main problem the peace movement in the Netherlands faced until 7 November 1985 when the Dutch cabinet decided to deploy cruise missiles on its territory.

(c) The scientific debate in the Netherlands

In contrast with Belgium, the scientific debate has been a constant factor in the Dutch public discussion on SDI. Early 1985 General G. C. Berkhof, a researcher at the Dutch Institute for International Relations Clingendael, published a book on the security dimensions of militarisation of outer space for Western Europe. As it was written in the Dutch language the book was soon to become the standard work in this field for the Netherlands and the Dutch-speaking part of Belgium.[46] Its publication immediately provoked a lively and polemical discussion in some newspapers, on militarisation of outer space and more specifically on SDI.[47] During the debate, General Berkhof came out as one of the main advocates of SDI. In an article published in May 1985, Berkhof argued that SDI can play an important role for West European security by diminishing the vulnerability of nuclear weapons.[48] SDI and its ATBM version for Western Europe could shift the balance of conventional forces, thereby reducing the prospects of a quick Soviet victory in Western Europe.

In the same journal Berkhof's thesis was countered by Mr G. Van Benthem Van Den Bergh, senior lecturer at the Institute for Social Studies in The Hague.[49] Van Benthem Van Den Bergh considers SDI dangerous because it creates the illusion that escape from the nuclear age is possible. Berkhof's vulnerability argument is hardly persuasive given the diversity and the extent of the existing nuclear weapons. Even in the long run it is higly unlikely that either party could launch a disarming first strike, which would be necessary to endanger the vulnerability of nuclear weapons. Berkhof's fear of Soviet superiority in conventional weapons is – according to Mr Van Benthem Van Den Bergh – based on the assumption that conventional and nuclear levels of violence can be distinguished in a military confrontation and that such a military conflict is politically controllable. The existing degree of uncertainty makes all strategic concepts which are based on these assumptions – such as SDI – dangerously illusive.

(d) The response of the industrial sector to SDI

Any Dutch industrial participation in the SDI research programme would most likely be in the electronic and aerospace industries. The companies most often mentioned are Signaal (a subsidiary of Philips), Fokker and Oldelft, all of which have an experience in military projects. If any SDI contracts are awarded to the Netherlands, they probably will go to one of these companies.[50]

However, as it has been emphasised in the Dutch Parliament, there exists a fear that the function of these companies might be reduced to that of subcontractors. For this reason, Philips has advocated a participation in the form of 'systems responsibility'. This would imply that the contractors would have total control over their part of the programme, even if it were very small. Philips regards this as a much more attractive form of participation than the role of a mere subcontractor. In order to protect their interests, potential Dutch industrial participants have organised themselves into a group, the Stuurgroep SDI, which is in official negotiation with the Dutch authorities and thus has a direct impact on the Dutch SDI position.

III THE REACTION TO THE SDI IN LUXEMBOURG

(a) The official Government reaction

On 30 April 1979, in his annual policy declaration to the parliament, Luxemburg's Defence Minister Mr Willy Bourg did not even mention SDI. It was only after a parliamentary interpellation in late June 1985 that the government made a statement on SDI in which the Luxembourg government supported the NATO policy that research on the feasibility of space-based defence systems was justifiable because of Soviet efforts in this field.[51] Nevertheless, this research should be accompanied by a study of the strategic consequences of such new weapons systems and its implications for arms control. As far as SDI was concerned, the government noted that there would be 'questions and doubts' expressed by leading politicians, both in the United States and in Europe.[52] The question of participation in the SDI research did not apply for Luxembourg, because it did not have any research centres that would meet such demands. However, according to the economic and free trade principles of the Grand Duchy, the government would not prevent bilateral contacts between enterprises in Luxembourg and the United States, if interest in such relations should be expressed.

The opinion of the ruling coalition partner, the Social Democratic Party (LSAP) is noteworthy. Its point of view on SDI does not directly contradict governmental policy, but it answers in a clear way the 'questions and doubts' the governmental declaration alluded to. The LSAP sees a threefold threat in the SDI: SDI would threaten the ABM Treaty, it would threaten the ongoing Geneva Talks, and it could become a threat to the NATO Alliance itself because it would decouple American and European security interests.[53]

(b) The position of the parliamentary opposition

The political opposition first introduced the SDI debate into parliament in late April 1985, following the defence policy declaration. The liberal party (*Demokratisch Partei*) favoured the SDI research 'because the Russians are advanced in their research and have already put some of the technology into practice'. According to the liberal party which held governmental responsibility from 1969 until 1984), the attitude of the government towards SDI should be positive because Luxembourg was interested in better defence systems.

The ecologist party (*Grèng-Alternative Partei*), rejected any form of militarisation of outer space 'whether SDI or just military satellites'.[54] The party claimed that SDI would be a destabilising project and a copy of the 'crazy High Frontier concept'.[55]

IV CONCLUSION

In developing their attitude towards SDI the government of the Benelux countries have made a clear distinction between a position on the concept of strategic defence in general, and a position on participation in the SDI research programme. On the concept itself they have agreed that a final decision about supporting the new strategic ideas of the American government must be made collectively by the West European allies within the WEU. An eventual final assessment may be linked to certain conditions such as Dutch demands that the SDI programme should constantly be evaluated during its process. The Luxembourg government has affirmed that there are 'questions and doubts' relating to the SDI. In this regard we can assume that at least some of the smaller European countries will take a different position on the SDI than they did in early 1985 at the Spring Session of the NPG in Luxembourg, where they unanimously supported SDI. This is a result of the internal political debates that have been conducted by the parties and parliaments ever since.

In principle, the Benelux countries are not against participation in the SDI research project. Those countries that would be able to participate – Belgium and the Netherlands – have not yet made a final decision, the preliminary decision being that any final decision must be based on the findings of the study groups which have been established in both countries for this purpose.

Among the ruling parties, no party has contradicted the chosen SDI policy in such a way that would endanger the existence of the government concerned. In the Dutch parliament, there is a majority against SDI because a wing of the CDA is opposed to it. However, until now this has not had any significant political implications. In Luxembourg the ruling Social Democrats have strongly criticised SDI in a public statement, but not in such a way that would contradict governmental policy. In Belgium such critical attitudes towards SDI within the ruling parties have not been observed.

The major opposition parties in all three Benelux countries have rejected any support for the strategic defence concept, and they have opposed any collaboration in SDI research as well. The arguments vary from emphasising the negative military strategic consequences to economic and technological arguments in favour of EUREKA. However, this political opposition must also be seen within the context of the still ongoing debate on cruise missile deployment, where for the first time in recent history a defence policy issue gathered all the attention of the opposition.

The extra-parliamentary opposition exists mainly within the organisations of the peace movement. The reaction here was relatively late in comparison with that of the political parties. The cruise missile debate had its influence. It has been their main issue, and their entire strategy is focused on it. Only very recently SDI – or 'Star Wars' as it is called – has appeared as a new 'hard' issue in their campaigns.

The French proposal for a civil research programme, EUREKA, has been generally perceived as a European 'alternative' to the technological challenge posed by SDI. EUREKA has had some influence on the decision-making process concerning SDI. The technological questions have been receiving more attention than the related military and strategic ones. The SDI debate has thus achieved an extra dimension that proves that SDI has now become a complex political problem.

In conclusion we can distinguish three considerations that play a major role in the decision-making process regarding a possible participation of the Benelux countries in the SDI. These considerations are concerned with foreign policy, military strategy and industrial technology and they often are presented as being related to each other. It has been recognised that the role of technology in economic development is becoming increasingly important and industrial–technological considerations seem to have become decisive as to whether or not any of the Benelux countries will participate in the SDI. What in an early phase of the SDI debate seemed a fairly simple political–military choice to be made has thus become a crucial industrial–technological choice that could determine economic future not only in the Benelux countries, but in Europe as a whole.

While in the BENELUX countries 1985 was characterised by a vivid debate on defence issues such as the cruise missile and SDI 1986, by contrast, was rather calm. No major changes occurred in the positions of governments, political parties and in the opposition towards SDI. No official government position was adopted concerning participation in the SDI, although that had been announced for late 1985. Several declarations, however, show that a formal agreement with the US, in the form of a Memorandum of Understanding, seems to be excluded. Contacts between SDIO and several of the above-mentioned industrial firms were confirmed, but no information is obtainable on whether or not this led to participation in the SDI. The peace movement continued to resist any participation, but it failed in making Star Wars an issue of massive mobilisation, as was the deployment of cruise missiles.

In 1986 several books on SDI were published in the Dutch language, an indicator of the growing scientific concern over SDI. In Belgium a comprehensive critical assessment was written by two Flemish authors.[56] The Dutch university of Nijmegen published a compilation of papers presented at an international conference on SDI held in Utrecht in April.[57] In December, the Dutch peace researcher Philip Everts edited a reader on the European consequences of SDI.[58]

Notes

1. Chambre of Representatives, *Belgisch Staatsblad. Vragen en Antwoorden* – question nr 25 from Mr Sleeckx (Brussels: Belgisch Staatsblad, 12 March 1985).
2. SDI interview of Leo Tindemans with W. Vandervorst, written version of the French-speaking radio news (RTBF), 12 March 1985.
3. Chambre of Representatives, *Belgisch Staatsblad. Beknopt Verslag* (Brussels: Belgisch Staatsblad, 28 March 1985).
4. *De Morgen*, 22 May 1985.
5. *De Morgen*, 22 May 1985.
6. See Chapter 9 by Alain Carton.
7. *Le Soir*, 20 April 1985.
8. Bernard Tuyttens, België en Star Wars, May 1985 (unpublished).
9. Belgian Ministry of Foreign Affairs, *Een Buitenlands Beleid ten Dienste van de Vrede, Veiligheid en Ontwapening* (Brussels: Department of Foreign Affairs, May 1985) p. 89.
10. Chambre of Representatives, *Belgisch Staatsblad. Vergaderingen van 12 mei 1985* (Brussels: Belgisch Staatsblad, 22 May 1985) pp. 983–7.
11. *La Libre Belgique*, 4 April 1985.
12. *De Morgen*, 22 May 1985.
13. Chambre of Representatives. *Belgisch Staatsblad. Vergaderingen van 22 mei 1985* (Brussels: Belgisch Staatsblad, 22 May 1985).
14. *Belgisch Staatsblad*, p. 987.
15. *Le Soir*, 20 April 1985.
16. *De Morgen*, 6 June 1985.
17. Literally translated: 'Another Way of Living'.
18. Chambre of Representatives, *Belgisch Staatsblad. Vergaderingen van 22 mei 1985* (Brussels: Belgisch Staatsblad, 22 May 1985) p. 987.
19. *Belgisch Staatsblad*, p. 985.
20. *Het Volk*, 10 June 1985.
21. *Technivisie*, 1 May 1985.
22. *De Morgen*, 4 April 1985.
23. *Kommentaar*, nr 3 (Antwerp: May 1985) pp. 6–7.
24. European Institute for Peace and Security (EIPS), *Star Wars Strategy: Implications and Consequences for the Europeans of the Atlantic Alliance*, vol. 1 (Brussels: EIPS, June 1984).
25. Michel Wautelet, *Aspects Scientifiques de l'IDS*, a Groupe de Recherche et d'Information sur la Paix (GRIP) publication, nr 81 (Brussels: GRIP, June 1985); André Dumoulin and Xavier Zeebroek, *Missiles Anti-Ballistiques et Traité ABM face à l'IDS*, a GRIP publication, nr 82 (Brussels: GRIP, June 1985); Rik Coolsaet, *L'Europe face à l'IDS*, a GRIP publication, nr 83 (Brussels: GRIP, June 1985).
26. International Association for University Peace Days, *Het Amerikaanse Strategische Defensie Initiatief*, document nr 1 (Brussels: IAUDP, August 1985).
27. According to a declaration by Steven De Batselier, a Flemish Socialist representative in the Chambre – expert in technology affairs, who attended both meetings – at a conference on SDI in Brussels on 6 June 1985.

28. Quoted in *Vrij Nederland*, 6 June 1985.
29. Vrij Nederland.
30. *NRC Handelsblad*, 5 February 1985.
31. Quoted in *Vrij Nederland*, 6 June 1985.
32. *De Volkskrant*, March 1985.
33. All quotations in this paragraph are literal citations from The Hague Speech as cited in *NRC Handelsblad*, 11 May 1985.
34. *Vrijheid en Demokratie*, May 1985.
35. Second Chamber, *Handelingen. Tweede Kamer der Staten-Generaal*, nr 33, 18–20 June (The Hague: Staatsuitgeverij, 1985) p. 5885.
36. *Tweede Kamer der Staten-Generaal*, p. 5879.
37. *De Volkskrant*, 12 March 1985.
38. *Tweede Kamer der Staten-Generaal*, p. 5880.
39. *Tweede Kamer der Staten-Generaal*, p. 5881.
40. *Tweede Kamer der Staten-Generaal*, p. 5880.
41. Partij van de Arbeid (PVDA), 'SDI, het concept positie over SDI en Eureka', The Hague, 11 June 1985 (unpublished).
42. *Tweede Kamer der Staten-Generaal*, ppp. 5889–91.
43. Pax Christi, *Advies Pax Christi aan de Nederlandse Bisschoppenkonferentie inzake de standpuntbepaling t.o.v. Star Wars* (The Hague: 17 April 1985).
44. *De Waarheid*, 18 June 1985.
45. *De Volkskrant*, 26 June 1985.
46. G. C. Berkhof, *Duel om de Ruimte* (The Hague: Institute Clingendael, 1985).
47. See for example *NRC Handelsblad*, 30 January 1985; 1, 5, 7 February 1985.
48. G. C. Berkhof, 'Het Amerikaanse "Strategic Defense Initiative" (SDI) en de Stabiliteit', *Internationale Spectator nr 5* (1985) pp. 303–10.
49. G. Van Benthem Van Den Bergh, 'Het Strategische Defensie Initiatief en het kernwapendebat', *Internationale Spectator nr 8* (1985) pp. 461–3.
50. Jan Wijkstra, "De Militaire of Civiele Route?", *Wetenschap en Samenleving 7/8* (1985) p. 63.
51. Chambre of Representatives, *Compte Rendu des Scéances Publiques, Chambre des Députés* (Luxembourg, 30 April 1985) pp. 3123–9.
52. The official Luxembourg view on SDI was given by Alphonse Berns in a letter to Marijke Van Hemeldonck, Member of the European Parliament and President of the International Association for University Peace Days. This letter is used as a source.
53. *Für die Schaffung eines Europas der Technologie*, a communiqué of the General Secretary of the LSAP, Luxembourg, 5 June 1985.
54. Chambre of Representatives. *Compte Rendu des Scéances Publiques*, Chambre des Députés, (Luxemburg, 30 April 1985) p. 3130.
55. *Compte Rendu des Scéances Publiques*, p. 3148.
56. Robert Berloznik, Patrick de Boosere, *Star Wars* (Berchem: Uitgeverij EPO, 1986).
57. J. Van Bentum (ed.), *Vrede Met Star Wars?* (Nijmegen, 1986).
58. Philip Everts (ed.), *De Droom der Ontwetsbaarheid. Het Amerikaanse Strategisch Defensie Initiatif en het belang van Europa* (Kampen: Kok Agora, 1986).

8 SDI – a Topic of Transatlantic Consultation and of Debate in International Parliamentary Bodies
Hans Günter Brauch

I INTRODUCTION

President Reagan's Strategic Defence Initiative (SDI) has become a major security issue both of transatlantic bilateral and multilateral consultations and of debate in three international parliamentary bodies: in the North Atlantic Assembly, in the Assembly of the WEU and in the European Parliament. However, the debate on a missile defence system (ABM) which includes Europe is not new. Twice within two decades the United States has initiated this debate without any prior consultation with its allies. In 1967, the then United States Defence Minister McNamara suggested a land-based terminal defence system (Sentinel) as a counter to the perceived Chinese ballistic missile threat, and sixteen years later President Reagan presented his radical vision of overcoming deterrence by deploying a multilayered BMD system. Many of the same arguments for and against such a system were already articulated in the first ABM debate from 1967 to 1972.

The signing of the ABM Treaty was applauded both by the two European nuclear powers (Great Britain and France) as a *stabilising* element of their national deterrent, and by the other non-nuclear European states as a major *symbol of détente and arms control*. The continued support for the ABM Treaty has become the lowest common denominator of all West European governments and of most West European political parties. If the United States should move in the early 1990s from research (SDI) to the development and testing of components, either modification of the ABM Treaty by negotiation or unilateral withdrawal would become necessary. A transatlantic conflict could become unavoidable.

II THE FIRST ABM DEBATE (1967–72)

In the early 1960s the first public debates over the deployment of ABMs began when both the United States[1] and the Soviet Union[2] were developing weapons with a limited capability of intercepting ICBMs.[3] In spite of the alarmist claims about the potential performance of the suspected early Soviet ABM systems the *Leningrad system* (Griffon missile), the *Tallinn system* (SA-5) and the *Moscow system* (Galosh), President Kennedy and President Johnson scrapped plans to install the United States *Nike-Zeus* and the *Nike-X* ABM systems, which were ready for deployment by 1964 and 1968 respectively.

Despite severe reservations concerning potential effectiveness, on 18 September 1967 the then Secretary of Defence Robert S. McNamara announced plans to install the Sentinel ABM system, which was designed to protect the United States population against a prospective limited Chinese ICBM capability in the 1970s and against a 'possible accidental launch of an ICBM by any of the nuclear powers'.[4] McNamara's announcement, only ten days before the second meeting of the newly established NPG of NATO, 'created considerable resentment among the allies', because 'the announcement had been made without sufficient consultation' and most Europeans felt 'that the United States had failed to honour its obligations to the NPG'.[5]

Raymond Garthoff has summarised well the alliance concerns that promptly emerged:

1. the absence of consultations;
2. an avowed anti-China orientation of the system;
3. suspicions that the anti-China rationale was a pretext for an eventual defence against Soviet missiles;
4. political concerns about the implications of defence of the United States but not of Western Europe;
5. uncertainty as to public reactions in Europe;
6. concerns about the effects on deterrence.[6]

In April 1967, at the first meeting of NATO's new NPG, its defence ministers discussed the 'technical and strategic (and financial) aspects of BMD' and began a study 'on ABM issues, including those involved in possible ABM defence in Europe'. However, the West European governments had not been consulted or informed prior to McNamara's announcement in San Francisco concerning ABM deployment while the Soviet leaders had been informed in advance.[7]

During the second NPG meeting in Ankara in late September 1967, McNamara's effort to downgrade the implications of the Sentinel decision for Europe was received rather critically by many West European defence ministers – Denis Healey being the most outspoken. In his general briefing to the NPG, McNamara tried to 'allay Allied doubts about the adequacy of the NPG as a means by which the United States would consult with their allies on strategic decisions which could affect their interests' and he committed the United States to keep its Allies informed on the planned arms control talks with the Soviet Union.[8]

In the course of the discussion of a NPG study on the possible deployment of an ABM system in Europe the British Defence Minister, Denis Healey, argued 'that a European system would not be particularly useful or viable, and that any attempt to develop such a system would have unfortunate consequences for the prospects for arms control and détente.'[9]

The reaction of the West European and especially of the West German press to McNamara's ABM announcement and to a possible European ABM system was highly sceptical.[10] Kurt Becker expressed the following reservations:

1. A missile defence on the national level would be useless and the prospects of a European solution would be hopeless.
2. The cost would be beyond the economic capabilities of the Europeans.
3. The security of Western Europe would be degraded if both the United States and the Soviet Union should deploy fully-developed ABM systems, given the technical difficulties of destroying Soviet SRBMs over the territory of the West European countries.[11]

The cost projections for a comprehensive ABM system for Western Europe were between 40 and 80 billion dollars, while the costs for a 'thin' ABM system to protect selected high-value targets in Western Europe were estimated between five and fifteen billion 1967 dollars. Lothar Rühl has summarised several of the concerns expressed during the Ankara NPG meeting:

1. an unpenetrable antinuclear umbrella appeared to be technically unfeasible;
2. such a system would require a permanent readiness that would not allow time for a political decision-making;
3. an ABM system could not protect Western Europe against SRBMs, cruise missiles and low-flying aircraft.[12]

Helmut Schmidt, the then leader of the parliamentary SPD, predicted an 'erosion of NATO and of the Warsaw Pact as a consequence of ABM systems in the Soviet Union and in the United States' Schmidt opposed a German ABM system for technical, fiscal, military and politcal reasons.[13]

In the meantime the study of ABM issues continued under the auspices of the NPG, and a final report was prepared for the third NPG meeting in The Hague in April 1968. According to Garthoff, who had participated in the NPG staff group on ABM issues in 1967–8, this study concluded that the 'light' United States ABM system would not be relevant to European defence:

> No feasible ABM defence for Western Europe could prevent the Soviet Union from 'inflicting catastrophic damage' on NATO Europe. In addition, it was recognised explicitly that European ABM deployment (by NATO or by European members of NATO) could have significant adverse political implications on East–West relations. The conclusion was that in the light of the current (and foreseeable) technological circumstances, the deployment of ABMs in NATO Europe was not 'politically, militarily or financially warranted'.[14]

In the final communiqué of the third NPG meeting the participating defence ministers stressed that under present conditions the deployment of a missile defence system in Europe would not be justified. However, development in this area should continue.[15] The Dutch Defence Minister, Mr den Toom, in summing up their conclusions stated 'that a European ABM would be too costly, not totally effective, and might compromise arms limitation discussions between the United States and the Soviet Union'.[16] A comprehensive ABM system for Europe would cost at least 40 billion dollars.[17]

The third NPG meeting put an end to the discussions about a potential European anti-missile screen until April 1984, when United States Secretary of Defense Caspar Weinberger reintroduced the same issue in the context of SDI during the 35th NPG meeting in

Cesme, Turkey. In April 1968 the seven NATO defence ministers coupled their rejection of a European ABM in their final communiqué with a reaffirmation of 'their hope that progress could be made in discussions with the Soviet Union toward a limitation of the strategic arms race'.[18]

The official criticism was shared by an expert conference organised by the Atlantic Institute in Paris in October 1968.[19] Lawrence W. Martin, then professor of war studies at King's College in London, summarised the conclusions concerning the potential effects of an American BMD deployment on United States–Soviet relations and on the NATO alliance:

> While it does not automatically follow that BMD entails a new arms race, it may well invite a further improvement of offensive weapons by which BMD might be overcome: there are strong arguments to suggest that BMD cannot offer anything approaching immunity. Furthermore, these new technologies might complicate attempts to limit the level of strategic armaments, make arms control measures still more difficult and at the same time breach the political requirements of equal sacrifice laid down by non-nuclear powers as a condition of their signing the non-proliferation treaty.

A year later, Martin made the following assessment:

> The European reaction to BMD has on the whole been unfavourable, the argument weighing most heavily against BMD being that it accelerates the arms race and thus it will worsen East–West political relations. Many Europeans regard it as an attempt to buy a superfluous margin of American security at the expense of Europe, at the same time introducing a new element of European inferiority into the Alliance.[20]

Johan Jørgen Holst, in his summary of the European reaction to the American ABM debate, refers to some additional strategic considerations:

> The general critical attitude does not differentiate between various alternative US BMD deployment configurations ... The introduction of thin BMD systems with area coverage in the United States may have another indirect impact on the deterrent mechanisms of the NATO defence posture. The deployment of tactical nuclear

weapons in Europe has, in the European view, in a significant degree contributed to the deterrence by constituting a link with the American strategic deterrent through a process of escalation. However, the introduction of controllable strategic nuclear forces and an associated doctrine of flexible and restrained use tended to decouple the tactical nuclear weapons from some of their deterrent value. Thus decoupled, tactical nuclear weapons in Europe may refocus ... European concerns on the possibility of the superpowers' confining an armed conflict to Europe and largely destroying the arena in the process without having to fear nuclear destruction in their own countries or having that fear significantly reduced.[21]

As far as the potential deployment of a BMD system in Europe is concerned Holst distinguishes among three possible alternatives: 'an American system deployed in Europe, a joint Euro–American system, or an American-built European system'. A European-built system would appear unlikely, and an American or a Euro–American system would raise serious problems of Alliance integration. While the Soviet threat to Western Europe would be larger, 'the relatively small defence area would, however, generate demands for fewer area coverage installations ... [which would] have to be for the defence of the population'. Even if the technical problems were surmountable, a European-based BMD 'would not close all the gaps. There remains also the issue whether BMD would be the most cost-effective means of damage limitation in Western Europe compared, for example, to civil defence measures ... and a reduction of the numbers of tactical nuclear weapons in Central Europe'.[22] For Holst there was also the danger 'that a BMD in Western Europe might tend to penetrate a posture and atmosphere of confrontation'.[23]

When in 1969 President Nixon decided to concentrate on the more sophisticated Safeguard system – a system utilising the nuclear-tipped Sprint (ER warhead) and Spartan (warhead in the MT range) missiles with a limited capacity to intercept Chinese and Soviet first-strike weapons against hardened Minuteman missile sites – the NPG took relatively little notice.[24] In Buteux's assessment: 'With respect to the Safeguard decision, the degree of consultation by the United States with her allies seems to have been adequate since it did not lead to open dispute'.[25] The initiation of SALT in 1969 and the prospects of negotiated constraints on ABM systems may also have contributed to the relative silence of the West European governments and experts as far as the effects of BMD on deterrence were concerned.

III EUROPEAN REACTIONS TO THE ABM TREATY AND TO SALT I (1972)

The ABM Treaty of 1972 and the subsequent Protocol of 1974 were welcomed by the European NATO allies. After an exemplary and close consultation process – according to Garthoff there were forty-five meetings or formal communications, including twenty meetings of the North Atlantic Council Permanent Representatives in Brussels on the SALT I negotiations – 'the ABM Treaty was seen as a major step toward stabilising the strategic arms race and removing any questions of differentiation in vulnerability between NATO Europe and the United States'.[26]

The ABM Treaty also helped to overcome the fears expressed by Michel Debré in 1970 and 1971 'that SALT might allow the USSR and the United States to deploy effective ABM systems behind which the two countries would retreat ... They feared that the continent might become the unshielded fighting ground for settling disputes with Europeans bearing the costs'.[27]

As a direct consequence of the ABM Treaty, the United States remained as vulnerable to ballistic attack as its Allies. The spectre of a 'Fortress America' was overcome. As a consequence of the constraints for the Soviet ABM systems, the continued credibility of the British and French nuclear systems was also enhanced. for this very reason Lord Carrington, then Great Britain's Secretary of State for Defence remarked in early 1973; 'We have a very direct interest ... in the continuation of the ABM Treaty'.[28] And for France the ABM Treaty 'provided a very welcome opportunity for the French to continue the expansion of their strategic nuclear force programme. The French government could have reasonable confidence that its deterrent's political utility would not be rendered ineffective without at least some advance warning through public abrogation of the ABM Treaty by either superpower, or through intelligence regarding clandestine Soviet research and development in ABM that might offer the Soviets an option of rapid ABM deployment'.[29]

Neither Great Britain nor France objected to the noncircumvention clause contained in Article IX, which prohibits the transfer of ABM technology to third countries. On 18 April 1972 the United States declared unilaterally that Article IX did not establish a precedent that would prohibit the transfer of offensive technologies as well.

The West German government interpreted the SALT process as a major pillar of East–West détente. The support for the continuation of the ABM Treaty regime is, as Yost correctly remarked, 'implicit in the FRG's support for the continuing SALT/START arms control process' which, 'according to its 1979 Defence White Paper is deemed "of permanent importance in all political efforts aimed at safeguarding peace and achieving stabilisation in the East–West balance of power"'.[30]

Gallis, Lowenthal and Smith have correctly summarised the general attitude of the other West European states, which

> view the ABM Treaty as an important sign that détente can yield significant agreements between the two superpowers. For this reason, discussion of new ABM systems often meets with scepticism and concern from the Allies. They have strongly supported the Treaty because in their view it serves to preserve known, and accepted, vulnerabilities. The Treaty thereby prevents a step into a new era which, in the view of some, could trigger a series of unknown countermeasures by each side, as the Soviet Union and the United States seek to develop new means and new weapons to penetrate an improving defence.[31]

Were the West European governments and the NATO Alliance adequately consulted prior to the most recent shift in the American attitude towards BMD and its potential implications for the ABM regime?

IV INTRA-ALLIANCE CONSULTATION

Within NATO the need for close consultation both in the area of military, – especially in nuclear – planning,[32] and in the arms control field[33] had repeatedly been stressed in political statements of NATO leaders and in the final communiqués of its meetings. According to Stanley R. Sloan, consultations in the NATO context

> include all those discussions that take place within formal NATO committees, groups and councils; for the purposes of this discussion, NATO consultations are interpreted more broadly to include bilateral and multilateral discussions that may take place outside formal NATO channels but that nonetheless affect relations among the allies.[34]

'Consultation', as Roger Hill put it, refers to the process of 'conferring, conversing, or otherwise communicating with each other in order to exchange information, give advice, or decide something'.[35] In Hill's formula, consultations are based on common purpose and goals and are therefore 'to be marked by trust, frankness, and a certain collegiality. Consultations on SDI within the official NATO structure consisted of the following elements:

1. *multilateral consultations* within the North Atlantic Council, the principal decision-making body of NATO; *consultations* in the Political Committee and the Senior Political Committee, the main source below the council level; the Defence Planning Committee the Nuclear Planning Group, the High-level Group that dealt primarily with deployment issues and the *Special Consultative Group* which was created as a special mechanism for co-ordinating NATO's approach to INF issues;
2. *bilateral consultations* taking place in Washington, in the respective national capitals, at NATO headquarters or during other international conferences on various political levels.[36]

(a) Multilateral consultations on SDI in the NATO context

With his 23 March 1983 speech[37] President Reagan inaugurated a major policy and technology review that led to the SDI. Neither the Pentagon or the United States Congress or the NATO Allies were adequately consulted or informed prior to this announcement. In order to counter the shock of the Allies the following sentences were added to the speech:

> As we pursue our goal of defensive technologies, we recognise that *our Allies* (my emphasis) rely upon our strategic offensive power to deter attacks against them. Their vital interests and ours are inextricably linked – their safety and ours are one. And no change in technology can or will alter that reality. We must and shall continue to honour our commitments.

The President concluded his major speech with the remark:

> Tonight, consistent with our obligations under the ABM Treaty and recognising the need for close consultation with our Allies, I am taking an important first step.

However, the official briefings and consultations with the allies did not start till February 1984, after the Fletcher and the Hoffman

Reports had been submitted in October 1983 and after President Reagan had signed the National Security Decision Directive 119 on 6 January 1984 that initiated the SDIO. According to the official Pentagon report of April 1985 on the SDI:

> Following President Reagan's decision on Allied consultations embodied in a Presidential Decision Directive, Administration briefing teams were sent to the capitals of Allies in Western Europe and the Pacific. The interagency teams were composed of officials from the Office of the Secretary of Defense, the Organization of the Joint Chiefs of Staff, the Department of State, and the Arms Control and Disarmament Agency. The European team briefed NATO Allies during early February 1984. The Pacific team visited Ottawa before traveling to Tokyo, Canberra and Wellington. As a preliminary step, Allied military attachés in Washington were presented with a briefing in order to ensure that they were informed and to provide them an opportunity to pass preliminary views to their governments in the interest of facilitating discussions with United States representatives.
>
> The Allies were presented a three-part briefing covering the scope of the Soviet efforts in both conventional ABM capabilities and advanced ABM technologies, the results of United States study of the policy implications of SDI and the dimensions of the United States technology research and development program.[38]

West European representatives at NATO reacted to this first comprehensive information on President Reagan's plan for a space-based BMD with a 'mixture of irritation, scepticism and concern'. According to a news analysis by Reuter, the implications of the SDI for arms control remained unclear. The main concern of Europeans participating in these briefings were whether such a BMD system could destroy ballistic missiles of shorter range as well. United States experts admitted in these off-the-record briefings that cruise missiles and nuclear bombers would create problems that could not be handled by such a defensive system.[39]

Until February 1986, SDI has been a major issue at the following ministerial sessions of the NPG: at its 35th meeting in Cesme on 3-4 April 1984, at its 36th meeting in Stresa on 10-11 October 1984, at its 37th meeting in Luxembourg on 26-27 March 1985, and at its 38th meeting in Brussels 28-29 October 1985.

At Cesme, according to the Pentagon report, 'Secretary Weinberger assured the Allies that the United States fully intended to continue consultations ... The Secretary stressed that he would welcome Allied technical participation as this could make significant contributions to the SDI programme'.[40]

However, according to reports both in the American and in the European press on the NPG meeting at Cesme, several Allied ministers indicated they were not altogether reassured after listening politely to Weinberger's explanation of United States plans 'to develop a comprehensive space-based missile defence and . . . not to seek a treaty with the Soviet Union banning antisatellite weapons'. Most outspoken in his criticism has been West German Defence Minister Wörner: 'My impression is that the Europeans were broadly united in their critical questions. I can't see that it would provide greater protection or stability. I can only hope it would give an incentive for arms control'. And according to the Dutch Defence Minister de Ruiter the discussion on SDI remained 'full of question marks. It has many aspects that can worry us'.[41]

According to a report in the *Wall Street Journal*, NATO defence ministers expressed 'uneasiness' over SDI and that there was a 'growing concern among Europeans' that the plans of the Reagan administration were 'leading America down an isolationist path that would leave Europe out in the cold'. 'Some Europeans fear', the newspaper said in describing the initial reaction of high-level European defence officials to the United States SDI briefing, 'if the United States concentrates on protecting itself, it would become less inclined to defend Europe'. One defence minister told the Wall Street Journal: 'there is concern that the US actions and proposals could set off a new kind of arms race between offensive and defensive systems'.[42]

Both Wörner and de Ruiter complained to the press at Cesme 'the US programme could do more damage to arms control at a time when nuclear arms reduction talks are stalled. West Germany', according to a report in the *Baltimore Sun*, 'believes the star wars effort, if implemented, could split the United States from its NATO allies, thereby leaving them more vulnerable to attack or to Soviet intimidation'.[43] And a few days after the NPG meeting in Cesme, Wörner followed up by telling a German daily 'that a US antimissile system could destabilise the East–West balance, "decouple" the US and Western Europe, and even lead to a splitting apart of the Western Alliance'.[44]

In July 1984 in Brussels, after the first public criticism by leading West European defence experts, officials of the United States government held a second set of detailed briefings both on the technological and the strategic aspects at a plenary session of NATO's Military

Committee. United States officials also met with the permanent representatives of the North Atlantic Council and NATO's High-Level Group to discuss 'the origin and purpose of SDI, implications for deterrence and arms control and the potential benefits for the Allies'.[45]

At the 36th ministerial meeting of the NPG in Stresa on 11–12 October 1984, Secretary of Defence Weinberger's briefing was received by his West European colleagues without additional comments on SDI. In the communiqué, the fourteen NATO defence ministers supported the readiness of the United States to initiate arms control talks including space weapons without any preconditions with the Soviet Union.[46] This support for the initiation of arms control talks including space weapons was also supported in the communiqués of the Ministerial Meeting of the Euro-Group on 3–4 December 1984 and of the North Atlantic Council on 13–14 December 1984.[47]

In February 1985 an interagency team of United States officials briefed the permanent representatives of the North Atlantic Council on SDI and on the Soviet efforts in research and development of defences against ballistic missiles.[48] In mid-February 1985, after a meeting of the Special Consultative Commission at which the forthcoming arms control talks in Geneva were discussed, the then Assistant Secretary of State for European Affairs, Richard R. Burt, admitted that it is 'no secret that there are different views' within NATO on 'many details, and even on the concept' of Reagan's plan to develop a space-based BMD.[49]

At the 37th Ministerial meeting of NATO's NPG in Luxembourg on 26–27 March, the European participants 'shelved their future concerns and embraced Reagan's proposed $ 26 billion research effort partly because of "what we know of Soviet capabilities and interest in the field"'. While Weinberger admitted in the press conference after the meeting that European officials voiced 'some reservations' about SDI[50], the communiqué of the NPG meeting supported 'the United States research programme into these technologies . . . This research, conducted within the terms of the ABM treaty, is in NATO's security interest and should continue. In this context, we welcome the United States' invitation for Allies to consider participation in the research programme'.

The defence ministers of the thirteen participating states 'noted with concern the extensive and long-standing efforts in the strategic defence field by the Soviet Union which already deploys the world's largest ABM and anti-satellite systems. The United States strategic defence programme is prudent', the communiqué stressed, 'in the light of these Soviet activities and is also clearly influenced by the treaty violations reported by the President of the United States'.[51]

On 22 May 1985, in a one-day ministerial meeting of NATO's Defence Planning Committee 'in the shadow of SDI', the NATO defence ministers emphasised their continued support for NATO's strategy of flexible response that would be fundamentally challenged if SDI should become reality. Ulrich Mackensen criticised the policy of harmony by excluding critical questions while Karl Feldmeyer observed an increasing difference of interests between the United States and its European allies on nuclear issues that had become obvious by the fundamental reorientation of strategy intended by the SDI.[52]

At the ministerial session of the North Atlantic Council in Estoril near Lisbon on 6–7 June 1985, at the insistence of France no endorsement of the SDI was included in the final communiqué. However the European members of the Council 'had been unanimous in stressing to Shultz that they wanted the United States to continue to abide by provisions of the unratified SALT II arms control agreement'. The European foreign ministers argued that an abrogation of the SALT II restraints 'would cause a backlash in West European public opinion and impede progress in the Geneva talks'.[53]

Denmark, Norway and Greece had shared the French reservations relating to the SDI. In order to minimise the differences that blocked an alliance endorsement for President Reagan's SDI, the sixteen NATO foreign ministers supported 'US efforts in all three areas of the Geneva arms control talks with the Soviet Union'.[54] The North Atlantic Council rejected 'military superiority' and in indirect criticism of the strategic change called for by SDI it emphasised 'Our strategy of deterrence has proved its value in safeguarding peace'; and 'The security of the North American and European allies is inseparable'.[55]

Although the defence ministers of NATO's NPG (with the exception of France, Iceland and Spain) had endorsed their support for the research phase of SDI, the foreign ministers of the North Atlantic Council only supported the efforts for limiting the deployment of weapons in outer space.

In Autumn 1985, both the multilateral and the bilateral consultations on the SDI and on the Geneva summit between President Reagan and the Soviet party leader Gorbachev intensified. In order not to document the differences that existed within NATO regarding SDI, the communiqués of the following meetings and conferences avoided any formal endorsement of SDI: the NPG meeting in Brussels on 29–30 October, the two special meetings of the North Atlantic Council on 15 October and on 21 November, the high-level conference of selected leaders of the major NATO countries in New York on 24 October, the Euro-group and the Defence Planning Committee meetings in Brussels on 2–4 December and the winter session of the North Atlantic Council on 12–13 December.

At the special meeting of NATO's foreign ministers on 15 October the NATO allies supported a narrow interpretation of the ABM Treaty as represented by Secretary of State George P. Shultz in the internal clashes within the Reagan Administration and they urged the United States to offer new arms control initiatives.[56] Once again, Shultz told his NATO colleagues 'that the US would only proceed with testing and production phases of the Strategic Defence Initiative after "extensive" consultations with its allies and negotiations with the Soviet Union'.[57]

Two weeks later, at the 38th meeting of the NPG in Brussels, United States Defence Secretary Weinberger tried to rally his counterparts – all NATO defence ministers with the exception of France and Iceland – behind the charge of major Soviet violations of arms control treaties, especially of the ABM Treaty.[58] In order to project a firm display of NATO solidarity, the European defence ministers 'noticed' the new American evidence, but they stopped short of joining the United States in explicitly accusing the Soviets of treaty violations. However, the ministers did not reiterate their former explicit support for the SDI after France, Norway, Denmark, the Netherlands and Greece had rejected any form of official participation.[59] After extensive negotiations at the working level the final communiqué of the NPG meeting mentioned the disputed SDI topic only indirectly and in the most general terms: 'We ... expressed strong support for United States' positions concerning intermediate, strategic, and defence and space systems'.[60]

In a communiqué issued on 3 December, after a meeting of NATO's Defence Planning Committee, NATO's defence ministers – once gain avoiding any direct reference to the SDI – 'expressed strong support for the United States' stance concerning intermediate range, strategic, and defence and space systems'.[61] At the same meeting, the West German Defence Minister, Manfred Wörner, launched the idea of EDI against short-range Soviet ballistic (SS-21, SS-22, SS-23) and cruise missiles that should be pursued independently of the SDI.[62]

At their winter meeting on 12–13 December 1985 in Brussels the foreign ministers attending the North Atlantic Council sessions, debated the Geneva summit and the implications of the SDI on NATO cohesion.[63] Several European foreign ministers, including Sir Geoffrey Howe and Hans-Dietrich Genscher, stressed that a second summit in 1986 'must produce more specific results'. Once more, the communiqué avoided any direct endorsement of President Reagan's SDI. The general support of 'US efforts in all three areas of negotiations' was interpreted by United States officials as a 'tacit approval of the US refusal to accede to Soviet insistence that Washington abandon the SDI programme'.[64]

Two years after the initiation of intensive briefings on behalf of the United States government and after numerous multilateral debates and consultations in the NATO framework, the Alliance could not agree on a joint position on SDI, while the Soviet Union had obviously failed to drive a wedge between the United States and their European allies, SDI nevertheless weakened both the WEU and Franco–German defence co-operation.[65]

In addition to the high visibility of the multilateral and bilateral consultations on the political level, there have been continuous United States efforts to meet with the Allies at the mid-level when representatives of Allied nations come to the United States. The bilateral consultations focused increasingly on the terms of a potential co-operation of Allied nations in the SDI research project.

(b) Bilateral consultations on the terms of participation of individual nations in the SDI research programme

In February 1985 United States Defence Secretary Weinberger invited Great Britain and West Germany to participate in the American SDI research programme. However, both German and British officials stressed that technological co-operation required a 'true two-way-street' that could be jeopardised by restrictive technology transfer rules.[66] On 26 March 1985 Secretary Weinberger sent a letter to the NATO members, Israel, Australia and Japan asking them 'as a first step, that you send me, within *60 days* (my emphasis), an indication of your interest in participating in the SDI research program of the areas of your country's research excellence that you deem most promising for this program'. Weinberger reassured his colleagues that the United States

> will work closely over the next several years with our Allies to ensure that, in the event of any future decision of defensive

weapons (a decision in which consultation with our Allies would play an important part), Allied as well as United States security against aggression would be enhanced...
The United States will, consistent with our existing international obligations including the ABM Treaty, proceed with co-operative research with the Allies in areas of technology that could contribute to the SDI research program. Pursuant to this policy, the United States is permitted – and is prepared – to undertake such co-operative programs on data and technology short of ABM component level as may be mutually agreed with allied countries.[67]

In a second mid-April letter, Secretary Weinberger tried to dispel any suspicion about the 60 days deadline that had been perceived by some Europeans as an ultimatum. The official SDI report of mid-April 1985 further specified the United States position on the co-operation issue:

With respect to SDI, the United States will not seek to arrange for the Allies to do for the United States what it cannot do under the [ABM] Treaty. Of course, exchanges with the Allies concerning defensive systems not covered by the ABM Treaty can continue as desired by the United States and its Allies.[68]

Since May 1985 many (often contradictory) statements on the terms of the potential co-operation of the Allies have been made by high United States officials. Secretary of Defense Weinberger and his Assistant Secretary, Richard N. Perle have indicated a tough position on sharing technologies with the European Allies and that it would require 'a major effort to keep Soviet spies from obtaining secret information'.[69] Returning from a NATO meeting in Brussels, Secretary Weinberger told reporters: 'I never felt a governmental response is required. What I wanted was to get their companies and scientific, educational institutions... all of them participating',[70] and in late June 1985 Richard Perle stressed in Cologne that 'a participation of European governments in the SDI programme was never intended'.[71]

The first NATO country to decline participation in the SDI programme was Norway, whose conservative coalition government did so on 15 April 1985, and on 11 May 1985 the Danish Parliament voted against Danish involvement.[72] During the Bonn summit in early May French President Mitterrand declined to accept the American invitation.[73] On 8 September 1985 the Conservative Canadian Prime Minister Brian Mulroney announced that participation in SDI was not in Canada's national interest.[74] Till December 1986, only Great Britain, West Germany, Israel, Italy and Japan had signed cooperative agreements with the United States on the terms of participation in the SDI research programme. Other governments had not reached a final position whether their governments should join the SDI research project by signing a government-to-government agreement.[75]

In a press conference on 26 August 1985, General Abrahamson, director of the SDIO, mentioned two minimal conditions for any such agreement: 'adequate government-to-government arrangement to ensure that classified information is protected' and that 'intellectual property rights are respected'.[76] On 5 December 1985 General Abrahamson told a Senate subcommittee that 'classified data would have to be transferred through government channels rather than directly from United States companies to their counterparts'.[77]

In order to specify the terms of such an agreement and to obtain additional information, several Western countries – Japan, Great Britain and West Germany – sent governmental delegations while the companies of several other countries, including France, sent representatives to the SDIO to discuss the possibilities of industrial participation in the SDI research programme.

In Great Britain some thirty companies have 'formed an informal industrial club . . . to study the best ways to respond to SDI requirements'. In West Germany, Dornier and Messerschmitt-Boelkow-Blohm, in France Aerospatiale and Matra, and a consortium of eight important Italian aerospace and electronic firms (CITES) whose goal is 'to be a major Italian partner in both the SDI and the EUREKA program',[78] have also indicated an interest in participating in SDI.

V THE INTERNATIONAL PARLIAMENTARY DEBATES ON SDI/EUREKA

President Reagan's strategic vision of overcoming deterrence and the SDI research programme have increasingly become an issue of debate in three international parliamentary bodies:

1. the *North Atlantic Assembly*, consisting of MPs from the sixteen NATO countries;
2. the *WEU* which consists of MPs from France, Great Britain, the Federal Republic of Germany, Italy, Belgium, the Netherlands and Luxembourg;

The Lefebvre report described the initial European reactions to the first United States briefings in February 1984 as 'cautious' and that of Great Britain and France as 'sceptical':

> The reason for this scepticism was primarily the possible erosion of the credibility of their deterrent forces though concern was also expressed about the possible undermining of deterrence and the increased likelihood of conventional war in Europe.[82]

While the Forrestal report tended to be rather sceptical of the implications of SDI for Europe, the Lefebvre report was more favourably inclined towards SDI in its conclusions:

> Should the United States decide in the future that the only means of maintaining stability and of promoting meaningful reductions in offensive nuclear weaponry is by deploying some form of BMD system, an arms control strategy must be formulated to accompany it. A reciprocal linkage between offensive and defensive weapons ... commends itself in this respect. A movement towards strategic stability and arms control might also be facilitated by sharing BMD technology with the Soviet Union as proposed by President Reagan. Also, it is felt that it would be militarily – and strategically – stabilising to deploy anti-tactical missiles in Europe to defend against short-range ballistic missiles, such a move would be politically prudent. This would reassure the European Allies that the United States is not seeking to protect only itself and would, perhaps, be a means of encouraging Alliance participation in elements of the SDI.[83]

In 1985 SDI became the major issue of debate and controversy among the nearly 200 parliamentarians from sixteen countries both at the spring session, held in Stuttgart from 17–20 May, and at the 31st annual session held in San Francisco, 10–15 October. Four draft reports dealt with different aspects of SDI or BMD: an updated version of the Lefebvre report on behalf of the Scientific and Technical Committee, a Draft Interim Report by John Cartwright for the Special Committee on Nuclear Weapons that focused on the alliance aspects, a Draft General Report on 'Alliance Political Developments' by Bruce George that discussed both the United States and the Soviet BMD programmes and EUREKA as issues and challenges for the Alliance, and a Draft General Report on 'The Future of Arms Control: Compliance and Verification Issues' submitted by Sir Geoffrey Johnson Smith on behalf of the Military Committee. An additional report on the 'Exploitation of Space' by Robert Banks included both military and private sector space activities.[84]

Addressing the Spring Session in Stuttgart, Chancellor Kohl – just two weeks after the clash between Reagan and Mitterrand during the Bonn summit on European participation in the SDI – indicated the European concerns and demands relating to SDI:

> For the North Atlantic Alliance, President Reagan's Strategic Defence Initiative constitutes *both an opportunity and a gamble*. We cannot today predict whether the SDI will prove to be an alternative means of preventing war and a way of reducing dependence on nuclear weapons and of eventually eliminating them. The following points remain particularly important for the North Atlantic Alliance:
>
> 1. The security of Europe must not be detached from that of the United States.
> 2. The NATO strategy of flexible response will remain valid as long as no more effective alternative means or preventing war has been found.
> 3. Specific research findings must lead to co-operative solutions.[85]

Speaking in the plenary debate Senator William Roth (Republican, USA) remarked that the key concept regarding the SDI was Alliance co-operation. He also rejected the argument that the SDI could lead to decoupling. Senator Roth felt that it would not be possible in the future to maintain popular support for the current strategy of deterrence based on the suicidal idea of MAD, and that SDI offered a possible escape from this. The United States wanted its European Allies to join in the SDI which would enhance their mutual security, and he hoped that co-operation in NATO would continue to grow in the new era just beginning.[86]

Replying to Senator Roth, the West German Social Democratic MP Karsten Voigt rejected the SDI on the grounds

> that it ran the danger of having a destabilising effect. It was totally illusory to believe that it could ensure a military deterrent. Nor could it solve economic problems. It could not, lastly, serve to strengthen peace. What was desirable was for East and West to arrive at an agreement in Geneva; for example, a moratorium on testing with verification provisions[87]

At Stuttgart the debate on the SDI took place largely in the Assembly's five committees and revolved around presentations by John Gardner, the Director of Systems at the SDIO, on the technological developments, on Soviet BMD efforts, on United States countermeasures against Soviet short-range missiles including SS-21, SS-22s and SS-23s, as well as on the potential spinoffs of the SDI.[88]

The Lefebvre Report of April 1985 had summarised the general mood of the transatlantic debate well:

> Support for the SDI has become, in a sense, the test of Alliance loyalty to the United States: a test of the United States' loyalty to the Alliance could be an agreement to offer genuine participation – and technology transfer – to those nations wishing to participate in the SDI.[89]

Bruce George, in his October 1985 report on 'Alliance Political Developments' touched both on the 'widespread concern on both sides of the Atlantic about the strategic implications of the President's vision' and on the 'fear of being excluded from the technological advances the specific programmes will entail and their commercial applications'. Based on past experience, George cautioned: 'The reluctance of the US to share high technology with Allies in the past does not augur well for broad, two-way technology transfer by participation in SDI'. For these very reasons, as George pointed out, 'The Allies ... have been increasingly nervous about committing themselves to a programme they find potentially destabilising in a strategic sense and disturbing in its technological implications. As a result, there was no support for SDI at the May Bonn Summit ... nor at the Lisbon ministerial meeting in late June, where France, Norway, Greece, and Denmark disagreed'. However, the European Allies have also been unable 'to agree on a collective European response to participation', for example, in several ministerial meetings of the WEU in April, July and in November 1985.[90]

Sir Geoffrey Johnson Smith noted in his discussion of compliance and verification issues 'considerable concern within the Atlantic Alliance about the ambitious Strategic Defence Initiative and the Soviet response to it. There are deepening fears that, in the long term, both programmes will seriously erode the ABM regime and that the net effect will be less, not greater, security for the West as a whole. The two Allied countries which have expressed the most serious concern about the threats to the ABM regime have been France and the United Kingdom'.

As the net beneficiaries, both Britain and France 'have an enormous stake in the ABM regime... Naturally, both countries are hostile to wholesale changes to the Treaty, whether carried out unilaterally or bilaterally, which might negate the effectiveness of their systems'. In the view of Sir Geoffrey's draft general report, 'there is a good case to be made that the ABM Treaty has... enhanced both strategic and crisis stability'.[91]

At the Annual Session of the North Atlantic Assembly in San Francisco the debate on the SDI evolved around a draft interim report submitted by the British Social Democrat John Cartwright and a special draft resolution on 'Strategic Defence and the Alliance' he had drafted with the Republican United States Senator Charles McC. Mathias, Jr.[92]

According to Cartwright, 'the SDI issue has... provoked deep apprehension and concern... within the Atlantic Alliance... Contradictory statements from Reagan Administration officials on the objectives of the SDI programme explain to a large extent the confused state of the debate in the Alliance. The early evidence strongly suggests that SDI carries with it a remarkable potential both to exacerbate existing transatlantic differences and create new tensions'. In Cartwright's view 'there is a real danger of a transatlantic crisis brought about by a conflict of nuclear philosophies'.[93] The cautious support for the research phase of the SDI which most West European Governments came to embrace in late 1984 and early 1985 represents, in the Rapporteur's view, 'agreement on short-term objectives while very serious doubts remain about the longer-term implications for Western security'.[94] Cartwright identified five specific aspects of the SDI on which particularly European apprehension has focused: (1) the future of strategic stability, (2) arms control, (3) decoupling, (4) implications for French and British nuclear forces, and (5) opportunity cost.

> As to strategic stability, Cartwright pointed to the danger that a substantial missile defence system, combined with an enhanced offensive capability, might well be construed by the other side as a development of a first-strike capability. In a crisis situation this could induce one side to launch a pre-emptive strike... From a purely European standpoint it has still to be demonstrated why a mixed offensive–defensive strategy would be better for European security than, for example, a strengthened ABM Treaty and deep cuts in strategic offensive weapons.[95]

In his view, the perennial European fear of decoupling has been rekindled as a result of the debate about strategic defences:

> Despite American assurances that the SDI programme is designed to offer protection both for the United States *and* its Allies, many remain sceptical. There is an underlying fear that should the two superpowers successfully achieve strategic defence, Europe would be much more exposed to a Soviet conventional attack.[96]

As to the opportunity cost, Cartwright pointed to the longer-term concern

> that pressures on the US defence budget (and budget deficit) as the SDI programme steps up its investment requests will dictate cuts elsewhere ... European apprehensions are ... that Congressional exasperation with European performance may generate a movement for reductions in the US contributions to NATO ... together with renewed Congressional threats to reduce US troops.

Cartwright concluded his critical evaluation of the SDI issue:

> The Strategic Defence Initiative is likely to pose major questions for the Alliance for several years, and its implications – strategic, political and economic – will be profound. Despite the limited agreement on short-term objectives, deep differences remain on the longer-term impact of such a dramatic change in philosophy. If not carefully managed, such differences will present the Soviet Union with a unique opportunity to create and exploit divisions between Alliance partners.[97]

The resolution on 'Strategic Defence and the Alliance' that was adopted at the 31st annual session of the North Atlantic Assembly held in San Francisco by 91 votes to 12 with 28 abstentions nevertheless urged the NATO governments: *'to support United States research into strategic defence consistent with the ABM Treaty'*. But it contained all the major reservations of the European governments as well:

1. to recognise the close relationship between strict adherence by all parties to existing arms control agreements and the construction of a framework for mutual restraint between East and West;
2. to support every effort to enhance Western security through

negotiated limitations and mutually verifiable reductions of offensive nuclear forces;
3. to encourage agreement between the Soviet Union and the United States on the technical definitions of the forms of research permissable under the terms of the ABM Treaty;
4. to ensure that any future arms control regimes covering strategic defence contain provisions which preclude circumvention by advanced ASAT development; . . .
6. to assess what forms, if any, of ballistic missile defence would allay fears about 'decoupling' and would contribute to the security of the Alliance as a whole;
7. to ensure that any Alliance participation in the Strategic Defense Initiative complies fully with the terms of the ABM Treaty; and
8. to ensure that the aim of extending the transatlantic two-way-street must also be pursued with regard to the research efforts by the United States.[98]

Many European concerns have also been incorporated into the preamble of this resolution: The Assembly: *reaffirmed* 'deterrence based on retaliatory offensive systems as an essential component of Alliance security'; *emphasised* 'that the existing strategy of flexible response must remain in force as long as there is no more efficient alternative'; *expressed the conviction* 'that the ABM Treaty continues to make a fundamental contribution to strategic stability' . . . ; *recognised* 'some similarities between ASAT and strategic defence technologies' that 'may undermine the ABM Treaty regime'; *supported* 'the United States' commitment to continue research into strategic defence technology within the provisions of the ABM Treaty'; *noted* 'that disagreement exists within the Alliance over the technical and financial feasibility, military utility and political desirability of deployment of large-scale ballistic missile defences; and *expressed its conviction* 'that decisions about the potential development and deployment of large-scale ballistic missile defences'; and *expressed its* partners have been consulted and negotiations with the Soviet Union on co-operative solutions have taken place, i.e. that there must be no automatic sequence of research, development and deployment'.[99]

This transatlantic compromise resolution emerged from six days of intensive and controversial debates both in the plenum and in the special committees. It came only after both United States Secretary of State George P. Shultz and Paul H. Nitze, the Reagan Administration's chief arms control adviser, had emphasised in their addresses to the Assembly that the United States intended to abide by a restrictive interpretation of the ABM Treaty. Shultz and Nitze reassured the delegates: 'Our research programme is and will continue to be consistent with the ABM Treaty'.[100] This allayed fears which had been created by Robert C. McFarlane, who had indicated 'that a new interpretation of the ABM treaty would allow the US wider latitude in developing and testing the space-based directed-energy weapons that are part of the SDI effort'.[101] Although he confirmed 'that a broader interpretation of our authority is fully justified', Shultz added, with applause from the delegates: 'This is, however, a moot point; our SDI research programme has been structured and, as the President has reaffirmed . . . , will continue to be conducted in accordance with the restrictive interpretation of the Treaty's obligations. Furthermore, any SDI deployment would be the subject of consultations with our Allies, and to discussion and negotiation, as appropriate, with the Soviets in accordance with the terms of the ABM Treaty'.[102] This restrictive interpretation was also emphasised by Paul Nitze a few hours before the vote: 'SDI is a research programme . . . The Programme is and will continue to be conducted in full conformity with the ABM treaty'.[103] Contradicting President Reagan's initial vision, which was still being upheld by his Science Advisor, George Keyworth, as late as 29 March 1985,[104] Nitze stressed 'that SDI is not designed to produce a regime that would replace deterrence, but rather to shift its means . . . Moreover, since out Allies benefit from the extended deterrence of the United States, a programme that enhances US security and US confidence in the viability of its strategic deterrent will increase the effectiveness of the US commitment to Europe as well'.[105] However, not all delegates were convinced, among them the British Labour MP, Kevin McNamara, who 'profoundly disagreed with US policy' and feared 'that the SDI programme would not end, but exacerbate the vicious spiral of the arms race'. McNamara 'noted the change in the US description of the SDI programme' and he 'also drew attention to the difficulty of finding a consistent voice within the United States Administration itself'. For him 'it was difficult to ascertain where the Administration really stood'.[106]

The Cartwright/Mathias draft resolution, which was finally adopted with several amendments, reflected the lowest common denominator between the enthusiastic SDI supporters and its harsh critics after a resolution by United States SDI supporters had been withdrawn and a critical draft resolution by the German delegate Karsten Voigt had been submitted too late. Voigt had called both for a drastic reduction of the nuclear offensive forces and for a mutual abandonment of strategic defences to be accompanied by a verifiable test stop for systems of strategic defence.[107] While all draft amendments by the European critics were voted down, it was only the German Christian Democrat Francke who succeeded in adding several of the European conditions for support of the SDI research programme. However, he withdrew the second part of paragraph 8, which had specified the European call for a transatlantic two-way-street: 'this includes free access to information and research results as well as unimpeded exchange and use of new technologies'.[108]

However, this transatlantic compromise is likely to erode and the conflicting assessments will become manifest as soon as the United States Administration drops its restrictive interpretation of the ABM treaty in order to start testing of SDI components.

San Francisco marked the end of the first phase of transatlantic controversy over SDI.

(b) The WEU and SDI

The WEU was established on 17 March 1948 by France, Great Britain, Belgium, the Netherlands and Luxembourg as a collective defence alliance. After a modification of the Brussels Treaty, Italy and the Federal Republic of Germany joined the WEU on 23 October 1954. Thirty years later, in October 1984, the foreign ministers and the defence ministers of the seven member countries met for the first time within the WEU framework. In the Rome Declaration of 27 October 1984 the foreign and defence ministers agreed that the WEU council should meet at the ministerial level regularly twice a year and that the ministerial organs should be reorganised in order to fulfil the threefold task of studying arms control and disarmament questions, dealing with security and defence problems and actively contributing the development of European armaments co-operation. The reorganisation should also contribute to the improvement of contacts between the Council and the WEU assembly, which consists of members delegated by the national parliaments of the seven WEU countries.[109]

The Assembly of the WEU prepares and discusses reports within the frameworks of the following committees: Defence Questions and Armaments; Scientific, Technological and Aerospace Questions;

Budgetary Affairs and Administration and the General Affairs Committee. After a report has been adopted by a committee, it is discussed in the plenary meeting. According to Article IX of the modified Brussels Treaty the Council of Ministers has annually to submit a report on 'matters concerning the security and defence of the member states . . . in particular . . . the control of armaments'.[110] According to the Rome Declaration, the Assembly 'should concentrate its activities on matters relating to every aspect of security policy and the defence of Europe'. In order to realise this commitment, a recent report by Mr van der Sanden on behalf of the General Affairs Committee on calls for 'more frequent participation by ministers of defence in Assembly sessions and their inclusion in the dialogue between the committees and the Council . . . While in Autumn 1984 this dialogue seemed to be starting in connection with the reactivation of the WEU it is to be feared', van der Sanden noted critically, 'that since the Rome Declaration the Council has stopped listening to the Assembly'.[111]

One test case for the willingness both of the Council and of the Assembly to upgrade the political role of the WEU could have become the formulation of a joint response to the American SDI.

The issue of 'the military use of space' was debated in the plenary meeting of the Assembly on 21 June and 4 December 1984, based on a report submitted by the conservative British MP Wilkinson on behalf of the Committee on Scientific, Technological and Aerospace Questions. The Wilkinson report discussed in detail the institutional framework of the European space effort, the Soviet space challenge, aspects of international space law, current initiatives and developments in defensive space systems, the long-term space strategy of the United States, the military satellite communications of the United Kingdom and a new European civil space programme for the 1980s and the 1990s. In the context of the United States long-term space strategy, the Wilkinson report briefly discussed SDI and BMD issues relating to Europe:

> No European NATO countries were involved in the SALT negotiations or of the ABM treaty . . . It is against this background that European NATO countries must consider their attitude to United States proposals for the development and deployment of strategic missile defences. They must decide whether any or all of them should contribute technically and financially to the development of projects recently approved by the Reagan administration which

could provide defence against ICBMs and IRBMs such as the SS-20 and SS-22. *There can be no doubt that it is in Europe's interest to do so* (my emphasis).[112]

In the context of his discussion of a European defensive space programme, Wilkinson suggested four major activities that should interest Europeans in order to improve the defence of their territories: telecommunications, military observation satellites, navigation satellites and attack satellites.[113]

During the debate of the Wilkinson Report in the WEU Assembly on 21 June 1984, several MPs stressed the need for independent European observation satellites. The French MP Fourré reminded his colleagues of the French proposal for an international satellite monitoring agency and for multilateral disarmaments talks in the Conference on Disarmament (CD) context to prevent ASATs and to place 'properly controlled restrictions on the new anti-ballistic technologies'. The German MP Scheer expressed these concerns clearly:

> The Europeans want space to be used for peaceful purposes only, secondly, ... they want to prevent a military arms race in space and, thirdly, we want to be less dependent on others for information. In other words, we need our own resources of information, and I should like to link this to the requirements of an arms control policy ... Europe urgently needs its own sources of information to enable it to undertake confidence-building measures in Europe without having to rely on the superpowers for all the necessary information.[114]

With few amendments the Assembly adopted a draft resolution on the military use of space that recommends that the Council

1. Urge the governments of member countries to do all in their power to secure negotiations between the United States and the Soviet Union so as to prevent the military use of space through the deployment of offensive space weapons systems ... ;
2. demand a larger European industrial involvement both in NATO telecommunications satellites and in NATO military satellite programmes ... ;
6. set clear European space policy objectives and priorities in the course of its politico-military consultations.[115]

The debate on the second part of the recommendations (that is, on the military use of space and on the implications for European security of American military developments in space) continued during the Assembly meeting on 4 December 1984. The British conservative MP and Rapporteur of the Space Committee Wilkinson stated:

> The SDI will continue and will have the full support of the United States Administration ... Our interests now should be to use [the ballistic missile] technology to create a more secure world, to enhance the opportunities for arms control, to reduce the risk of pre-emptive attack, to increase the value of our deterrents for the Alliance as a whole, and to try to make the world a safer place.[116]

Wilkinson's second report was openly criticised, primarily by German, British and Dutch MPs. Both Scheer and Tummers noticed various contradictions with his first report. The Dutch Labour MP de Vries pointed to major deficits and to a lack of balance:

> The Rapporteur was obviously so enthralled by the technical possibilities that he sees in the American initiatives that he has not mentioned anyone who was critical of that approach. The Assembly must be anxious to do that ... If one thing is clear from the report, it is the Rapporteur has no notion of the different perceptions of Americans and Europeans in such matters.[117]

The Spring sessions of the WEU Council in Bonn on 22–23 April 1985 in Bonn and of the Assembly from 20–23 May 1985 in Paris were dominated by the debate on the SDI, although neither could agree on a joint and coordinated position.[118] At the Bonn meeting, the West German Foreign Minister, Mr Genscher, and his French colleague, Mr Dumas, failed in their effort to formulate a joint position in response to the United States invitation to participate in the SDI research programme. Sir Geoffrey Howe, the British Foreign Minister, and Michael Heseltine, the former British Defence Minister, were instrumental in deluting the final communiqué both regarding a joint WEU position on SDI and on EUREKA. The ministers could only agree: 'to *continue their collective consideration* (my emphasis) in order to achieve as far as possible a co-ordinated reaction of their governments to the invitation of the United States to participate in the research programme'.[119]

At the Spring session of the WEU Assembly, several reports that touched upon the controversial SDI issue were to be debated, for example a Report by the British MP, Mr Hill on 'United States–European Co-operation in advanced technology'[120] and another by the French parliamentarian, Mr Fourré on 'The Military Use of Computers'.[121] A third report on 'Emerging Technologies and Military Strategy',[122] which had been drafted by the Dutch MP, Mr van den Bergh, for the Assembly's Committee on Defence Questions and Armaments[123] on 20 May 1985, had been postponed by a majority of the Conservative and Christian Democrat members to the December session. By this procedural vote a formal debate within the Assembly both on SDI and on Eureka was taken from the agenda of its Spring meeting.

In presenting the 30th annual report of the Council, its Chairman-in-Chief, West German Foreign Minister Genscher, indicated his scepticism about the SDI by stressing the need to stick to NATO's strategy of flexible response. However, after turning to the global competition in the high-technology area, he emphasised:

> Europe cannot afford to be decoupled from developments in high technology. It must pool its technological capabilities in order to remain a partner of the United States on equal terms. We must arrive at a European response to the global technological challenge ... France and the six other WEU members hold the same position: Europe must strengthen its own technological capabilities with a view to creating a technological community.[124]

In replying to a question by the Dutch Liberal MP Blaauw whether a European aerospace defence initiative would bring together all the European technologies, Genscher limited his remarks to the need for a European observation satellite. Without committing himself on SDI, the German Foreign Minister also alluded to the very important strategic effects of SDI. And on the potential competition between SDI and EUREKA, he replied that 'European technical co-operation is necessary regardless of SDI. Europe's top scientists should not be led to believe that research in high technology was the monopoly of the United States. The question is not only whether to participate in SDI but how'.[125]

The Dutch Liberal MP Blaauw, in pointing to the need for an ATBM, suggested that a multilateral West European study group should be set up 'to work out a conceptual framework' for a defence against SS-21, SS-22 and SS-23. 'That would form the basis for a joint co-ordinated United States–Western European concept. We now have the agency to carry out this work. Western European security aspects could then be fully incorporated into the overall SDI project ... In this context, the opportunities for Western European companies would be more promising ... Instead of duplicating Alliance research efforts, which should be the venue of EUREKA, we would reach a true division of labour with mutually supportive programmes. There is a more general feeling that Western European co-operation in armaments is needed. A Western European effort on SDI-related research', in the view of the Dutch MP 'could be an effective catalyst in this respect ... That leads to the Western European answer to the American SDI offer – that is, for instance, the establishing of a European aerospace defence initiative (EADI). This can be seen as the European component in the overall Alliance defence initiative in which American and European technology will work together and share their achievements'.[126]

The opposite view, however, was taken in the draft report of the Dutch Labour MP, Mr van den Bergh. In a preliminary draft resolution Mr van den Bergh had suggested to the Assembly to call 'upon the Congress of the United States, pending the outcome of current arms control negotiations, to make budgetary provision only for a modest SDI research programme limited to the projects and scale of that of previous administrations'.[127]

In order to prepare for the December debate on SDI, the Committee on Scientific, Technical and Aerospace Questions of the WEU Assembly organised a colloquy 'on the space challenge for Europe', held in Munich on 18–20 September 1985.

Franz Josef Strauss, the Minister President of the Free State of Bavaria, although supporting German participation in the SDI, objected to any inferior role of the European space industry. Europe should develop its own aerospace projects, even in competition with the United States. He supported EUREKA as well, and he did not foresee any direct competition between the two programmes.[128] The Co-ordinator for German Aerospace Policy, the Parliamentary State Secretary Martin Grüner, was more cautious, stating: 'European participation in the American SDI ... can only be successfully implemented on the basis of mutual technology transfer'. Grüner indicated his preference for:

> technology programmes aimed directly at civilian applications with positive effects in as many sectors of the national economy as possible and without taking the indirect route of meeting expensive military requirements. One of these solutions can be Eureka. But

Eureka will not allow us to dispense with further considering the needs of our security.[129]

Less reserved in his scepticism as far as European participation in SDI is concerned was the French Minister for Research and Technology Hubert Curien, who supported space activities in a defence context, in national frameworks and then in co-operative frameworks relating to 'non-aggressive applications of space technology: communications gathering, information, navigation, etc. which allow the defensive potential of present means to be maintained without becoming involved in a new arms race'. In Curien's view an ABM with a strong space element 'can but revive the arms race and . . . it is therefore undesirable. The concept of an absolutely impenetrable barrier is admittedly attractive, but it is not certain that it is feasible and the erection of a partial barrier could but lead to a strengthening of offensive means.

Curien supported a major European space programme with both civil and defence components to meet the twofold space challenge posed by the United States:

> [Europe] is also preparing to respond in a broader sense to the technological challenge of the end of the century: the Eureka programme seeks to mobilise European energies and competence for specific purposes, through programmes with clearly-defined aims in the five key technological sectors. Moreover, the space sector has a major role to play at the side of Eureka by offering its experience of how technological progress can be used to good purpose . . . Conversely, certain results of work on Eureka will undoubtedly be of considerable use in the development of the space systems of tomorrow'.[130]

Within the WEU Assembly the debate on the military–strategic, political and technological aspects during the 31st session on 2–5 December 1985 was based on three reports which had been drafted by the Dutch Labour MP, Mr. van den Bergh, on 'WEU and the Strategic Defence Initiative – The Strategic Defence Initiative (Defence Aspects)'[131] on behalf of the Committee on Defence Questions and Armaments; by the French Socialist, Mr Berrier on: 'WEU and the Strategic Defence Initiative – The European Pillar of the Atlantic Alliance'[132] submitted on behalf of the General Affairs Committee; and by the German Christian Democrat, Mr Lenzer on 'WEU and the Strategic Defence Initiative – Guidelines Drawn from the Colloquy on the Space Challenge for Europe (Proposals)[133] submitted on behalf of the Committee on Scientific, Technological and Aerospace Questions. While the WEU Council could not agree on a joint position of the seven member countries during a joint meeting of the foreign and defence ministers in Rome on 14 November 1985,[134] the WEU Assembly approved three resolutions on 4 December, after two days of intensive and controversial debate in the plenary.[135]

The van den Bergh report provided a critical assessment of the history of BMD. It took into consideration both the Soviet ABM systems as well as President Reagan's initiative. The public presentation of SDI, its implications on arms control and the European reactions were detailed and the debate on possible European participation was reviewed. Van den Bergh pointed specifically to the contradictory American assessments of the size of the Soviet ABM programme[136] and to the controversial evaluations of existing arms control agreements, especially to the open dispute which surfaced in October 1985 among President Reagan's former security adviser, Mr McFarlane, George P. Schultz and Paul Nitze about the interpretation of the ABM Treaty.[137]

After quoting at length the reservations and questions about SDI held by the British Foreign Secretary, Sir Geoffrey Howe and his German counterpart, Mr Genscher, and after pointing to the French arms control proposals of 12 June 1984 and August 1985 relating to the military use of space, van den Bergh summarised the European positions:

> With the reservations..., the European countries have generally supported in public the United States decision to conduct research into the possibilities of ballistic missile defence, within the terms of the ABM treaty, but on the understanding that that treaty does not permit testing or development of such weapon systems.[138]

Van den Bergh reminded his colleagues that the WEU Council, in its reply to Recommendation 413 on 11 April 1985, adopted a clear position concerning the restrictions imposed on SDI by the ABM treaty and stressed the importance of not eroding the latter:

> The Council notes that the strategic defence initiative ... is no more than a research programme and hence does not contravene the provisions of the 1972 ABM Treaty ... Relevant tests or deployments will have to be a matter for negotiation, under the

terms of the ABM treaty. In view of the contribution of this treaty to stability, the Council stresses the importance of preventing its erosion.[139]

The van den Bergh report concluded by noting 'the reservations on SDI expressed by European circles', by criticising the Weinberger letter as a 'quite inappropriate way to handle relations between Allied governments', and by endorsing 'the French proposal that a treaty to ban high-level anti-satellite weapons and to protect satellites of third parties should be negotiated in the Conference on Disarmament in Geneva'.[140]

The Berrier report which had been adopted by the General Affairs Committee by 13 votes to 0 with one abstention focused primarily on the relationship between the Atlantic Alliance and the defence of Europe, on the reactivation of the WEU and on tightening institutional links between the Assembly and the other WEU organs. Section III dealt specifically with 'Europe and President Reagan's Strategic Defence Initiative'. Berrier pointed especially to five measures that will become necessary within the next few years:

(a) Progress achieved by the United States and the Soviet Union in the use of space must be followed closely . . . ;

(b) [Ideal] participation of European firms in the American programme [does] not lead to a brain drain . . . towards the United States but provides work for firms in Europe itself and allows them to improve their knowledge of the technology they use . . . ;

(c) Considering that American firms are . . . the main beneficiaries of the SDI . . . , it is essential for Europe to make a research and development effort which will allow its own firms to remain competitive [e.g., EUREKA] . . . ;

(d) In the military field proper, Europe must take full account of the fact that:

1. the SDI will not lead to deployment for [many] years and that it is not sure it ever will;
2. it is not yet possible to assess the degree of security it will offer;
3. it is not at all sure that systems deployed as a result of the SDI will effectively be able to destroy medium and short-range missiles, i.e. to ensure as much security for Europe as possibly for the United States . . .

(e) If the protection of the United States and Western Europe is to be ensured by different means, in increasingly different conditions and in accordance with even more different strategic hypotheses, there will be a greater risk of Europe's territory being uncoupled from that of the United States, with all the drawbacks this entails for Europe's security.[141]

Berrier noted the advantageous use of space for defence purposes, especially communications and observation satellites. In order to avoid an increased militarisation of space the Berrier report emphasised that 'the agreements concluded ... in the ABM treaty must be extended'. It suggested that the French disarmament proposals in this respect should 'be studied by the United States' European allies with a view to preparing a European position not only towards participation in SDI but also towards measures for avoiding the balance of forces in the world being disturbed too quickly'.[142]

The Lenzer report, after reviewing the aerospace policies of the member states and discussing the space station and the issue of exploitation and commercialisation of space knowhow, dealt briefly with 'the strategic defence initiative and European co-operation' by pointing to the different assessments by government officials, industrialists and scientsits during the Munich colloquy. In his view three points should be emphasised:

> First, WEU, but also Europe, must grasp the fact that at present there is no longer any real major technical obstacle to the development and deployment of first-generation SDI systems ... Second, no European country alone can carry out such an undertaking. But Western Europe can. However, it has a choice between doing nothing, co-operating with the United States at the risk of becoming irreversibly dependent and, finally, providing itself with the means of acceding to independence ...
> Third, whatever decision is taken, the technological and economic spin-off will be considerable ... Europe will or will not be present in this implacable competition whose real stake is its place, and perhaps even its survival, in the twenty-first century.[143]

'All SDI proposals contain the risk of splitting the European nations', the French Minister for External Affairs, Mr Dumas, reminded the members during the December session of the Assembly in Paris.[144] The two days of intensive debate reflected the positions the members of the defence and foreign affairs committees had previously taken in their national parliaments.[145] While the critics – primarily of the left and some liberals – stressed the need for strengthening the ABM regime and for maintaining strategic stability and were concerned that SDI might increase the risks of war, the SDI supporters, primarily British Conservatives, German and Italian Christian Democrats and Dutch liberals interpreted SDI as a response to Soviet BMD activities. In their view, SDI could be pursued in the framework of the ABM treaty to the advantage of the European technological development.[146] While most French critics stressed the need to maintain deterrence, the British Labour MP Mr Freeson disagreed, and called for a different Western defence policy:

> The SDI expensively sidetracks us from the fact that reliance on the MAD policy is increasingly unacceptable. We must not loose sight ... of the five fundamental principles of defence which, in combination, can be the basis for reshaping our defence policy effectively. First, defence must be effective ... Secondly, defence must be non-provocative ... Thirdly, defence must be non-nuclear ... Fourthly, defence must be legitimate ... Fifth, defence should lead to comprehensive disarmament and world security.[147]

In the subsequent vote on the resolution submitted by Mr van der Bergh most of the amendments introduced by Mr Hill on behalf of the Conservatives were adopted while those of the critics, introduced by Mr Gansel, were rejected. In the final Resolution 428, the WEU Assembly recommends that the Council:[148]

1. agree a common response to the United States' strategic defence initiative or, if that seems impossible, specify Europe's own interests in this area by harmonising as far as possible the answers of the seven WEU member countries which should:

    (a) stress the importance of avoiding an arms race in space;
    (b) accept research compatible with existing arms control agreements and of a nature and scale which will enhance stability and security;
    (c) permit European industry to participate in all areas of SDI research on terms providing a genuine exchange of technology;
    (d) ensure that the answers of members of WEU to the American invitation do not jeopardise the development of Europe's technological capability and encour-

age the development of this capability, in particular through the early implementation of the Eureka programme;

2. give priority and special emphasis to a joint European programme for defence and arms control purposes, including observation and communications satellites, and to promoting civil technological research of EUREKA type within ESA and the European Communities;
3. request all countries concerned to ensure that no obstacles will be placed in the way of balanced and verifiable agreements limiting strategic and intermediate-range nuclear weapons and encourage the pursuit and success of the Soviet–American negotiations in Geneva on the limitation of armaments in the three areas covered;
4. emphasise the need, when the results become available, for the United States and its European partners to discuss the political as well as the military and strategic implications of research on SDI;
5. instruct the new agency for the study of arms control and disarmament questions to report annually on the arms control impact of the SDI;
6. Ensure maintenance of the nuclear deterrent capability of the Atlantic Alliance as long as Europe's security is not effectively guaranteed by other means and consider the question of the case for adequacy in conventional defence capacity, both in the present situation and in regard to the development of the strategic defence initiative.

In Recommendation 429, prepared by Mr Berrier, the WEU Assembly recommends with respect to the SDI that the Council: 'have the appropriate agency conduct a continuing study of the strategic consequences of the development of new weapons, whatever the results of its efforts to co-ordinate the answers of member countries to the American proposal that they take part in the strategic defence initiative'.[149] Recommendation 430, prepared by Mr. Lenzer recommends that the Council urge member governments:

3. to define the co-operation framework in which the defence aspects of European space activities can be discussed and determined;
4. to accept non-aggressive applications of military space tech-

nology such as communications, surveillance, navigation and the use of satellites for crisis management and treaty verification to strengthen strategic stability in relations between NATO and Warsaw Pact countries as indicated in the NATO statement of 8th January 1985;
5. to pursue jointly research on a European anti-missile system independently or as part of SDI.[150]

These three resolutions reflect the positions of those parties that form the Conservative governments of Great Britain, West Germany, Belgium, the Netherlands and the coalition government of Italy. Nevertheless, the conditions that have been added to the general support for a European participation in the SDI research programme indicate the areas of conflict with the United States in the years to come. While the Soviet Union failed to drive a wedge between the United States and its European allies, SDI has succeeded however, as Mr Dumas emphasised in his address, in splitting the Europeans. The hopes that had been created by the Declaration of Rome of October 1984 had become the first victims of the SDI. The official rhetoric of strengthening the European pillar of NATO and of perspectives for a Europeanisation of European defence has lost much of its credibility during the agonising debate on the SDI within the Council of the WEU.[151]

(c) The reaction to SDI and EUREKA by the European Parliament

The supranational EC and its directly elected European Parliament representing twelve member countries – all but Ireland being members of NATO, and all but Denmark, Greece, Spain, Portugal and Ireland being members of the WEU as well – has no direct authority to deal with defence-related issues. At the Stuttgart summit in Spring 1983 the intergovernmental European Political Co-operation (EPC) agreed to deal also with the political and economic aspects of security policy. Irrespective of these legal constraints, the Political Committee of the European Parliament, after its second direct election in June 1984, established a subcommittee on 'security and disarmament'. SDI and EUREKA-related issues have been dealt with by this Subcommittee on Security and Disarmament and by the Committee on Energy, Research and Technology. At its October session in 1985, the European Parliament debated both SDI and EUREKA-related issues in some detail.

In a working document by the Rapporteur of the Subcommittee on Security and Disarmament on 'Arms Control Disarmament and its Relevance for the European Community' of 30 May 1985, Sir Peter Vannick summarised many of the political concerns that have been raised by West European governments against SDI.[152]

The Committee on Energy, Research and Technology ordered several reports on the 'Creation of a European Research Area' (W. Münch),[153] which dealt with problems of technology co-ordination and technology transfer; on the 'European Reaction to the Technological Challenge' (Poniatowski),[154] on 'European Space Policy' (Toksvig)[155] and on the 'Differences in the Level of Technological Development in the Member Countries of the European Community' (Longuet).[156] However, none of these reports dealt with SDI, and they only touched on the EUREKA project in general terms.

Several draft resolutions on behalf of the European People's Party and of the Socialist Group as well as by individual MPs dealt with SDI and 'Star Wars'. The draft resolution of the Conservative Parties calls on the foreign ministers to formulate in the context of the EPC a common European position relating to the SDI, to include the French proposals for a space shuttle and observation satellite in a comprehensive Western concept and to enable the European partners in NATO to contribute to the strategic and technological planning of the project.[157] The draft resolution of the Socialist Group emphasised the principle of the peaceful use of space; it stressed that the common survival could not be achieved by military and technological systems but only by political solutions; it rejects all steps that could lead to an intensification of the arms race between the United States and the Soviet Union in outer space; it calls upon the member states to formulate a common position on SDI in the EPC context and it asks the Political Committee to prepare a report on the European reaction to the American SDI programme.[158]

A group of British European MPs called upon the Commission and on the European member countries to make any treaty relating to SDI dependent on progress in the Geneva talks and on the signing of a treaty prohibiting the deployment and testing of offensive weapons in space.[159]

In a second draft resolution on the SDI research programme, the Socialist Group opposed any participation in the SDI research programme, any contribution to it from financial resources and research capacities of the EC and any militarisation of outer space. It supported instead peaceful research in the ESA context, a European effort for an independent observation satellite to verify arms control agreements, and it supports a European research programme against world hunger.[160]

Several other draft resolutions by Ms Charzat, Mr Glinne and Mr Saby[161] called for an end to the arms race and to the militarisation of outer space; by Mr Linkohr,[162] who objected to any participation of the EC in the SDI; by Mr Meten, who called for an enlargement of the activities of the European Community in the peaceful use of Outer Space[163]; by Ms Lizin, who called for the opinion of the European institutions on the SDI and for a debate in the European Parliament[164] and by Mr Ulburghs[165] on the necessity to reject the SDI and to transform Europe into a nuclear weapons-free zone. One resolution submitted by members of the Socialist Group supported the initiative of the French government for EUREKA and called upon the Commission to accept this proposal and on the Council to submit an outline which follows the example of the ESPRIT programme.[166]

During its first October session, from 7–11 October 1985, the European Parliament organised a symposium on 'Europe in the year 2000 – The technological challenge',[167] and debated the technological challenge for the EC posed by the United States and Japan in general, and by the SDI research programme in particular. Four general issues were mentioned in the reports by the French Liberal MP Mr M. Poniatowski: 'On Europe's response to the modern technological challenge',[168] which had been adopted by the Committee on Energy, Research and Technology on 27 September 1985 by 20 votes to 0 with 4 abstentions, and in a report by the Dutch Labour MP Mr Metten on 'Technology Transfer',[169] which had been unanimously adopted in the same Committee on 16 September 1985: the concept of a European technological Community,[170] EUREKA, SDI and the associated issues of technology transfer.

During the plenary debate on the Poniatowski report, a small majority of 152 to 147 deleted the following reference in its draft resolution: 'rejects European participation in the SDI project on political, economic, scientific and ethical grounds', which had been backed by most French members and by the parliamentarians of the Socialist Group.[171] No resolution or reference in support of the SDI had been put for a vote, but both the Conservative and the Christian Democratic groups opposed any rejection of SDI by the European Parliament. EUREKA, on the other hand, was welcomed by a majority of the European Parliament 'both as a non-military European response to the SDI programme and as a means for an aggressive European technological and industrial policy'. However, according to the wishes of the European MPs 'the EUREKA project [is] to be incorporated into the European Community' and 'the Commission [is] to have a vital role in the development of the EUREKA project'. The resolution, which was adopted along with the Poniatowski report by 197 to 85 with 12 abstentions, appealed to 'all European scientists and research workers to commit themselves to developing a European technological community for non-military purposes'.[172]

The Poniatowski report described the present American strategy in R&D terms: Its goal is 'to maintain American competitiveness, particularly in regard to Japan, across the board and retain military supremacy with regard to the USSR in all sectors'.[173] From a European perspective, a major concern has been the tightened United States limits on technology transfer not only to the rivals in the East but to the Allies in the West as well. The Metten report[174] analysed in detail the United States technology transfer controls in relationship with its European Allies and the implications of the increasing influence of the Pentagon over the transfer of civilian technologies for the European industry.

In his appended draft resolution, Metten called on the governments of the member countries to counter United States efforts to apply its technology transfer rules against citizens and firms within the EC.

The increased dependence of European companies on the national technology transfer rules as a direct consequence of their participation in the SDI research programme has been a major concern not only to members of the European Parliament[175] but to the European Commission as well. The Vice President of the European Commission, the German Christian Democrat Mr Narjes, expressed his deep concern in a letter he sent on 5 December 1985 to the twelve members of the Research Council of the EC. As a consequence of a heavy participation of European companies in the SDI research programme, the existing shortage of highly specialised research manpower could be depleted. American technology transfer rules could have a major impact on technology transfer within the EC. These restraints could possibly prevent European co-operation in the area of science policy and technological co-operation. These problems would apply both to European companies and to European governments. Mr Narjes was concerned about the implications of these rules for the technological co-operation within the EC. The Vice President of the European Commission urged the ministers of science and technology of the Twelve to avoid any restrictions which would prevent intra-European technological co-operation.[176]

Many members of the European Parliament and of the Commission were also concerned about the development of EUREKA at the Hanover meeting on 5–6 November 1985. A resolution tabled by Mr Linkohr *et al.* on behalf of the Socialist Group, which was adopted by the European Parliament during its December meeting, expressed regret that the instruments and the institutions of the EC would not be utilised and that parliamentary control of the EUREKA programme would not be guaranteed.[177]

SDI was a controversial issue in all three international parliaments during 1984 and 1985, and is likely to remain a major topic for political, strategic, economic and technological debates for the years to come. The European technological, economic, political and military response to the major challenge posed by SDI is gradually emerging. The parliamentary debates are just one public aspect of the intensive consultations within the European governments in the framework of the EC, of the WEU, and of the North Atlantic Alliance.

VI CONCLUSION

By the end of 1985 no common West European position on SDI had developed. Instead, SDI has become the *most divisive element* in East–West relations in general (and especially in the United States–Soviet arms control negotiations in Geneva), but also in transatlantic relations between the United States and its Allies, in the relationship among the member governments of the WEU and the EEC and in Franco–German cooperation. The challenge posed by the SDI research programme has become both a *catalyst* for the EUREKA technological programme and a *brake* for the intra-European defence co-operation that was proclaimed in the Rome Declaration of the WEU in October 1984.

SDI took the European Allies by surprise; there were no prior consultations. The aim of later consultations between the United States and its Allies has been 'persuasion' in order to gain political support for the internal budgetary fights in Congress. The official United States interpretations of the goals of SDI have remained contradictory, especially when one compares the rhetoric applied at the home front (strategic revolution overcoming deterrence) with the selling strategy used abroad (a complement to deterrence).

During 1986, SDI remained a controversial issue in the deliberations of the three transnational parliamentary bodies. With the enlargement of the European Community (Spain and Portugal joined on 1 January 1986), the European Parliament tended to be more critically inclined towards the SDI.

The debate about the Reykjavik summit was a major issue of the meetings of the European Parliament in late October, of the Northatlantic Assembly in its meeting in Istanbul in November, and of the WEU Assembly in Paris in December 1986. The European Parliament expressed its disappointment about the failure of the summit and emphasised that SDI must not become an impediment for concrete disarmament measures.[178]

The acronym 'SDI' has replaced 'INF' in the headlines, in scientific deliberations, in national and international parliamentary debates and in official consultations. It will stay with us for many years, regardless of whether or not this vision based on the imagination of science fiction writers is ever realised.

Notes

1. Benson D. Adams, *Ballistic Missle Defense* (New York: American Elsevier Publishing Co., 1971); Ernest J. Yanarella, *The Missile Defense Controversy – Strategy, Technology, and Politics, 1955 – 1972* (Lexington: University Press of Kentucky, 1977); Graham T. Allison and Frederic A. Morris, 'Armaments and Arms Control: Exploring the Determinants of Military Weapons', in Franklin A. Long and George W. Rathjens (eds), *Arms, Defense Policy, and Arms Control* (New York: W. W. Norton, 1976) pp. 99–130.
2. Lawrence Freedman, *US Intelligence and the Soviet Strategic Threat* (Boulder, Co.: Westview Press, 1977); John Prados, *The Soviet Estimate, US Intelligence Analysis & Russian Military Strength* (New York: Dial Press) 1982.
3. Paul E. Gallis, Mark M. Lowenthal, Marcia S. Smith, *The Strategic Defense Initiative and United States Alliance Strategy* (Washington: Congressional Research Service, 1 February 1985) p. 3ff.
4. 'Text of McNamara Speech on Anti-China Missile Defense and US Nuclear Strategy', *New York Times*, 19 September 1967.
5. Paul Buteux, *The Politics of Nuclear Consultation in NATO 1965–1980* (Cambridge: Cambridge University Press, 1983).
6. Raymond L. Garthoff, 'BMD and East–West Relations', in Ashton B. Carter and David N. Schwartz (eds), *Ballistic Missile Defense* (Washington: Brookings Institution, 1984) p. 281.
7. Garthoff, 'BMD and East–West Relations', pp. 282–3.
8. Buteux, *The Politics of Nuclear Consultation*, pp. 79–80.
9. Buteux, pp. 80–1.
10. Pierre Simonitsch, 'Ein atomares Riesen-Feuerwerk über Europa? Plan für Raketenabwehrsystem hat kaum Chancen/Politisch und

finanziell zu teuer', *Frankfurter Rundschau*, 27 September 1967; Kurt Becker, 'Der allzu teure Schirm. Ein Raketen-Abwehrsystem: für Europa unerschwinglich', *Die Zeit*, 29 September 1967.
11. Becker, 'Der allzu teure Schirm'.
12. Lothar Ruehl, 'Schutz gegen Tiefflieger ist vordringlicher als Raketenabwehr. Die Diskussion in der NATO über ein ABM-System geht weiter', *Die Welt*, 9 October 1967.
13. Rolf Breitenstein, '"Anti-Raketen durchlöchern Pakte" – Schmidt: Abwehrsysteme machen die nuklearen Garantien unglaubwürdig', *Frankfurter Rundschau*, 11 October 1967.
14. Garthoff, 'BMD and East–West Relations', pp. 283–4.
15. Bundesministerium der Verteidigung, 'Kommuniqué zur NPG-Sitzung in Den Haag', *Mitteilungen an die Presse*, V/37, (Bonn: 19 April 1968); Robert C. Doty, 'Despite US Decision – NATO Nuclear Group Decides ABMs Aren't Justified Now', *International Herald Tribune*, 20 April 1968.
16. Buteaux, *The Politics of Nuclear Consultation*, p. 85.
17. 'Die Frage eines europäischen Raketenabwehrsystems – Abschluss der NATO-Tagung in Den Haag', *Neue Zürcher Zeitung*, 21 April 1968.
18. Doty, 'Despite US Decision'.
19. 'Raketenabwehr-System für Europa wenig sinnvoll – Politischer Schutz durch Washington im Vordergrund der Sicherheits-Politik/ Überlegungen westlicher Fachleute', *Frankfurter Allgemeine Zeitung*, 22 October 1968.
20. Laurence W. Martin, *Ballistic Missile Defence and the Alliance* (Paris: Atlantic Institute, 1969) p. 4–5.
21. Johan J. Holst, 'Missile Defense: Implications for Europe', in Johan J. Holst and William Schneider, Jr (eds), *Why ABM? Policy Issues in the Missile Defence Controversy* (New York: Pergamon Press, 1969) pp. 190–7.
22. Holst, 'Missile Defence', pp. 200–1.
23. Holst, p. 201.
24. Abram Chayes and Jerome B. Wiesner (eds), *ABM – An Evaluation of the Decision to Deploy an Antiballistic Missile System* (New York: Signet, 1969).
25. Buteux, *The Politics of Nuclear Consultation*, p. 115.
26. Garthoff, 'BMD and East–West Relations', p. 285.
27. Gallis et al. *The Strategic Defense Initiative*, p. 6; Michel Debré, 'France's Global Strategy', *Foreign Affairs*, April 1971, p. 403.
28. Lawrence Freedman, 'The Small Nuclear Powers', in Ashton B. Carter and David N. Schartz (eds), *Ballistic Missile Defence*, p. 258.
29. David S. Yost, 'Ballistic Missile Defense and the Atlantic Alliance', *International Security* 7 (1982) p. 147.
30. Yost, *Ballistic Missile Defense*, p. 151.
31. Gallis, *The Strategic Defense Initiative*, p. 12.
32. Buteaux, *The Politics of Nuclear Consultation*.
33. Stanley R. Sloan, 'Arms Control Consultations in NATO', in Hans Guenter Brauch and Duncan L. Clarke (eds), *Decisionmaking for Arms Limitation – Assessments and Prospects* (Cambridge, Mass.:

Ballinger, *Decisionmaking* 1983) pp. 219–36; Richard E. Darilek, 'Separate Processes, Converging Interests: MBFR and CBMs', in Brauch and Clarke (eds), pp. 237–58; Simon Lunn, 'Policy Preparation and Consultation within NATO: Decisionmaking for SALT and LRTNF', in Brauch and Clarke (eds), *Decisionmaking*, pp. 259–74.
34. Sloan, 'Arms Control', p. 219.
35. Roger Hill, *Political Consultations in NATO, Wellesley Papers 6/78* (Toronto: Canadian Institute of International Affairs, 1978) p. 11.
36. Sloan, 'Arms Control', p. 229.
37. President Reagan's speech of 23 March 1983, *Weekly Compilation of Presidential Documents 12*, 28 March 1983 (Washington DC: United States Government Printing Office, 1983) pp. 423–66.
38. US Department of Defense, *Report to the Congress on the Strategic Defense Initiative* (Washington, DC: United States Department of Defence, 1985) pp. A2, A3.
39. 'Im Hintergrund: US-Vertreter verwirrten Europäer', *Frankfurter Rundschau*, 29 February 1984.
40. *Report to the Congress*, p. A3.
41. Fred Hiatt, 'US Antisatellite Plan Draws Fire. NATO Ministers Reportedly Express Skepticism, Anxiety', *Washington Post*, 4 April 1984, p. 18.
42. Robert S. Greenberger and Metin Demisar, 'NATO Allies Question "Star Wars" Plan, Pressure Netherlands to Deploy Missiles', *Wall Street Journal*, 4 April 1984, p. 33.
43. 'NATO nations challenge cost, practicality of US "star wars"', *Baltimore Sun*, 24 April 1984.
44. Elizabeth Pond, 'Europe fears "star wars" may destroy, not defend West', *Christian Science Monitor*, 12 April 1984.
45. *Report to the Congress*, p. A3.
46. 'Weitere Vorbehalte gegen Weltraumrüstung. Europäische NATO-Staaten erinnern Weinberger an ihre besonderen Sicherheitsinteressen', *Süddeutsche Zeitung*, 11 October 1984; Horst Schreitter-Schwarzenfeld, 'NATO-Partner halten still – Verbündete haben keine Einwände gegen Weltraumrüstung', *Frankfurter Rundschau*, 12 October 1984; for the text of the communiqué of the 36th NPG meeting, see *NATO Review*, 5/ 1984.
47. For texts see *NATO Review*, 6/1984.
48. *Report to the Congress*, p. A5.
49. 'NATO Mutes "Star Wars" Criticism', *Hartford Courant*, 14 February 1985.
50. Michael Weisskopf, '"Star Wars" Research Supported – NATO Ministers Call Plan "Prudent"', *Washington Post*, 28 March 1985, p. 29; David Fouquet, 'US allies support "star wars" research,', *Christian Science Monitor*, 28 March 1985, p. 1.
51. 'NATO Group Supports US SDI Research (Text: Final NPG communiqué)' *Wireless Bulletin from Washington*, 58/1985, 28 March 1985.
52. For the text of the communiqué see *NATO-Review* 3/1985; Ulrich Mackensen, 'NATO bleibt bei "flexibler Antwort" – Bisherige Strategie soll nicht geändert werden/Distanzierungen von SDI?', *Frankfurter*

Rundschau, 23 May 1985; Karl Feldmeyer, 'NATO-Beratung im Schatten von SDI', Frankfurter Allgemeine Zeitung, 28 May, 1985.
53. John M. Goshko, 'NATO Support for SDI Blocked by France', Washington Post, 7 June 1985, p. 11.
54. 'NATO Backs US Efforts At Geneva', International Herald Tribune, 8 June 1985; '"Star Wars" strikes out in NATO', Chicago Tribune, 8 June 1985; John M. Goshko, 'NATO Backs All Areas of US Geneva Efforts', Washington Post, 8 June 1985; 'NATO Now Cooler to Space Weapons', New York Times, 7 June 1985.
55. 'NATO Strongly Supports US Arms Control Efforts (Text: NATO communiqué and extracts of minutes)', Wireless Bulletin from Washington, 104/1985, 10 June 1985.
56. William Drozdiak, 'Arms Offer Urged – US Allies Press Pre-Summit Initiative', Washington Post, 16 October 1985, 'Shultz Briefs NATO on SDI', Wireless Bulletin from Washington, 191/1985.
57. Michael Feazel, 'Shultz Affirms European Role in SDI Test, Production Aspects', Aviation Week & Space Technology, 21 October 1985, pp. 21–2.
58. 'Weinberger Briefs NATO Ministers On US Charge of Arms Violations', International Herald Tribune, 30 October 1985.
59. 'NATO pressed on Soviet Cheating', Washington Times, 29 October 1985, p. 5; 'President backed by NATO for summit', Washington Times, 31 October 1985, p. 7; William Drozdiak, 'NATO Backs US on Arms Charges', Washington Post, 30 October 1985, p. 8; Ray Moseley, 'NATO supports "Star Wars" plan', Chicago Tribune, 31 October 1985, p. 5.
60. For text see Nato Review, 6/1985; 'Die Nato verdeckt ihre Uneinigkeit. Das Kommuniqué vermeidet Erwähnung von SDI', Frankfurter Allgemeine Zeitung, 31 October 1985.
61. 'NATO Will Continue Conventional Force Improvements (Text: Defense Planning Committee communiqué)', Wireless Bulletin from Washington, 4 December 1985, p. 18.
62. 'Bemühungen Bonns um verbesserten Schutz gegen Angriffe aus der Luft – Trainingszentrum für Tiefflüge, Nato-Herbsttagung beendet', Frankfurter Allgemeine Zeitung, 4 December 1985; 'Wintertagung der Nato in Brüssel', Neue Zürcher Zeitung, 5 December 1985.
63. Jan Reifenberg, 'Nach Genf geht es um das Kleingedruckte: Der "Geist" allein reicht nicht', Frankfurter Allgemeine Zeitung, 11 December 1985; John M. Goshko, 'Shultz Says Allies Value Defense Genscher sieht Auswirkungen auf Zusammenhalt der Allianz', Frankfurter Rundschau, 13 December 1985; 'Genscher fordert eine SDI-Diskussion der Nato', Frankfurter Allgemeine Zeitung, 13 December 1985.
64. 'Allies Ask US Effort on Arms', International Herald Tribune, 13 December 1985; John M. Goshko, 'Schultz Says Allies Value Defense Over Arms Pact', International Herald Tribune, 14 December 1985; 'Nato-Ausblick auf ein Jahr der Verhandlungen – Suche nach Stabilität der Ost–West-Beziehungen', Neue Zürcher Zeitung 16 December 1975.
65. See especially the disappointment of France as far as Franco–German military co-operation is concerned in the French press.

66. Hans Günter Brauch, 'From Strategic to Tactical Defense? – European political, strategic, technological reactions to the "Star Wars" vision and the Strategic Defense Initiative: Eureka and/or EDI', in John McIntyre (ed.), *International Space Policy* (New York: Greenwood, 1987).
67. Upi, NATO text, 26 March 1985.
68. *Report to the Congress*, p. A4.
69. Michael Weisskopf, 'US Might Not Share All Its "Star Wars" Secrets With Europeans', *Washington Post*, 2 May 1985, p. A32; Christopher Thomas, 'US fear on sharing secrets', *The Times*, 21 May 1985; 'Sharing SDI data risks spying, Weinberger says', *Christian Science Monitor*, 6 June 1985, p. 2.
70. James Gerstenzang, 'NATO Caution on "Star Wars" Won't Affect Program, Weinberger Says', *Los Angeles Times*, 24 May 1985, p. 4.
71. 'USA für Firmenkooperation bei SDI. Perle: Europäische Regierungsbeteiligung niemals geplant', in *General-Anzeiger*, 29 June 1985.
72. 'Norway bars "star wars" role', *Washington Times*, 19 April 1985, p. 6; 'Norway refuses space-defense role', *Boston Globe*, 18 April 1985, p. 20; 'Storting Gives Conditional Support to US SDI', *Current News*, special edition, Strategic Defense Initiative, 16 July 1985.
73. See chapter 5 in this volume.
74. 'Government Role in SDI is Refused By Canada', *International Herald Tribune*, 9 September 1985.
75. Brauch, in McIntyre (ed.), *International Space Policy*.
76. 'US Has Minimal Conditions For Allied Cooperation in SDI (Transcript: Abrahamson press conference)', *Wireless Bulletin from Washington*, 29 August 1985.
77. ' "Star Wars" Gain Seen With Allies', *New York Times*, 6 December 1985, p. D-3.
78. David A. Brown, 'European Industry Begins to Seek US SDI Contracts', *Aviation Week & Space Technology*, 16 December 1985, pp. 12–15.
79. NAA, *Texts Adopted by the North Atlantic Assembly at its 29th Annual Session, The Hague, 2–7 October 1983*, Doc. AA 330, SA(83) 32, (Brussels: North Atlantic Assembly, October 1983) p. 17; NAA, *Official Report of the Second Sitting, The Hague, 6 October 1985*, AA 302, SA/CR 3 (83) 21 rev.1 (Brussels: North Atlantic Assembly, November 1983) pp. 10–12.
80. NAA, *30th Annual Session, Official Report of the Second Sitting, Brussels, 15 November 1984*, Doc. AB 258, SA/CR 3 (84)15, rev.1 (Brussels: North Atlantic Assembly, December 1984).
81. NAA, Military Committee, *General Report on Alliance Security, Mr Michael Forrestall (Canada) General Rapporteur* (Brussels: North Atlantic Assembly, November 1984), p. 23.
82. NAA, Scientific and Technical Committee, *General Report, Mr Thomas-Henri Lefebvre (Canada), General Rapporteur* (Brussels: North Atlantic Assembly, November 1984), pp. 27–8.
83. *Lefebvre report*, p. 31.
84. NAA, Scientific and Technical Committee, *Draft General Report presented by Mr Thomas-Henri Lefebvre, General Rapporteur*, Doc.

AC 70, STC(85)2 (Brussels: North Atlantic Assembly, April 1985); NAA, *Draft Interim Report of the Special Committee on Nuclear Weapons presented by Mr John Cartwright (United Kingdom) and Mr Julian Critchley (United Kingdom), Co-Rapporteurs*, Doc. AC 67, CS/AN(85)3 (Brussels: North Atlantic Assembly, April 1985), NAA, Special Committee in Nuclear Weapons, *Draft Interim Report, Mr John Cartwright (United Kingdom), Rapporteur* (Brussels: North Atlantic Assembly, October 1985); NAA, Political Committee, *Draft General Report on Alliance Political Developments, Mr Bruce George (United Kingdom), Rapporteur*, Doc. AC 180, PC(85) 7 (Brussels: North Atlantic Assembly, October 1985); NAA, Military Committee, *Draft General Report on The Future of Arms Control: Compliance and Verification Issues, Sir Geoffrey Johnson Smith (United Kingdom), General Rapporteur* (Brussels: North Atlantic Assembly, October 1985), Doc. AC 175, MC (85) 11; NAA, Scientific and Technical Committee, *Draft Special Report on the Exploitation of Space presented by Mr Robert Banks (United Kingdom), Special Rapporteur*, Doc. AC 121, SA/CR 1 (85) 2, (Brussels: North Atlantic Assembly, 1985).

85. NAA, *Official Report, of the Spring Plenary Sitting, held in the Landtag of Baden-Württemberg, Stuttgart, Monday 20 May, 1985*, Doc. AC 121, SA/CR 1 (85) 3, (Brussels: North Atlantic Assembly, June 1985), p. 6.
86. NAA, *Spring Plenary Sitting*, p. 13.
87. NAA, *Spring Plenary Sitting*, p. 19.
88. 'NAA Spring Session', in *News – The North Atlantic Assembly*, June–August (1985) 52–53, p. 3.
89. *Lefebvre report*, pp. 17–18.
90. *George report*, pp. 23–4.
91. *Johnson-Smith report*, pp. 25–7.
92. NAA, *Draft Resolution on Strategic Defence and the Alliance, presented by Mr John Cartwright (UK), Senator Charles Mathias (United States)*, Doc. AC 256 (85) 19 rev. 1 (Brussels: North Atlantic Assembly, no date).
93. *Cartwright report*, pp. 8–9.
94. *Cartwright report*, p. 12.
95. *Cartwright report*, pp. 14–15.
96. *Cartwright report*, p. 16.
97. *Cartwright report*, p. 17.
98. NAA, *Policy Recommendations, 31st Annual Session, San Francisco 10–15 October 1985*, Doc. AC 262, SA (85) 25 (Brussels: North Atlantic Assembly, October 1985), res. 170.
99. NAA, *Policy Recommendations*.
100. NAA, *31st Annual Session, Official Report of the Opening Ceremony and the First Plenary Sitting, 14 October 1985*, Doc. AC 254, SA/CR 3(85) 17 (Brussels: North Atlantic Assembly, Nov 1985).
101. Richard G. O'Lone, 'North Atlantic Assembly Declares Support for SDI After Lobbying Effort', *Aviation Week & Space Technology*, 21 October 1985, pp. 18–19.
102. Shultz, NAA, AC 254, SA/CR 3(85) 17, p. 10.

103. 'Address by Ambassador Paul H. Nitze', NAA, *31st Annual Session, Official Report of the Second Plenary Sitting, 15 October 1985*, AC 255, SA/CR 4 (85) 18 (Brussels: North Atlantic Assembly, November 1985) pp. 5–7.
104. George Keyworth, 'The President's Defense Initiative', remarks to the SDIO University Review Forum, 29 March 1985, excerpted in US Congress, Office of Technology Assessment, *Ballistic Missile Defense Technologies*, OTA-ISC-254 (Washington, DC: United States Government Printing Office, September 1985) p. 302.
105. Nitze, AC 255, p. 7.
106. McNamara, in NAA, *31st Annual Session, Second Plenary Sitting*, pp. 23–4.
107. Draft resolution by Karsten Voigt in German, 26 September 1985 in private files of the author; Fritz Wirth, 'Skepsis und Zustimmung für SDI in San Francisco, NATO-Parlamentarier uneins über Reagans Initiative', *Die Welt*, 12 October 1985.
108. NAA, *Amendments nos 17–20 to the Draft Resolution on Strategic Defence and the Alliance presented by Mr Klaus Francke (FRG)*, Doc. AC 256, SA(85) 19 rev.1, (Brussels: North Atlantic Assembly, October 1985).
109. Assembly of the WEU, *The outlook for WEU. Reply to the 30th Annual Report of the Council, Report on behalf of the General Affairs Committee by Mr van der Sanden, Rapporteur, Document 1012* (Paris: Western European Union, 3 May 1985), pp. 12–13.
110. Assembly of the WEU, 3 May, p. 7.
111. Assembly of the WEU, 3 May, pp. 8–9.
112. Assembly of the WEU, *The military use of space, Report submitted on behalf of the Committee on Scientific, Technological and Aerospace Questions by Mr Wilkinson, Rapporteur, Document 976* (Paris: Western European Union, 15 May 1985), p. 190.
113. Asembly of the WEU, 15 May, pp. 195–6.
114. Assembly of the WEU, *30th Ordinary Session (Second Part), Official Report of the 8th Sitting, 4 December 1984, Doc. A/WEU (80)* Cr 8 (Paris: Western European Union, 1984).
115. Assembly of the WEU, 15 May, pp. 183–4.
116. Assembly of the WEU, 4 December, pp. 10–15.
117. Assembly of the WEU, 4 December, pp. 17–18.
118. 'Europäische Wünsche und amerikanische Bedenken vor der WEU-Tagung in Bonn', *Frankfurter Allgemeine Zeitung*, 20 April 1985; 'Westeuropäische Union tagt in Bonn. Genscher: Europas Rolle aufwerten', *Süddeutsche Zeitung*, 23 April 1985; Anna Tomforde, 'Europeans plan joint SDI response', *Boston Globe*, 24 April 1985; William Drozdiak, 'Europeans to Act Jointly on "Star Wars"', *Washington Post*, 24 April 1985, p. 27; Tyler Marshall, '7 NATO Allies Fail to Agree on US Bid to Join "Star Wars" Research', *Los Angeles Times*, 24 April 1985, p. 19; 'Space Arms Unity Eludes Europeans', *New York Times*, 24 April 1985, p. 5.
119. 'Communiqué issued at the close of the ministerial meeting of the Council of the Western European Union', *Document 1011* (Paris, Western European Union: 24 April 1985).

120. Assembly of the WEU, *30th Ordinary Session (Second Part), United States–European co-operation in advanced technology, Report submitted on behalf of the Committee on Scientific, Technological and Aerospace Questions by Mr Hill, Rapporteur, Document 992* (Paris: Western European Union, 8 November 1984).
121. Assembly of the WEU, *31st Ordinary Session (First Part), The Military use of computers. Reply to the 30th annual report to the Council, Report submitted on behalf of the Committee on Scientific, Technological and Aerospace Questions by Mr Fourré, Rapporteur, Document 1007* (Paris: Western European Union, 11 April 1985).
122. WEU Assembly, Committee on Defence Questions and Armaments (30th Session of the Assembly), *Emerging Technology and Military Strategy, Draft Report submitted by Mr van den Bergh, Rapporteur, Document A/WEU/DA (85) 4* (Paris: Western European Union, 9 May 1985).
123. 'Divided WEU puts off Star Wars discussion', *The Times*, 22 May 1985; 'Unmut bei der WEU-Versammlung. Das Ziel eines "europäischen Pfeilers der NATO" wurde nicht erreicht', *Süddeutsche Zeitung*, 21 May 1985; 'Die WEU legt sich noch nicht fest', *Frankfurter Allgemeine Zeitung*, 22 May 1985.
124. 'Address by Mr. Hans-Dietrich Genscher', in Assembly of the WEU, *31st Ordinary Session (First Part), Official Report of the 5th Sitting, 22 May 1985, Doc. A/WEU (31) CR 5* (Paris: Western European Union, May 1985).
125. Assembly of the WEU, 22 May, p. 5.
126. Assembly of the WEU, 22 May, pp. 23–6.
127. 'WEU-Versammlung verschiebt Debatte über SDI-Projekt', *Frankfurter Allgemeine Zeitung*, 24 May 1985; 'Genscher als Anwalt der "Eureka"-Initiative – Rede vor der WEU-Versammlung in Paris', *Neue Zürcher Zeitung*, 24 May 1985; Diana Geddes, 'Genscher applauds Eureka', *The Times*, 23 May 1985.
128. Peter Schmalz, 'Strauss: Nicht mit Abfall begnügen. Für Eigenständigkeit in Luft- und Raumfahrt. Europäer zu Einigkeit ermahnt', *Die Welt*, 19 September 1985; Rudolf Metzler, 'Europa uneins über SDI-Beteiligung. Bonn setzt auf zivile Anwendungsmöglichkeiten. Paris weiterhin dagegen', *Süddeutsche Zeitung*, 19 September 1985.
129. 'Statement by Parliamentary State Secretary Dr Martin Grüner, Co-ordinator for German Aerospace Policy', WEU Symposium on *The Space Challenge for Europe*, Munich, 18 September 1985, p. 5.
130. Assembly of the WEU, Committee on Scientific, Technological and Aerospace Questions, *Colloquy on The Space Challenge for Europe*, Munich, 18–20 September 1985. Paper submitted by Mr Herbert Curien, 13 A/P 11660 (Paris: Western European Union, September 1985).
131. Assembly of the WEU, *31st Ordinary Session (Second Part), WEU and the strategic defence initiative. The strategic defence initiative (defence aspects). Report submitted on behalf of the Committee on Defence Questions and Armaments by Mr van den Bergh, Rapporteur,*

Document 1033 (Paris: Western European Union, 4 November 1985).

132. Assembly of WEU, *31st Ordinary Session (Second Part), WEU and the strategic defence initiative – The European pillar and the Atlantic Alliance. Report submitted on behalf of the General Affairs Committee by Mr Berrier, Rapporteur, Document 1034* (Paris: Western European Union, 5 November 1985).

133. Assembly of the Western Economic Union, *31st Ordinary Session (Second Part), WEU and the strategic defence initiative – guidelines drawn from the colloquy on the space challenge for Europe (Proposals), Report submitted on behalf of the Committee on Scientific, Technological and Aerospace Questions by Mr Lenzer, Chairman and Rapporteur, Document 1036* (Paris: Western European Union, 6 November 1985).

134. See Assembly of the WEU, The Counsellor in charge of Press Affairs (Paris: Western European Union, 19 November 1985) p. 1.

135. Assembly of the WEU, *31st Ordinary Session (Second Part), Texts Adopted* (Paris: Western European Union, December 1985).

136. *Van den Bergh report*, p. 9.

137. *Van den Bergh report*, pp. 9–10.

138. *Van den Bergh report*, p. 16.

139. *Van den Bergh report*, p. 18.

140. *Van den Bergh report*, p. 18.

141. *Berrier report*, pp. 9–10.

142. *Berrier report*, p. 11.

143. *Lenzer report*, p. 14.

144. 'Address by Mr Roland Dumas, Minister for External Affairs of the French Republic', Assembly of the WEU, *31st Ordinary Session (Second Part), Official Report of the 10th Sitting, 4 December 1985, Doc. A/WEU (31) CR 10* (Paris: Western European Union, December 1985) p. 22.

145. See chapters 4–7 in this volume.

146. See note 135; Assembly of the WEU, December session 1985, 8th to 11th Sitting, *Doc. A/WEU (31) CR 8–11* (Paris: Western European Union, December 1985).

147. Assembly of the WEU, *31st Ordinary Session (Second Part). Official Report of the 9th Sitting, 3 December 1985, Doc. A/WEU (31) CR 9* (Paris: Western European Union, December 1985) p. 7.

148. See note 135, *Recommendation 428 on WEU and the strategic defence initiative. The strategic defence initiative (Defence aspects) Doc. A/WEU (31) PV 11* (Paris: Western European Union, December 1985).

149. See note 135, *Recommendation 429 on WEU and the strategic defence initiative – The European pillar of the Atlantic Assembly*, Doc. A/WEU (31) PV 11 (Paris: Western European Union, December 1985) pp. 4–6.

150. See note 135, *Recommendation 430 on WEU and the strategic defence initiative – guidelines drawn from the colloquy on the space challenge*

for Europe (Proposals), Doc. A/WEU (31) PV 11 (Paris: Western European Union, December 1985), p. 6.
151. 'WEU-Versammlung berät über SDI', *Frankfurter Allgemeine Zeitung*, 30 November 1985; 'WEU für Beteiligung an SDI-Programm', *Frankfurter Allgemeine Zeitung*, 6 December 1985; 'Die WEU-Versammlung für "Star Wars"', *Neue Zürcher Zeitung*, 7 December 1985.
152. Europäisches Parlament, Politischer Ausschuss, Unterausschuss Sicherheit und Abrüstung, *Arbeitsdokument über Rüstungskontrolle, Abrüstung und ihre Bedeutung für die Europäische Gemeinschaft, Berichterstatter Sir Peter Vanneck*, PE 97 992 (Luxembourg: European Parliament, 30 May 1985).
153. Europäisches Parlament, Ausschuss für Energie, Forschung und Technologie, *Arbeitsdokument zur Schaffung eines europäischen Forschungsraumes (Dok. 2-1183/84), Berichterstatter Herr Werner Münch*, PE 96.715 (Luxembourg: European Parliament, 13 May 1985).
154. Europäisches Parlament, Ausschuss für Energie, Forschung und Technologie, *Bericht über die Antwort Europas auf die technologische Herausforderung, Berichterstatter Herr Poniatowski*, PE 98 542 (Luxembourg: European Parliament, 30 Sept 1985).
155. Europäisches Parlament, Ausschuss für Energie, Forschung und Technologie, *Bericht über die Europäische Weltraumpolitik, Berichterstatter Herr Toksvig*, PE 95.639 (Luxembourg: European Parliament, 30 September 1985).
156. Europäisches Parlament, Ausschuss für Energie, Forschung und Technologie, *Bericht über den unterschiedlichen Stand der technologischen Entwicklung in den Mitgliedstaaten der Europäischen Gemeinschaft, Berichterstatter Her Gérard Longuet*, PE 99 802 (Luxembourg: European Parliament, 30 September 1985).
157. Europäisches Parlament, Politischer Ausschuss, Unterausschuss 'Sicherheit und Abrüstung', *Mitteilung an die Mitglieder, Entschliessungsanträge* (Luxembourg: Europäisches Parlament, 6 September 1985) see Dokument B 2-80/85.
158. Europäisches Parlament, Sitzungsdokumente 1984-85, *Entschliessungsantrag von den Abgeordneten Hänsch u.a. im Namen der Sozialistischen Fraktion... zur Strategischen Verteidigungsinitiative (SDI) der Vereinigten Staaten*, PE 96 456/rev. (Luxembourg, European Parliament, 11 March 1985); Europäisches Parlament, Sitzungsdokumente 1985-86, *Entschliessungsantrag von den Abgeordneten Hänsch u.a. im Namen der Sozialistischen Fraktion zur Strategischen Verteidigungsinitiative (SDI) der Vereinigten Staaten*, Dokument B 2-95/85 (Luxembourg: European Parliament, 9 April 1985).
159. Europäisches Parlament, Sitzungsdokumente 1985-86, *Entschliessungsantrag von den Abgeordneten Tongue u.a. zum Krieg der Sterne*, Dokument B2-193/85 (Luxembourg: European Parliament, 22 April 1985).
160. Europäisches Parlament, Sitzungsdokumente 1985-86, *Entschliessungsantrag der Abgeordneten Linkohr u.a. im Namen der Sozialisti-

schen Fraktion zum SDI-Forschungsprogramm, Dokument B 2-240/85 (Luxembourg: European Parliament, 30 April 1985).
161. Europäisches Parlament, Sitzungsdokumente 1984–85, *Entschliessungsantrag eingereicht von Frau Charzat u.a. zur Beendigung des Wettrüstens und des Rüstungswettlaufs bei der Militarisierung des Weltraums*, Dokument 2-1553/84 (Luxembourg: European Parliament, 1 February 1985).
162. Europäisches Parlament, Sitzungsdokumente 1985–86, *Entschliessungsantrag eingereicht von Herrn Linkohr zur europäischen Forschung im Hinblick auf das weltraumgestützte Abwehrsystem (SDI)*, Dokument B 2-346/85 (Luxembourg: European Parliament, 20 May 1985).
163. Europäisches Parlament, Sitzungsdokumente 1985–86, *Entschliessungsantrag eingereicht von Herrn Metten zur Nicht-Beteiligung der EWG an der Forschung über Weltraumwaffen*, Dokument B2-358/85 (Luxembourg: European Parliament, 22 May 1985).
164. Europäisches Parlament, Sitzungsdokumente 1985–86, *Entschliessungsantrag von Frau Lizin zur Strategischen Verteidigungsinitiative*, Dokument B2-597/85 (Luxembourg: European Parliament, 20 June 1985).
165. Europäisches Parlament, Sitzungsdokumente 1985–86, *Entschliessungsantrag von Herrn Ulburghs zur Notwendigkeit einer Ablehnung der Strategischen Verteidigungsinitiative (SDI)*, Dokument B 2-251/85 (Luxembourg: European Parliament, 2 May 1985).
166. Europäisches Parlament, Sitzungsdokumente 1985–86, *Entschliessungsantrag der Abgeordneten Fuillet u.a. zu der Initiative der französischen Regierung zur Schaffung einer europäischen Koordinierungs- und Forschungsstelle (EUREKA)*, Dokument B 2-353/85 (Luxembourg: European Parliament, 22 May 1985).
167. Europäisches Parlament, 'Symposium: Europa 2000', *die Woche im EP* Tagungswoche vom 7–11 Oktober 1985 (Luxembourg: European Parliament, October 1985) pp. 16–25.
168. European Parliament, Working Documents 1985–86, *Report drawn on behalf of the Committee on Energy, Research and Technology on Europe's response to the modern technological challenge, Part B: Explanatory Statement, Rapporteur M. M. Poniatowski*, Document A 2-109/85/B, PE 98.542/fin./B (Luxembourg: European Parliament, 30 September 1985).
169. Europäisches Parlament, Sitzungsdokumente 1985–86, *Bericht im Namen des Ausschusses für Energie, Forschung und Technologie über den Technologietransfer, Berichterstatter Herr Alman Metten*, Dokument A2-99/85, PE 95.409/endg. (Luxembourg: European Parliament, 30 September 1985).
170. Commission of the European Communities, *Memorandum Towards a European Technology Community*, COM(85), 350 final (Brussels: Commission of the European Communities, 25 June 1985); Commission of the European Communities, *Implementation of the Commission's Memorandum 'Towards a European Technological Community'*

(Communication of the Commission to the Council) COM (85) 530 final (Brussels: Commission of the European Communities, 30 September 1985).
171. European Parliament, Working Documents 1985–86, *Report drawn on behalf of the Committee on Energy, Research and Technology on Europe's response to the modern technological challenge, Part A: Motion for a resolution, Rapporteur Mr M. Poniatowski*, Document A2-109/85/A, PE 98.542/fin./A (Brussels: European Communities, 30 September 1985); M. G. Möhnle, 'Parlament in Strassburg lehnt Nein zu SDI ab', *Die Welt*, 10 October 1985; 'Europa-Parlament bei SDI gespalten', *Süddeutsche Zeitung*, 10 October 1985.
172. See *European Parliament, Working documents*, A2-109/85/A/5; 7; 19.
173. See *European Parliament, Working documents* A2-99/85, p. 12.
174. *Europäisches Parlament, Sitzungsdokumente* A2-99185.
175. Europäisches Parlament, Sitzungsdokumente 1984–85, *Entschliessungsantrag von Herrn Linkohr zu den Beschränkungen des internationalen Technologietransfer durch die USA und deren schädlichen Auswirkungen auf die industrielle Entwicklung der Europäischen Gemeinschaft, Dokument 2-721/84* (Luxembourg, European Parliament, 15 October 1984).
176. 'EG-Kommissar Narjes warnt vor SDI', *Süddeutsche Zeitung*, 18 December 1985.
177. Europäisches Parlament, Sitzungsdokumente 1985–86, *Entschliessungsantrag der Abgeordneten Linkohr u.a. zur Eureka-Konferenz in Hannover und zur Europäischen Technologiegemeinschaft*, Dokument B2-1338/85 (Luxembourg, European Parliament, 10 December 1985).
178. 'Gipfel von Reykjavik: Bedauern und Hoffnung – Grossmächte müssen Verhandlungen fortsetzen – SDI führt in Sackgasse', in *Europäisches Parlament*, 9/1986, pp. 1–2.

9 EUREKA: a West European Response to the Technological Challenge posed by the SDI Research Programme
Alain Carton

I THE AMERICAN AND THE JAPANESE CHALLENGE FOR WESTERN EUROPE IN THE FIELD OF HIGH TECHNOLOGY AND THE ORIGINS OF EUREKA

The European nations do not have many choices in the face of the growing American and Japanese dominance in the fields of high technology:

1. They can undergo joint technological progress following the successful precedence of some joint European projects such as ARIANE, AIRBUS, ESPRIT and JET.
2. If not, they can act separately and be confronted with gradual technological and economic decline.

The joint European research project EUREKA is conceived as a means of facing the technological challenge posed by the American SDI. However, it would not be correct to present EUREKA as a direct 'response' to SDI. For the United States, SDI may lead to substantial progress in all fields of advanced technology, The Europeans must act today on these developments if they do not want to let this unique opportunity of overcoming a technological gap (and preventing it widening) pass them by. This does not imply that EUREKA should imitate the American initiative. Neither its objectives nor its organisation are the same.

The surprisingly quick start of EUREKA after the initial invitation by the French Foreign Minister, Mr Roland Dumas, to the other European governments in mid-April 1985, and its successful pre-phase, are to a great extent due to the fact that a month earlier the

Europeans had been invited to participate in the American SDI research programme but did not want to become 'subcontractors' in an enterprise of which they had neither control nor influence and which did not correspond to their own needs.

(a) The American and the Japanese challenge

New developments in the area of high technology will dramatically change the structure of the world's industry, especially in the fields of space technology, information, electronics, and the production of new materials. Many of our present manufacturing techniques and products will become obsolete. This is not a new situation – nor is the European effort to counter the challenge. The situation will, however, intensify as a consequence of the SDI programme and of the Japanese projects sponsored by *MITI (Ministry of International Trade and Industry)*. In FY85 SDI will make up only 10 per cent of the overall United States military research budget. However, over the next five years, it will inevitably have an impact on the competitiveness of American civilian technology.

The present rapid expansion of the telecommunications industry – a result of American deregulation and the fusion of telecommunication with computer technology – has intensified competition between Japan and the United States. America has been forced to take second place in the electronic components industry since the Japanese have claimed more than 30 per cent of the world market. The American government has made several allegations concerning collaboration between government and industry in Japan. However, the considerable influence the American government has in the area of technological and industrial development, in the form of military spending, must not go unmentioned.

The original demand for integrated circuits was from the military and NASA. Texas Instruments, for example, provided guidance systems for the Minuteman II missile and Fairchild for the Apollo programme. In 1963, defence and NASA contracts constituted 85 per cent of total production. This share decreased gradually – in 1965 to 75 per cent and in 1972 to 55 per cent – as new applications for semiconductor technology appeared. The most recent example of the role the public market plays is the Defense Department's SHSIC (Super High Speed Integrated Circuit) programme, for which $300 million have been allocated for the period 1979–89.

In Japan, MITI plays a decisive role in defining the directions of industrial development. However, the future depends on the strategies in the fields of high technology of the individual companies concerned. Especially important is the fusion of sectors which have hitherto been separate, such as computer science and telecommunications.[1]

In coming years, competition from Japan and the United States will be a serious challenge for European ventures. A major factor here is the treaty signed by President Reagan and Prime Minister Nakasone in November 1983 concerning the transfer of Japanese civilian-oriented high technology to the United States for military application. This will have two major consequences:

1. Japanese companies will be inspired to produce highly sophisticated military goods.
2. Japan's market position in the United States will be strengthened.

If no major changes occur today, Europe will become an industrial satellite of the Japanese–American oligopoly.

A direct consequence of the Japanese technological offensive was the founding of the European Community of Technology. The Japanese challenge should not be underestimated. The first major MITI project was started in 1966; the duration of these projects has tended to grow, and they currently concentrate on the most advanced technologies. These projects have been initiated and controlled by MITI and the Agency for Industrial Sciences and Techniques. Current project areas include optronics (1979–86), super-computers (1981–3), new components (1981–91) and software development (1979–96). The projects were planned by MITI in close consultation with the respective industries. Furthermore, selected companies are co-operating within a research association which is closely linked to MITI. Considerable sums are being invested in R&D by industry, with government acting as only a partner.

The development of a super-computer has recently been planned which should be able to reach a speed of ten thousand million flops (floating points per second). This would be ten times faster that the American CRAY-1 computer. This is part of the artificial intelligence project 1982–91.

It must however be said that in Japan publicly-financed research accounts for only 27 per cent of total research spending. In the Federal Republic of Germany it is 42 per cent and it is 50 per cent in the United States. In France publicly-financed research is 53 per cent of the total.

(b) Issues of technology transfer and competition

The member countries of the Atlantic Alliance admittedly hold divergent opinions concerning commercial relations with eastern bloc countries. COCOM has been established to co-ordinate export policies to these countries, and more recently the United States has imposed limitations on the export of certain high-technology products to the West European Allies.[2] One example in 1981 was the refusal to sell the CRAY computer system to France for its defence requirements, which forced France into developing its own ISIS and BULL systems. Britain has also been restricted in its development of software originally from the United States.

These examples make co-operation between Europe and the United States difficult and perhaps they indicate the true nature of the American offer to the Europeans of participation in future research programmes. For Western Europe, technological co-operation with the United States has often meant playing the role of subcontractor. The best example for this has been the Spacelab programme, which was developed by the ESA.

The possibility of closer co-operation between Europe and the United States raises two important questions:

1. Will the Americans be able to profit from such co-operation? This may very well be the case, if they can use European experience in advanced technologies and if, by reducing duplicate research, they can spread the costs of research and development. However, in view of the failure of the 'two-way-street' in military matters, many Europeans are not at present very enthusiastic about the prospects of such co-operation.
2. In what ways could such co-operation be best expanded? The first step would be the revision of agreements concerning technological transfer within the Alliance, especially in view of technologies with possible military applications. Another possibility would be an overhaul of COCOM. Priority should also be given to the application and exploitation of technology rather than to the export policies concerning it.

Many European governments are concerned that Europe has been losing its competitive footing in the world high-technology market. In 1978 Europe's trade balance for high-technology products showed a surplus of $500 million. By 1982, this had dropped to a deficit of ten thousand millions. Western Europe controls only 10 per cent of the world market; Japan and the United States control the rest.

The European problems concerning advanced technology are the result of two factors:

1. obstacles to inter-European co-operation in research and development programmes – a result of nationally-oriented policies concerning standards and concerning import limitations;
2. inability to turn ideas into innovations – a result of the lack of entrepreneurial spirit and capital, the limitedness of the domestic markets and the poor technological transfer between research institutes and industry.

II EUROPEAN CO-OPERATION IN THE FIELD OF ADVANCED TECHNOLOGY – FROM THE VERSAILLES SUMMIT (1982) TO THE EUROPEAN CONFERENCE ON TECHNOLOGY (1985)

(a) The background of European co-operation

Europe has already achieved remarkable results in various fields. Examples are AIRBUS, ARIANE, JET and ESPRIT. Other projects such as BRITE, RACE, and STIMULATION are in progress. Many people believe that the time has now come to organise more systematic co-operation, based on past experience, between the European countries in all fields of advanced technology.

At this point the role of the EEC and the European Commission should be mentioned. One of the most innovative elements of the European co-operative effort has been the creation of an information technology task force within the frame of the ESPRIT[3] programme. The task force, which is in charge of managing ESPRIT, is largely autonomous and its costs are shared. It reports directly to the committee responsible for industrial issues within the Commission. The EC finances 50 per cent of the project, making it an instrument of support for the computer industry. In 1985 ESPRIT was concerned with the following areas: microelectronics software technology, advanced information processing, office systems and computer-assisted production.[4]

The activities of the task force in the field of computer science are rounded off by the BRITE and RACE programmes. BRITE (Basic Research in Industrial Technology for Europe) aims at promoting the information industry in Europe and making it more competitive by co-ordinating the concentrated actions in the field of basic research. These projects are defined by a series of contracts signed by industrial organisations, research institutes and universities in the EC member countries. Research effort is to focus on the following areas: laser technology, production of new materials, catalysis and particle beam technology.

The RACE programme (Research and Development in Advanced Communication Technology for Europe) involves the creation of vast broadband integrated communication networks (IBC). The decision to initiate the definition phase was taken by the ministers of research and development from the ten EC countries on 4 June 1985.[5] The RACE programme is considered a complement to the ESPRIT programme. The Commission estimates that the EC's contribution will be $400 million, spread over ten years.

The European effort is equally significant in matters concerning outer space.[6] At its ministerial level meeting in Rome (30–31 January 1985), the board of ESA stressed the common will of its member countries to develop a 'coherent and balanced' European space programme. In particular, the members decided to incorporate the ARIANE V launcher and the COLUMBUS projects within the agency programmes. France considers these projects a stage towards the development of a European space station. The agency board also took note of the French decision to launch the HERMES space shuttle programme, which will provide Europe with its own means of transport and service in outer space. The preliminary studies phase has already begun.

(b) The genesis of EUREKA – the work group Technology, Growth and Employment

The establishment of the work group on Technology, Growth and Employment (TGE) after the Versailles Summit of highly-industrialised nations in 1982 can be seen as the origin of EUREKA. The question of how to set up a co-operative programme for the development of a means to stimulate growth and employment was discussed for some time. The work group submitted a list of proposals for co-operation under which each member country could determine its own interests for specific projects. A number of scientific and technological issues were reviewed by the work group in the attempt to determine where additional international collaboration might contribute to knowledge and improve social and economic conditions. The eighteen proposals submitted by the workgroup were approved at the Williamsburg Summit in May 1983. There is a wide range of co-operation already under way in the following areas:

1. conquest of space,
2. renewable sources of energy,
3. safety research for light water reactors,
4. computer and artificial intelligence.

Several of the subjects which are part of these programmes have been proposed again in the context of the EUREKA programme.

The general secretary of TGE, Mr Yves Stourdze who is also the present director of *CESTA (Centre d'Etudes des Systèmes et des Technologies Avancée)* in Paris, has written in this respect:[7] 'The feeling prevailed that the traditional forms of technological co-operation have in most cases resulted in failure. With the exception of a few exemplary cases such as ARIANE, AIRBUS, and CERN, such projects have generally perpetuated themselves into weighty, inefficient and costly bureaucratic structures. The initiation of a precisely working programme therefore appeared to be more urgently needed than a text stating the general principles of an international agreement'.[8] The experiences of TGE were to prove important for the EUREKA programme.

III EUREKA – A NEW EUROPEAN PERSPECTIVE

In April 1985 the President of the French Republic, Mr François Mitterrand, launched the idea of a 'Europe of Technology'. French officials had been calling on their European partners for years to intensify their efforts in this field.

A number of propositions had already been made by Prime Minister Fabius and the Foreign Ministers Cheysson and Dumas. These propositions had been received positively among the other EC countries. In addition, the EC Commission in Brussels had submitted a working paper concerning the strengthening of the technological base and competitiveness of European industry to the European Council at its Spring meeting in Rome (29–30 March 1985). Mr Dumas's letter to other EC countries foreign ministers before the WEU ministerial meeting in Bonn in late April 1985 was thus not an initiative *a nihilo*.

The talks with Mr Genscher, the German Foreign Minister, and Mr Genscher's speech in Saarbrücken on 13 April 1985 concerning European industrial co-operation inspired Mr Dumas to propose to his counterpart the speedy creation of a Europe of Technology. He also sent letters to all the foreign ministers of the EC and of Spain and Portugal and to the President of the European Commission, Mr Jacques Delors. In his letter, Mr Dumas proposed the establishment, based on past experience, of a better system of co-operation. This would mean the creation of an appropriate, flexible structure of counselling among governments, research institutes and private industry. European R&D would thus be given a stimulus comparable to that given by the Pentagon in the United States or by MITI in Japan.

The starting point of EUREKA was the ministerial session of the WEU in Bonn on 23 April 1985. The final communiqué of the session stresses the necessity of a co-ordinated response to the American offer to the Europeans of participation in the SDI research programme. The call for 'co-ordinated response' in fact concealed the different opinions which existed among the Western European countries concerning the American initiative. The positive element was the expression of the political will to realise a unified European effort in the field of technology. However, a debate among the European countries concerning the organisation of this effort was also provoked.

(a) Description of the EUREKA programme

The European partners have increasingly spoken of a *Community* of Technology rather than of a European *agency* as France had originally conceived it.[9] It has therefore been necessary to remove all ambiguity. The other European countries and the EC Commission interpret 'community' as relating to actions with which the Commission is associated under various statutory and institutional arrangements. SME (*Serpent Monétaire Européen*) or JET (Joint European Torus) are examples. For France, 'community' implies a common policy, which would operate completely within the framework of the EC Commission and be subject to the EC rules concerning common policies. It is therefore important to realise that the EUREKA project would not exist as a common policy incorporated within the budget of the EC, requiring unanimous decision and subject to the authority of the Commission and the EC Parliament. On the contrary, EUREKA would have to operate as a specific community action. The flexibility of such community actions has shown that problems concerning non-member countries can be dealt with as they appear.

The EUREKA project is based on three main objectives:

1. the establishment of genuine European co-operation in advanced technologies such as microelectronics, optronics, new material production and laser technology;
2. the detailed definition of common programmes in the areas of space, energy and oceanography;
3. the extension of technological innovation to traditional sectors, where research has been limited and which could therefore become vulnerable.

Once the Eureka programme is established, it will have three major functions:

1. It will be the basis for political collaboration and decision at the ministerial level: this would involve the definition of major objectives based on the conclusions of several consultative work groups from various sectors and the creation of policies concerning the opening of public markets, standardisation of private industry and stimulation of the economy.
2. It will stimulate R&D. This would involve the strengthening of the contact network among the various European laboratories and universities in the fields of information technology, new material production and biotechnology.
3. It will help co-ordinate market-oriented action. A permanent co-ordination work group has to be created in order to maximise the effects of efforts towards specific objectives which could stimulate the relevant sectors of private industry. It would be an *ad hoc* organisation – a 'co-ordinative authority' rather than an agency – which would have the task of defining and managing the programmes in close

co-operation with the respective companies and research institutes.

The EC Council voiced its support for the French-led EUREKA programme at the meeting in Milan (28–29 June 1985), and it also supported the EC Committee proposals, which are similar. The Council also approved extending participation to European companies in non-EC member countries. It stressed, however, that Community prerogatives would have to be maintained concerning the following points:

1. the unification of the domestic EC market;
2. the enhancement of the Community's technical and financial means, especially the services of the EIB (*European Investment Bank*), which would be immediately available.

The operational phase of EUREKA was drafted at the European Conference for Technology held in Paris (17–19 July 1985).[10] During the meeting, the major project areas were agreed upon and the problem of financing was discussed.

Within this wider context, the four largest European electronic concerns have planned joint ventures in both the civilian and military sectors. These companies include the Siemens AG of the Federal Republic of Germany, General Electric Co. of Great Britain, NV Phillips of the Netherlands and the French state-owned Thomson concern. The industrial agreements deal with advanced electronics such as microprocessors, microwave components, high-density memory components, flat screens, and so on.

The French delegation presented a working paper, the EUREKA White Book, to the conference in order to provide information concerning market-oriented programmes.[11] Five specific areas of research were approved at the meeting: computer technology, telecommunication, processors for industry, new materials production and biotechnology. Related to these five areas are the five specific projects below:

EUROMATIC (European information science)
This programme consists of the following projects:

1. a Franco–German super computer by 1992,
2. artificial intelligence projects in both hardware and software technology and the creation of multilanguage data banks,
3. the establishment of a software engineering centre.

EUROCOM (European telecommunications)
This consists of:
1. development of a vast European numerical communications system,
2. creation of research computer networks for European laboratories.

EUROBOT (European robotics)
Three programmes have been given priority here:
1. the development of robots for dangerous work,
2. agricultural processors,
3. fully automated factories (based on the earlier ESPRIT and BRITE programmes).

Eurobiot (European biotechnology)
Agricultural bioengineering and research into genetic engineering.

Euromat (European materials)
The development of new materials for industrial turbines.

(b) Questions about EUREKA

Some of the ambiguities that plagued EUREKA during its gestation phase have been overcome in its pre-phase. The major problems EUREKA faces are the following:

1. Is EUREKA a civilian or a military project?
2. What will be the institutional framework of EUREKA?
3. How will it be financed?
4. In what ways are the SDI and the EUREKA programmes related?
5. How do the various EEC member countries view EUREKA?

EUREKA – a civilian or a military project?
Given the fact that advanced technologies are crucial from all points of view, it is hardly possible to make a clear distinction between purely civilian and military projects. It is also difficult to estimate the spin-off of the various programmes described. However, the EUREKA project involves the mobilisation of the energies of both private and public organisations. There might be some military applications, but of a peaceful nature, such as observation and

communication. The EUREKA programme will not involve any co-operation and research in the fields of armaments, and this is especially true for the offensive arms race in outer space. The French Defence Ministry has planned the development of observation and communication satellites for the 1986 military budget and it believes that the definition of military objectives in space will encourage the European EUREKA effort, even if the EUREKA programme is conceived much more widely, not limited to military subjects.[12]

The institutional framework of EUREKA
EUREKA requires a flexible form including both organisational bodies within the EC to work together with the countries participating in various projects and organisational bodies independent of the EC for the specific 'à la carte' programmes. A narrow framework based on community policies is not envisaged. The EUREKA advanced technologies programme will rather be based on the existing procedures of, for example, the nuclear research (Euratom) project.

The EC will also play a part in the framework of the EUREKA programme. However, there have in the past been some weaknesses in Council decisions relating to research programmes, and these are partly due to institutional problems. This is closely related to the fundamental question of institutional reform in the EC itself. The best solution would probably be a small secretariat to co-ordinate decision-making among the governments, private industry and the EC. Another possibility would be the establishment of a board of management consisting of representatives from the various research and technology ministries and of high civil servants responsible for specific projects.

The present chairman of the European Commission, Mr Jacques Delors, has stated that a modification of the Rome Treaties abrogating the unanimous decision stipulation is necessary to ensure a large domestic market for the EC member countries.[13]

The funding of EUREKA
From the financial point of view, one of the first steps to be taken within the framework of the EUREKA programme will have to be the elimination of duplicate R&D projects existing in the various European countries. As far as funding is concerned, governments, financial institutions, banks and private industry will be called upon to finance different projects. Contributions by the various countries should supplement other types of funding, especially for the 'à la carte' projects and when Community austerity is necessary.

Funding of EUREKA should be similar to that of the JET programme or of other EC R&D programmes.

The relationship between SDI and EUREKA
EUREKA is neither an 'alternative' to SDI nor is it a reaction to the American invitation to participate in SDI research. The SDI programme has, however, provoked European awareness concerning advanced technologies. The main obstacle preventing immediate acceptance of the American invitation to participate in SDI has been the problem of *technology transfer*. In the view of many managers of high-technology firms, the SDI programme would mean an important transfer of technology in the next few years, financed by the military. However, the SDI programme would be entirely under control of the Americans and the role of the Europeans would be reduced to that of a subcontractor.

Moreover, the effect of space research on the economy must not be exaggerated. The resources allocated to the SDI programme will be low in comparison to former projects such as the Apollo programme. European companies may expect to obtain contracts worth one thousand million dollars in the first five years – 5 per cent of the total SDI research budget. Divided among the many European high-technology companies, the SDI programme becomes much less attractive.

In some fields, Europe must maintain control over the whole technological process. EUREKA could provide the appropriate framework for this and for the ideal form of co-operation among the three types of participants – scientists, private companies and public authorities.[14]

The EUREKA and the SDI debates have influenced the political climate in Europe. One result of EUREKA has been that in Europe the strategic aspects of SDI have become less important in view of the economic and technological aspects. Furthermore, EUREKA has generated a technological dynamism of its own. A good example is the agreement of the four largest electronic companies of Western Europe to co-operate in developing new electronics ventures in both civilian and military sectors. In late June 1985, company executives of Siemens AG of West Germany, General Electric Co. of Britain, NV Philips of the Netherlands, and Thomson, a state-owned French company, agreed to study developing joint research projects in advanced sectors of electronics, primarily covering new generation microprocessors, gallium arsenide integrated circuits, microwave components, high-density memories, flat screens and a wide range of sensory devices.[15]

The main aspect of EUREKA has become the pursuit of market-oriented projects, while taking into account European misgivings about the military applications of advanced technology. The projects allow the various European countries to make independent decisions concerning military applications of knowledge gained in joint projects in which they have participated.

The EC members' view of EUREKA
Since its beginnings, EUREKA has provoked varying reactions among the EC member countries.[16] The programme was officially approved at the meeting of the EC foreign ministers in Luxembourg on 29 April 1985 and again at the Milan Summit on 28 June 1985.

The Foreign Minister of the Federal Republic of Germany, Mr Genscher, has declared his support of the 'Europlean Community of Technology'. The French government has adopted detailed positions on both the EUREKA and SDI programmes. Italy also favours the EUREKA programme, provided it does not involve the creation of a new agency. The same is true for Great Britain, whose major concern has been the emergence of a costly bureaucratic system. In addition, the Benelux countries, Denmark, Ireland, Greece, Spain and Portugal have all expressed their interest in EUREKA.

Outside the EC, Norway, Switzerland, Sweden and Austria have indicated interest in participation.

However, the fact remains that some countries - Germany and Great Britain for example - have interpreted EUREKA as a form of SDI participation rather than a European research initiative.

Finally, the European Commission in Brussels has approved the French initiative in general terms. Mr Karl Heinz Narjes, Vice Chairman and Commissioner in charge of research issues would have preferred an organisational body within the framework of the EC. However, the President of the EC, Mr Jacques Delors, stated 29 April 1985 that the EUREKA project was in fact on the same lines as his own proposals.

(c) The results of the second EUREKA meeting

The decisions taken during the Hanover meeting in November 1985 mean that the EUREKA programme has entered the operative phase. Much of the anti-SDI aspect of the project has been lost, but the conviction has been gaining ground that EUREKA can be a viable framework for mobilising Western Europe's high-technology industry. The objectives are civilian and commercial, but there might be some military fallout – something completely different from the so-called 'spin-off effect'.

The EUREKA project now exists, but a large European home market must yet be developed in order to ensure its success.

The resolutions adopted in Luxembourg run along these lines. The EC is becoming more and more convinced that a unified European market is a prerequisite for technological innovation. There is today in Europe a strong commitment to strengthening the multilateral trade system, to promoting new technologies, and especially to creating new fields of employment.

Ten 'Category I' projects were adopted at the Hanover meeting, which means that companies in at least two EC countries have agreed to co-operate and that the proposals have been approved by the governments of the countries involved. The projects are:

1. production of a standard microcomputer for education and domestic use;
2. production of a new type of computer chip made of amorphous or uncrystallised silicon;
3. development of a high-speed computer;
4. development of a laser for cutting cloth in the textile industry;
5. development of membranes for filtering water;
6. development of high-power laser systems;
7. development of a system for tracing air pollutants;
8. development of a European research computer network;
9. development of a diagnosis kit for sexually transmitted diseases;
10. development of advanced optic electronics.

In addition, there are about twenty 'Category II' projects – meaning either that the companies are still in negotiation or that their governments have not yet given formal endorsement.

The Ministerial Conference will be the co-ordinating organ of the EUREKA programme. It will be composed of representatives of the member countries and of the EC Commission. Under this commission, a secretariat will be created to increase the transparency and efficiency of the EUREKA programme.

(d) The developments in 1986: the results of the third EUREKA meeting in London and of the fourth EUREKA meeting in Stockholm[18]

EUREKA was expanding steadily both in terms of membership and in the number of agreed joint projects and in committed funding. In

1986, both the Commission of the EC and Iceland were admitted as full members and several other countries from Eastern Europe and from Canada, China, Japan and Brazil expressed an interest in participating in EUREKA projects.[19]

During 1986, 102 new projects were added to the initial list of 10 projects.

- (a) on 22–23 January 1986, civil servants agreed on 16 collaborative projects worth a total of up to £398 million[20];
- (b) on 29–30 June 1986, at the third ministerial meeting in London, 40 ministers from 19 EUREKA countries added 62 new projects (including those already agreed upon in January) with a value of 2 billion European currency units (ECU)[21];
- (c) on 17 December 1986, at its fourth ministerial meeting in Stockholm an additional 40 collaborative research and development projects were adopted.[22]

At its London meeting, the EUREKA member states agreed to establish a small secretariat in Brussels to be headed by the French official Xavier Fels with a small staff of 12–15 experts and assistants that started to operate on 1 November 1986.

During 1986, both the British and the West German governments announced their interest to co-sponsor several Eureka projects, e.g. the West German Minister of Science and Technology, Mr Riesenhuber, announced that his government had approved EUREKA subsidies for German companies of up to 518 million marks over a eight-year period.

In March 1986, the European Commission published plans to double the spending on EC research programmes over the next five years from 3.75 billion Ecu (1984–7) to 9 billion Ecu (1987–91).[23] The discussions among Western Europe's three largest semiconductor manufacturers (Siemens, Philips and Thomson) intensified[24] and talks among European bankers were initiated to establish a 'Eureka Financial Roundtable'.[25]

IV CONCLUSION

EUREKA's success depends on the engagement of all the parties concerned and especially on the political will of the various governments. France, in initiating the EUREKA programme, has taken a special responsibility upon itself, but the impetus must come from the EC, and this may well involve a revision of the Rome Treaties.

It is important that EUREKA is not understood as a competitor to the SDI. It may very well be that some European companies participate in projects of both programmes; this would not be a conflict.

By December 1986, relatively few SDI contracts had been awarded to European companies, primarily in areas where they had a leading edge. With the increasing criticisms in the US Congress, both with respect to the SDI programme and with the funding for competitors of US companies in particular and with the increasing protectionism as far as exports of European high-technology products to the United States are concerned, more companies shifted from SDI to an intra-European co-operation in the EUREKA context.

SDI may have provided the necessary economic challenge for European countries and companies to overcome non-tariff barriers, to avoid duplication in similar research and development projects, and to foster closer co-operation in the realm of science and technology. With a market of 360 people – larger than the population of the United States and Japan – the 19 EUREKA member nations do have a good chance of competing successfully both with the American military–industrial complex and with the Japanese technological offensive.

Notes

1. Ministry of Industrial Trade and Industry, *Features of the Industrial Development of Japan* (Tokyo: MITI, March 1983).
2. NAA, Scientific and Technical Committee, *Interim Report of the Sub-Committee on Advanced Technology and Technology Transfer by Mr Lothar Ibrügger, Rapporteur* (Brussels: North Atlantic Assembly, November 1984).
3. At the Community level, the ESPRIT programme (European Strategic Programme for Research and Development in Information Technologies) represents a public- and private-sector investment to a total of 1.5 million ECU for the period 1984–5. See, for example, Commission of the European Communities, *Memorandum Towards a European Technology Community*, COM (85) 350 final (Brussels: European Commission, 25 June 1985).
4. A. Danzin, 'Esprit et vulnerabilité technologiques de la CEE', *Revue de Defense Nationale* (February 1985) pp. 7–32.
5. See the following memoranda of the European Commission to the Council of the European Communities – for 1984 nos 215, 230, 271 and 608 and for 1985 nos 84, 140, 112, 113 and 145.
6. ESA, *Annual Reports* for 1984 and 1985 (Paris: European Space Agency, 1984 and 1985).
7. CESTA published the 'White Book on EUREKA' in June 1985.
8. Yves Stourdze, 'Vers une nouvelle co-operation scientifique et indus-

trielle – l'ambition d'Eureka', *Le Monde Diplomatique* (August 1985), p. 17.
9. 'EURECA' (European Research Co-ordination Agency) was changed to 'EUREKA', thereby avoiding the confusion with EURECA, a programme of the ESA.
10. 'Document du Quai d'Orsay', *Bulletin d'Information*, 139/85, 23 July 1985.
11. CESTA (ed.), *Eureka, la chance technologique de l'Europe, propositions françaises* (Paris: CESTA, June 1985).
12. Ministre de la Défense, *Rapport au Parlement sur l'exécution et la réévaluation de la loi de programmation militaire 1984–1988* (Paris: Defense Ministry, June 1985) p. 10.
13. Press conference with Jacques Delors in Brussels, 26 June 1985.
14. Interview with Etienne Davignon, the former Vice President of the EC for Research and Industry, *Le Monde*, 25 June 1985.
15. Axel Krause, 'Electronic Plans Set in Europe – 4 Firms Draft Eureka Project for $1 billion', *International Herald Tribune*, 27 June 1985; Elizabeth Pond, 'W. Europe decides to pull together on defense research – "Eureka" would keep some "star wars" type research in Europe', *Christian Science Monitor*, 28 June 1985, p. 9; Bernard Brigouleix, 'Les quatre grands de l'électronique européenne s'engagent dans Eurêka', *Le Monde*, 27 June 1985; 'Europäische Elektronikunternehmen kooperieren bei Eureka-Projekten', *Handelsblatt*, 27 June 1985.
16. 'Europäische Antwort auf die SDI-Initiative – Informelles Treffen der EG-Forschungsminister in Rom', *Neue Zürcher Zeitung*, 24 April 1985.
17. 'Abschluss der Ministerkonferenz in Hannover – Grundsatzerklärung zu Eureka verabschiedet – Schaffung eines Sekretariats', *Neue Zürcher Zeitung*, 8 November 1985.
18. The developments in 1986 were added by Hans Günter Brauch.
19. Axel Krause, 'France Supports Eureka Role for Non-Europeans', *International Herald Tribune*, 27 February 1986.
20. Fiona Thompson, 'Eureka projects worth up to £398m announced', *Financial Times*, 24 January 1986.
21. 'The Eureka joint project list', *Financial Times*, 1 July 1986.
22. Hannes Gamillscheg, 'Osteuropa an Eureka vorerst nicht beteiligt – Technologie - Programm wird um 40 Forschungsvorhaben erweitert/ Konferenz in Stockholm', *Frankfurter Rundschau*, 18 December 1986.
23. Quentin Peel, 'Commission call to double R&D spending', *Financial Times*, 7 March 1986.
24. Guy de Jonquières, 'Europe's microchip leaders discuss joint research for 1990s', *Financial Times*, 25 June 1986.
25. 'Riesenhuber: Fortschritte bei Eureka', *Frankfurter Allgemeine Zeitung*, 18 December 1986.

Part III
Assessments of the Strategic Consequences of SDI for Europe

10 The Implications of BMD for Britain's Nuclear Deterrent
Paul B. Stares[1]

I INTRODUCTION

Britain has many reasons to be concerned about the long-term consequences of the United States Strategic Defence Initiative (SDI), not least the stimulus it will doubtless give to Soviet BMD research. Like France, Britain relies on the absence of significant Soviet missile defences to ensure the effectiveness (and with it the credibility) of its relatively small deterrent force. If such defences become a permanent feature of the strategic landscape, then Britain will clearly have to reconsider its options for staying in the nuclear business.

In many respects, Britain finds itself in a situation not dissimilar to that of the late 1960s. Having decided in December 1962 to modernise its independent nuclear deterrent with Polaris missiles procured from the United States, Britain was confronted soon afterwards with unambiguous evidence that the Soviets were constructing an ABM system around Moscow. The likelihood of further ABM deployments made some seriously question whether Britain's small force of Polaris missiles would be able to penetrate these defences in sufficient numbers if called upon to do so. Despite this concern, the Polaris programme was continued and as Lawrence Freedman notes, 'Britain's refusal to be panicked by the trends of the late 1960s toward [BMD] was rewarded in 1972 with the signing of the US–Soviet ABM treaty'.[2] By placing a ceiling on the deployment of ABM systems, the Treaty effectively guaranteed the continued viability of small 'minimum deterrent' forces.

Today, in the mid-1980s, Britain has once again embarked on a major strategic force modernisation programme. The aging Polaris missile-carrying submarines are to be replaced and armed with Trident D-5 missiles, which are also being purchased from the United States. As will be discussed shortly, this represents a considerable increase in Britain's strategic capabilities. Although the choice of

Trident was heavily criticised at the time for being excessive to Britain's deterrent needs, it will doubtless now be considered as the minimum essential to hedge against possible Soviet BMD deployments.

So far, however, there has been little public discussion of the potential implications of the SDI for the long-term future of the British independent deterrent and the credibility of the Trident force in particular. Several reasons account for this. Having already weathered a storm of controversy over the Trident decision, the British government is clearly reluctant to give the public a new reason to doubt their choice. On the international stage they are likewise wary of being seen to place national concerns before such wider issues as strategic stability, Alliance cohesion, and arms control. The opponents of the Trident programme also find themselves in somewhat of a dilemma. To use the possibility of Soviet BMD deployments as a further argument in their campaign to defeat Trident could rebound on them. For instance, it might be interpreted as an endorsement for the SDI or even further expenditures to buttress the credibility of the deterrent.

Moreover, like the 1960s, there are good reasons for Britain not to be panicked by the latest 'trend' towards BMD; the Trident system promises to be a potent strategic instrument that is unlikely to be undermined either very easily or very soon by Soviet missile defences. Also, there is still considerable uncertainty over where the SDI will eventually lead, if indeed it is not traded for reductions on offensive forces as part of a 'grand deal' on arms control. Nevertheless Britain should not be complacent and expect the SDI to be just another episode in the superpowers' periodic flirtation with BMD. Towards the end of its projected lifetime, it is not inconceivable that the British Trident force will be facing significant Soviet BMDs. Prudence suggests, therefore, that Britain should begin to think about the implications of this well in advance of its possible occurrence. Before turning to a more detailed discussion of the available options for maintaining the credibility of the Trident force, it is useful to review briefly the British debate about the SDI.

II BRITAIN AND SDI

The British government's initial reaction to both the 'Star Wars' speech and the subsequent announcement of the SDI can be characterised as 'wait and see'; first to see whether Reagan's rhetoric represented a serious United States policy shift, and second to see whether it would be translated into a serious R&D effort. Furthermore, there was (and still is) considerable scepticism over whether the SDI would ever reach the deployment stage. Many of the technologies required to perform the more ambitious goals of the programme are still immature and even with an immense injection of funds there is no early prospect of deployment. Even the supporters of the SDI recognise that fielding of a relatively modest 'point defence' BMD system designed to protect United States ICBM silos and key military installations is 15-20 years away from fulfilment, while more comprehensive defences are at the very least 20-30 years into the future. With these time scales, observers can be forgiven for questioning whether the necessary political and financial commitment can be maintained. This in part explains one of the earliest official British statements on the SDI. When questioned in the House of Commons on whether the prospects for a space-based BMD system would cause the government to change its mind on Trident, the then Secretary of State for Defence Michael Heseltine replied:

> That space-based system is very much a research project and nobody knows whether it is technically feasible or will enter service. We cannot base a defence strategy on that hypothesis.[3]

With the re-election of President Reagan in November 1984 and his subsequent reaffirmation of the SDI, British officials became more concerned about the possible consequences of the United States initiative. Their concerns were made known privately to the United States when Prime Minister Margaret Thatcher visited President Reagan at Camp David on 22 December 1984. Just prior to this meeting, the Foreign Office and the Ministry of Defence under Cabinet Office supervision completed their own internal assessment of the implications of the SDI. In addition to questioning whether the United States could overcome the formidable technological obstacles, the review apparently expressed serious doubts about the wisdom of the enterprise. Above all else the Foreign and Commonwealth Office and Ministry of Defence were worried about the implications of the SDI for superpower stability, strategic doctrine, NATO cohesion conventional military budgets and – last but not least – the prospects for arms control.[4]

Although Thatcher voiced all these worries to Reagan, the overriding concern was to avoid a major United States–European rift over the SDI. With both sides still licking their wounds after the deployment of Cruise and Pershing missiles, nobody wanted another NATO crisis on their hands. As a result the agreed statement following the meeting was, as one British observer noted, 'a masterpiece of ambiguity'.[5] In return for Mrs Thatcher's less than wholehearted blessing of the SDI, President Reagan agreed to four policy guidelines presented to him by the British delegation:

1. the United States' and Western aim is not to achieve superiority but to maintain balance, taking account of Soviet developments;
2. SDI-related deployment would, in view of treaty obligations, have to be a matter for negotiation;
3. the overall aim is to enhance, not undermine, deterrence;
4. East–West negotiation should aim to achieve security with reduced levels of offensive systems on both sides.[6]

Mrs Thatcher's carefully qualified support for the SDI was reiterated when she addressed a joint session of Congress in February 1985. She also expressed her hope that British scientists would participate in some of the research.[7]

Behind the facade of Anglo–American consensus, the British government (and in particular the Foreign Office) still harboured serious concerns about the SDI. In March 1985, before the audience at RUSI, Foreign Secretary Sir Geoffrey Howe fired a broadside of criticism thinly disguised as a series of questions about the SDI. In the process he called for caution and restraint from the Americans.[8] Yet, what remains the most comprehensive official statement of British views on the SDI made no reference to its possible repercussions for Britain's independent deterrent, although it did warn of potential 'offsetting developments on the Soviet side'. It is to this issue that we now turn our attention.

III TRIDENT AND SOVIET BMD

One of the principal reasons for Britain's decision in 1980 to replace its ageing Polaris force with the Trident I missile system was the need to provide a hedge against potential Soviet ABM advances. Indeed, the importance of this factor in British strategic planning was amply demonstrated with the earlier decision to modernise the existing Polaris A-3 warhead under the extremely costly Chevaline programme. This has apparently improved the ability of the warheads to penetrate Soviet ABM defences.[9] The new system designated Polaris A-3TK is now being retrofitted onto the current British force of four SSBNs. However, as the government admitted to the House of Commons Expenditure Committee in 1980:

> Though the Chevaline programme will keep our Polaris missiles able to penetrate anti-ballistic missile defenses into the 1990s, continuing Soviet effort in research and development, allowed by the 1972 ABM Treaty, might in time reduce our assurance of this ... [Trident I's] MIRV capability and long range give excellent margins of long term insurance against further advances in Soviet ABM and ASW capability.[10]

Michael Quinlan of the Ministry of Defence went even further in stating that the Trident I (C-4) missile was 'a system which has within it, and could have to a further degree, a very considerable capability to counter an ABM defence even if that were *significantly extended* beyond what is currently allowed under the ABM Treaty' (my emphasis).[11]

The margin of insurance has been further increased – at least in theory – by the decision in 1982 to procure the improved Trident II (D-5) missile, although ensuring commonality with the United States was the prime consideration in this case. While the D-5 missile has the capacity to launch fourteen warheads compared to the C-4's eight, the British government acknowledged in 1982 that 'the move to Trident D-5 will not involve any significant change in the planned total number of warheads associated with our strategic deterrent force in comparison with the original intention for a force based on the C-4 missile'.[12] The 1984 *Statement on Defence Estimates* also noted that 'the Trident force will not involve using the full capability of the system'. This decision, as will be discussed shortly, may be re-evaluated in light of Soviet BMD activities.

On the basis of these statements, it is fair to assume – certainly for the time being – that each of British D-5 missiles will have eight warheads. Four SSBNs each with 16 tubes are to be built to accommodate these missiles. This will make for a total force of 64 missiles with 512 warheads. The first Trident boat is expected to be operational by 1993.[13] With most probably a two-year interval between each of the subsequent boats entering service as the remaining Polaris submarines are decommissioned, the evolving capability of the British strategic force during the 1990s should be as follows:

Year	Trident SSBNs	Polaris A-3(TK)*	Available Warheads
1993	1	3	272
1995	2	2	352
1997	3	1	432
1999	4	—	512

*Assumes three warheads per missile.

On the assumption that each Trident SSBn will have a lifetime of at least its Polaris *Resolution* class predecessor – i.e., 30 years – the *full* complement of Trident boats will be operational between 1997 and 2023. During this period at least one boat (with 128 warheads) will be on station all of the time, with two some of the time (with 256 warheads). Given adequate warning, Britain could possibly generate three boats on station bringing the total warheads available to 384. However, this cannot be relied on for contingency planning purposes.

Will this force provide a credible deterrent over its projected lifeime? The answer to this question is quite obviously impossible to ascertain with any certainty. Moreover, it should not be forgotten that the credibility of the Trident force is not just determined by Soviet BMD deployments but also by Soviet ASW capabilities, which can also be expected to improve. Leaving aside the ASW question, the British government appears to be confident (though not totally unconcerned) about maintaining the effectiveness of the Trident system in the face of Soviet BMD deployments. This was apparently far from the biggest item on the 'worry' list presented to the Cabinet before Mrs Thatcher left for Camp David.[14] Not long after, the former British Chief of Defence Staff, Lord Lewin, also stated – if rather unclearly:

> The United States Strategic Defence Initiatve, aimed at testing such an ABM system, is irrelevant to Trident. If the Soviet Union developed one, it could not be deployed until long after Trident was operational.[15]

Lewin, however, was almost certainly referring to a space-based system that could intercept ballistic missiles during the 'boost-phase' of flight and therefore before the separation of their multiple warheads. Against Britain's relatively modest force, a BMD system of this kind would prove especially potent. While Lewin's predictions of the likely deployment timescale for such a system seems intuitively correct, he fails to consider what the Soviets might deploy in the interim. If one believes official United States estimates of what the

Soviet Union could do in the absence of ABM Treaty constraints over the next twenty-five years the British Trident force may have to adapt in significant ways to remain an effective deterrent.

In particular, United States analysts are concerned with the 'breakout' potential of current Soviet BMD activities. Since the mid-1970s the Soviets have been upgrading the country's ABM-1 system around Moscow consisting of Galosh nuclear-tipped exoatmospheric interceptors, Hen House detection and tracking radars, Try Add guidance and engagement radars, and Dog and Cat House target tracking radars. The new system, known in the West as ABM-X-3, is expected to be operational by 1987. The Galosh missiles are now being replaced with an improved version (the SH-04) and an entirely new system of endoatmospheric interceptors (the SH-08), to be housed in hardened silos, is also expected to enter service in the near future. In addition to the relatively recent deployment of large phased-array early warning radars around the Soviet Union which will improve long-range target acquisition, two new types of ABM radars – the Flat twin for re-entry phase tracking and the *Pawn Shop* for missile guidance – will provide the dedicated battle management support for the ABM-X-3 system.[16] Of particular concern to Western analysts is the speed at which the two radar systems could be deployed, the Flat Twin can apparently be erected on a prepared site in approximately 6–8 weeks while the Pawn Shop is housed in a van-sized container and therefore potentially transportable.[17]

With all these developments in mind the CIA reported to Congress in 1985:

> The Soviets have the major components for an ABM system that could be used for widespread ABM deployments well in excess of ABM Treaty limits . . . The potential exists for the production lines associated with the upgrade of the Moscow ABM system to be used to support a widespread deployment. We judge they could undertake rapidly paced ABM deployments to strengthen the defenses at Moscow and cover key targets in the western USSR, and to extend protection to key targets east of the Urals, by the early 1990s.[18]

Although the same report acknowledged the existence of an active Soviet directed-energy weapon research programme, a ground-based laser system for BMD is not expected to be operational 'until after the year 2000'.[19]

Clearly then, towards the end of its projected lifetime the British Trident force could be confronted with significant Soviet BMD deployments. Before we consider what Britain might do in response, some important points need to be stressed.

First, the British strategic force may still be able to perform its primary mission with some impunity as Soviet BMD deployments are likely to be directed at the principal threat namely the United States. For example, if the Soviet Union concentrated its efforts on the defence of its ICBM silos this would probably not affect Britain's capacity to inflict horrendous damage on Soviet cities.

Second, while Britain has to plan on the assumption that it might have to act alone, penetration of Soviet missile defences will most likely increase if the attack occurs in conjunction with, or after a strike by, either American or French nuclear forces.

Third, as the Soviet Union is unlikely to have *total* confidence in its ability to counter a British strike even after extensive BMD deployments, the deterrent value of the Trident force will not diminish completely.

As Soviet defences increase, Britain can adopt a variety of corresponding measures to ensure the penetrability of the Trident force. These include first, the development of new penetration aids (penaids). As the United States will be developing its own countermeasures to Soviet missile defences, Britain should maintain its fullest possible co-operation to avoid a recurrence of the Chevaline fiasco. New attack strategies – such as deliberately depressing the trajectory of the warheads to reduce Soviet warning time and with it BMD target acquisition – is another possible option.

Second, Britain could make strategic targeting responsive to Soviet BMD deployments. If the defence of Moscow does improve significantly, Britain would have to balance the importance of the Soviet capital as the centre of state power against the more assured capacity to inflict unacceptable damage on other less well defended but less important Soviet cities.[20] Such a policy, arguably, would not seriously reduce the deterrent properties of the Trident force.

Third, the Trident force could be either augmented or complemented. Additional Trident missiles might be acquired if new Soviet BMD deployments made it necessary and politically acceptable though this appears the least attractive option. Alternatively, Britain might consider buying or developing Cruise missiles which could be deployed on a number of different platforms to complicate further the calculations of Soviet defence planners.[21]

Finally, there could be greater collaboration in the development and procurement of new strategic systems. With France, for example,

Britain could collaborate on the development of new cruise missiles or, perhaps, go further and co-ordinate targeting with the objective of optimising the effectiveness of their respective deterrent forces.

In conclusion, it would be optimistic for Britain to assume that its independent nuclear deterrent can continue to benefit indefinitely from the 'free ride' it receives with the ABM Treaty. Unless there is a sharp reversal of the current trend, the growing superpower interest in BMD will surely erode the foundations of that Treaty. While Britain can exert diplomatic pressure to prevent or delay this, it should also anticipate a world without the ABM Treaty. As discussed above, this may come sooner than many believe, possibly before the expected operational life of the Trident force has ended. Depending on the extent of Soviet BMD deployments the Trident force may have to adapt in ways that are unlikely to be cost-free. By the time a follow-on system to Trident is considered – around the year 2010 – the costs of continuing with an independent nuclear deterrent in the face of potential BMD developments will almost certainly be too expensive for Britain. Moreover, by then the United States may no longer be willing to provide Britain with new strategic systems if President Reagan's goal of increasing the emphasis on defensive weapons (SDI) whilst reducing offensive force levels with the Soviet Union takes effect. This could force Britain to collaborate with the French on nuclear weapons or, more likely, get out of the nuclear business altogether.

Notes

1. A shortened version of this paper appeared in *Defense Analysis* 1 (June 1985).
2. Lawrence Freedman, 'The Small Nuclear Powers', in Ashton B. Carter and David N. Schwartz (eds), *Ballistic Missile Defense* (Washington, DC: Brookings Institution, 1984) p. 258.
3. Quoted from 'Revised Trident Cost Coming Next Spring', *The Times*, 27 June 1984.
4. John Barry, 'Will Star Wars Split the West?' *The Times*, 24 January 1985.
5. *The Times*, 24 January.
6. See text of speech by the Secretary of State: 'Defense and Security in the Nuclear Age', 15 March 1985 before the RUSI. Interestingly, the United States State Department drafted a telegramme for American

missions quoting these four points as policy guidance. However, the Defense Department refused to clear it despite the approval of the President. See John Newhouse, 'The Diplomatic Round', *The New Yorker* (22 July 1985).

7. 'Thatcher Tells Hill She Backs Reagan on "Star Wars" Plan', *Washington Post*, 21 February 1985, p. 1.
8. *Washington Post*, 21 February.
9. For description of the Chevaline programme and system see Freedman, 'The Small Nuclear Powers', p. 262 and Robert Hutchinson, 'Chevaline: UK's Response to Soviet ABM System', *Jane's Defence Weekly* 15 December 1984, pp. 1068–9.
10. Quoted from David Yost, 'BMD and the Atlantic Alliance', *International Security*, 7 (Fall 1982) p. 150.
11. Quoted from Freedman, 'Small Nuclear Powers', p. 267.
12. Freedman, p. 266.
13. *Statement on the Defense Estimates 1984* (London: HMSO, 1984) p. 24.
14. *Jane's Fighting Ships 1984–1985* (London: Jane's, 1984) p. 593.
15. Lord Lewin, 'Why We Need Trident,' *The Times*, 6 February 1985, p. 12.
16. For details of current and expected Soviet BMD deployments see Thomas K. Longstreth, John E. Pike and John B. Rhinelander, *The Impact of US and Soviet Ballistic Missile Reference Programs on the ABM Treaty*, A Report for the National Campaign to Save the ABM Treaty (March 1985) pp. 19–22; see also chapters 12, 15 and 16 in this volume.
17. Longstreth *et al*.
18. *Soviet Strategic Force Developments*. Testimony Before a Joint Session of the Subcommittee on Strategic and Theater Nuclear Forces of the Senate Armed Services Committee by Robert M. Gates and Lawrence K. Gershwin (26 June 1985) pp. 5–6.
19. *Soviet Strategic Force Developments*, p. 8.
20. The British government has already indicated its flexibility on this issue. See Lawrence Freedman, 'British Nuclear Targeting', *Defense Analysis* 1, (June 1985) p. 94.
21. See David Hobbs, *Alternatives to Trident*, Aberdeen Studies in Defence Economics 25 (Summer 1983) for a comprehensive review of alternative strategic platforms.

11 The Implications of SDI for French Defence Policy
Alain Carton

I INTRODUCTION

At the Bonn Economic Summit in May 1985, when President Francois Mitterrand issued his famous refusal to the United States invitation to participate in SDI research,[1] he made it clear that it was not the idea of building military systems in space to which France was objecting but rather to the prospects of becoming what he called 'subcontractors' to the United States in such a venture.[2]

This distinction is very important. France believes that the SDI programme could in fact revive the arms race and lead to a strengthening of offensive weapons capacities. On the other hand, the contribution of space technology to French deterrence strategy must be considered in view of French defence policy. Here, space-related ventures are necessary in consideration of both purely national interests, as well as the interests of the Western Allies, which still have to be defined.

The French believe that such activities should relate to the non-aggressive applications of space technology: communication, observation, navigation, and so on. The present defence potential could be maintained without risking involvement in a new arms race.

The question remains whether the implications SDI is having on Europe will mean that France must react on a purely national level or whether a co-ordinated response with the European Allies is possible.

A major issue for France is the problem of maintaining its nuclear deterrence capability and responding to the continual need of modernising its forces in both a qualitative and quantitative way. France fears that a major BMD system would trigger Soviet counteractions and thus weaken the credibility of the French nuclear forces. The SDI programme might well change France's part in maintaining strategic balance in Europe and with it the nature of French solidarity with its Allies. The current definition of France's vital interests is at stake.

The ambiguities concerning the EUREKA programme must be removed. France is determined to play a leading role in the development of European technology, but it has insisted on the civilian character of the European response to the SDI challenge.[3] Nevertheless, there is still a number of possibilities for European military activities in space. This might be in the form of bilateral, trilateral or multilateral co-operation.

All these subjects play a role in the debate on the so-called EDI, the European extension of SDI. The French decision to become a major space power on its own or in co-operation with other countries is in fact related to this debate, although it might not seem so at first glance. This debate is under way in a few European countries, and it seems necessary to clarify a few details.

II FRENCH DEFENCE POLICY AND THE ABM DEBATE

(a) Independence, credibility and sufficiency – the three pillars of the French deterrent

The French decision in 1965 to develop an *independent* 'Force de Frappe' and the 1966 decision to withdraw from NATO's integrated military command provoked a considerable discussion on the need for an autonomous response to aggression.[4] In a speech to the French National Assembly on 17 June 1965, Premier Georges Pompidou declared that no nation could exist without an independent policy, its own defence and its own powers of decision. He went on to say that in defending France's independence, the French were defending that of Europe to which they belonged.[5]

Almost twenty years later, in the military planning law 1984–88 adopted by the Socialist majority in parliament, a similar reference to national independence is contained: a great nation is forever in the position of having to make decisions – military decisions must be made independently. The defence system must therefore be kept a purely national affair, and the domestic armaments industry must be able to meet the needs of the armed forces. Export and co-operation with other countries are necessary to keep the costs of the programmes at acceptable levels.[6]

The French armaments industry is characterised by its dependence on the public sector. This is either in the form of purely state institutions (army and navy armaments) or in partnership with the private sector (aeronautics). The private sector plays an important role in the area of advanced technology, but in consequence of the nationalisation programme which began in 1981, the state now controls over 75 per cent of the armaments industry.[7]

The issue of *credibility* is related to the concept of vital interests. The protection of vital interests is not only a matter of strict military logic; the continuous modernisation of the deterrent forces is necessary in order to ensure the government's liberty of decision and action in retaliating with the full power and flexibility necessary for all situations.

The concept of *sufficiency* is a strategic concept, not a tactical one. It is a dynamic concept closely related to NATO's flexible response strategy. France's strategy doctrine is determined by its possession of nuclear forces of sufficient potential to deter any aggression against the national sanctuary. Quantitative or qualitative modifications of the nuclear forces – the increase of nuclear warheads on sea-based nuclear missiles, for example – do not reflect a change in this doctrine, but an extension of the concept of vital interest. 'Vital interest' is not a geographical, but rather a political concept. The multiplication of warheads is linked with the threat of nuclear response any time France's vital interests are in danger. This must hold true even in the face of a 'shield' against nuclear missiles in the form of land- or space-based systems.

According to French deterrence doctrine, nuclear response must be massive and based on the 'countercity' strategy. In view of technological advances, France also possesses the capability of selected nuclear strikes against military, industrial and other vital objects of the aggressor to protect its own vital interests. This is not new. Since 1960 military policy has been based on the so-called 'weak to strong' strategy: flexibility in responding to various levels of aggression. It would be erroneous to assume that the modernisation of the French nuclear forces is a result of changes in strategic doctrine.

(b) The role of the ABM Treaty in the French nuclear doctrine

As the ABM Treaty was being ratified in 1972, France was carrying out the third Military Planning Law (1970–75). For the nuclear forces, this involved the creation of a sea-based component, the first nuclear submarine with SLBMs becoming operational in January 1972, and the initiation of a land-based, tactical nuclear programme.[8]

The ability of France to retaliate after suffering a first strike is guaranteed by the structure of its nuclear forces. Moreover, France has been conducting research and testing since 1972 in order to modernise its strategic nuclear weapons (MIRV, penetration aids, and so on).

The ABM Treaty is seen by France very positively as one of the few genuine arms control agreements. It enhances the credibility of both the American and the French deterrents.

Under the terms of the ABM Treaty research is allowed. There would be no means to control and prevent this. However, France holds the position that the ABM Treaty must be modified to prevent 'unfair' deployment of ABM systems.

Post-war French military doctrine has been greatly influenced by its Second World War experiences and especially by the failure of the static Maginot Line defence. A 'nuclear Maginot Line' has been rejected and instead a doctrine which emphasises the threat of extreme and decisive escalation has been formulated. This nuclear threat '*du faible au fort*' (weak-to-strong strategy) must be sufficient to deter all forms of aggression.

The ABM Treaty is thus not only beneficial to the two superpowers, but it is also very important to France as an independent nuclear power.

The first 'Defence White Book' was issued in 1972 under the then Minister of Defence Michel Debrè, one of General de Gaulle's oldest companions. It dealt with the modernisation of the nuclear forces and with their protection. Some of the points in the chapter on anti-air defence made in view of the ABM debate of the early 1970s are still relevant:[9]

1. France is interested in space satellites for military observation, but does not consider them indispensable. The deployment of aggressive space weaponry can only result in mass destruction.
2. France is against an ABM system based on the concepts that had been developed by the United States and by the USSR. Such a system could not be impermeable and could be realised only at a tremendous cost.
3. Aggression is made vain and improbable at a much lower cost by ensuring the survivability of second-strike, retaliatory forces.

III THE MODERNISATION OF FRENCH NUCLEAR FORCES AND THE ROLE OF SPACE TECHNOLOGY IN FRENCH DEFENCE POLICY

Even if SDI does not actually threaten the basic principles of French nuclear doctrine, it makes it more difficult for France to maintain the conditions of its deterrent. SDI might well have a destabilising effect on the present relationship between the two superpowers.[10] The retaliatory forces cannot remain efficient in the face of two conflicting strategies; the one based on ABM systems and the other based on the threat of 'going nuclear', signalling to an aggressor that vital interests will be protected by nuclear retaliation.

It is perhaps possible that the United States and the Soviet Union will be able to develop defensive space systems before the turn of the century. However, because of the inability of such a system to deal with short-range weapons, their respective allies would remain unprotected. Europe should rather devote its energies to developing systems suitable for their own protection. France, however, cannot find comfort in the dream of making nuclear weapons obsolete by deploying SDI.

(a) The effects of the technological aspects of SDI on French military policy

The belief that technological progress will overcome the threat of nuclear war is nothing new. The SDI programme can be linked with plans to increase the conventional deterrent by using new weapons technology. The principle is the same in either case: complete trust is placed in high technology in order to deal with a political problem. This point has been emphasised by the new French Defence Minister, Mr Paul Quiles. He has posed the question of whether technological advances will require a reassessment of strategic doctrine and armaments planning.[11]

As far as the technological challenge is concerned, the French answer is well known. Since the Hanover Summit in November 1985, the EUREKA programme is on a firm foundation.[12] It is the goal of this programme to mobilise and concentrate European energies by functioning as the framework of specific, well-defined projects. The main interest areas in the SDI programme and the EUREKA programme can be compared.

The SDI programme encompasses the following major space technology areas:

1. radar technology, especially image-sensitive space radar;
2. optical technology and image-sensitive lasers;
3. boost-phase and mid-course surveillance systems;
4. new weapons;
5. battle management and communications;
6. space power systems and related logistics;
7. robot technology.

The priorities of EUREKA are:

1. information science (more powerful and faster computers);
2. robot technology;
3. artificial intelligence;
4. communications (wideband communication systems and components);
5. biology (biotechnology and genetic engineering);
6. new materials.

It would not be correct to assume that these projects are competitive. However, in view of various political considerations, compatibility might prove difficult.

Scientific and technological collaboration will nevertheless become increasingly necessary in the future. Such co-operative efforts, when based on equality and mutual gain, can greatly serve the common interests of both sides of the Atlantic.

The EUREKA programme has been set up in accordance with the financial resources of Western Europe and some outer-European, neutral countries. The EUREKA projects are in the areas of industry where a strong impetus is needed.

Given its size and goals, the SDI programme can be realised only with the financial resources the United States can offer. Furthermore, many objectives of the SDI ideology will serve only United States national interests, not those of its European Allies.

(b) Strategic modernisation and the role of space technology

As far as the military challenge is concerned, Defence Minister Quiles has stressed that the Soviet response to the American initiative will be that the Soviets will greatly strengthen their own defence systems. He has pointed out that the result of the superpowers' increased efforts in the fields of their respective strategic defence programmes will be that the penetration capability of the French deterrent is put into question.[13]

At the end of 1972 the decision was taken to equip French nuclear submarines with a new generation of missiles by 1985. The sixth strategic nuclear submarine, the 'inflexible' SNLE (*Sous Marin Nucleaire Lanceur d'Engin*) was launched in 1985, and was the first to carry these new MIRVed missiles.

The main tasks in developing this missile were reducing the size of the warhead and increasing its penetration ability. Penetration is aided by hardening the warheads enough to assure that an ABM explosion would destroy at the most one warhead. A megaton ABM system could thus be saturated and exhausted, assuring penetration of the surviving warheads. The new M-4 missile will triple the efficiency of the French strategic sea-based component.

The modernisation programme has been accelerated in the Military Planning Law 1984–88, which was presented by the former Defence Minister, Mr Charles Hernu, to parliament in July 1985.[14] All the components of the strategic force have been included. Of special importance is the modernisation of four first-generation SNLEs by 1992. These submarines will be equipped with M-4 missiles carrying six warheads each.

In addition, the detection and identification avoidance capabilities of the submarines will be improved. There are a number of features of submarines which can be detected: optical discontinuity with its surroundings, noises, electromagnetic disturbances, heat, and so on.[15]

Funding for the development of a new SNLE generation has been appropriated in the 1986 military budget. This submarine will have a new nuclear boiler especially designed to avoid detection by enemy ASW systems. It will go into service by 1994 and be equipped with missiles carrying very small nuclear warheads – 'almost invisible' as the Defence Minister has said. the new warheads, despite their small size, will be able to survive megaton nuclear explosions as well as the extreme physical conditions of the re-entry phase. In the Ministry of Defence, providing the missiles with similar protection is being considered in order to assure their survivability against laser weapons.[16] In November 1985 it was also announced that a new weapons system would be developed which could survive SDI-type systems as soon as the first results of the SDI research programme were known. This would most likely be in the form of cruise missiles or the already planned SX system (land- or air-based mobile missiles).

The French defence policy concerning space and the ongoing strategic modernisation are closely related. France intends to continue its peaceful activities in space. This position was clearly expressed in August 1985 by the French delegation to the Conference on Disarmament in Geneva. The French government doubts whether it will be possible to ban and liquidate all types of space weapons.[17] The term 'space weapon' cannot even be clearly defined. The French delegation in Geneva has argued that an international ban on space weapons might not even be able to ensure the security of space objects. Furthermore, it will not be easy to define the criteria for distinguishing between passive and active military space activities. Consequently, only verifiable agreements have been proposed by the French delegation in Geneva. Their primary objective is a ban on ASAT systems.

In June 1984 the French proposed improving the system of registration of all objects launched into space. In this way it should be possible to agree on 'rules of the road' for different types of space orbital systems.

The Defence Minister has stated that the main space interests for France are communication and surveillance for crisis prevention and management.[18] Several steps have been taken in this direction:

1. 800 million Francs have been invested in the SPOT observation satellite programme;
2. the SYRACUSE I telecommunications satellite is already in service and is conducting surveillance over the Mediterranean, the Middle East, Africa and the Atlantic;
3. the more sophisticated SYRACUSE II will be put into service in 1992;
4. the first funding for the HELIOS military satellite has been appropriated in the 1986 military budget; it should be put into service by 1992;
5. finally, a study group has been set up within the defence department to analyse defence requirements in various areas such as ocean surveillance, electromagnetic detection and new satellite generations with radar and infrared systems.

IV A EUROPE OF TECHNOLOGY AND EUROPEAN SECURITY: THE EDI CONCEPT

In an interview with *Le Monde* Defence Minister Quiles has commented in connection with the so-called EDI that only a land-based system is conceivable. He added that a great number of units would be necessary to protect the civilian population: For Europe, only a system protecting military targets is feasible.[19]

However, Daniel Pichoud, engineer in charge of military space affairs, recently said at a seminar organised by the French School of Administration on French space and defence policies that Europe possesses the technological ability to undertake the most sophisticated military space activities.[20] The ARIANE V being developed within the framework of the ESA and the European shuttle programme HERMES will put Europe in the position to launch a military telecommunications satellite into a geostationary orbit.

General Pierre Gallois, influential in formulating French nuclear doctrine in the 1960s, has been very sceptical about the EUREKA project. In a book published in Autumn 1985 he wrote that France had again chosen to ignore developments in the concept of strategy by investing only in civilian- and commercially-oriented research. He supports a genuine European system of strategic defence. General Gallois believes it wrong to assume that the challenge posed to France by SDI can be responded to on a purely civilian level.[21]

It is true that the EUREKA programme is a purely civilian response to the technological challenge posed by SDI, but the military aspect has not been forgotten. As has been argued above, the military modernisation now under way is an adequate response to this military challenge. France's answer has been to separate these two aspects.

Defence Minister Quiles has described the areas of high technology which will prove important for the military:[22]

1. communication, telemetry and missile guidance applications of lasers;
2. computer science for command and control systems, naval systems, satellite systems, and so on;
3. space technology for the purposes of observation, communication and reconnaissance;
4. robot technology.

The R&D expenditures in the French military budget are being constantly increased. Quiles has claimed that France is second only to the United States among Western countries in devoting resources to defence research; the French are competitive in all the major fields of high technology, including those relevant to SDI research. In his view, France is the most advanced in Europe in the field of high-power lasers, both in their military and their metallurgic applications, and he considers his ministry to have been one of the major factors in bringing this about.

V CONCLUSION

For the past three decades space has been used for military purposes. Reconnaissance, early warning and communication satellites have been sent into orbit and ASAT systems, ballistic missiles and ABM systems have been tested. In addition, the trajectories of all modern ICBMs and IRBMs and their warheads enter space.

Many of these systems play a major role in ensuring peace. In particular, reconnaissance, early warning and communication satellites are of vital importance for arms control and crisis prevention.

It is the French position that there are many beneficial defensive uses of outer space. The military use of space can contribute to deterrence but should not be allowed to become a new arms-race dimension.

Former Research and Technology Minister Curien said at the WEU conference on space matters that participation in SDI would not give Europe the necessary degree of independence in military space matters; European R&D would be completely dictated by the Pentagon: 'It is first necessary to master your own defence system before you can determine your own defence and foreign policies.'[23]

Notes

1. Francois Mitterrand, Press Conference after the Economic Summit in Bonn, 4 May 1985, *Bulletin d'information du Quai d'Orsay* (088/85) 9 May 1985.
2. The same position was taken by the French delegation to the WEU Assembly during its December 1985 session.
3. See chapter 9 in this volume on EUREKA.

4. For more details see David Yost, *France's Deterrent Posture and Security in Europe. Part I: Capabilities and Doctrine, Adelphi Paper 194* (London: International Institute for Strategic Studies, 1984/1985); Yost, Part II: *Strategic and Arms Control Implications, Adelphi Paper 195* (London: International Institute for Strategic Studies, 1985).
5. Georges Pompidou, 'Fusion de certaines institutions européennes', 1st sitting, *Journal Officiel*, 17 June 1965, p. 2212.
6. *Annexe de la Loi de Programmation Militaire 1984–1988* (Paris: Ministry of Defence, June 1983).
7. See WEU, Standing Armaments Committee, *The Armaments Industry in the WEU member countries* (Paris: Western European Union, 1982).
8. See *Compte Rendu sur le programme d'équipement militaire*, presented by the government (Paris: Government Printing House, 1975).
9. See Document of the French Ministry for External Affairs, *SDI* position francaise (Paris: Ministry of External Affairs, 8 May 1985).
10. Paul Marie de la Gorce, 'L'Europe et la guerre des étoiles', *Herakles* 27, March/April 1985, p. 8.
11. Paul Quiles, address to the Institut des Hautes Études de Défense Nationale (Paris: Ministry of Defence, 12 November 1985).
12. See Document du Quai D'Orsay (220/85) 'Réunion de Hanovre 5–6 Novembre (Paris: Ministry of External Affairs, November 1985).
13. Quiles, address, 12 November.
14. *Rapport au Parlement sur l'éxécution et la réévaluation de la loi du 8 juillet 1983 portant approbation de la programmation militaire pour les années 1984–1988, Troisieme partie, La programmation des Forces 1986–1988* (Paris: Ministry of Defence, June 1985).
15. See Michel Cordelle, 'La détection magnétique des sous-marins, Acte du colloque', Science et Défense (Paris: Ministry of Defence 1983).
16. Quiles, address, 12 November.
17. Intervention de la délégation française devant le Comité ad hoc de la Conférence du desarmement sur la prévention d'une course aux armements dans l'espace extra-atmosphérique (Paris, 5 August 1985).
18. Paul Quiles, interview in *Le Monde*, 18 December 1985.
19. Quiles, interview 18 December.
20. 'La France et l'Europe peuvent accéder à léspace militaire', *Air et Cosmos*, 14 December 1985, pp. 64–5.
21. Pierre Gallois, *La guerre des cents secondes* (Paris: Fayard, 1985) pp. 102–3.
22. Paul Quiles, Allocution du Ministère de la Défense en cloture de la journéc nationale, Science et Défense (Paris, 5 December 1985).
23. Assembly of the WEU. *The Space Challenge for Europe Colloquy*, Munich 18–20 September 1985, Official Record (Paris: Office of the Clerk of the Assembly of WEU, 1985) pp. 28–31. For the most recent strategic debate in France, see Report of the National Assembly, document 432, 1st ord. sess., 6 November 1986: *Project de loi de programme d'equipement militaire pour les années 1986–1991*, presented by the Government (Paris: Government Printing House, November 1986) pp. 14–16.

12 The Geostrategic Risks of SDI
Peter A. Wilson

A new phase in the protracted process of negotiating constraints on nuclear arms began with the joint announcement by the United States and the Soviet Union that their delegations would meet in Geneva in March 1985 to begin the 'umbrella' talks. Unlike the SALT and START/INF negotiations, these talks include space-based offensive and defensive weapons as explicitly part of the agenda. They consist of three separate negotiation groups which include the elements of START, INF and space armament. Although the negotiations are not *de jure* merged, the Soviets clearly during the Reykjavik summit signalled that implicit and explicit trade-offs between the negotiations will have to be made. In particular, they have argued that 'satisfactory' results in the negotiations on space arms will have a significant impact on the two negotiations dealing with nuclear offensive armament – START and INF. The stage has now been set for the inclusion of President Reagan's Strategic Defence Initiative (SDI) in what has become the most complex arms negotiations heretofore conducted by the two superpowers.

Since President Reagan made his March 1983 'Star Wars' speech in which he declared the ultimate objective of making nuclear-armed ballistic missiles 'impotent and obsolete', the issue of strategic defence has dominated the arms control scene. Proponents of BMD, including the Administration, have argued that the current dependence upon the deterrent effect of an offensive-dominated nuclear military environment is fraught with the danger of failure and may be considered in the final analysis 'immoral'. It is striking that the President has strongly suggested in a national magazine that the current posture of mutual vulnerability is of dubious moral standing.[1] He has strongly advocated that the United States transform the current nuclear balance with the Soviet Union currently dominated by the overwhelming superiority of offensive nuclear weapons. In turn his critics, both domestic and foreign, have been quite harsh in opposing the SDI. In essence, most of the critics within the National

Security Community have focused on the technical unworkability of any strategic defence concept which provides very high levels of protection for the population and industry of the United States from a massive ballistic missile attack.

Spokespersons for the Administration reply that the SDI is but a major R&D effort to lay the groundwork for a policy decision to be made at some future date as to whether the United States should press ahead with a major BMD programme with the President's ultimate goal in mind. On the other hand, some advocates of BMD believe that an intermediate goal of protecting selected military forces such as the United States fixed land-based ICBMs is more immediately desirable and feasible. A recent example of this position is the article, 'Defense in Space is Not "Star Wars"' by Zbigniew Brzezinski, Robert Jastrow, and Max M. Kampleman.[2] They call for the early by the mid-1990s deployment of a limited ground- and space-based defence specifically designed to protect United States land-based ICBMs from a Soviet countermilitary attack with its current generation of ICBMs. This type of advocacy suggests that pressure may build for a rather early decision to press ahead with the deployment of a limited BMD with the intention of creating a more complete defence at a future date.

Critics of SDI have attacked the idea of a limited defence by arguing that the system can be overwhelmed with countermeasures or a further expansion of Soviet offensive forces. Proponents in and out of the Administration reply that the problem of a Soviet offensive 'reaction' can be managed through the arms control process. In fact, President Reagan has clearly stated that under certain circumstances the United States would be willing to share with the Soviets the fruits of its new BMD technology. Further, the Administration has argued that both the Soviet Union and the United States have a mutual interest in designing an arms control agreement which allows both sides to deploy BMD systems in a fashion which does not destabilise the current deterrent system based on offensive nuclear weapons.

Superficially, the idea of a grand strategy of nuclear 'co-operation' between the two superpowers appears very attractive. If appropriately packaged this concept might have great domestic political appeal. Unfortunately, there are major security concerns which must be considered before the United States even hints that it is considering a revision if not revocation of the ABM Treaty in collusion with the Soviet Union.

I A SOVIET INTEREST IN SDI?

Currently, the Soviet public position on the subject of the SDI is one of hostility. Like many American critics, the Soviets have argued the case that a high-confidence population defence is technically not feasible. Further, some spokesmen have threatened that the Soviets would merely overwhelm any United States BMD with a major expansion of their offensive forces. The reason for this uniform hostility usually given by Western observers is the Soviets' fear of the severe strain that a space weapon race would place on their military industrial complex and overall economy. The Soviets do face a deteriorating 'correlation of forces' with their military facing major current and future forces posture requirements. The simplest explanation may thus be the most powerful – that the Soviet leadership has concluded that a competition in advanced space weapons will lead to an overall deterioration of their military position *vis-à-vis* the United States. Yet there may be circumstances rather near at hand which could lead to a Soviet interest in a major revision of the ABM Treaty with objectives not in the interest of the United States.

From a geostrategic perspective, the Soviet Union faces a very different nuclear threat than the United States, which includes three nuclear-armed potential opponents in Eurasia: the United Kingdom, France and the People's Republic of China (PRC). Since the Nixon Administration initiated a rapproachment with the PRC, the United States has viewed the Soviet Union as the only nuclear armed power of concern. Now China is considered as an important military counterweight to the Soviet Union in Asia.

All three of these third-party nuclear forces are undergoing major expansion and modernisation programmes. The most visible are the programmes of the United Kingdom and France.[3] Recent public calculations indicate that either force could cause great damage to European Russia. These analyses indicate that if the French and United Kingdom forces were used in a co-ordinated retaliatory strike, Soviet industrial assets would suffer damage close to the assured destruction criteria set down by former Secretary of Defense McNamara some twenty years ago.[4]

Although less noticed by public commentary, the PRC's nuclear forces have been and are likely to be expanded and modernised. Of note is the launching of a multiple satellite, a precursor technology for MIRV, by their large space booster and their 1982 test of a solid propellent SLBM. Although the Chinese leadership has clearly signalled that the modernisation of their military establishment is subordinated to the demands of agricultural and industrial modernisation, they have not precluded substantial investments in nuclear weapon programmes. More specifically, Minister of Defence, Zhang Aiping, made an important policy statement in March 1983 which stated that Chinese military's modernisation priorities would focus on nuclear weapons and their delivery systems. This is precisely the priority followed by the Soviet Union during the Khrushchev era and the French during the buildup of their nuclear forces. The Soviet Union will thus face three nuclear forces which may well have the independent capability of carrying out a retaliatory strike which can cripple European Russia. These potential threats will probably mature by the early 1990s.

In the face of these third-party threats, both actual and potential, the Soviets have tried to enlist the acquiescence of the United States to some form of compensation during the nuclear arms negotiations which began with SALT I. In the past the Soviets have argued that the United States should compensate the Soviet Union for the presence of the United Kingdom and French nuclear forces by allowing for asymmetric limits on both superpower strategic and/or intermediate offensive systems which favour the Soviet Union. An example was the position that the Soviets adopted in the INF negotiations on December 1983. At that time they claimed that the United States should allow the Soviets the right to deploy substantial theatre nuclear forces to offset United Kingdom and French nuclear forces while precluding any United States deployment of Pershing II and GLCM. The United States throughout the nuclear arms negotiating process has held to the position that both superpowers' offensive forces should be placed under numerically equal limits.[5]

Might not the Soviets now find a different mechanism to gain 'compensation' for these lesser nuclear threats?

Under pressure to produce operational results, the United States SDI community could press for a modification of the ABM Treaty. A United States intiative could be designed to modify the ABM Treaty to allow for more ambitious R&D tests and/or the deployment of a defence of United States ICBM sites. The Soviets could seize upon this initiative as an opportunity to satisfy several political–military goals. First, a modified ABM Treaty might allow the Soviets to deploy a rather robust ground-based BMD to protect European Russia against third-party nuclear attacks. The Soviets could thus gain compensation by deploying BMD systems which have an asymmetric geostrategic effect. Currently, the Soviets have successfully developed a nuclear-armed ABM system similar in performance to the Sentinel/Safeguard missile defence which the Nixon Administration negotiated away in SALT. Of note is the rationale given by former Secretary of Defense McNamara for the Sentinel ABM system in 1967. It was advocated as a means to protect the United States from a threat of a Chinese nuclear attack. Second – with or without a successful renegotiation of the ABM Treaty – a broader, if not more significant, political–military goal can be served. This is the powerful disruptive potential that an ABM Treaty renegotiation would have between the United States and its two nuclear-armed allies and China.

A central objective of Soviet foreign policy is to disrupt – if not destroy – the political–military coalitions that the United States has created in Eurasia to contain the Soviet Empire. In Europe, the Soviet Union has worked tirelessly to bring about a reduction of the United States political–military commitment to its Allies without stimulating the creation of a powerful united Western Europe, one above all dominated by Germany. This objective has been pursued with varying degrees of vigour since the mid-1950s. More recently, since the early 1970s, the Soviets have had to face the spectre of increasing political–military collaboration between China and the United States. By advancing the option of revising the ABM Treaty in the name of revolutionising the relationship between the nuclear offence and defence, or for the less grand goal of protecting a fraction of its strategic deterrent, the United States provides the Soviet Union with a diplomatic weapon of considerable potential.

Nothing would inflame more a virulent form of Gaullism – whether of a French or Chinese type, than if London, Paris and Beijing were informed that the United States was prepared to collude with the Soviet Union in fundamentally altering the geostrategic environment. If the Soviets were permitted under a revised ABM Treaty a wide-area BMD to cover European Russia, then the deterrent power of the nuclear third parties could well be substantially compromised.[6] Even if the actual operational capability of the permitted BMD under the new ABM Treaty regime were less than satisfactory, the overall geostrategic effect would be quite favourable for the Soviets.

No doubt the Soviets would press during the course of a renegotiation of the Treaty more restrictive non-transfer agreements to inhibit the United States from providing defence penetration technologies to the nuclear-armed third parties.

Given these risks of renegotiating the ABM Treaty, the United States should move with great care at the umbrella talks in Geneva. What is at stake is more than a change in the rules of the road for the nuclear arms competition. The United States may now be on the verge of benefitting from a profound and favourable change in the geostrategic balance – that is, the 'correlation of forces'. A strategic mismanagement of the SDI by this Administration could vitiate some of these positive trends.

II A FAVOURABLE GEOSTRATEGIC SHIFT?

From the perspective of the evolving correlation of forces since the Second World War, we may be witnessing a period in which the Soviet position has and will undergo a serious deterioration. Unlike many Western national security commentators who overemphasise the narrow military aspect of the East–West balance, the Soviets true to their Marxist–Leninist ideological tradition have considered the state of the international struggle in a broader context. More specifically they have given great weight to the overall aspect of the political–ideological, social and economic factors which are a part of the correlation of forces. From the perspective of the Moscow leadership, the global environment has undergone a serious turn for the worse since the late 1970s. To make this point, several major changes in the geostrategic balance can be highlighted.

(a) The decline of Marxist–Leninism

On 7 December 1984 the Chinese Communist Party released an authoritative statement which rejected the universal and historically continuous validity of Marxism as an economic development model. This statement is a reflection of the overall global decline of belief in the Soviet ideological model as the 'wave of the future'. Many other examples can be cited such as the rise and decline of Eurocommunism, the intellectual shift within France away from Marxism, the rise of Solidarity in Poland, and the fading support for the Soviet model in the less-developed world. Compared to the sense of momentum that was symbolized by Khrushchev's public ideological optimism twenty-five years ago, the current ideological power of Marxist–Leninism seems positively decrepit. This is not to gainsay that Marxist–Leninism ideology does not have force in selected areas of the Third World such as parts of Latin America, where there are great political, social and economic inequities. Overall, from the Soviet perspective, the idea of a Communist International has become a distant historic memory. Only in Eastern Europe can the Soviets impose a degree of ideological conformity, but this has been clearly demonstrated as a conformity derived by the threat of force and not conviction.

Central to this decline has been the recent deterioration of Soviet economic performance. For approximately three decades after the Second World War Soviet economic performance was quite respectable when compared to the industrial democracies. Since the mid-1970s that performance has been poor, and is the economic aspect of the correlation of forces which must be one of greatest concern for the Soviet leadership.

(b) The decline of the Soviet economy and the resurgence of the 'West'

Since the mid-1970s the Soviet economy has become at best a mediocre performer with growth rates declining from 4 per cent per annum to approximately 2.5 per cent per annum during this decade.[7] The reasons for this historic slowdown have been well documented elsewhere. It is sufficient to note that the reasons for this decline are multiple and negatively reinforcing. Such factors as the decline in the birth rate, the reduced availability of cheap resources in European Russia and declining return on capital investment have been cited as causes for this slowdown. Furthermore, the social environment has shown a serious deterioration with the rise of infant–adult male mortality, alcoholism and corruption. Evidence accumulates that the Soviet labour force is operating on the philosophy that 'they pretend to pay us, so we pretend to work'. Western observers have argued whether the Soviet leadership is capable of reaching effectively to this slow motion crisis. Currently, the consensus of these observers is that the economic difficulties of the Soviet Union are not sufficiently stark for any likely leadership in Moscow to run the risks of losing political and economic control in the name of radical economic reform. In essence, the Stalinist model for economic growth and political control was too successful for too long.

If the Soviet leadership does not choose major reform, then they face the prospect that the economic power of the Soviet Union will decline relative to the United States, Japan and Germany for a period which could well reach to the turn of the century. Simultaneously with the Soviet economy deteriorating, the United States is undergoing a major economic transformation which could well lead to a period of sustained growth. Central to this transformation is that the economically debilitating effect of the multiple petroleum crises is over. The industrial democracies may be entering a period of moderate growth with low inflation stimulated by the collapse of OPEC's power to dictate petroleum prices.

For the United States a robust economic future is somewhat uncertain. Much will depend upon the nature of the political compromise between the President and the Congress over the continuing fiscal crisis. The pessimistic Washington observer will conclude that nothing much will be done until the economic crisis is upon us. A more sanguine view is that a compromise on spending and taxes is possible. Although the jury is still out, there is a reasonable chance that a sufficiently robust fiscal compromise is possible which would allow the United States economy to continue through the decade on a course of moderate growth with modest inflation.

From the perspective of the Soviet leadership, the overall economic performance of the United States is likely to remain impressive, even if the budgetary impasse is not completely resolved. Of great worry to Moscow is the intense technoeconomic competition raging between the United States and Japan. Both countries are in a sustained and diverse struggle over the development of advanced technologies for industrial and commercial purposes. In this technology race, only selected elements of European science and industry are in the competition. The Soviets are not even in the game. Some of the more potentially dramatic advances are likely to take place as a result of the commercial exploitation of new microelectronic, biotechnology, materials and communications research. Of particular note is the competition between Japan and the United States for supremacy in the fifth-generation computer technlogy. Most ominous is the transformation of the United States, European and Japanese education, industrial and service sectors with the mass diffusion of microcomputers and their associated communications infrastructure. As vividly reported elsewhere, the Soviet system of political and social control precludes the free-flowing diffusion of this technology through Soviet society.

Aside from prospect of falling behind the industrial democracies, the Soviets now face the prospect that some of the major new industrial countries such as Brazil, India, South Korea, and China may rapidly outpace Soviet growth rates for the rest of the century. The invidious comparison that the rest of the world will make between the Soviet Union and these 'lesser' economic powers is bound to create further political difficulties for Moscow. Most dramatic is the possibility that their 'main enemy' (the Chinese) may be very successful with their 'socialism with Chinese characteristics' and launch a period of sustained agricultural and industrial growth.[8] Already the performance of their decollectivised farming system has been quite dramatic.[9] Might not the Chinese now offer many third world 'socialist' countries a new model which further undermines the relevance of the Soviet economic model?

Aside from the change which appears to be under way in the non-military aspects of the correlation of forces, the narrower military aspect has and will undergo a deterioration from the Soviet perspective.

III THE CHANGING MILITARY ENVIRONMENT

Since the late 1970s, the United States has been making a historically unprecedented peacetime increase in military investment. Since 1979, the defence budget has been rising in real terms on an average of approximately 6 per cent per annum. This investment growth rate which appears to have been outpacing the Soviets during this period is leading to an across-the-board modernisation of all aspects of United States military capability.[10]

Although this investment has not been the most efficient or the product of a coherent grand strategy, it is (or will be) producing substantial results. From the Soviet perspective, the most worrying is the modernisation of the United States strategic and theatre nuclear strike forces with their associated C^3I upgrades. In particular, the United States is likely to deploy by the early 1990s a robust capability to attack the full range of Soviet hard targets. This capability will be reinforced with a large improvement in manned and missile aerodynamic strike forces. These, in the face of perceived inadequacies of Soviet air surveillance and defense capabilities, may in Soviet eyes further increase the United States potential for pre-emptive nuclear warfare. In conjunction with these offensive forces, the Soviets may well view the United States improvement in global surveillance capabilities as most ominous. By means of these surveillance systems, the United States may gain an improved capacity to acquire and attack mobile and imprecisely-located targets. These targets can include the new generation of Soviet land-mobile ICBMs and submersibles.

A second worry, since the May 1984 *Red Star* interview of the former Chief of Staff Ogarkov is valid, will focus on United States attempts substantially to modernise its theatre warfighting potential. Unlike many sceptics in the United States and Europe, the Soviets seem quite impressed with the potential that new technologies can bring to bear to the threatre war equation. Certainly United States standing forces in Europe are undergoing a major modernisation with the introduction of new tactical aircraft, battle tanks, infantry fighting vehicles, rocket artillery, sensors and C^3.

These forward-deployed assets will now be backed by reinforcement capabilities which are in the process of major expansion which includes the buildup of the United States sea and airlift potential along with expanded equipment pre-positioning programmes. Further strengthening this theatre reinforcement potential is a rapidly modernising navy and marine corps.

Obscured by the recent political controversy in Europe over the deployment of the Pershing II and GLCM theatre missiles is the simple fact that the overall state of the military component of the Atlantic Alliance has undergone a major renovation. In particular there has developed a *de facto* integration of the French military within the NATO command structure. Although the French have continued to give emphasis to their nuclear force modernisation, their conventional theatre forces have been reconfigured to participate in any early conventional defence of Germany.[11] As indicated, the French nuclear programme is about to undergo a major increase in capability with the deployment of MIRVed SLBMs.

Simultaneously the United Kingdom continues to modernise its conventional forces while pressing ahead with its Trident II SLBM programme. Although there are indications that the United Kingdom defence budget will be constrained for the rest of the decade, the outlook may not be so gloomy if the United Kingdom and continental Europe undergo a period of sustained growth with low inflation. Clearly improved economic prospects will allow the major European contributor to NATO's conventional power the Federal Republic of Germany to sustain its modernisation programme.[12]

In contrast to the relative health of the military structure of the Atlantic Alliance, the Warsaw Pact on its thirtieth anniversary is facing sustained difficulties. The ongoing political and economic crisis in Poland has severely vitiated the overall military effectiveness of the

Pact. The Soviet General Staff must have grave doubts about the reliability of the Polish military establishment. Although the Solidarity Union has been suppressed, the ideals it represents have not.

No doubt the Soviet political and military leadership can count on some elements of the East European military establishment during a hypothetical war with NATO. Yet, these are likely to be a relatively small number of elite units which, like the Polish Zomo, will continue to fight for the preservation of their communist regimes. From the Soviet perspective this means that the overwhelming bulk of the East European armies can at best be expected to perform defensive tasks during a NATO–Warsaw Pact war. Many may refuse to participate at all or may go so far as to actively support NATO in the event of open hostilities. The static measures of the NATO–Warsaw Pact balance which so worries many commentators may thus give the Soviet military leadership little comfort.[13]

To the south, Moscow faces a greater Middle East area where American influence remains strong. Although it suffered a loss of prestige with its failed intervention policy in Lebanon, the United States' political military position remains substantial. The United States continues its active military supply relationship with Egypt and Saudi Arabia. In the latter case, the United States operates a forward deployed presence with its AWACS aircraft which continue to monitor the Iran/Iraq war. Infrastructure construction continues in Saudi Arabia and Oman which would facilitate any rapid deployment of air and ground forces to support a USCENTCOM contingency. All of this has continued even with an expanded (albeit somewhat tempestuous) security relationship with Israel.

In contrast, the Soviets can be satisfied only with their increased military presence in Syria and the deployment of advance air defence systems. Even the stability of that position is endangered by the uncertain health of Hafez Assad and the likely succession struggle that will follow his death. The Iran–Iraq war continues as a bloody stalemate with the Iraqis expanding their military sources of supply from such diverse states as France, Egypt, Brazil and China. The latter's new entry into the international arms market with the sale of Soviet-designed hardware must be worrying for Moscow. As for Iran, the scenario of a post-Khomeini disintegration of the government seems less and less plausible as the Shiite revolutionary regime consolidates its domestic grip by physically destroying all forms of opposition. This has included the regime's efforts to crush the Tudeh Party.

In South Asia, the occupation of Afghanistan remains a stalemate. Now the Soviets face the prospect that India under the leadership of Rajiv Gandhi may tilt more to the West. Indira's son and his young cadre of advisors show little interest in the ideology of socialism; rather, they seem fascinated with the prospect of exploiting new technologies and the dynamism of private entrepreneurship. For all of their sustained effort to build a political–military relationship with India, the Soviets must have uncertain confidence of its strength and endurance.

In East Asia, the long-term security problem for the Soviets has been aggravated by the prospect of China undergoing a fundamental economic transformation resulting in a much more effective competitor for the Soviets in the 1990s. As noted, Chinese reforms in agriculture may become the model for many Third World countries which are moving away from Marxist–Leninism without totally rejecting the ideology of 'socialism'.

In Southeast Asia, the Socialist Republic of Vietnam remains a militarily useful but economically expensive ally. As in the case of Cuba, the Soviets have a bastion which gives their military an expanded global reach. Yet the price of this both in economic and political terms may well continue to rise. Vietnam is bogged down in Cambodia, is a source of major concern to its ASEAN neighbours and is another element on the negative agenda between China and the Soviet Union. Similar to their expanded military presence in Mongolia, the Soviets have gained a regional military advantage *vis-à-vis* China. Yet this military presence guarantees the long-term character of the rivalry between both powers.

IV A NUCLEAR DETERRENT GRIDLOCK?

From the prospect of the Soviet military, all of these Eurasian security demands will likely grow simultaneously with the United States military buildup. By the early 1990s the Soviet General Staff will face the prospect of wartime contingency planning requirements which involve a global conventional/nuclear conflict with the United States while facing three other nuclear armed states. A form of nuclear deterrent gridlock will be in place. Although the Chinese nuclear forces will remain technically more primitive than either of the superpowers', its deterrent potential should not be underestimated. If cleverly deployed in concealed locations, this relatively small force might prove very difficult to disarm by a pre-emptive attack by either superpower. The leverage of this force will loom even larger in the context of any Soviet contingency plan involving war with the United States. In the event of a NATO–Warsaw Pact war, the Soviet General Staff will have a central question to resolve: What should be done about China's military potential? In the event of a global nuclear war with the United States, the surviving Soviet political–military establishment, after suffering grave nuclear damage to its strategic and regional military forces, would face the prospect of dealing with undamaged Chinese nuclear and conventional forces in the Far East. The Soviet leadership would face the prospect of a 'stab in the back' even if China had remained studiously neutral during a United States–Soviet nuclear conflict.[14]

In Europe, the Soviet military face an analogous planning dilemma. In the event of a major conventional campaign in Europe, the Soviet leadership will have little confidence of being able to forecast accurately when that campaign would shift to nuclear warfare. In that theatre, they face three independent nuclear-release decision authorities. By the early 1990s, the French will have the option of launching a large (e.g. several hundred weapon) theatre-wide strike against the Red Army which could still be configured for conventional operations. They could make this decision unilaterally before a more detectable collective decision to do so was made by the NATO political/military authorities.

These uncertainties about escalation control and plausible war outcomes must loom larger and larger as the decade progresses. This suggests that whatever confidence the Soviet military has – and this seems modest – in solving its warfighting requirements *vis-à-vis* the United States, the presence of these increasingly capable third nuclear forces further undermines any plausible strategy that leads to a war outcome that can be described remotely as 'victory'.[15]

V WHAT OF SDI IN THIS CONTEXT?

It is in the context of this favourably shifting correlation of forces that a coherent policy toward BMD should be devised. To press ahead with the SDI with hope of an early deployment runs serious risks to US national security objectives and may rise substantially if the United States attempts to renegotiate the ABM Treaty in collusion with the Soviet Union. Some of the likely adverse consequences are as follows:

1. A disruption of the Atlantic Alliance – a United States decision to renegotiate the ABM Treaty will strike at the heart of the rationale of the French and British independent nuclear deterrents. The United Kingdom might acquiesce with the promise of BMD countermeasures which the Soviets will try to block. For France, the reaction might be very bad with the revival of a virulent form of anti-American Gaullism. The current *de facto* acceptance of NATO's doctrine of flexible response by France's political–military elites would collapse.
2. Damage relations with China – evidence of superpower collusion would ignite an anti-American reaction within the Chinese political–military elites for reasons similar to an adverse French response.
3. Undercut the rationale for nuclear force modernisation – if the SDI grew substantially, especially if stimulated by an early deployment decision, then there would be budgetary pressures to divert resources to this enterprise. Some of the Administration's nuclear force programme could be at risk. The pressure to cut offensive programmes could be intensified if the Administration persists in characterising the current deterrent posture based upon offensive weapons as 'immoral'.[16]
4. Facilitate Soviet BMD breakout – left unstated by advocates of a limited space-based BMD is how a modified ABM Treaty would deal with the risk of a Soviet unilateral decision to deploy a much larger system. By allowing for a limited deployment, the United States facilitates the Soviet development and deployment of a BMD mobilisation base for a complete breakout of the ABM Treaty.

VI SDI AND ARMS CONTROL

United States arms control strategy should be consistent with United States broad national security objectives. As indicated in this essay, the United States should appreciate the favourable shift in the correlation of forces with the Soviet Union. To paraphrase Mao Zedong, we should respect the Soviets' military power but despise their long-term economic and political power. This strongly suggests several basic political–military objectives for the United States. First, the United States should continue to encourage the maintenance of the military stalemate in Eurasia. Second, the United States should encourage the continued development of a more pluralistic global society which has and will become more resistant to Soviet imperial designs.

In this broad context, what should be the objectives of the Geneva negotiations? The first is the resolution of the issue of the SDI. Clearly, the President's proposal was of value, if only to lever the Soviets back to the arms negotiations. Now the United States will have to consider whether it is in the United States geostrategic interest to allow both superpowers the right to deploy BMD beyond those permitted by the ABM Treaty. The answer is very likely to be negative; therefore, the status quo of nuclear balance dominated by offensive weapons will have to be preserved. Thus, the United States should continue the ABM Treaty. In fact, the language might well be tightened to make it more difficult for the Soviets to use their new generation of SAMs as part of a BMD breakout scenario.

Second, a multiyear ban on all space-based weapons should be a part of a future Treaty. A ban could be monitored in space by techniques already demonstrated by some of the more recent satellite recovery missions of the Space Shuttle. Further, serious consideration should be given to the option that such a Treaty would allow both superpowers the right to deploy a land-based ASAT to enforce any space weapon ban. This provides both superpowers additional insurance against a space weapon breakout scenario.[17]

Both sides would continue considerable R&D efforts in space weapon-related technologies. Much of this activity could be concealed and therefore not made credibly verifiable as part of a restraint regime. Furthermore, it is clearly not in our interest to constrain many elements of the R&D related to space weapons. After all, much of the new technology in space-based sensors is relevant toward improving United States space-based global warning capabilities.

As for offensive weapons in the START/INF baskets of these talks, the focus should be on gaining gross aggregate reductions on the payload potential of both sides' missile and bomber forces.[18] Directly or indirectly these limits would combine the forces in the separate START and INF categories. Our primary concern should be to reduce the overall first-salvo firepower potential of the Soviet missile forces. Instead of worrying about the threat of accurate MIRVed weapons to our land-based ICBM forces, we should be more concerned about a future Soviet capability to place our mobile strategic weapons at risk. If unconstrained by arms control, the Soviets might have sufficient missile firepower to carry out barrage attacks against our mobile assets. Such high firepower attacks may become operationally feasible by the mid-1990s with the development and deployment of new space-based target acquisition sensors.[19]

It will not be in the United States interest to constrain the development and deployment of these sensors for the same reasons cited above. Such sensor innovation may be impossible to monitor and much of this technology is critical for the development of a new generation of strategic–tactical warning sensors. The countermilitary threat to imprecisely-located systems should thus be constrained indirectly by placing limits on the nuclear barrage potential of both superpowers.

No doubt the Soviets will strongly resist any calls for substantial reductions in their strategic and theatre nuclear delivery forces so long as the issue of nuclear armed third parties is finessed by the United States. This suggests that a linkage between deep START/INF reductions and third-party constraints can be designed to serve United States and Allied interests. One possibility is to place upper bounds on third-party nuclear force modernisation programme in conjunction with substantial superpower reductions, for example, START/INF aggregates of approximately 1,500 launchers/vehicles. In a fashion analogous to the Washington Naval Agreement, limits could be placed on third-party forces in terms of a ratio of their individual forces to a superpower limit which would be at numerical parity.

Conducting the Geneva talks as a multinational effort would be a strategy designed for stalemate and failure. Rather, it might be possible to design a United States–Soviet reduction agreement as a multistage process. For example, the United States and the Soviet Union could agree to a combined START/INF aggregate agreement of 1,800 launchers/vehicles without gaining consent of constraint by any of the third parties. Further reductions to aggregates in the 1,500 range would require the acceptance by the United Kingdom and France of a constraint regime. The politically more difficult task of including the Chinese forces in an agreement could wait for a third reduction phase below the 1,500 aggregate regime.

VII CONCLUSION

A multinational nuclear arms agreement may seem to many observers as positively utopian. Yet an approach to nuclear arms control and strategic defences as suggested by this chapter seems better designed to serve the United States broad geostrategic interest of supporting a global evolution toward increased pluralism while containing Soviet military power. On the other hand, the idea of rendering nuclear weapons 'impotent and obsolete' in collusion with the Soviet Empire seems worse than utopian, it is fraught with substantial peril.

Notes

1. 'The Alternative Is So Terrible', *Time*, 28 January 1985, p. 29.
2. Zbigniew Brzezinski, Robert Jastrow, and Max M. Kampleman, 'Defense in Space is Not "Star Wars"', *New York Times Magazine*, 27 January 1985. Max Kampleman is the head of the United States negotiating teams in Geneva.
3. See chapter 10 by Paul Stares and chapter 11 by Alain Carton in this volume.
4. In an article published in September 1986 in *Scientific American*, 'British and French Nuclear Forces', John Prados, Joel Wit and Michael Zagurek, Jr examine the mid-1990s retaliation potential of the United Kingdom and French nuclear forces.
5. This was a central issue that was resolved only at Vladivostok in 1974. The Soviets agreed to equal strategic delivery vehicle/launcher aggregates, albeit at a high level of 2,400. Equal aggregates were successfully negotiated by the United States in SALT II. However, the Soviets did signal that further reductions would have to be 'compensated'.
6. Prados, Wit and Zagurek, 'British and French Nuclear Forces' explores the implications of an expanded 1990s Soviet BMD for the United Kingdom and French nuclear forces.
7. See Gertrude Schroeder 'The Slowdown in Soviet Industry', *Soviet Economy* (Winter 84–85).
8. Dwight Perkins wrote an essay which makes the case about China's capacity to 'catch up' with the Soviet economy, 'The Economic Background and Implications for China', in Herbert T. Ellison (ed.), *The Sino–Soviet Conflict – A Global Perspective* (Washington, DC: University of Washington Press, 1982).
9. See the two articles 'Peasants Rising', p. 11 and 'In Praise of Peasants', p. 86, *Economist*, 2 February 1985.

10. See Richard F. Kaufman, 'Causes of the Slowdown in Soviet Defense' *Soviet Economy* (Winter 1984–5) for an interesting discussion of the slowdown in Soviet military investment during the late 1970s and early 1980s.
11. For an excellent short article on the changing French defence posture and policy, see Giovanni de Briganti 'Independence Marks Fifth Republic's Approach' *Armed Forces Journal* (November 1984) which quotes French Defence Minister Charles Hernu to the effect that France's conventional forces are being reconfigured to provide the option of early theatre support to NATO's central front.
12. See Hans Günter Brauch, 'Official and Unofficial West German Alternatives for Reducing the Reliance of European Security on Nuclear Weapons – A Plea for a Pragmatic Approach', in Frank Barnaby and Terrence Hopmann (eds), *Rethinking the Nuclear Dilemma in Europe* (London: Macmillan, 1987).
13. The economic difficulties of the East European countries have had a notable effect on the rate of modernisation of their armed forces. For example, they have been very slow to replace their obsolete T-55 series tank force with the more modern T-72 series. Modernisation of their tactical air forces has also moved at a slow pace.
14. This deterrent interaction has already worked to the benefit of the Chinese. Uncertainties about a United States reaction to a Soviet attack on China have helped delay a decision by Moscow to go to war, an idea strongly considered during the 1969 border crisis. In the future, Soviet military planners will have to deal with the problem of deterring an undamaged United States and Europe following a major nuclear conflict with China.
15. Former Chief of Staff Nikoli Ogarkov, in a May 1984 interview in *Red Star*, stated. 'On the one hand, it would seem a process of steadily increasing potential for the nuclear powers to destroy the enemy is taking place, while on the other there is an equally steady and, I would say, even steeper reduction in the potential for an aggressor to inflict a so-called 'disarming strike' on his main enemy. The point is, with the quantity and diversity of nuclear missiles already achieved, it becomes impossible to destroy the enemy's systems with a single strike'. This statement clearly reflects a Soviet military view which suggests that something approximating 'saturation parity' exists between the two superpowers. 'Ogarkov Interview in Krasnaya Zvezda', in *FBIS Daily Report*, III (91) article 054, p. R14.
16. Former Secretary of Defense Schlesinger has criticised the Administration for calling into question the morality of the current offensive dominated deterrent posture. See E. Ulsamer, 'The Battle for SDI', *Air Force Magazine* (February 1985). Anxiety about the SDI programme taking resources away from other USAF programmes is seen in J. Canan, 'Protecting the Priorities', *Air Force Magazine* (February 1985).
17. As part of an arrangement to permit both superpowers an ASAT system, there may be an opportunity to allow both parties the right to deploy a *very thin* nationwide ABM system. A recent test by the United States Army of their infrared Homing Overlay Experiment (HOE)

vehicle suggests the feasibility of a dual ASAT–ABM system. A limited ASAT–ABM could be deployed in a small number of fixed sites to provide protection against that low-probability, but catastrophic event – an accidental launch of a nuclear-armed ballistic missile. One could imagine a revision of the ABM Treaty which allowed both superpowers the right to deploy a very limited (few tens) ABM–ASAT force. Over the longer run it may be possible to include third parties as they join a multinational nuclear arms agreement to provide them protection against this type of accident.

The value of such accident insurance would have to be weighed against the difficult negotiating task of designing an agreement which allowed an ABM deployment different from that permitted by the ABM Treaty, while protecting broader United States strategic interests, for example, a BMD breakout scenario.

18. A restraint on only missile launchers or aircraft may not be sufficient. It has been argued that a direct constraint on missiles would lead to a more effective agreement by limiting refires. On the other hand, a direct constraint on missile inventories will require intensive onsite inspection arrangements, a very difficult negotiation task. Also it is not clear that constraining the multiple-salvo potential of both sides is in United States best interests. The main United States future counter-force worry is the large *first* salvo potential of the Soviet missile force. See Glenn A. Kent, Randall J. DeValk, Edward L. Warner III *A New Approach to Arms Control* (Santa Monica: RAND/R-3140-FF/RC) for a discussion of this issue.

19. See Theodore H. Moran and Peter A. Wilson, 'A New Strategic Arms Agreement: Taking the Long View', *SAIS Review* (Winter 1984) for a more detailed discussion of the concept of the first salvo barrage attack as a counterforce threat of the 1990s.

Part IV
The ABM Treaty – From SDI to EDI

13 The ABM Treaty – its Evolution, Interpretation and Grey Areas, and an Official Attempt at Reinterpretation
John B. Rhinelander[1]

I INTRODUCTION

This chapter draws on the experience of the author as the legal advisor of the American SALT I delegation from 1971 to 1972. The official United States SALT I negotiating record is massive but still classified. The author has not had access to it since 1972 and he has never seen the official ACDA history of SALT I, which is also classified. This chapter is consistent with a chapter the author wrote in 1973 which *has until recently been the standard unclassified reference in government on the interpretation of the ABM Treaty*.[2] In analysing the most recent efforts of the Reagan Administration to reinterpret the ABM Treaty this chapter employs statements the author made before the Senate Foreign Relations Committee on 12 November 1985 and responses the author made on 11 January 1986 to written questions submitted to him by the House Foreign Affairs Committee.[3]

After a brief survey of the evolution of the ABM Treaty and its major components, this chapter will focus on the three remaining grey areas of this treaty, on ASAT, ATBM and radars, on official statements and interpretations from the ABM Treaty by successive American administrations from Nixon to Reagan, and on the most recent efforts of the Reagan Administration to reinterpret the ABM Treaty. The chapter will then turn to the Soviet attitude towards the Treaty and to their official reactions to the American debate provoked by the reinterpretation efforts. The chapter concludes with a few precise suggestions how the ABM Treaty may be strengthened, a demand that has been made by various West European governments in their reaction to the American SDI research programme.

II EVOLUTION OF THE ABM TREATY[4]

Concurrent with the debate over the merits of ABMs, the United States and USSR agreed to enter into negotiations seeking to limit their strategic nuclear forces. The process was delayed due to the Soviet invasion of Czechoslovakia in the summer of 1968, but negotiations finally began under the Nixon Administration in Helsinki, Finland, in November 1969.

In the initial sessions of the first Strategic Arms Limitation Talks (SALT I) negotiators undertook wide-ranging discussions about the interaction between offensive and defensive systems. Was it more important to limit defensive systems first, as the United States had originally proposed during the Johnson Administration, or offensive and defensive systems simultaneously, as urged by the Nixon Administration? Each side searched for a better understanding of how the other perceived the offence–defence relationship.

With the completion of the ABM Treaty and the Interim Agreement on offensive nuclear forces in May 1972, the superpowers implicitly acknowledged the overwhelming strategic reality: however seductive in theory, nationwide defences against missile attack were not feasible because the destructiveness of nuclear weapons gave offensive systems an insurmountable advantage.

In addition, the deployment of large-scale missile defences would lessen the stability of the strategic balance. A competition in building ABM systems would inevitably instigate an uncontrolled buildup in offensive nuclear forces, as each sought to ensure its ability to penetrate its opponent's defensive shield. Conversely, as later summarised in the Preamble to the ABM Treaty, 'The limitation of anti-ballistic missile systems . . . would contribute to the creation of more favourable conditions for further negotiations on limiting strategic arms'.[5]

It was recognised that such an unregulated offensive–defensive arms competition could also have adverse consequences on 'crisis stability', and therefore increase the risk of nuclear war. Crisis stability exists when each side is secure in the knowledge that it possesses the capacity to threaten the other side with devastating retaliation even if struck first with nuclear weapons. Both superpowers spend considerable funds and effort to maintain this 'second-strike' capability.

The projected cost of an unregulated offensive–defensive competition was also an inducement for controls. It was well understood that the dual pursuit of both nationwide ABM systems and new technologies for offensive systems to penetrate defences would be prohibitively expensive.

For all of the above reasons, the United States and the USSR decided at the SALT I negotiations to conclude an agreement of unlimited duration banning the large-scale deployment of ABM systems and placing strict limitations on the development of ABM capabilities. They also reached a five-year interim agreement limiting ICBM and SLBM launchers and agreed to continue negotiations toward greater curbs on offensive weapons. The SALT II Treaty (signed in June 1979 for a term through 31 December 1985 but never ratified) represented the first comprehensive agreement limiting strategic ballistic missile launchers and heavy bombers, and requiring some reductions in offensive systems.[6]

The SALT I and SALT II negotiations were premised on the assumption that limitations on strategic offensive forces would not be possible without extensive constraints on strategic defences. The collapse in November 1983 of the Strategic Arms Reduction Talks (START) and negotiations to reduce intermediate-range nuclear forces (INF),[7] coupled with renewed official interest in strategic defences, has resurrected the same fundamental issues that defined the strategic debate in the late 1960s, and that many believed had been resolved with the signing of the ABM Treaty in 1972.

This was reflected in the framework for arms control negotiations which resumed in Geneva in 1985.[8] In 1967, the United States proposed negotiations on ABM systems, while then Soviet Premier Kosygin resisted.[9] Now in 1986 this situation has reversed, with President Reagan arguing that defensive systems would be both 'moral' and 'stabilising' and the Soviet leaders countering that they would lead to a dangerous escalation of the arms race.[10]

(a) Personal recollections of the ABM Treaty negotiations

In April 1971, while I was serving in Washington as a Deputy Legal Advisor at the Department of State, Ambassador Gerard Smith asked me to come to the fourth negotiating session of SALT I in Vienna and prepare drafts of an ABM Treaty and an Interim Offensive Agreement. In March the Soviet delegation had tabled a draft ABM Treaty. The United States delegation believed it was appropariate to begin to prepare formal texts of agreements on defensive and offensive strategic weapons limitations. I spent most of that session reviewing the record of the negotiations, familiarising myself with United States position papers and the technical characteristics of the weapons and in discussions with members of the United States delegation. I also prepared rough first drafts of texts. The negotiating session ended in May 1971, shortly after the '20 May understanding' between the United States and Soviets was announced.

During the remainder of May and in June in Washington I prepared successive drafts of an ABM Treaty and in Interim Agreement after input from others. I recollect that the Soviet draft of Article III was permissive – it stated both sides may deploy a single fixed land-based ABM system with 100 ABM launchers and no limits on ABM radars within the deployment areas. This text was vague, imprecise and, among other things, an invitation to pursue and deploy both stand-alone components, such as long lead-time ABM radars and ABM systems based on 'exotic' technologies. In my drafts I turned Article III around into a form eventually agreed upon and also tightened it.

Article III, as drafted, prohibited deployment of any ABM system or components *except* those in the deployment areas and as limited quantitatively, qualitatively and geographically. The text of Article III, standing alone, prohibited the *deployment* of fixed land-based 'exotic' ABM systems and components because only systems utilising ABM launchers, ABM missiles and ABM radars could be deployed. This raised the 'exotic systems' question directly for inter-agency consideration.

The other substantive Articles always referred to 'ABM systems' and to 'components' to make clear the United States position that components were limited and not just entire systems.

The Soviet draft of March 1971 contained prohibitions on testing and deployment of 'space-based' in what is now Article V(1), as did my drafts of May–June 1971 which, I believe, added 'develop'. The gist of this article was derived from the 4 August 1970 proposal by the United States for bans on production, testing and deployment of all mobile-type ABM systems.

The drafts of May–June 1971 were reviewed by members of the SALT delegation while in Washington. Some of them had sharply differing views on 'exotic systems' and other questions.

The fifth negotiating session began in Helsinki in early July 1971. After taking into account the President's written instructions, the delegation revised my drafts, cabled texts of an ABM Treaty and an Interim Agreement to Washington and sought authorisation to table them in a plenary session. On the 'exotic systems' questions the delegation was split. Gerard Smith[11] wrote that he and Harold Brown supported a broad ban; Paul Nitze concurred except for sensors; but General Allison and Ambassador Parsons favoured no restraints at all on 'exotic systems'.[12]

The delegation was subsequently authorised to table the text of both agreements, which it did on 27 July, but with the article in the ABM Treaty covering space-based systems omitted. The Verification Panel in Washington was still analysing the 'exotic systems' question. Eventually, the President rejected an ABM ban which Ambassador Smith had urged, but about the same time he approved a White House staff compromise to the basic Smith–Brown position on 'exotic systems' which would prohibit (1) the deployment of fixed land-based systems and (2) the development, testing and deployment of all other basing modes. The Joint Chiefs of Staffs were particularly interested in preserving the option to develop and test fixed land-based lasers. The President's decision preserved this option, as does the ABM Treaty itself.

The United States delegation filled in the blank Article in its ABM Treaty in mid-August 1971. The Soviets initially balked at discussing, let alone agreeing to any limitations on, 'exotic systems'. They were probably without any instructions on this issue and may have felt the United States was on an intelligence-fishing expedition. Progress was soon made nevertheless. Various working groups and a drafting group were set up to seek agreement issue by issue. Joint Draft Texts of the Soviet and United States drafts of the Treaty were prepared with disagreed language, which at first was extensive, in brackets. Before the conclusion of the fifth negotiating session in September 1971, the Graybeal–Karpov[13] working group which focused on Article V agreed, *ad referendum* to the two delegations, that current Article V(1) covered 'current' as well as 'exotic' technologies. The United States delegates agreed that the Americans on this working group had carried out the President's instructions. The brackets around 'develop' in that paragraph in Article V in the Joint Draft

Text were subsequently removed in the drafting group during the sixth negotiating session after both delegations had noted their approval. The Administration now contends that *either* the Soviets never agreed with the United States interpretation *or* that the Soviets later modified their agreement or changed their interpretation during negotiations over Agreed Statement D.

The major sticking point then, and through late into the sixth negotiating session, was on fixed land-based systems. United States instructions were to preserve the right to develop and test, but not to deploy, fixed land-based lasers. Accordingly, the United States delegation insisted that Article III should authorise deployment of only ABM systems and components which are based on 'current' technology. Further, development and testing, whatever the technology, of fixed land-based systems and components could be carried out only at ABM test ranges. The Soviets resisted any limitations on *fixed land-based* 'exotic systems'. As John Newhouse wrote

> Back in the summer, Moscow's attitude, as reflected by its delegation, had been sympathetic. Then, in the autumn, it hardened, probably under pressure from the military bureaucracy. Washington was accused of injecting an entirely new issue. Moscow would not agree to a ban on future defensive systems, except for those that might be space-based, sea-based, air-based, or mobile land-based. The United States delegation persisted and was rewarded. Land-based exotics would also be banned. The front channel had produced an achievement of incalculable value.[14]

The Article III issue was not resolved until late in the sixth (Vienna in November 1971–February 1972) negotiating session. It was handled principally in the Garthoff and Kishilov or Grinevsky working group and also in the Garthoff/Parsons–Kishilov/Grinevsky group.[15] The United States proposed the 'currently consisting of' phrase which was agreed upon for Article II to make clear that the Treaty was *not* limited to 'traditional' technology. The United States proposal for the 'except that' formulation for Article III was accepted, which made clear that fixed land-based 'exotic' systems could not be deployed. The ban against a nationwide defence or 'base' for such a defence in Article I(2), which was a Soviet initiative intended in part to deal with 'exotic systems', was agreed. In each case, agreement was *ad referendum* to the delegations. Together, these textual provisions completed all the key words in the Treaty relating to 'exotic systems'.

An agreed interpretation tied to Articles I and III was first proposed by Garthoff in mid-December 1971. I distinctly recollect advising that no supplementary interpretation was technically necessary. The United States effort, therefore, was to reinforce the clear meaning of the specific Article III and the more general Article I(2).

The United States had originally proposed a paragraph for the Treaty in August 1971. It stated: 'Each party undertakes not to deploy ABM systems using devices other than ABM interceptor missiles, ABM launchers or ABM radars to perform the functions of these components'.[16] The Soviets balked at any such Treaty language and, subsequently, the initial United States proposals for an agreed interpretation. Eventually the Soviets proposed a counterdraft. This was modified several times at United States insistence (including the insertion of the opening phrase 'In order to insure fulfillment of the obligation not to deploy ABM systems and their components except as provided in Article III of the Treaty . . .'). The reference to Article XIV in Agreed Statement D indicated that the Treaty would have to be amended before a fixed land-based 'exotic', such as a laser, could be deployed. The final compromise language was proposed by Garthoff to the Soviets in late January 1972, and early in February Kishilov informed Garthoff of Soviet agreement. This was eventually noted in a United States plenary statement. Agreed Statement D and the other Agreed Statements were initialled on 26 May 1972 by Ambassadors Smith and Semenov.

Agreed Statement D refers to and interprets only Article III, although the references to 'other physical principles' and 'components capable of substituting for' are equally applicable to Article V(1). While the language is admittedly opague, the United States has always understood that Agreed Statement D reinforced Articles I(2) and III and reinforced the prohibition on deployment of *fixed land-based* 'exotic systems' unless and until the Treaty is amended. *Finally and most importantly, Agreed Statement D certainly does not diminish or amend Article V(1) and the other substantive Articles such as I(2), IV, V(2) and IX.*

During the seventh negotiating session I prepared detailed, still classified, memoranda on both the ABM Treaty and the Interim Agreement intended to serve four distinct purposes:

1. inform the delegation on what was agreed with the Soviets and what was not;
2. suggest whether the United States should consider seeking

one or more Agreed Statements to provide more specific interpretations;
3. indicate what types of weapons programme, current and future, were prohibited and permitted;
4. serve as the basis for the eventual transmittal documents to Congress and background for the Congressional hearings.

Successive drafts of my memoranda were shared within the delegation. Where there was any doubt that a matter was agreed, the proposition was enclosed in brackets. The brackets were removed only after I and others on the United States delegation were satisfied. I constantly revised the drafts as issues were reviewed. The draft memoranda were never made 'final'.

In certain cases the United States delegation sought and achieved Agreed Statements. In others, it did not seek them. Some matters were judged agreed, while others were not. To the best of my recollection, the United States delegation never sought an Agreed Statement confirming that Article V(1) covered 'exotic systems'. We probably felt that seeking further specific agreement was unnecessary and would not be productive. In any event, I am absolutely certain that my *contemporaneous advice* to the United States delegation on the scope of Articles III and V(1) with respect to 'exotic' systems was clear and that *none of the delegates nor their advisors (State, ACDA, JCS and OSD) disagreed with that advice*. I recall no indication that the Soviets thought otherwise.

During SALT I the United States delegation and particularly Washington, did not insist on the kind of precision reached in the SALT II Treaty with its 98 Agreed Statements and Common Understandings. The Soviets stubbornly resisted the level of textual detail the United States had initially sought at SALT I. Neither the President nor Henry Kissinger cared much for detail. During the final three negotiating sessions, the United States delegation made *ad hoc* decisions on how the Presidential instructions should be sought and recorded on many issues, how hard to push for additional clarity and what sufficed, but it constantly reported to Washington.

In one instance relating to the ABM Treaty ('current' Soviet ABM test ranges) the United States delegation identified the two United States ABM test ranges and the Soviet test range at Sary Shagan. The Soviet response noted that national technical means permitted the identification of test ranges. The United States delegation noted immediately that the Soviets did not respond to the United States identification of Sary Shagan as their ABM test range, but the delegation believed the Soviet response reflected extreme Soviet insensitivity to any discussion of their test range. However, in the mid-1970s the Soviets claimed a second 'current' ABM test range at Kamchatka based on the presence of an old radar. Paul Nitze has referred to this negotiating technique as unworthy of bazaar traders. I agree. The United States eventually accepted the Soviet claim in 1978 because there was a factual basis for it, but it learned from this example, and particularly from the Moscow Summit negotiations on the Interim Agreement, that explicit agreement and written precision are important. The SALT II documentation reflects this learning.

SALT I, however, did not have this benefit of latter-day hindsight indicating the need for precision and detailed Agreed Statements and Common Understandings reflected in the SALT II Treaty. Some of the SALT I underlying understandings are reflected in formal plenary statements, others in the less formal mini-plenary statements and some in working documents, memoranda of conversations ('memcons') and reporting cables. Agreement was reached *ad referendum* in one or more working groups, approved by the two delegations, referred to the drafting group, to the interpreters, etc. On many points there will not be simple, clear documentation. In addition, the United States government (but not the Soviets) has lost most of its SALT I historical memory.

Some interpretation matters were in 1972, and remain today, ambiguous and need clarification. The dividing line between permitted 'research' and prohibited 'development' and 'testing' is *not* clear, nor is the related meaning of 'component' in the broad prohibitory context of banning 'exotic systems' under article V(1). There is no Agreed Statement on either issue. The former was discussed in a formal statement delivered by Harold Brown and a general understanding, although not a fully documented record, was reached. I do not recall any discussion of the latter with the Soviets.

The SALT delegation remained in Helsinki until agreement on open points, primarily the Interim Agreement, was reached at the Moscow Summit. When the delegation returned to Washington and the transmittal documents and Congressional statements were being prepared under White House control, Henry Kissinger directed that all 'understandings' be culled from the negotiating record and made public to refute criticism of secret arrangements. This was the derivation of the SALT I Common Understandings.[17] The search of the files for Common Understandings was limited. It did not cover all the myriad of agreed understandings reached in less formal ways during the negotiations.

The hearings before the Senate Foreign Relations Committee[18] and particularly before the Senate Armed Services Committee[19] led to a much fuller public record on many of the nuances. Some of the initial testimony of officials was not clear, but the record was frequently supplemented. This includes the statement for the record of the Senate Armed Services Committee, prepared after inter-agency review of reporting cables, on the difference between research and development for purposes of Article V. It includes explicit confirmations submitted by Secretary of Defence Melvin Laird, Under Secretary for Defence Research and Engineering John Foster and Acting Chief of Staff of the Army General Palmer, all to Senator Jackson, that development and testing, as well as deployment, of space-based 'exotic systems' were prohibited. Senator Jackson (Democrat, Washington), who was a sharp critic of SALT I but voted *in favour* of the ABM Treaty, understood this point clearly. He was probably the most knowledgeable Senator on the impact of SALT I on weapons programmes. Finally, Senator James Buckley, (Republican, New York) stated on the Senate floor on 3 August 1972 that he opposed the ABM Treaty and would vote *against* it largely because of this prohibition. He said:

> Thus the agreement goes so far as to prohibit the development, test or deployment of sea, air or space based ballistic missile defence systems. This clause, in article V of the ABM treaty, would have the effect, for example, of prohibiting the development and testing of a laser type system based in space which could at least in principle provide an extremely reliable and effective system of defences against ballistic missiles. The technological possibility has been formally excluded by this agreement.[20]

The vote in favour of advice and consent to ratification was 88–2. I resigned from government in June 1972 after the transmittal documents had been sent to Congress and before the hearings.

In 1972–3 while in private practice, I co-edited a book on SALT and wrote chapter 5 on 'The SALT I Agreements'; which summarises my recollection of the advice I gave to the United States delegation which was the basis for the Executive position before, during and after the ratification process. Prior to publication of the book, I informally cleared my chapter with government officials to ensure both accuracy and non-disclosure of sensitive information.[21]

(b) A personal comment on the October 1985 reinterpretation efforts

Over the past eight years I have been informally asked about various issues by officials at the JCS, OSD, State, ACDA and the CIA. One question in the late 1970s was whether there were any *deployment* limits on *fixed land-based* 'exotic systems'. This question had been reopened in OSD, sharply debated with JCS supporting the traditional United States position and then correctly resolved. This question also involved Agreed Statement D, but in this case the OSD argument was that there were no *deployment* limits under Article III on fixed land-based 'exotic systems' and only an obligation to discuss. This is almost the exact reverse of the Reagan Administration's 'reinterpretation' which now claims the limits in Agreed Statement D *prohibit only deployment* of systems referred to in Articles III and V(1).

To the best of my knowledge the challenge to Article V(1) within the Executive arose only recently, although the Heritage Foundation circulated a Backgrounder dated 4 April 1985 rejecting the traditional interpretation.[22] A footnote stated that it was authored by an unnamed government official.

In my judgement the FY85 Arms Control Impact Statement prepared by the Reagan Administration correctly states the agreement reached with the Soviets in 1971–2 on the meaning of Article V(1). It provides:

> The ABM Treaty bans the development, testing and deployment of *all* ABM systems and components that are sea-based, air-based, space-based or mobile land-based ... The ABM Treaty prohibition on development, testing and deployment of space-based ABM systems, or components for such systems, *applies to directed energy technology (or any other technology used for this purpose)*. Thus, when such directed energy programs enter the field testing phase they become constrained by these ABM Treaty obligations (my emphasis).[23]

The SDI Report to Congress (April 1985), especially Appendix B, is consistent with this statement.[24]

III THE ABM TREATY – A BRIEF SURVEY

The ABM Treaty bans the deployment of nationwide systems to defend against ballistic missile attack. The explicit purpose of the ABM Treaty is to preclude the type of advanced nationwide missile defence system that President Reagan envisions and the less sophisticated one that some claim the Soviet Union is developing. Deployment of such systems would clearly violate the basic objective and explicit terms of the Treaty.

(a) Provisions of the Treaty[25]

As amended by the 1974 Protocol, the ABM Treaty limits the United States and the USSR to one ABM site of 100 interceptors and 100 launchers either around the national capital or an ICBM field. The Treaty also places strict limits on the number of ABM radars at the permitted ABM site. It allows research on all types of ABM systems and components. Advanced development, testing and deployment of certain types of ABM systems and their components are banned. This ban on advanced development of specific types of ABM systems and components is particularly relevant to the early stages of the President's SDI as outlined in the following section.

The ABM Treaty consists of a Preamble and sixteen Articles. In the course of the negotiations, agreement was reached on some interpretations related to the Treaty. These 'Agreed Statements' and 'Common Understandings' are an integral part of the Treaty and help to clarify some elements of its text. Additional Protocols and Agreed Statements were later reached in the Standing Consultative Commission (SCC) established by the Treaty as the forum to discuss Treaty-related issues. Some key definitions and concepts used in the Treaty were not clarified by agreed interpretations during or subsequent to SALT I.

The Preamble sets forth the common premises and objectives of the United States and the Soviet Union which are the basis for entering into the ABM Treaty. Most other provisions of the Treaty are summarised below. (For the complete text of the ABM Treaty and Protocols, see Appendix.)

Article 1: bans the deployment of ABM systems which would provide a defence for the entire territory or a base for such a defence. It also bans the defence of individual regions except as permitted by Article III.

Article II: defines an ABM system as one designed to counter strategic ballistic missiles or their elements in flight. Current types of components are described as ABM missiles, ABM launchers, and ABM radars.

Article III: (as amended by the 1974 Protocol) allows one fixed, land-based ABM site in each country, of a radius of 150 kilometres, either to defend the national capital or one ICBM field. Each site is limited to no more than 100 ABM launchers and missiles. Each site has limits on ABM radars which are somewhat different. A site to defend ICBM silos may have no more than two large ABM radars and eighteen smaller ABM radars. A site to defend the national capital is permitted no more than six radar complexes, each complex having a radius of no more than three kilometres.

Article IV: permits development and testing of ABM systems and components at mutually agreed upon test ranges, which may have no more than fifteen ABM launchers.

Article V(1): bans the development, testing or deployment of ABM systems or components which are sea-based, air-based, space-based or mobile land-based.

Article V(2): bans the development, testing or deployment of ABM launchers for multiple launch or rapid reload of ABM interceptors.

Article VI(a): bans giving non-ABM systems ABM capabilities (i.e., capabilities to counter strategic ballistic missiles in their flight trajectory), or testing such non-ABM systems 'in an ABM mode'. Although these non-ABM systems are not defined by the Treaty, they could include air defence or anti-tactical ballistic missiles, strategic offensive missiles, or anti-satellite weapons.

Article VI(b): requires that ballistic missile early-warning radars be located along the periphery of the national territory and oriented outward.

Article VII: permits modernisation and replacement of ABM systems and components subject to other Treaty provisions.

Article VIII: directs that excess or banned ABM systems or components be dismantled within the shortest possible time.

Article IX: prohibits the transfer to other states, and the deployment outside each party's national territory, of ABM systems or their components.

Article XIII: establishes SCC as the forum for discussion of future ABM Treaty issues.

Article XIV: establishes that each Party may propose amendments to the Treaty.

Article XV: provides that the Treaty is of unlimited duration, but permits either side to withdraw on six months' notice if its supreme interests are jeopardised.

Agreed statements

D: establishes that future ABM systems based on other physical principles will be subject to discussion. Their deployment in a fixed land-based mode requires Treaty amendment.

E: bans the development, testing, or deployment of MIRVed ABM interceptors.

F: exempts LPARs used for tracking objects in outer space or for verification from other restrictions on LPARs.

G: specifies that Article IX's prohibition on the transfer of ABM systems or components to other states includes technical descriptions and blueprints.

Common understandings

B: specifies that non-phased-array radars for range safety and instrumentation are permitted outside of ABM test ranges. The Soviets further elaborated that such radars are not limited by the Treaty.

C: specifies that 'mobile' ABM systems and components include those that are 'not permanent fixed types'.

1974 ABM Protocol
In 1974 the United States and the Soviet Union signed a Protocol that reduced the allowed number of ABM sites on each side from two to one. The Protocol also allows each side to change its original choice of an ABM site (defending either its national capital or an ICBM field) but it can do so only once and advance notice must be given.

Additional protocols and agreed statements
A number of confidential agreed statements and protocols have been reached within the SCC by the United States and the USSR.

1974 Protocol on Procedures for ABM systems: establishes procedures for replacement and dismantling of ABM systems.

1976 Supplementary Protocol on ABM Procedures: establishes additional procedures for replacement and dismantling of ABM systems.

1978 Agreed Statement: defines test ranges for ABMs and identifies current ranges; specifies criteria for the Article VI term 'tested in an ABM mode'; and refines criteria for permitted and prohibited activities of air defence components at ABM test ranges.

(b) Activities permitted and prohibited by the ABM Treaty

The basic approach of the ABM Treaty is that anything which is not prohibited is permitted. Research on all types of ABM systems and components is permitted. The following section outlines specific permitted and prohibited activities.

Fixed land-based ABM systems and components
Research, development, testing and deployment of fixed, land-based ABM systems and components is permitted, provided such activity meets the geographic, quantitative, and qualitative constraints of Article III (as modified by the 1974 Protocol) and utilises 'current components', namely: ABM missiles, ABM launchers and ABM radars as outlined in Article II.

Fixed, land-based systems are limited, however, by Article V(2) and Agreed Statement E. Article V(2) bans the development, testing or deployment of 'ABM launchers for launching more than one ABM interceptor missile at a time'. It also bans the development, testing or deployment of systems for 'rapid reload' of ABM launchers. Agreed Statement E clarifies that the provisions of Article V(2) prohibit the development, testing or deployment of more than one independently-guided warhead on each ABM interceptor missile.

Research, development and testing (but not deployment) of fixed, land-based components utilising kinetic-energy (which destroy their targets by high-speed impact or with shrapnel explosives) and directed-energy systems (such as lasers or particle beams) is permitted (see Article III and Agreed Statement D). However, testing of such systems and their components must be conducted at designated ABM test ranges (see Article IV).

Space-, air, sea- and mobile land-based ABM systems and components
Research and preliminary development of space-based, air-based, sea-based or mobile land-based (including transportable or otherwise

not of a permanent, fixed type) ABM systems and components is permitted. Research, development, testing and deployment of anti-satellite weapons is not prohibited by the Treaty, but such weapons may not be given the capability to intercept strategic ballistic missiles or tested in an ABM mode (Article VI(a)).

Radars
The deployment of ballistic missile early-warning radars is permitted but limited to those 'at locations along the periphery' of the national territory and 'oriented outward' (Article VI(b)). This restriction limits the ability of early-warning radars to function as ABM radars by reducing their proximity to missile fields that they could be used to protect. Further insurance is provided by the fact that radars located at the periphery are also, themselves, quite vulnerable to direct attack.

Space-tracking radars, as well as radars used for arms control treaty verification (national technical means), are not limited by the ABM Treaty (see Agreed Statement F). Therefore, deployment of large phased-array radars (LPARs) without regard for location or orientation for these purposes is permitted. Non-ABM radars for test instrumentation or range safety purposes are permitted outside of agreed ABM test ranges (see Common Understanding B). There is no agreed statement distinguishing by technical characteristics an ABM radar from a non-ABM radar or one type of radar from another (e.g., an early-warning from a spacetrack radar).

Lasers and directed-energy weapons
The development, testing and deployment of any type of space-based, air-based or mobile ground-based ABM system or component, whether based on present or future (e.g., one utilising kinetic-energy or directed-energy weapons) technologies, is banned by Article V of the ABM Treaty. The American negotiators were aware of the problem of technological innovation, and finally achieved Soviet agreement to provisions which severely constrain the development and testing of these future types of technologies.

Each year the Director of the Arms Control and Disarmament Agency (ACDA) submits a report to Congress, on behalf of the President, assessing the arms control impact of various weapon programmes. The 1984 Arms Control Impact Statement (ACIS) stated, 'The ABM Treaty prohibition on development ... applies to directed energy technology ... When such directed energy weapons enter the field testing phase, they become constrained by these ABM

Treaty obligations.' This statement correctly reflects the definition of 'develop' prepared by the United States during the ratification of SALT I.[26]

Agreed Statement D, which supplements Article III of the Treaty, covers future fixed, ground-based ABM sytems based on other physical principles (e.g., lasers and particle beam weapons) and including components 'capable of substituting for ABM interceptor missiles, ABM launchers, or ABM radars'. Agreed Statement D, in conjunction with Articles III and XIV, requires prior consultation and an amendment to the ABM Treaty before a fixed, ground-based directed-energy or kinetic-energy ABM system can be deployed.

In summary, while Article V severely limits the development and testing of ABM systems and components, directed-energy or otherwise, the development and testing of *fixed, land-based* directed-energy systems or components is permitted. However, as stated in the 1984 ACIS, 'although the Treaty allows the development and testing of fixed, land-based ABM systems and components based on other physical principles ... the Treaty prohibits the deployment of such fixed, land-based systems and components unless the parties consult and amend the Treaty'.[27]

Role of the SCC

The SCC was established to deal with perceived ambiguities in the ABM Treaty's provisions. The SCC has been suitable to resolve questions of interpretation and it can adopt Agreed Statements to overcome disputes and different assessments of treaty terms.

In 1974 and 1976 the SCC agreed on two Protocols governing the dismantlement of ABM systems. In 1978 and 1985 the SCC negotiated agreed statements relating to 'test in an ABM mode' and related issues. The texts of each remain classified.

The United States has had a series of excellent Commissioners to the SCC. General Ellis, the current Commissioner, is one of the best of all. But the United States Commissioner is an instructed agent. His instructions are set on an inter-agency basis and he is not free to act on his own. The SCC Commissioners – United States and Soviets – follow instructions from the respective capitals.

If the United States had legitimate doubts whether the Soviets agreed to the historic United States interpretation of 'exotics' in the ABM Treaty and had sought to resolve the issue, the matter should have been raised in the SCC. It would have been a simple matter to get Soviet agreement.[28]

(c) Critical definitional issues

Both the United States and the USSR are involved in many military development efforts with potential application to an ABM role. Over the next several years, the SDI programme as presently structured will focus on developing and testing the necessary technologies and components for a modern ABM system. Such tests are a critical prerequisite to any decision on deployment.

President Reagan has stated that all research and development programmes will be 'carried out in a manner consistent with our obligations under the ABM Treaty'. Secretary of State Shultz, likewise, insisted to Soviet Foreign Minister Gromyko in their talks last January that 'SDI is a research programme fully consistent with the ABM Treaty.' This point was further emphasised in the White House report 'The President's Strategic Defence Initiative,' released in January 1985.[29]

It is probably true that the first four years of the SDI will be generally consistent with the ABM Treaty, at least through 1988. This policy was determined in NSDD 119 of January 1984. However, starting in 1988 this situation will change. A close look at what is now known about the SDI indicates that beginning in 1988 the programme as presently planned will come into conflict with the Treaty's limits. Because the SDI research, development and testing efforts are scheduled to be carried out from now through the early 1990s, compliance with the ABM Treaty will be affected well before any decision is taken to deploy a layered strategic anti-missile system.

Supposed imprecision in Treaty language does not provide any basis for claiming adherence to the Treaty's terms by the SDI after 1988. A careful examination of the Treaty and its negotiating history indicates that many of the demonstrations planned under the SDI, starting in 1988 are almost certainly inconsistent with the terms of the Treaty.

Some of the current uncertainty with regard to permitted and prohibited activities under the ABM Treaty stem from the different possible interpretations of certain Treaty terms. During the SALT I negotiations the two Parties did not reach an agreed interpretation of the term 'develop', although there were numerous exchanges on its meaning. The term 'ABM component' was defined with respect to current technologies but not future technologies. The phrase 'tested in an ABM mode' was not defined at all, although the United States did issue a unilateral statement on the matter.

While this lack of agreement has led to some differing interpretations of certain activities that occurred after the Treaty was ratified, some terms and phrases were further clarified following the signing of the Treaty. The following sections provide a detailed examination of how the ABM Treaty defines prohibited activities related to missile defence and outlines the negotiating history that led to these provisions.

Definition of 'Test in an ABM Mode'
While no agreement was reached with the Soviets in 1972 on a definition for 'test in an ABM mode', in 1978 the United States and the USSR reached an Agreed Statement in the SCC specifying the criteria for applying the term as it is used in the Treaty to refer to missiles, launchers and radars.

Although the exact text of the 1978 Agreed Statement remains classified, it is apparently similar to, with one significant change, the United States unilateral statement attached to the Treaty which provides the United States interpretation of the definition. In this statement the United States indicated that it would regard a missile to be 'tested in an ABM mode' if the missile were:

> flight tested against a target vehicle which has a flight trajectory with characteristics of a strategic ballistic missile flight trajectory ... or is flight tested to an altitude inconsistent with interception of targets against which air defences are deployed.[30]

The parties were unable to agree to a precise definition of the altitude that marks the difference between air defence and missile defence interceptions, and this part of the definition was not adopted. However, lack of a common understanding on the definition of a 'strategic ballistic missile' is relevant to several current compliance issues, particularly the controversy over Soviet testing of their new SA-12 air defence missile against a target based on the SS-12 tactical ballistic missile.

Definition of 'Component'
The Treaty defined ABM components as 'currently consisting of ... interceptor missiles constructed and deployed for an ABM role, or ... tested in an ABM mode; ... launchers constructed and deployed for launching ABM interceptor missiles; ... and radars constructed and deployed for an ABM role or ... tested in an ABM mode.'[31]

During the course of the negotiations, discussions were held on the difference between a 'component', which would be limited by the Treaty, and an 'adjunct' which would not be limited by the Treaty. One example of an 'adjunct' which was mentioned during these discussions was a small optical telescope that might be used in conjunction with an ABM radar, perhaps to provide assistance in calibrating the radar. Since such an adjunct would not have meaningful ABM capabilities, these discussions suggested that 'adjuncts' would not be subject to Treaty limitations. However, a precise definition of the distinction between a 'component' and an 'adjunct' was never formally sought in the negotiations. There has not been an authoritative statement on this subject by American officials (as was the case with the definition of the term 'develop'). There has also been no agreed definition reached in the SCC.

Definition of 'Develop'

Certain provisions of the ABM Treaty deal primarily with development and testing, as separate from deployment, of ABM systems and components. Article IV allows the development and testing of fixed, land-based ABM systems and components at agreed test ranges. Article V prohibits either side from undertaking 'to develop, test, or deploy ABM systems or components which are sea-based, air-based, space-based or mobile land-based.' Also banned by Article V is the development, testing, or deployment of multiple, independently-guided warheads for ABM interceptor missiles, systems for launching more than one ABM interceptor missile at a time from each launcher, or automatic, semi-automatic, or similar systems for rapid reload of ABM launchers. Article VI prohibits either party from giving non-ABM (e.g., air defence, strategic offensive, or anti-satellite) missiles, launchers, or radars 'capabilities to counter' strategic ballistic missiles or from testing such systems 'in an ABM mode.'

The ABM Treaty language does not provide a definition of the term 'development' (as used in Article IV) or 'develop' (as used in Article V). The meaning of 'development' and 'develop', however, was discussed between the American and Soviet delegates during the SALT I negotiations.

American negotiators proposed to Soviet negotiators that development is that stage which follows research and that research includes the conceptual design and laboratory testing which precedes field testing. While development often overlaps with research, it is usually associated with the construction and testing of one or more prototypes of a weapon system or its major components. It therefore made sense for the Treaty to ban development of those systems where testing and deployment were prohibited.

Article V of the Russian text of the Treaty uses the verb 'to create' for the English word 'develop'. In response to the proposed United States definition of 'develop', a Soviet negotiator apparently indicated that there was little difference between 'develop' and 'test' as proposed by the United States and that the dividing line should be where national technical means could identify specific systems as ABM-related.

The most explicit and authoritative American formulation on this matter was in response to a question from Senator Henry Jackson to Ambassador Gerard C. Smith (chief negotiator of the ABM Treaty) during Smith's testimony before the Senate Armed Services Committee on 18 July 1972. This written submission for the record, which was prepared by the executive branch after reviewing the SALT delegation's reporting cables to Washington, states:

> The obligation not to develop such systems, devices, or warheads would be applicable only to that stage of development which follows laboratory development and testing. The prohibitions on development contained in the ABM Treaty would start at that part of the development process where field testing is initiated on either a prototype or breadboard model. It was understood by both sides that the prohibition on 'development' applies to activities involved after a component moves from the laboratory development and testing stage to the field testing stage, wherever performed. The fact that early stages of the development process, such as laboratory testing, would pose problems for verification by national technical means is an important consideration in reaching this definition. Exchanges with the Soviet Delegation made clear that this definition is also the Soviet interpretation of the term 'development'.

The submission went on to state:

> Article V ... places no constraints on research and on those aspects of exploratory and advanced development which precede field testing. Engineering development would clearly be prohibited.[32]

However, this sensitive and now critical unilateral definition, which appears to be drafted in terms of the Department of Defense categorisation of the research and development process (see below), was never reduced to an agreed statement during the negotiations or subsequently under the auspices of the SCC. No Soviet official has publicly stated whether they agree with the definition prepared during the ratification process. Privately, the Soviets have indicated they are under no obligation to comment on documents prepared for United States ratification purposes.

Recently some Soviets have asserted that, as to development, the Treaty bans even the earliest stages of research conducted for the purpose of creating systems or components limited by Article V. It is not known whether this is the official Soviet view.

In other 1972 testimony before the Senate Armed Services Committee, then Director of Defence Research and Engineering, Dr John Foster, Jr, elaborated on Ambassador Smith's submission. Dr Foster's submission stated that:

> Constraints imposed by the phrase 'development and testing' would be applicable only to that portion of the 'advanced development stage' following laboratory testing, i.e. that stage which is verifiable by national means. Therefore, a prohibition on development – the Russian words is 'creation' – would begin only at the stage where laboratory testing ended on ABM components, on either a prototype or a breadboard model.[33]

For the purpose of categorising programmes in the defence budget, the Pentagon both in 1972 and now divides the research and development process into five stages:

6.1 Basic research: efforts directed toward the expansion of specific knowledge of natural phenomena, but not tied to a specific programme.

6.2 Exploratory development: efforts directed toward the expansion of technological knowledge and the development of materials and components with potential application to new military weapons and equipment. Emphasis on exploring the feasibility of various approaches to military problems up to the point of breadboard and prototype fabrication.

6.3 Advanced development: efforts directed toward the development of experimental hardware for technical or operational testing of its suitability for military use.

6.4 Engineering development: efforts directed toward the

development of a particular system engineered for service use, but which has not yet been approved for production and deployment.

6.5 *Operational systems development*: efforts directed toward the continued test, development, evaluation and design improvement of projects which have already entered (or have been approved for) the production–deployment stage.

A careful reading of Ambassador Smith's submitted statement and the Department of Defense categorisation of the research and development process suggests that the Article V limitations on space-based and other mobile systems and components would permit basic research (6.1) and those aspects of exploratory (6.2) and advanced (6.3) development which preceded field testing. Article V would ban field testing as part of exploratory (6.2) and advanced (6.3) development, as well as engineering (6.4) and operational (6.5) systems development. The SDI is funded under the 6.3 advanced development budget category.

Current plans of the Administration for the SDI, if carried through during the 1988–1993 time period, would be inconsistent with the limits of Article V of the Treaty as explained in the 1972 Senate Armed Services Committee hearings. The *1984–88 Five-Year Defense Guidance*, signed by Secretary of Defence Caspar Weinberger, states that the United States plans to initiate 'the prototype development of space-based weapons systems ... so that we will be prepared to deploy fully developed and operationally ready systems'.[34]

IV THREE GREY AREAS OF THE ABM TREATY – ASAT, ATBM AND RADARS

Another prominent threat to the viability of the ABM Treaty results from Soviet and American development efforts in other weapons technologies which have an application for missile defence. The three most difficult such 'grey areas' are ASAT weapons, ATBMs and LPARs.

Each nation maintains that its programmes in these areas are in compliance with Treaty terms as narrowly interpreted. It is possible, therefore, that neither side will take the step of formally abrogating the agreement. Each may simply continue to undertake activities that undermine the Treaty, steadily eroding its restrictions until the Treaty has lost its significance.

Indeed, this threat of erosion of the Treaty emerged in early 1985 as one of the prominent themes of the American negotiating posture at the renewed strategic arms negotiations in Geneva. However, it is unlikely that American concerns about Soviet erosion of the Treaty will be eased unless Soviet concerns are also addressed. Both sides must begin to acknowledge the symmetry of the problem of the erosion of the ABM Treaty if the process is to be halted.

(a) ASAT weapons[35]

ASAT systems and components are not directly covered by the ABM Treaty. An ASAT system employing interceptor launchers, missiles, and radars is prohibited by the Treaty only if, as provided in Article VI(a), it is given the capability to intercept strategic ballistic missiles or is tested in an ABM mode.

Both the United States and the USSR have active ASAT development programmes. In 1984 the United States Air Force began testing its miniature homing ASAT vehicle, designed to be launched from an interceptor aircraft, collide with and destroy an orbiting satellite. It is scheduled to become operational in 1988. The Soviets already have a less capable ASAT system in operation. They have pledged, however, not to conduct further tests of their system as long as the United States does not test its new ASAT. The Soviets ceased testing their ASAT system in June 1982 and have not indicated whether the American ASAT tests have invalidated their test moratorium offer.

The technologies and components necessary to destroy satellites and ballistic missile objects overlap considerably. For example, because the ABM Treaty does not explicitly limit the development or deployment of ASAT systems or components, many of the technologies for BMD including sophisticated sensing, tracking, and non-nuclear kill devices, could be tested and even deployed under the guise of ASAT systems.

One consultant to the Administration on these issues, Keith Payne, has written:

> A space-based system (laser or conventional) could be tested as satellites that resembled ICBM or SLBM upper stages. Such space objects simulating ICBMs/SLBMs in free flight would not have flight trajectories comparable to those of ballistic missiles launched

on sub-orbital trajectories. Such simulations would be useful for testing pointing and tracking, and in regard to lasers, beam jitter and deposited power levels at various distances.[36]

Dr Keyworth seemed to confirm the usefulness of this approach in July 1983, when he talked about near-term research milestones for the Star Wars plan:

> One is a geosynchronous anti-satellite capability ... It may not necessarily be the best way for the ASAT mission, but a geosynchronous anti-satellite capability is important to test the technology to destroy missiles in each of the three layers.[37]

In the early 1990s the Administration plans to test space-based weapons and sensors that are part of the SDI against satellite targets that will simulate the characteristics of ballistic missile components. These tests will demonstrate ABM capabilities but are being justified by the Administration as Treaty-compliant because they involve satellite targets rather than ballistic missile targets.

In the absence of an agreement limiting ASATs, their development, testing or deployment would undermine the prohibitions in Article V(1) to the extent that such ASAT systems or components were dual-capable for ABM use or aided the development of space-based ABM systems.

(b) Anti-tactical ballistic missiles (ATBM)[38]

Air defence systems rely primarily on high velocity SAM interceptor missiles to destroy attacking enemy aircraft. These systems have reached a level of sophistication such that they have some capability against both Cruise and ballistic missiles. Weapons capable of destroying medium and intermediate-range ballistic missiles (MRBMs and IRBMs) are called anti-tactical missiles (ATMs) or anti-tactical ballistic missiles (ATBMs) more or less interchangeably.

Article VI(a) of the ABM Treaty dealt with this emerging problem by prohibiting either side from giving air defence (or other) missiles, launchers, or radars the capability to counter strategic ballistic missiles or their elements (re-entry vehicles) while in flight and not to test them 'in an ABM mode.'

The problem is now becoming more acute. Both the United States and the USSR are deploying air defence systems that with some modifications could be upgraded to provide some ATBM capability. These include the United States Patriot and Improved Hawk and the Soviet SA-10 and SA-12 'Gladiator' systems. The United States is presently spending hundreds of millions of dollars per year on various programmes that directly or indirectly are related to the ATBM mission. The Army research and development budget contains the bulk of this funding, including several classified programmes. The Soviet ATBM programme, is described in detail in other sections of this chapter.

Systems with an ATBM capability might also be effective against SLBMs, which fly at a similar speed, flight trajectory and re-entry angle to MRBMs and IRBMs. Since SLBMs are included as strategic ballistic missiles in the SALT I and SALT II agreements on offensive weapons, deployment of ATBMs with such a capability would be inconsistent with Article VI(a) of the ABM Treaty.

The American ATBM programme began in the late 1950s, with the Plato programme, and the Field Army Ballistic Missile Defense Systems (FA-BMD) Project, which ran from 1959 through 1962. Subsequent efforts focused on enhancing the anti-missile capabilities of the SAM-D anti-aircraft missile, which, after many delays, is now entering service as the Patriot.

Over the past decade most American ATM work has been focused on upgrades to the existing Patriot and Improved Hawk anti-aircraft missiles. Enhancements to the radars and battle management computers of these systems, as well as upgrades to the missiles themselves, could have provided a limited anti-missile capability for the late 1980s. However, these upgrades did not seem to be a particularly promising solution to the problem, and funding has been severely cut back by the Congress.

In parallel with these efforts the Army is also pursuing an alternative, non-interceptor, approach to dealing with the Soviet missile threat, the highly classified Grass Blade programme. Grass Blade is developing techniques to disguise and hide potential targets of Soviet missiles, as well as means of locating Soviet theatre missile launchers so that they can be destroyed before launch.

In recent years the Army has also initiated work in conjunction with the Navy on a more advanced Anti-Tactical Missile Programme, which could use an improved version of the Patriot missile in conjunction with a ground mobile version of the Navy's SPY-1 Aegis radar. However, in 1984 this effort failed to receive Congressional support due to budget, schedule and technical uncertainties. The Navy continues to examine the use of the SM-2 nuclear tipped SAM in conjunction with the Aegis radar system for the ATBM mission[39]

Much of the American ATBM effort is now concentrated in the SDI, where the Small-Radar and Extended-Range Homing Intercept Technology (SR-HIT and ER-HIT) programmes are intended to support ATM applications. In the longer term the SDI systems that are part of the Terminal System Demonstration, the HEDI interceptor, and the TIR and AOS sensors, would be used for the ATM mission.[40]

The Future Security Strategy Study Team (Hoffman Panel), appointed by the President 'to assess the role of defensive systems in our future security strategy', was enthusiastic about the potential for ATM and recommended major additional investment in this area. The unclassified summary of the Hoffman Panel's report stated that:

> Deployment of an anti-tactical missile (ATM) system is an intermediate option that might be available relatively early . . . The advanced components, though developed initially in an ATM mode, might later play a role in Continental United States (CONUS) defense . . . We can pursue such a programme *within ABM Treaty constraints* (emphasis in original).[41]

The Hoffman Panel thus concluded that a large-scale ATM programme would be a less provocative way to develop, test and even deploy advanced missile defence concepts in the near-term. The Hoffman Panel and other ATBM advocates are particularly enthusiastic about the potential role of ATBMs in Europe. Later these systems could be adapted to large-scale strategic defence of the United States. However, large deployments of Soviet ATBMs could affect the ability of United States medium-range missiles in Europe to reach their targets, as well as erode the penetrability of United States, British and French SLBMs.

Despite the Hoffman Panel's interpretation of what the Treaty permits, the development and deployment of such systems raises direct questions of compliance with a number of provisions of the ABM Treaty. The Hoffman Panel's Summary Report implies that ATBMs are not constrained by the ABM Treaty. Such a conclusion is unfounded. The unconstrained development and deployment of ATBMs by both the United States and the USSR would circumvent both the purpose and letter of the ABM Treaty. The application of such systems to a strategic ABM role over the long term appears quite likely if specific limits on their development and deployment are not agreed upon in the near future.

(c) Large phased–array radars

During the negotiation of the ABM Treaty it was recognised that establishing useful restrictions on radars would be extremely difficult. The United States was very concerned about the ABM potential of the existing Soviet Hen House radars and sought strict limitations in this area. The Soviets were reluctant to agree to any limits on radar deployments that might interfere with their large air defence system, with its many radars designed to track aircraft and other air-breathing vehicles. However, after extensive discussions on the problem, the two sides agreed to some limits in the Treaty on ABM and other types of large radars.

As radar technology has advanced, the problem has taken on a new dimension. Today's modern and sophisticated large phased-array radars (LPARs) can serve many functions. They can provide early warning of missile or bomber attack. LPARs can track satellites and other objects in space and observe missile tests to obtain information for monitoring purposes. They are also an essential component of present-generation ABM systems, providing initial warning of an attack and battle management support, distinguishing re-entry vehicles (RVs) from decoys and guiding interceptors to their targets.

The Treaty recognised that LPARs can be used for tracking satellites, verification and early warning, and made provisions for such activities. Unfortunately, distinguishing an LPAR designed for one of these functions from one designed for an ABM role is very difficult.

This issue is particularly important to clarify because of the importance of LPARs to any ABM system. Because of their size and complexity, LPARs can take a number of years to construct. In a sense, they are the 'long lead-time' for an ABM system. The restrictions on these radars imposed by the ABM Treaty were based on the recognition that it might be possible for a country to build the large radars needed for an ABM system over a long period of time, and then quickly deploy a force of ABM interceptors and smaller site radars which require less elaborate preparations.

During the late 1970s the Soviets began work on a new type of very large phased-array radars, known as the Pechora type, after the locality of the first radar of this type to be identified. By the mid-1980s the Soviets were building a nationwide network of about half a dozen of these radars, providing almost complete coverage of potential missile attack trajectories.

Given the inherent ABM potential of early warning radars, the United States first raised the questions about these radars in the SCC in the late 1970s. The most disturbing of these radars is the one under construction at Abalakova (also referred to as 'near Krasnoyarsk'). The Abalakova radar is discussed in the section on Soviet compliance questions.

The Soviets have raised several analogous questions about American compliance with ABM Treaty provisions concerning radars. Of particular note are the two new Pave Paws SLBM early-warning radars under construction in Georgia and Texas which will become operational in 1986 and 1987, respectively. Two other Pave Paws radars in Massachusetts and California have been operational since the late 1970s. Three of the four Pave Paws radars will receive radar power upgrades and two computer upgrades. These upgrades will significantly enhance their capabilities.

The Pave Paws radars provided a 240-degree field of coverage. Initial plans for the deployment of the two new radars resulted in a field of coverage that included almost two-thirds of the CONUS. The final deployment plan apparently will reduce this coverage, but it still includes greater portions of the United States than were covered by the first two radars of this type. Hence, these sites may be inconsistent with the requirement in Article VI(b) that such radars be 'oriented outward'. In addition, they also might be construed as providing a base for the defence of territory, which is prohibited by Article 1. These radars are similar in performance to the Perimeter Acquisition Radars (PARs) planned for the Sentinel/Safeguard ABM system in the late 1960s and early 1970s.

In 1978 the Soviets raised within the SCC the question of whether the first two Pave Paws radars were consistent with the undertaking in Article I of the ABM Treaty 'not to provide a base' for ABM territorial defence. They have also publicly complained about the two new sites. United States officials insist that all four Pave Paws radars are for space tracking and early warnings of SLBM attack, and thus are fully consistent with ABM Treaty limits.

In 1975, the Soviets complained at the SCC that the United States Cobra Dane LPAR at Shemya Island, Alaska, was an ABM radar and forbidden under Article III of the Treaty. The United States explained that the radar was for verification purposes (national technical means), space tracking and early warning.

The United States is now upgrading the existing Ballistic Missile

Early Warning System (BMEWS) by replacing the present fixed-array and mechanically steered radars with phased-array radars that are somewhat smaller than Pave Paws, though with a power aperture product large enough to give them some ABM capability, and subject them to the limits of the ABM Treaty. This replacement process is nearly complete at the Thule, Greenland site, and is under consideration for the Clear, Alaska site. The Fylingdales Moor site in the United Kingdom will be upgraded in the next few years. It will be getting a phased-array radar of the Pave Paws type, but with three faces (and 360-degree coverage) instead of two. The Fylingdales LPAR may become operational in 1989. All three sites are receiving new data processing equipment and missile attack assessment computers.

The installation of new phased-array radars in Greenland and in the United Kingdom could be regarded as inconsistent with the Article VI commitment 'not to deploy in the future radars for early warning of strategic ballistic missile attack except at *locations along the periphery of the national territory* and oriented outward' (my emphasis). During the ratification hearings, Senator Percy submitted a written question on this subject, to which Ambassador Smith responded:

> *Question*: What was the rationale behind agreeing to Article VI of the treaty which prohibits the future deployment in third countries of early warning radars?
>
> *Answer*: Neither the United States nor the USSR believed that it is necessary to deploy future radars for early warning of strategic ballistic missile attack in third countries in order to obtain sufficient warning of such attack. Therefore, consistent with the goal of placing tight constraints on systems which could contribute to ABM capabilities beyond those envisioned in Article III of the treaty, the sides agreed to prohibit such deployments. Article VI does not affect existing ballistic missile early warning radars.[42]

This statement was prepared for the ratification record. It is not clear the extent to which the specific issue of modernising a 'grandfathered' early warning radar was discussed during negotiations.

This Congressional response confirms the explicit understanding that the then-existing BMEWS radars in Greenland and the United Kingdom are permitted. It leaves unclear whether deployment of new types of radars at these sites would be permitted. The Administration's position is that since the new phased-array radars are at the same location as the existing radars, they are a modernisation permitted under Article VII of the Treaty. Article VII is not relevant, and this position appears inconsistent with the language of Article VI(b) and Agreed Statement F. Further, public release of information on the Thule upgrade was delayed by more than a year, apparently to avoid association of this activity with the controversy over the Soviet radar in Siberia.

V ATTITUDES OF THE REAGAN ADMINISTRATION ON THE ABM TREATY

During the early days of the SDI there were some within the Administration who argued that the ABM Treaty would permit all the planned SDI test programmes to go forward to the 1993 deployment decision point. However, this interpretation was difficult to reconcile with a careful examination of the Treaty or with the actual characteristics of the demonstrations planned under the SDI, some of which would have come into conflict with the Treaty as early as 1987.

In NSDD 85 (dated March 1983) and NSDD 119 (dated January 1984) President Reagan directed that, during the 1984–88 period, the SDI would remain in compliance with ABM Treaty. However, testing of the Airborne Optical System, scheduled for 1988, would be inconsistent with the ABM Treaty. Some Administration officials now appear to accept that the Treaty, as understood prior to the 6 October 1985 reinterpretation, would have to be amended or abrogated in order for many of the demonstrations planned from 1991–1993 to take place. In testimony before the House Armed Services Committee on 27 February 1985 General Abrahamson stated that while the United States can currently test SDI-related components in the laboratory, 'at some point, however, we would have to depart from the Treaty.' He also went on to say that that would be necessary 'about the turn of the next decade'.[43] Consequently, an affirmative decision would be required to restructure (or cancel) these demonstrations in order to remain compliant with existing Treaty provisions.

(a) Interpretation of 'component'

Many in the Administration have argued that a device would not be an ABM component unless it could perform the complete function of, or substitute on a 'stand alone' basis for, an ABM component as defined in Article II of the Treaty. If a device could only perform part of the function of an ABM radar, launcher, or interceptor, then it would not be constrained as an ABM component under their interpretation.

These same Administration officials have maintained that the technology which will be demonstrated in many of the planned tests is not sufficiently mature to be integrated into a workable ABM system. Therefore, they argued, these experiments would not violate article V of the ABM Treaty because ABM systems or components will not be developed or tested. A violation would only occur, according to these officials, when and if technical experiments become part of a total system through integration, at the testing stage, with command and control elements, interceptors, and the other necessary elements of an ABM system.

This position was outlined by presidential science advisor Dr George Keyworth II in a speech delivered 29 February 1984. According to Dr Keyworth:

> As it's emerging, the Strategic Defence Initiative would move towards a series of progressive demonstrations of evolving subsystems. Each of these demonstrations would test out a piece of military meaningful technology. These would be building blocks from which an eventual system could be designed, but in and of themselves would not constitute a weapons system. Such activity would be fully within the provisions of existing treaty limitations.[44]

Other official documents suggest a similar interpretation of permitted activities.

The summary of the SDI programme released by the Department of Defense in April 1984 states that the SDI programme involves 'several component technology development programmes which culminate in hardware demonstrations.'[45] Another Department of Defense report, released in March 1984, mentions 'near-term feasibility demonstrations that could be developed into elements of a total ballistic missile defence system.'[46]

Under this interpretation, the Treaty was understood to permit development and testing of assemblies and subassemblies for space-based, air-based and mobile land-based systems which do not constitute full ABM 'components'. For instance, the Administration argues that the Airborne Optical System (which would provide initial target tracking data) is merely an adjunct to the Terminal Imaging Radar, which would provide direct guidance information to ground based interceptors. In this view, the Airborne Optical System would have to perform all sensor and battle management functions in order to be a 'component'. This would require this single sensor to search for attacking warheads, acquire (identify) individual warheads, discriminate these warheads from decoys and other objects, track the warhead, assign an interceptor to the warhead, track the interceptor during its flight and provide updated guidance information to it, assess whether the interception was successful, and to repeat the process if needed.

This line of reasoning ignores the history of the Treaty negotiations, which clearly suggest that ABM sensors do not have to perform the full spectrum of ABM battle management functions in order to be subject to the limitations of the Treaty. This line of reasoning also seems to rest on an extremely limited conception of the nature of the components that constitute an ABM system, in which there is a single sensor, such as a radar, that performs all of the tracking and battle management functions for the interceptor.

Although there are some missile defence systems with a single sensor (e.g., the previously proposed Site Defence system) they are the exception, rather than the rule. In practice, most missile defence systems have more than one sensor component, each of which plays some role in the management of the battle.

For example, the early Nike-Zeus system had not one or two, but *four* separate types of radars, for target acquisition, decoy discrimination, target tracking and interceptor tracking. Under this interpretation of the difference between a 'component' and an 'adjunct', all of these radars would be considered to be adjuncts to one another, and one of them would be considered to be a component.

Yet, the Airborne Optical System performs a role similar to that of the Perimeter Acquisition Radar (PAR) in the Sentinel/Safeguard system. Radars such as the PAR were clearly considered to be ABM components, and subjected to strict limitations in the Treaty. The United States went so far as to make a unilateral statement concerning the limitations on the Soviet's Hen House radars, even though these radars could play even less of a role in ABM battle management than that played by the PAR.

One danger in adopting the Administration's logic is that the distinction between a component and an adjunct would be impossible to achieve within current verification capabilities. A very detailed understanding of the performance capabilities of components, and of the complex interaction of components with each other, would be required to determine whether a particular activity was fully in compliance with Article V. This would clearly be beyond the capabilities of existing national technical means of verification.

For example, in the case of the Airborne Optical System, national technical means could observe that a large infrared telescope was mounted on an aircraft which was being used in conjunction with strategic missile testing. This activity would, on a *prima facie* basis, raise questions of compliance with Article V. Distinguishable and observable external characteristics, such as the performance of the aircraft, and particularly the aperture of the telescope, would indicate the suitability of the sensor for the ABM role, making a more definitive determination possible. Indeed, one of the enunciated rationales for United States ABM research is to provide the technical data needed for making such assessments about Soviet systems on the basis of information derived from national technical means of verification.

Some in the Administration would add the further criterion that the detailed characteristics of the sensor hardware and the computer software of the device be sufficiently capable that they could effectively perform the ABM mission in practice. This would, of course, be impossible to determine by national technical means, and it is difficult to imagine any means to make such a determination short of observing the successful operation of the system as a whole.

By that time the component would be perfected, the development process completed and the Treaty circumvented. The clear intention of Article V was to limit the development of new types of ABM technology at the earliest possible stage, that is, at the time that they would become detectable by national technical means.

Modifications to planned demonstrations under the SDI have further reduced the applicability of this interpretation of the Treaty put forward by the Reagan Administration. In particular, the demonstrations of the Airborne Optical System (AOS) and the Talon Gold pointing and tracking experiment, which were scheduled for 1987 prior to the initiation of the SDI, have been restructured in order to enhance the ABM capabilities of the demonstrated technologies. In the case of the AOS, this includes the addition of a laser range finder and on-board computers that will enable AOS to perform virtually the entire range of ABM battle management functions. This has resulted in a delay of over one year in the AOS demonstration, as well as a decision that the parts of Talon Gold will initially be demonstrated on the ground.

(b) The Reagan Administration and the ABM Treaty

In seeking to justify United States programmes as lawful under the ABM Treaty, senior Administration officials do not appear to be concerned that these types of ABM development efforts fundamentally compromise United States compliance with the ABM Treaty. On 8 April 1984 Defense Secretary Caspar Weinberger stated that 'I've never been a proponent of the ABM Treaty'.[47] Weinberger has also made many erroneous statements regarding the Treaty. For example, on 13 April 1981 Weinberger mistakenly told an interviewer, 'The treaty limiting anti-ballistic missiles expires in 1982.[48] (The ABM Treaty is of unlimited duration.) On 24 March 1983 he stated, 'The treaty only goes to block deployment.'[49]

Assistant Secretary of Defence Richard Perle, a chief architect of Reagan's arms control policy, voiced the opinion of many Reagan officials in 1982 testimony before Congress. Stated Perle, 'I believe that this review in 1982 of the ABM Treaty is an appropriate occasion to raise some questions about the underlying logic of that treaty because the preclusion of strategic defense as that treaty entails is, in my judgement, destabilising. It was a mistake in 1972 and the sooner we face up to the implications of recognising that mistake the better.'[50]

Some in the Reagan Administration have focused on the Krasnoyarsk radar as a means to justify United States 'breakout' of the Treaty's provisions. In testimony before the Senate Armed services Committee in March 1984 Richard Perle said that one of the responses the Reagan Administration was considering as a result of the Soviet radar was to deem ourselves no longer bound by the ABM Treaty. Other Presidential advisers share Perle's feelings about the ABM Treaty and would like to abrogate it whenever it becomes politically convenient to do so.[51]

(c) Efforts since October 1985 to reinterpret the ABM Treaty

A new version of the ABM Treaty was introduced by National Security Advisor Robert McFarlane on a nationally televised interview programme on 6 October. McFarlane asserted that research, development and testing of defensive systems[52] involving new physical concepts ... are approved and authorised by the treaty. Only deployment is foreclosed.' According to this new interpretation, sea-based, air-based, space-based and mobile land-based 'exotic systems and components',[53] such as those being pursued in the SDI, may be developed and tested, but not deployed, consistent with the ABM Treaty. Unfortunately, the result of McFarlane's announcement is much more than a legalistic squabble. Unless Congress intervenes and limits strategic defence expenditures according to the traditional United States interpretation, wherein development and testing of space and other mobile basing modes are prohibited, all restraints on Reagan's Star Wars plan may come unleashed. In the words of its chief negotiator, Gerard Smith, the ABM Treaty would be rendered a 'dead letter'.

Prior to the summit in early October 1985 the Administration had concluded that because *the Soviets never agreed to the United States position at SALT I, the Soviets cannot be held to abide to it today and, therefore, the United States is not legally bound*.

On 11 October the President decided that he agreed 'in principle, but not in practice' with this 'reinterpretation'.[54] Based on a Presidential directive, Secretary Shultz announced on 14 October in a speech before the thirty-first Annual Meeting of the North Atlantic Assembly that 'a broader interpretation of our authority is fully justified', but SDI 'will be conducted in accordance with a restrictive interpretation of the treaty's obligations'.[55]

This leaves the United States *legally* free to return to the 'reinterpretation' whenever the President and his advisors deem advantageous.[56] Department of Defense officials do not admit that they have yet lost the argument, and stress that Secretary Shultz did not state how long the Administration would continue to abide by the *new* 'restrictive interpretation', which represents presidential policy rather than a matter of law.

The legal rationale for the reinterpretation revolves around Article II(1) and Agreed Statement D. Article II(1), which defines ABM systems, includes the phrase 'currently consisting of' immediately before the definitions of 'traditional' ABM components. The administration argues that 'currently consisting of' would be better understood if the comma in the text were deleted and the text read 'and only those consisting of'. Therefore, the administration argues (1) the treaty, and particularly Article V(1), constrains only 'traditional' ABM technology (ABM interceptor missiles, ABM launchers and ABM radars). (2) The treaty permits development, testing and deployment of 'exotic' ABM systems and components, however based. (3) Agreed Statement D implicitly amends Article V(1) and Article III to prohibit *deployment* of 'exotic' systems and components whatever their basing mode.

This rationale is absurd as a matter of policy, intent and interpretation. If the Administration sticks with it as the best legal interpretation of the Treaty, then the Administration has effectively repudiated the ABM Treaty as a legal instrument. If the truncated Treaty remains in effect, then both the United States and Soviets can develop and test, *without quantitative or geographic limits*, any sea-based, air-based, space-based or mobile land-based ABM system or component provided they are based on 'exotic systems and components'.

But the result could be even more far-reaching. Because the Administration's new interpretation is that Article V(1) and other Articles of the Treaty do not apply to 'exotic systems' and Agreed Statement D blocks only their deployment, then the necessary consequences are that the limits on 'ABM systems or components' throughout the Treaty do *not* include 'exotic systems'. This results in:

1. the deployment bans on a *nationwide* ABM defence, a *base* for such a defence, and a *regional* ABM defence (except as permitted by Article III) in Article I(2), which were fundamental statements of the Treaty's scope, are all limited to 'traditional' ABM technology, (i.e., ABM launchers, ABM missiles and ABM radars) and do not apply to 'exotic systems';
2. the words 'currently consisting of' in Article II(1), intended to make clear that the Treaty applied to all ABM technologies and not just 'traditional' ones, are rendered devoid of meaning;
3. because Article IV dealing with ABM test ranges explicitly refers back to Article III (which authorises limited deployments of *fixed, land-based* ABM launchers, ABM missiles and ABM radars), the geographic, quantitative and implicit qualitative limits in Article IV on ABM tests do not apply to *tests* of any type of mobile or space-based 'exotic systems';
4. Article V does not apply to all or almost all SDI programmes, and HOE (which was a kinetic-energy test with a

single intercept mechanism) could have been tested in a MIRVed configuration; and

5. the prohibitions in Article IX against transfers of ABM systems or their components to other States, and deployment outside national territories, apply only to 'traditional' technology and not to 'exotic systems'.

The consequences of this 'reinterpretation' are dramatic when one considers that the principal United States concern has historically been with Soviet 'breakout' capability based on 'traditional' or 'low-tech' systems. These remain tightly constrained notwithstanding the 'reinterpretation'. On the other hand, most of SDI is now 'legally' unconstrained by the Treaty.

With particular respect to the Soviets and their emphasis on 'traditional' systems: (a) ABM deployment is limited to the one area surrounding Moscow; (b) ABM tests must be limited to their two ABM test ranges; (c) the development, testing and deployment of land-mobile 'traditional' ABM systems and components is prohibited; and (d) the ban on the 'upgrade' of SAM systems remains in full force. However, under the 'reinterpretation' the Soviets now legally could place in the field an unlimited number of *mobile* land-based lasers (the Soviets have an active laser programme) across the Soviet Union provided they were labelled for 'test' purposes.

With particular respect to the United States, it is now free to exploit Western technology in the full pursuit of Star Wars. A full-scale, operational orbiting system with accompanying ground stations and including as many as 100 to 400 killer satellites and related sensors could now be 'legally' put in place as an extensive 'test programme' to test the new technology in a BMD *system* configuration. United States allies would be free of any Treaty restraints to participate in *two-way* transfers of most SDI technology, with the only 'legal' constraints on 'West–West' SDI technology transfers under the Munitions Control an Export Administration Acts.

This result is absurd. Unbeknown to the United States SALT I delegation, the SALT I backstopping apparatus in Washington, the Nixon Administration and each of its successors and Congress, the United States would now be in the most one-sided Treaty relationship imaginable.

Of course, it could not last for a minute. Arms control agreements are viable only as long as they are in the *net interests* of each party. Secretary Shultz had spoken of the need to 'prevent the erosion of the

ABM Treaty', but Defence Secretary Weinberger, Under Secretary Iklé, and Assistant Secretary Perle have repeatedly stated that they have no use for the ABM Treaty and the sooner the United States is rid of it the better.[57] Secretary Weinberger's 13 November 1985 letter to the President, leaked to the press on the eve of the Geneva summit, reinforces the view that treaty commitments that impinge United States programmes are of little relevance to the Office of the Secretary of Defence.[58] Unless the President or Congress repudiates this self-defeating step, OSD officials can claim that any action they wish to take short of full-scale final deployment is legally permitted under the Treaty.

The timing of the announcement of the initial 'reinterpretation' remains obscure. The Department of Defense has known, of course, that under the historic interpretation the evolution of SDI research into development and testing would have to be stopped somewhere between 1988, as I believe, and 1990 as even the Department of Defence officials have privately conceded unless either the Soviets agree to amend the Treaty or the United States formally withdraws. From a policy and political point of view, six weeks before the Summit, the 'reinterpretation' by the United States with respect to a legally-binding treaty could have been a disaster. The first concrete United States response to the Soviet proposal (admittedly lopsided) to cut offensive forces by 50 per cent was to repudiate the ABM Treaty which, as both had agreed last January, is interrelated to any offensive limitations.

One of the political reasons for the Administration's initial 'reinterpretation' at this time may have been the Department of Defence's attempt to encourage more Allies to support SDI by participating in co-operative SDI 'research'. Foreign corporations, particularly in the United Kingdom and West Germany, might be encouraged by the 'reinterpretation' because co-operation might be extended from ABM 'research', which is all that is permitted under the historic United States interpretation, to include now 'development and testing' with full sharing and two-way transfers.

The actual effect on United States Allies was the reverse, because the political fallout of this full sharing in SDI technology directly associated with ABM systems or components would have been the implicit or explicit ratification by Allied governments of the repudiation of the ABM Treaty. That is a role that none is prepared to accept or condone, including the United Kingdom. Margaret Thatcher earlier had achieved at Camp David the President's agreement to four basic principles relating to SDI. Compliance with the ABM Treaty was one of them. The political cost in West Germany and the Netherlands might be much higher for their governments and NATO as a whole.

The most immediate consequence of President Reagan's having agreed in principle, but not in practice, with this reinterpretation is uncertainty, because the President's 'policy' of following the new interpretation could be reversed at any moment. OSD will certainly urge the President to abandon this 'policy' restraint before the end of his term, probably arguing that this would be a fitting response to the Soviet radar at Krasnoyarsk.

More fundamental, there are now no legal or Treaty constraints on developing or testing 'exotics'. OSD has succeeded in repudiating a fundamental purpose of the Treaty while, at the same time, urging rejection of the extension of the 'no undercut' policy relating to offensive agreement.

The net effect is an abandonment of arms control past and certain frustration of arms control future, unless the President or Congress reject this course.

(d) Comments on the Department of State's 29 October 1985 publication – 'Analysis of US Post-Negotiation Public Statements Interpreting the ABM Treaty's Application of Future Systems'[59]

The Administration is using two arguments supporting its 'reinterpretation': (1) the Soviets never agreed to the historic (now labelled restrictive) interpretation[60] and (2) that the restrictive view was the *minority* view in 1972.

It is not clear whether the latter argument, set forth in the State Department's Analysis, reflects incredibly shabby research and analysis done in haste or whether it represents a studied and disingenuous attempt to rewrite history. The *Sofaer* (legal adviser, State Department) prepared statement *rejects* a functional' approach to Article II, but the State Analysis *omits* the clear and definitive statement in the submittal letter from Secretary of State Rogers to President Nixon which emphasises that the Treaty *adopts* a functional approach.

The State Analysis states that the 'first indication' of the historic (or restrictive) interpretation 'came in a book written by John Rhinelander in 1974'.[61] If the authors of the State Analysis had contacted me, which they did not, they would have learned that:

1. I wrote detailed, still classified memoranda for the SALT delegation in the spring of 1972 which included the meaning of Article I(2) and the broad ban of Article V(1), and that *no* representative on the United States SALT delegation (State, ACDA, OSD or JCS) disagreed with my conclusions;
2. the memoranda served as the basis for the United States public positions before, during and subsequent to the 1972 hearings on the ABM Treaty;
3. I wrote the book chapter in 1972–3, *less than a year after leaving government*, and cleared it with government officials to ensure both accuracy and non-disclosure of sensitive information. I wrote the chapter at the urgings of government officials to provide a public document, in summary form, on what the ABM Treaty and other three SALT I agreements mean.

The State Analysis (p. 1) relies upon and purports to suggest that the summary in the periodic ACDA compilation, *Arms Control and Disarmament Agreements – Texts and Histories of Negotiations*, accurately reflects the United States position. It does not, and never has. In seeking to compress the material, the compilation necessarily skims or omits issues.[62]

The State Analysis fails to comprehend that the major focus in the public statements subsequent to the signing of the Treaty on 26 May 1972 was on *fixed land-based* 'exotic' systems because the President's instructions to the United States delegation in July 1971, reflected in the final text, was to *permit* the development and testing, but *not* the deployment, of fixed land-based 'exotics'. The JCS and the Senate Armed Services Committee wanted to preserve the right to pursue the development of fixed land-based lasers. The major argument with the Soviets had been on fixed land-based 'exotics'.

The question of spaced-based 'exotics' was not a contentious issue since the United States were not pursuing them at that time. The Soviets had agreed to the ban on space-based 'exotics' by September 1971, less than two months after the issue was first raised. No one on the United States delegation, including those who had originally opposed any limits on 'exotics', suggested there was any doubt as to what the Treaty as negotiated meant on the 'exotic' issues or that the Soviets had agreed.

The State Analysis (pp. 3–5) stresses Ambassador Smith's oral comments in Moscow and statements of Secretary Rogers and

Ambassador Smith before the Senate, and draws inferences from them. *Nowhere, however, does the State Analysis suggest that any witness stated that 'exotic' space-based systems were not broadly banned.* The State Analysis relies on the imprecision of certain statements, which implicitly were focused on *fixed land-based* systems, to reach a conclusion on *space-based* ones. This is a *non sequitur*. If the State Analysis were even half complete, it would have noted that Ambassador Smith's comments in Moscow, the 10 June 1972 submittal letter, and various statements suggested just the opposite conclusion in discussing *Article II(1)*.

For instance, as Ambassador Smith stated in Moscow at the news conference on 26 May 1972, 'Now, *Article II* defines what we are talking about and *has a very important bearing on the whole question of what we call future systems* (my emphasis)'.[63] The State Analysis quotes this sentence, but does not comprehend the reference to *Article II*, which contains the definitions for the entire Treaty. The State Analysis (p. 2) notes that Ambassador Smith failed to refer to *Article V*, but this was unnecessary.

More explicitly, the submittal letter dated 10 June 1972 from the Secretary of State to the President includes the following:

> Article II defines an ABM system as 'a system to counter strategic ballistic missiles or their elements in flight trajectory.' It *indicates* that such systems *currently consist of* ABM interceptor missiles, ABM launchers and ABM radars (my emphasis).[64]

With respect to development and testing, the submittal letter states:

> Article V limits development and testing, as well as deployment, of certain types of ABM systems and components. Paragraph V(1) limits such activities to fixed, land-based ABM systems and components by prohibiting the development, testing or deployment of ABM systems or components which are sea-based, air-based, space-based or mobile land-based.[65]

Nowhere is there any suggestion in the submittal letter that these blanket prohibitions exclude 'exotics'. With respect to the discussion in the immediately following section under the heading Future ABM Systems, the submittal letter notes that:

> Article II(1) defines an ABM system *in terms of its functions* as a 'system to counter strategic ballistic missiles or their elements in flight trajectory', *noting* that such systems *'currently'* consist of

ABM interceptor missiles, ABM launchers and ABM radars (my emphasis).[66]

While this sentence is set in a paragraph basically discussing fixed land-based systems, it is evident from this text and the submittal letter as a whole that the Treaty adopts a *functional* approach which is applicable to Article V and every other substantive Article (i.e., Articles I(2), III, IV, V, VI and IX).

It is mindboggling to the extreme to understand how the Sofaer prepared statement, but particularly the State Analysis, can now claim that the Treaty does *not* adopt a *functional* approach. The only answer is that by *omitting* these *key* sentences and quoting other relevant material out of context the intent of the State Analysis was not to be forthright. Further, while the Sofaer prepared statement distorts the entire Treaty because of Agreed Statement D and falsely states that other physical principles were not well understood in 1972, the submittal letter sets Agreed Statement D in the context of reinforcing Article III.[67]

Finally, the State Analysis (pp. 5–8) of the Oral Testimony before the Senate Armed Services Committee is shabby if not intentionally misleading. It suggests that testimony of Defence witnesses is of lesser consequence than that of Secretary Rogers and Ambassador Smith (p. 8), but even that incredible approach does not explain the *omission* of the written response by the Department of Defense to one of the questions submitted to *Secretary of Defense Laird* by Senator Goldwater. The State Analysis also tries to finesse what General Palmer said absolutely clearly, by referring to what he had said *earlier* in the exchanges.

It is important to understand the setting of the Senate hearings. The Senate Armed Services Committee began its hearings on 6 June, a week *before* the transmittal letter of 13 June from the President and the submittal letter of 10 June from the Secretary of State were released by the White House. The third of three 'Assurances' set by the JCS as conditions for JCS support of SALT I was 'Vigorous Research and Development Programmes'.[68] A statement by Senator Jackson had said that a laser contract for defensive purposes had been cancelled because of the Treaty. On 6 June 1972 Secretary Laird denied to Senator Jackson that a contract had been cancelled.[69] Ambassador Gerard Smith denied to Senator Goldwater at a later date that development of lasers was prohibited.[70] His response was of course true, but had nothing to do with Articles II(1) or V(1).

The Department of Defense written answer for Secretary Laird in response to Senator Goldwater agreed with the historic interpretation. The response, cleared in the inter-agency process which almost certainly involved Paul Nitze and General Allison as well as Ambassador Smith, provides[71]

Question: The ABM bit does not bother me too much, although I have not seen the fine print. For my money, we should have long since moved on the space based systems with boosting phase, destruction with shot, nuces, or lasers. I have seen nothing in SALT that prevents development to proceed in that direction. Am I correct?

Answer: With reference to development of a boost-phase intercept capability of lasers, there is no specific provision in the ABM Treaty which prohibits development of such systems.
There is, however, a prohibition on the development, testing, or deployment of ABM systems which are space-based, as well as sea-based, air-based or mobile land-based. The US side understands this prohibition not to apply to basic and advanced research and exploratory development of technology which could be associated with such systems, or their components.
There are no restrictions on the development of lasers for fixed, land-based ABM systems. The sides have agreed, however, that deployment of such systems which would be capable of substituting for current ABM components, that is, ABM launchers, ABM interceptor missiles, the ABM radars, shall be subject to discussion in accordance with article XIII (Standing Consultative Commission) and agreement in accordance with article XIV (amendments to the treaty).

This authoritative response on behalf of the Secretary of Defense *at the beginning of the hearings* should put the question to rest once and for all. Each following statement was consistent with it. While some did not clearly limit the response to fixed land-based system, the answers are clear in context.

The exchanges between Senator Margaret Chase Smith and Dr John Foster, and between Senator Jackson and Dr Foster on 22 June 1972[72] clearly and correctly concluded that: (1) an 'adjunct' to a fixed land-based 'traditional' component is not limited, whether the 'adjunct' is 'traditional' or 'exotic'; (2) a fixed land-based 'exotic' can be developed and tested, but not deployed; and (3) a space-based 'exotic' cannot be developed, tested or deployed. The State Analysis (p. 6) misquotes Foster at one point by changing 'deployment' to 'development'. The key exchanges with Foster, first on fixed land-based and then on space-based 'exotics', are:

> *Senator Smith*: Mr Secretary, would you tell the committee what progress is being made in the research and development of the laser [deleted].
> *Dr Foster*: Certainly, Senator Smith. The United States today has a research and development effort in lasers that totals a little above [deleted].
> *Senator Smith*: Thank you, Mr Secretary. Is there anything in the agreements that would prevent us from continuing our effort along this line?
> *Dr Foster*: There is nothing in the agreements, Senator Smith that prevents us [deleted]. The agreement does forbid the replacement of the currently allowed defence, that is, interceptor missiles, by a laser system.
> *Senator Smith*: In other words, the laser, if it was developed to the ultimate, could not be used at one of the two sites?
> *Dr Foster*: Yes, its deployment would be prohibited by the treaty. A laser could be used as part of an auxiliary designator system but it could not be used in substitution for a prime detector, that is, the ABM radar, or interceptor missile component.
> *Senator Smith*: But that will not slow us up or slow us down on continued research and development of the laser, will it?
> *Dr Foster*: No Senator, it will not. [Deleted.]

* * *

> *Dr Foster*: Senator Jackson, I support the treaty that has been signed; I share with you the concern about the survival on Minuteman but I have an even greater concern and that is for the long-range future in the face of a growing and already superior Soviet research and development effort.
> Quite frankly, I was shocked when the House Armed Services Committee reported out a bill that cut the proposed research and development for fiscal 1973 by $482.9 million. I don't see how the United States can keep its current momentum in research and development. There is nothing more important to the security of this Nation than our will to survive and the maintenance of our superiority in research and technology.

Senator Jackson: I couldn't agree with you more, but, you see, the problem is that we are creating a climate of confusion. I want to – before we conclude here – is there anything in these agreements that impinge on our right to research those areas that bear on both our defence and on defense capability? Specifically, there is a limitation on lasers, as I recall, in the agreement and does the SAL agreement prohibit land-based laser development?
Dr Foster: No, sir; it does not. [Deleted.]
Senator Jackson: [Deleted.]
Dr Foster: [Deleted.] What is affected by the treaty would be the development of laser ABM systems capable of substituting for current ABM components.
Senator Jackson: I am saying offence and defence, now.
Article V says each party undertakes not to develop and test or deploy ABM systems or components which are sea-based, air-based, space-based or mobile land-based.
Dr Foster: Yes, sir; I understand. We do not have a programme to develop a laser ABM system.
Senator Jackson: If it is sea-based, air-based, space-based, or mobile-land based. If it is a fixed land-based ABM system, it is permitted; am I not correct?
Dr Foster: That is right.
Senator Jackson: What does this do to our research –
I will read it to you: section 1 of article V – this is the treaty: 'Each party undertakes not to develop' – it hits all of these things – 'not to develop, test or deploy ABM systems'. You can't do anything: you can't develop; you can't test and finally, you can't deploy. It is not 'or'.
Dr Foster: One cannot deploy a fixed land-based laser ABM system which is capable of substituting for an ABM radar, ABM launcher, or ABM interceptor missile.
Senator Jackson: You can't even test; you can't develop.
Dr Foster: You can develop and test up to the deployment phase of future ABM system components which are fixed and land-based.
My understanding is you can develop and test but you cannot deploy. You can use lasers in connection with our present land-based Safeguard system provided that such lasers augment, or are an addendum to, current ABM components. Or in other words, you could use lasers as an ancillary piece of equipment but not as one of the prime components either as a radar or as an interceptor to destroy the vehicle.

Senator Jackson: The way I read this – but I may be wrong; it depends upon the interpretation here – but it says each party undertakes not to develop, test or deploy ABM systems or components which are sea-based, air-based, space-based, or mobile land-based.
Dr Foster: That is correct.
Senator Jackson: Now, it could well be read into this that even though you are conducting research you have not deployed it, that you cannot do that either. The way I read this, Mr Chairman – you might take a look at it – I think it raises a real question here whether you can actually engage in research.
[The information follows:]
Article V prohibits the development and testing of ABM systems or components that are sea-based, air-based, or mobile land-based. Constraints imposed by the phrase 'development and testing' would be applicable only to that portion of the 'advanced development stage' following laboratory testing, i.e. that stage which is verifiable by national means. Therefore, a prohibition on development – the Russian word is 'creation' – would begin only at the stage where laboratory testing ended on ABM components, on either a prototype or breadboard model.
Dr Foster: No.

John Foster was the senior technical official at Department of Defence. While the State Analysis (p. 6) avoids the point, John Foster's submission for the record on the dividing line between permitted 'research' and prohibited 'develop' was important and makes sense only if space-based 'exotics' are banned at the development stage.

The exchanges between General Palmer, Senator Goldwater and Senator Jackson on 19 July, if read in full,[73] leave no doubts that the two Senators and this JCS representative, the Acting Chief of Staff of the Army had responsibility for ABM programmes understood and agreed. The key and conclusive colloquy is:

General Palmer: On the question of the ABM, the facts are that when the negotiation started the only system actually under development, in any meaningful sense, was a fixed, land-based system. As the negotiations progressed and the position of each side became clear and each understood the other's objective better, it came down to the point where to have agreement it appeared – this is on the anti-ballistic missile side – this had to be confined to

the fixed, land-based system. The Chiefs were consulted. I would have to go to a closed session to state precisely the place and time. They were consulted on the question of qualitative limits on the ABM side and agreed to the limits that you see in this treaty.
Senator Jackson: Even though it can't be monitored?
General Palmer: Yes.
Senator Jackson: I just want that; so the Chiefs went along with the concept here that involved –
General Palmer: A concept that does not prohibit the development in the fixed, land-based ABM system. We can look at futuristic systems as long as they are fixed and land-based.
Senator Jackson: I understand.
General Palmer: The Chiefs were aware of that and had agreed to that and that was a fundamental part of the final agreement.[74]

In brief, the State Analysis is a sorry product. It would be amusing, if the issue were not serious, to suggest as it does that the views of the Secretary of Defense, the Undersecretary of Defense for Defense Research and Engineering, and the Army representative of the JCS on the implications of the ABM Treaty on United States weapons systems count for nothing. Senator Jackson clearly understood the 'exotic' issues as he repeatedly led the witnesses. Richard Perle sat in many of the key hearings, including the day General Palmer testified.[75]

Senator Buckley clearly understood, also. He had testified on an early date before the Foreign Relations Committee. One of the issues he stressed on 29 June in his oral remarks and his written statement was that the ABM Treaty prohibited the development and testing of a space-based laser.[76] Senator Cooper stated he disagreed with Buckley's opposition to the Treaty but commended him on his detailed knowledge.[77] Senator Buckley stressed this ban on space-based lasers on the Senate floor before voting against the Treaty on 3 August 1972.[78]

The vote in favour of advice and consent to ratification was 88–2. The Administration's support for its reinterpretation rests on the implicit argument that testimonies of Pentagon witnesses before the Armed Services Committee are of little weight in assessing the Senate's understanding of the ABM Treaty.

A fair and dispassionate review of the 1972 position of the United States would have led to a far different result from the Sofaer prepared statement and the State Analysis – the Executive stated in

June–July 1972, and the Senate understood, that Article II(1) was functional and that Article V(1) banned space-based 'exotics'. It is clear that the current Administration has repudiated the basic scope and purpose of the ABM Treaty thirteen years after ratification by the Senate.

(e) Major weaknesses and contradictions of the Reagan Administration's reinterpretation of the ABM Treaty

The Administration's 'reinterpretation' as advocated in the prepared Sofaer statement and his response to question is, as a technical legal issue, egregiously flawed in at least four accounts:

(a) It rests, in effect, on deleting the comma, inserting an 'and' in its place before the phrase 'currently consisting of' in Article II(1), and rewriting the text to read 'currently utilised physical principles'.[79] This twists a *functional* definition into one of *limitation*. Article II(1), including the comma and the words 'currently consisting of', were approved by the *entire* United States delegation to make clear the Treaty was based on a *functional* approach.

(b) It also rests on a new canon of construction, never before heard of, that the *unambiguous* text of a treaty should be distorted to give an agreed interpretation an independent and amendatory role.

(c) It totally ignores the Vienna Convention on the Law of Treaties and the ALI Restatement of the Foreign Relations Law of the United States which stress the importance of subsequent *practice* in interpreting a treaty.[80] Subsequent practice, including statements, of *both* the United States and the Soviets, *reinforce* the historic interpretation of the ABM Treaty.

(d) Finally, it is internally inconsistent. The 'reinterpretation' is based on a construction of the entire Treaty, including Article III, limiting 'ABM' systems and their components to 'traditional' technology. Therefore, if the Treaty is so read, Agreed Statement D is meaningless because it is explicitly linked to fulfilling the obligations of *Article III*. Accordingly, the only logical interpretation of the Treaty, under this distorted reading of Article II(1), is that *there are no deployment limits on 'exotic' systems under either the*

Treaty or Agreed Statement D. Nevertheless, the Sofaer statement abandoned legal analysis at some point and interpreted the 'spirit' of Agreed Statement D to decide that it *amends* Articles III and V(1) and prohibits deployment of 'exotics'.

In 1972-3, within a year after leaving government, I wrote a chapter in a book which *has been until recently the standard unclassified reference in government on the interpretation of the ABM Treaty*. To the best of my knowledge, it accurately reflects the detailed classified analysis of the ABM Treaty which I prepared while a member of the SALT I delegation. This chapter includes the statement that:

> The future systems ban applies to devices which would be capable of substituting for one or more of the three basic ABM components, such as a 'killer' laser or a particle accelerator. Article III of the treaty does not preclude either development or testing of fixed land-based devices which could substitute for ABM components, but does prohibit their deployment. Article V, on the other hand, prohibits development and testing, as well as deployment, of air-based, sea-based, space-based, or mobile land-based ABM systems or components, which includes 'future systems' for those kinds of environments. The overall effect of the treaty, therefore, is to prohibit any deployment of future systems and to limit their development and testing to those in a fixed land-based mode. Certain devices, such as telescopes, which are simply adjuncts to, not substitutes for, present ABM components are not covered.
>
> These constraints on future ABM systems, which the US proposed, are an attempt to censure a long-term, effective limitation on strategic defensive systems in an age of changing technology. Neither laser technology nor any other kind of device now appears to have a significant ABM potential; for instance, a land-based laser would not be effective in the event of cloud cover, and the platform for an air-based laser would itself be a vulnerable target. However, the failure to include a broad proscription against future devices might have led to a significant effort by either or both the US and USSR to develop esoteric ABM systems. The prohibition in the treaty thus serves the same preventive purposes as the prohibition on placing weapons of mass destruction in orbit which is contained in the Outer Space Treaty of 1967.

* * *

The effect of paragraph 1 of Article V is to limit ABM systems or ABM components which may be developed, tested, or deployed under Articles III and IV to those which are fixed land-based. While paragraph 1 of Article V expressly refers to sea-based, air-based, space-based, or mobile land-based ABMs, the intention was to foreclose ABM components except fixed land-based; therefore ABM components are prohibited not only from the bottom of the sea but also on or in lakes or rivers. The prohibitions in Article V apply to each of the three basic ABM components, as well as the system as a whole, and would therefore prohibit a sea-based ABM radar linked with land-based ABM launchers and interceptor missiles. It would also prohibit future ABM systems, such as an airborne killer laser, as a substitute for a fixed land-based ABM interceptor missile, or a space-based sensor that, coupled with land-based components, was capable of substituting for one or more current ABM components. There are, however, grey areas since the treaty does not limit deployment of sensors or satellites which provide early warning of missile launches.[81]

More recently this interpretation has also been supported in the FY85 Arms Control Impact Statement prepared by the Reagan Administration which correctly states the agreement reached with the Soviets on the meaning of Article V(1).[82]

The April 1985 SDI Report to Congress, which provides the justification for the SDI programme under the Treaty, is essentially consistent with this analysis.[83]

VI SOVIET ATTITUDES TOWARD THE ABM TREATY AND SOVIET REACTIONS TO AMERICAN EFFORTS TO REINTERPRET IT

In the Soviet parliamentary ratification deliberations, the First Deputy Minister of Foreign Affairs, Vasily V. Kuznetsov, 'on behalf of the Soviet Government', gave the Presidium of the Supreme Soviet the official Soviet position on the ABM Treaty. He said that, 'The sides pledge themselves not to create or develop ABM systems or components emplaced in the sea, the air or space or of a mobile ground type'. He presented this as a clear obligation of the Treaty as a whole.[84]

Based on my review of available documents, the Soviets had not explicitly tied this interpretation to 'exotic systems' in public until recently although their statements *implicitly* supported this as the only interpretation of the entire Treaty, including Article V(1).

From the best of my recollection and after discussions with former SALT I colleagues now out of government, there is no doubt that the Soviets understood that the text of the ABM Treaty prohibited the development and testing of space-based 'future' or 'exotic' systems based on the language the United States sought and obtained in Article II(1) and Article V(1).

The Soviets accepted the United States insistence that 'currently consist of' be included in the text of Article II(1). This was agreed (*before* the final negotiations on Agreed Statement D) in order to reinforce the United States approach to Articles III and V(1). This and other questions of interpretation were carefully reviewed by me with other members of the delegation during the seventh and final negotiating session (Spring 1972).[85]

The Gorbachev interview with *TIME*, which occurred a month *before* the 'reinterpretation' was announced, is specific and unambiguous: 'In our view, it [SDI] is the first stage of the project to *develop a new ABM system prohibited under the Treaty of 1972* (my emphasis).[86]

Specific Soviet responses to the United States 'reinterpretation'[87] include the translation of a TASS commentary and an English-language TASS article. The former, as translated by FBIS, includes:

> According to the CBS television company, one of the latest administration reports contains the 'conclusion' that the antimissile defence treaty, which strictly restricts the development [sozdaniye] of antimissiles, allegedly does not restrict the development [razrabotka] and testing of 'exotic' types of weapons – laser and beam weapons – at all. It is quite clear which way such 'interpreters' are taking the matter. Having just the other day tested land-based laser installations, the United States is now planning to site a laser weapon on board a spacecraft and test it directly in space.
> It would evidently not be inappropriate to remind some people in Washington yet again that the antimissile defense treaty (Article V) prohibits both the development [sozdaniye] and testing of space-based antimissile defence systems or components. The treaty provisions relate to any systems designed, as defined by Article II, for fighting against strategic ballistic missiles or their elements on

flight trajectories. Since the antimissile defense components being created within the 'star wars' programme are designed for precisely this purpose, that is, are intended to replace the antimissiles mentioned in the treaty (or to act together with antimissiles), all provisions of the treaty relate to these, regardless of the degree of 'exoticness' of their principles of operation. It is high time the irresponsible 'interpreters' [tolkovateli] from Washington gave up their useless and dangerous occupation, listened to the voice of the world public, which they are trying to delude, and directed their efforts to positive goals. And they do have something to think over: The set of Soviet initiatives offers broad scope for constructivism.

Finally, Marshall Sergei Akhromeyev, the Chief of the Soviet General Staff, made lengthy comments in an article in *Pravda*. He said the ABM Treaty 'unambiguously bans' the development, testing and deployment of space-based ABM systems.[88]

The article by Marshall Akhromeyev, the Chief of General Staff, states:

> In order to justify the militarisation of outer space, Washington alleges that the United States work on the so-called 'Strategic Defence Initiative' (SDI), in fact a Star Wars programme, is something quite legitimate and even allowed by the 1972 Soviet–American treaty limiting antiballistic missile defenses. A 'new interpretation' of the treaty has been offered, according to which it purportedly allows the development, testing and creation of antimissile weapons sytems on the basis of 'other physical principles', that is, laser, particle-beam and other types of weaponry, both land and space-based.
>
> Thus, the President's national security advisor Robert McFarlane, when appearing on NBC television programme on 6 October, distorted the essence of the ABM Treaty. Trying to substantiate the 'lawfulness of experiments' within the framework of the infamous Strategic Defence Initiative, he contended that the treaty supposedly sanctions tests of any ABM systems if only they are based on other principles of physics. The ABM Treaty is negated also by the so-called 'new' confidential study prepared by the Pentagon concerning limitations envisaged by the treaty. It is contended in the study that the provisions of the treaty supposedly can be applied only to radars and antimissiles but not to the

development and testing of 'exotic' ABM systems (lasers, beam weapons).

Such 'interpretations' of the ABM Treaty, to put it mildly, are deliberate deceit. They contradict reality. Article V of the treaty absolutely unambiguously bans the development, testing and deployment of ABM systems or components of space or mobile ground basing, and, moreover, regardless of whether these systems are based on existing or 'future' technologies.

In accordance with the agreed-upon statement 'D' to the treaty, to which the Administration now refers so often, the conduct of research, development, and testing of ABM systems or their components based on other physical principles is allowed in areas that are strictly limited and defined by the treaty and only using permanent land-based ABM systems (as they are defined in Article III of the treaty). Moreover, if either side wants to deploy this type of new system in these limited areas, it cannot do so without preliminary consultations with the other side and without introducing the appropriate agreed-upon amendments to the treaty.

Only this and no other interpretation of the key provisions of the ABM Treaty, which was initiated in its time by the United States itself, was worked out and adopted by the two sides in the course of talks on this treaty. The present aim of the United States Administration is clear: to prepare a 'legal base' for carrying out all the practical stages of work within the framework of the SDI programme, that is, the development, testing and deployment of space strike systems.

The ABM Treaty is becoming an obstacle to the United States in the fulfillment of Star Wars plans. In his striving to clear the road to the militarisation of outer space, Secretary of Defence Caspar Weinberger, when speaking in the National Press Club in Washington, threw aside the subterfuges of McFarlane and others and bluntly stated that the United States should study the possibility of really withdrawing from the ABM Treaty. This is the actual position of the United States.

The Soviet Union is of a diametrically opposite opinion. The termless ABM Treaty is of fundamental importance for the entire process of nuclear arms limitation. Even more, it is the basis on which strategic stability and international security rest. We are convinced that everybody, including the United States, will stand to lose from a violation of this treaty. The USSR is strictly observing

> all commitments under the treaty and is not doing anything that would contradict its provisions.
> The Soviet stand on space strike arms was clearly formulated by Mikhail Gorbachev. It is necessary for a ban to embrace every phase of the inception of this new class of arms. This, however, does not deny the right and possibility to conduct basic research in outer space. But it is one thing to conduct research and studies in laboratory conditions and quite another thing when models and prototypes are created and samples of space arms are tested. This is always followed by deployment of arms. It is precisely such a line, backing it up accordingly with propaganda, that the United States Administration is pursuing as regards the Star Wars programme. The USSR views as impermissible any out-of-laboratory work connected with the development and testing of models, pilot samples, separate assemblies and components. Everything that is being done for the subsequent designing and production of space strike systems should be banned.
> The Soviet Union's approach is substantiated and realistic also from the point of view that out-of-laboratory work can be verified by national technical means. If this process is cut short at the initial stage of research, the possibility of development space strike arms will vanish.[89]

Marshall Akhromeyev *explictly* confirms the historic United States position of the ban on space-based 'exotic systems'. The Soviet position is identical to the United States position in the last four Administrations that tight constraints on strategic defensive systems are a *precondition* to limits on offensive systems. The United States would *never* consider deep cuts on its offensive forces if the Soviets were deploying a nationwide ABM defence.

In my judgement there is no possibility whatsoever of the Soviets accepting reductions on offensive forces while the United States insists on the 'reinterpretation' of the ABM Treaty.

It is not clear to me that the Soviets will agree to discuss in detail deep cuts on offensive forces before or unless, the United States agrees to strengthen the ABM Treaty, including a repudiation of the 'reinterpretation' and negotiate a separate ASAT Treaty. This is apparently what the Soviets mean by 'Space Strike Arms.'[90] Sound military logic is on the Soviet side, as well as the United States insistence in 1967–72 that the precondition to any offensive limitations was tight limits on strategic defensive systems.

Some sophisticated analysts in the United States, who strongly support arms control, now suggest the Soviets should agree to phased, annual reductions on offensive systems and finesse SDI because domestic support for SDI would wither away. While the argument is clever, it is inconveivable to me that the Soviet government would opt for an approach that did not tightly constrain ABMs *by agreement*, as had been agreed in 1972.

If the Soviets increase their offensive capabilities (ICBM and SLBM warheads, penetration aids, heavy bombers, SLCMs and ALCMs), any strategic United States strategic defence could be overwhelmed or bypassed, as even SDI supporters admit.

The Administration's justification for its 'reinterpretation' is that the Soviets cannot be held to comply with the historic United States position. Instead of reinterpretating the clear text of the 1972 Treaty based on a selective review of the *classified United States negotiating records*, the better approach would have been to ask the Soviet negotiators in private in Geneva whether or not the Soviet Union agrees that Article V(1) bans the development, testing and deployment of 'exotic systems'. If the private Soviet response had been 'no', then the Administration's 'reinterpretation' would have been justified.

If the private Soviet response in Geneva were 'yes', as one would expect from their public statements since 1972, then the 6 October 'reinterpretation' and the 13 October recanting by the Administration would have been unnecessary. Agreed Statements on the basic points could have been quickly negotiated if deemed necessary for clarity.

If a private but positive Soviet response in Geneva were now rejected by the United States as 'too late' because the United States wanted to keep open the option of reasserting its 'reinterpretation', then the OSD motive behind the initial change in United States position – to erode immediately and eventually destroy the ABM Treaty – would be clear. This now appears to be the case. The question was not asked before the reinterpretation was announced on 6 October because the Soviets would have agreed, which is not what OSD wants.

If arms control is to remain an element of United States security policy, the challenge will be to strengthen the ABM Treaty through specific, mutual and verifiable Agreed Statements and Common Understandings. This challenge can be met if the political will exists.

VII CONCLUSION AND SUGGESTIONS

The primary issue of the renegotiation effort has been whether Article V(1) of the ABM Treaty prohibits the development and testing of space-based and other mobile-type 'exotic systems' (e.g., space-based lasers). The secondary issue has been whether *any* of the Treaty's substantive constraints on 'ABM systems or components' in Articles I(2), IV, V and IX apply to space-based 'exotic systems'. The answers are four-fold:

1. *the prohibitions are clear from the text of the Treaty*, particularly Article V(1) which states, 'Each Party undertakes not to develop, test or deploy ABM systems or components which are sea-based, air-based, space-based or mobile land-based';
2. the negotiating history, as interpreted in 1972 by the SALT I delegation and the backstopping representatives in Washington, supports the broad ban on space-based 'exotic systems' as the only permissible interpretation;
3. this has been the interpretation of the Executive, accepted and relied upon by Congress, since 1972;
4. any other result is patently absurd and would frustrate the stated premise of this Treaty of indefinite duration – to prohibit the deployment of nationwide ABM systems or a 'base' for such a system.

The Soviets accepted this interpretation during the negotiations, reflected it in their ratification proceedings, and have taken no actions and have not made any official statements inconsistent with this interpretation. This is the only conclusion one can draw from their public statements which sometimes deal with the issue implicitly and elliptically rather than explicitly.

The reinterpretation was the central reason for the collapse of the Reykjavik summit in October 1986. While the Soviets adopted an interpretation of the Treaty stricter than agreed in 1972, the radical and unilateral reinterpretation by the United States undercut the essential understanding that limitations on defensive systems are a precondition to those on offensive systems.

The challenge is now to agree on the Treaty as negotiated in 1972, and eventually to strengthen the ABM Treaty through specific, mutual and verifiable Agreed Statements and Common Understandings. Six former Secretaries of Defense endorsed the importance of the Treaty and the need to strengthen it *before* this controversy broke.[91]

Chapter 15 in this book contains a series of specific recommendations which were intended to start a constructive process consistent with Secretary Shultz's stated goal of reversing the erosion of the ABM Treaty.

In conclusion, let me suggest approaches I propose for three Agreed Statements based upon, and entirely consistent with, my recollection of the SALT I negotiating record which would clarify the overall scope of the ABM Treaty, particularly Article V(1);

1. *First Agreed Statement to Article II(1)*. As used in this Treaty, 'ABM systems', 'ABM systems or components', 'ABM systems and components and 'ABM systems or their components' include ABM interceptor missiles, ABM launchers, and ABM radars as defined in Article II(1) and any devices based on other physical principles which are capable of substituting for or performing the functions of ABM interceptor missiles, ABM launchers, or ABM radars.
2. *First Agreed Statement to Article V(1)*. Article V(1) applies to ABM components and any devices based on other physical principles which are capable of substituting for or performing the functions of ABM components, any of which are sea-based, air-based, space-based or mobile land-based.
3. *Second Agreed Statement to Article V(1)*. As used in Article V(1), 'develop' refers to that stage of the research and development cycle at which field testing, observable by national technical means, is initiated on ABM components or on any devices which are capable of substituting for or performing the functions of ABM components.

The third suggestion is obviously incomplete. It points out the compelling need to begin the difficult process of resolving some of the ambiguities inherent in the ABM Treaty. The SCC was established with this as one of its assigned tasks. The SCC has been underutilised. The SCC could easily, and quickly, also review and revise Agreed Statement D to make its intended meaning clearer. A starter in replacing Agreed Statement D could be:

1. *First Agreed Statement to Article III*. Article III prohibits the deployment of fixed land-based devices based on other physical principles which are capable of substituting for or

performing the functions of fixed land-based ABM systems or components as defined in Article II(1).
2. *First Agreed Statement to Article IV*. Fixed land-based devices based on other physical principles which are capable of substituting for or performing the functions of ABM components as defined in Article II(1) may be developed and tested at ABM test ranges described in Article IV.
3. *First Agreed Statement to Article XIV(1)*. Any obligation in this Treaty may be discussed in accordance with Article XIII and an amendment adopted in accordance with Article XIV.

The six suggested Agreed Statements do not even touch on the question of distinguishing a 'component', or device capable of substituting for or performing the function of a component, from a 'subcomponent', assembly, adjunct, etc. or the equally difficult question of distinguishing ABM-related space-based sensors from space-based sensors for early warning or for other purposes. Counting rules, presumptions, and *ad hoc* approaches will all be necessary. These challenges will be truly difficult even with the best of intents.

Before constructive steps can start, however, and assuming the Soviets are prepared to negotiate and not just posture, the President should publicly repudiate the *legal* advice he has recently received from his advisors on a narrow scope of Article V(1) and other critical Articles of the ABM Treaty. The United States Senate could have a critical role since senators have been reviewing the classified negotiating record since September 1986. Their affirmation of the historic interpretation could lead to progress in Geneva.

This whole sorry business could lead to a constructive ending if the United States and Soviets were to agree privately, before or at a 1987 summit in the United States, on the First Agreed Statements that I have suggested. This should be only the first of many steps needed to avoid further erosion of the ABM Treaty of 1972.

Notes

1. The author would like to thank Thomas K. Longstreth and John E. Pike to use a part of the booklet he co-authored with them: *The Impact of US And Soviet Ballistic Missile Defense Programs on the ABM Treaty* Washington, DC: National Campaign to Save the ABM Treaty, 3rd

edn, March 1985) especially pp. 3–10, 23–41. The author acknowledges the permission of the Arms Control Association to use a few quotes from his article: 'Reagan's "Exotic" Interpretation of the ABM Treaty, Legally, Historically, and Factually Wrong', *Arms Control Today* 15 (October 1985) pp. 3–6. This article consists in large parts of material previously published in the joint *Impact* study and in Congressional testimony. See especially US Congress, House of Representatives, Committee on Foreign Affairs, Hearing, *ABM Treaty Interpretation Debate*, 22 October 1985 (Washington: United States Government Printing Office, 1986). See here also for the statement by Abraham D. Sofaer (referred to as Sofaer prepared statement).
2. John B. Rhinelander, 'The SALT I Agreements', in Mason Willrich and John B. Rhinelander eds, *SALT The Moscow Agreements and Beyond* (New York: The Free Press: 1974) pp. 125–59.
3. Statement before the Senate Foreign Relations Committee, 12 November 1985 on ' "Exotic Systems" and the ABM Treaty'; 'Responses by Mr Rhinelander to additional questions submitted by Chairman Fascell', in US Congress, House, Committee on Foreign Affairs, Hearings: *ABM Treaty Interpretation Dispute* (Washington, DC: United States Government Printing Office, 1986) pp. 171–99.
4. Gerard Smith, *Doubletalk – The Story of the First Strategic Arms Limitation Talks* (New York: Doubleday, 1980); John Newhouse, *Cold Dawn. The Story of SALT* (New York: Holt, Rinehart and Winston, 1973); Raymond L. Garthoff, *Détente and Confrontation – American–Soviet Relations from Nixon to Reagan* (Washington, DC: Brookings Institution, 1985).
5. 1980 edition: *Arms Control and Disarmament Agreements – Texts and Histories of Negotiations* (Washington, DC: United States Arms Control and Disarmament Agency, August 1980) pp. 137–47.
6. Thomas W. Wolfe, *The SALT Experience* (Cambridge: Ballinger, 1979); Strobe Talbot, *Endgame. The Inside Story of SALT II* (New York: Knopf, 1979).
7. Hans-Henrik Holm and Nikolaj Petersen (eds), *The European Missiles Crisis. Nuclear Weapons and Security Policy* (London: Frances Pinter, 1983).
8. For details see *Arms Control Reporter. A Chronicle of Treaties, Negotiations, Proposals* (Cambridge, Mass.: Institute for Defense and Disarmament Studies) vol. 4 (1985).
9. For details see Raymond L. Garthoff, 'BMD and East–West Relations', in Ashton B. Carter and David N. Schwartz (eds), *Ballistic Missile Defense* (Washington, DC: Brookings Institution, 1984) pp. 275–329.
10. 'Appendix K: Excerpts from Soviet Statements on BMD', in US Congress, Office of Technology Assessment, *Ballistic Missile Defense Technologies*, OTA-ISC 254 (Washington, DC: United States Government Printing Office, September 1985) pp. 312–15.
11. Smith, *Doubletalk*, pp. 263–5.
12. Newhouse, *Cold Dawn*, pp. 230–1; 237; Garthoff, *Détente and Confrontation*, chapter 5.
13. Sid Graybeal was later the US Commissioner to the Standing Consulta-

tive Commission and Victor Karpov has been in 1986 the head of the Soviet delegation in Geneva.
14. Newhouse, *Cold Dawn*, p. 237.
15. Raymond Garthoff and Nicolai Kishilov were the executive secretaries of the respective delegations.
16. Smith, *Doubletalk*, pp. 265; 343–4.
17. See Common Understanding E: for text see *Arms Control and Disarmament Agreements*, p. 144, and Appendix to this volume, p. 565.
18. US Congress, Senate, Committee on Foreign Relations, Hearings on *Strategic Arms Limitation Agreements* (Washington, DC: United States Government Printing Office, 1972).
19. US Congress, Senate, Committee on Armed Services, Hearings on *Military Implications of the Treaty on the Limitations of Anti-Ballistic Missile Systems and the Interim Agreement on Limitation of Strategic Offensive Arms* (Washington, DC: United States Government Printing Office, 1972).
20. See *Congressional Record*, Senate, 3 August 1972.
21. Rhinelander 'The Salt I Agreements', pp. 128–9; 134.
22. 'U.S.–Soviet Arms Accords are no Bar to Reagan's Strategic Defense Initiative', *Heritage Foundation Backgrounder*, 421 (Washington, DC: Heritage Foundation, 4 April 1985).
23. *Fiscal Year 1985 Arms Control Impact Statements* (Washington, DC: United States Government Printing Office, 1984) pp. 251–2.
24. *Report to the Congress on the Strategic Defense Initiative 1985* (Washington, DC: Department of Defense, April 1985) Appendix B.
25. For text see note 5 and appendix to this volume.
26. *Fiscal Year 1984 Arms Control Impact Statements* (Washington, DC: United States Government Printing Office, 1983).
27. *Fiscal Year 1984*, pp. 131–5.
28. See Rhinelander, 'The Salt I Agreements'.
29. *The President's Strategic Defense Initiative* (Washington, DC: The White House, January 1985).
30. See for text note 5.
31. See note 5.
32. Smith in (note 19), *op. cit*, p. 377.
33. Foster in (note 19), *op. cit*, p. 275.
34. For *1984–1988 Five-Year Defense Guidance* see Richard Halloran, 'Pentagon Draws Up First Strategy for Fighting a Long Nuclear War', *New York Times*, 30 May 1982.
35. Paul B. Stares, *Space Weapons & US Strategy: Origins and Developments* (London: Croom Helm, 1985).
36. Keith Payne, 'Introduction and Overview of Policy Issues', in Keith B. Payne (ed.), *Laser Weapons in Space, Policy and Doctrine* (Boulder, Co.: Westview Press, 1983) pp. 9–13.
37. See the personal files of John Pike (Federation of American Scientists) and Tom Longstreth (Legislative Assistant of Senator Kennedy) in Washington, D.C.
38. Hans Günter Brauch, *Antitactical Missile Defense – Will the European Version of SDI Undermine the ABM-Treaty?* AFES Papier 1, (Stutt-

gart: AG Friedensforschung und Europäische Sicherheitspolitik, Institut für Politikwissenschaft, July 1985).
39. See chapter 14 by Brauch in this volume.
40. See chapter 16 by John Pike in this volume.
41. Fred S. Hoffman, *Ballistic Missile Defenses and US National Security. Summary Report*. Prepared for the Future Security Strategy Study (FSSS), (Washington, DC: Department of Defense, October 1983) p. 2.
42. Smith, *Doubletalk*, p. 53.
43. General Abrahamson, Statement before the House Armed Services Committee, 27 February 1985.
44. 'Proposed Remarks of Dr G. A. Keyworth to the Brookings Forum on the Future of Ballistic Missile Defense', *Ballistic Missile Defense: Current Issues* (Washington, DC: Brookings Institution, 29 February 1984).
45. US Department of Defense, *Defense against Ballistic Missiles: An Assessment of Technologies and Policy Implications* (Washington, DC: United States Government Printing Office, April 1984).
46. US Department of Defense, *Strategic Defense Initiative – Defense Technologies Study* (Washington, DC: United States Government Printing Office, March 1984).
47. Caspar Weinberger, as reported in ABC's 'This Week with David Brinkley' (Washington, DC: ABC, 8 April 1984).
48. Caspar Weinberger, see the files of the Arms Control Association in Washington for the *Impact Study*.
49. See *Impact Study*.
50. Richard Perle, as quoted in US Congress, Senate, Armed Services Committee, Hearings, *Department of Defense Authorization for Appropriation for Fiscal Year 1983*, part 7, (Washington, DC: United States Government Printing Office, 1982) p. 5001.
51. Richard Perle, Statement before the Senate Armed Services Committee, see in the files of the Arms Control Association for the *Impact Study*.
52. Richard McFarlane, Statement in *NBC, Meet the Press*, 6 October, 1985.
53. Don Oberdorfer, 'White House Revises Interpretation of ABM Treaty', *Washington Post*, 9 October 1985, p. 21.
54. Don Oberdorfer, 'Reagan Claims "Star Wars" Progress Does not Violate Terms of ABM Pact', *Washington Post*, 13 October 1985, p. 11.
55. Don Oberdorfer and David B. Ottaway, 'U.S. Clarifies ABM Pact View – "Restrictive Interpretation" Set for Space Defense Plan', *Washington Post*, 15 October 1985, p. 1.
56. Don Oberdorfer, 'White House Revises Interpretation of ABM Treaty', *Washington Post*, 9 October 1985, p. 21.
57. Anthony Lewis, 'This Switch Puts the Summit at Risk', *International Herald Tribune*, 15 October 1985.
58. 'Weinberger Letter to Reagan on Arms Control', *New York Times*, 16 November 1985.

59. Department of State, 'Analysis of US Post-Negotiation Public Statements Interpreting the ABM Treaty's Application of Future Systems' (Washington, DC: Department of State, 29 October 1985. This document will be quoted as State analysis in the subsequent text with the pages given in brackets.
60. See 'The ABM Treaty and the SDI Program', *Current Policy*, 755 (Washington, DC: US Department of State, Bureau of Public Affairs, October 1985) pp. 1–3.
61. Rhinelander, 'The Salt I Agreements'.
62. Arms Control and Disarmament Agreements, pp. 136–8 and Rhinelander, 'the Salt I Agreements', pp. 125ff.
63. See *Weekly Compilation of Presidential Documents*, 5 June 1972 (Washington, DC: United States Government Printing Office, 1972) p. 931.
64. See *Hearings on Military Implications of the Treaty*, p. 79.
65. *Hearings*, p. 81.
66. *Hearings*, p. 79.
67. *Hearings*, p. 81.
68. *Hearings on Strategic Arms Limitation Agreements*, p. 70.
69. *Hearings on Military Implications of the Treaty*, p. 30.
70. *Hearings*, p. 306.
71. *Hearings*, pp. 40–1.
72. *Hearings*, pp. 222; 274–5.
73. *Hearings*, pp. 437–44.
74. *Hearings*, p. 443.
75. *Hearings*, p. 399.
76. *Hearings on Strategic Arms Limitation Agreements*, pp. 256; 262.
77. *Hearings*, p. 269.
78. *Congressional Record*.
79. 'Analysis of US Post-Negotiation Public Statements'.
80. See ALI, Tentative Draft 6, para. 325 (12 April 1985).
81. Rhinelander, 'The Salt I Agreements', pp. 128–9; 134.
82. See *Fiscal Year 1985*.
83. See *Report to the Congress on the Stategic Defense Initiative 1985*.
84. See *Pravda*, 30 September 1972, translated in *FBIS* 3 October 1972.
85. See Section II(a) above.
86. See TIME (9 September 1985) p. 24 (my emphasis); see also note 10.
87. See for references *FBIS* 9 and 10 October 1985.
88. Philip Taubman, 'Soviet Military Chief Accuses US of Distorting Terms of ABM Pact', *New York Times*, 19 October 1985, p. 3.
89. See for reference *FBIS* 21 October 1985; see also in ABM Treaty hearing (FN.1), p. 305–311.
90. Committee of Soviet Scientists for Peace Against the Nuclear Threat, *Space-Strike Arms and International Security* (Moscow: Novosti, October 1985).
91. See 'Statement in Support of the ABM Treaty' (Washington, DC: National Campaign To Save the ABM Treaty, June 1984).

14 From SDI to EDI – Elements of a European Defence Architecture
Hans Günter Brauch

I FROM SDI TO EDI AND EXTENDED AIR DEFENCE

SDI has been a controversial political, strategic, technological and arms control issue in the domestic politics of most West European countries – within coalition governments (most visibly between Chancellor Kohl and Foreign Minister Genscher), among the NATO allies and between the United States and the Soviet Union.[1] The call for an EDI,[2] for a TDI[3], for an ATBM[4] or more recently for an extended air defence has become the lowest common denominator within the West German coalition government,[5] between the West German and French defence ministers[6] and among those defence experts who have stressed 'damage limitation'[7] as a preferable goal in relation to 'mutual assured destruction' (MAD).[8]

American rhetorical commitments and publicly-expressed concerns by European defence ministers[9] have contributed to the conceptual development of a BMD system for Europe against those tactical Soviet SRBMs and IRBMs targeted against Western Europe which could not be countered by a space-based multilayered SDI systems architecture.

In his so-called 'Star Wars' speech President Reagan indicated that the new defence technologies which he called on scientists to develop 'to give us the means of rendering these nuclear weapons impotent and obsolete' should reduce any incentive 'that the Soviet Union may have to threaten attack against the United States *or its allies*'. (my emphasis).[10]

Fred S. Hoffman, the study director of the Future Security Strategy Study (FSSS) in his summary report on 'Ballistic Missile Defenses and U.S. National Security', suggested several 'intermediate options' as the preferred path to achieving President Reagan's goal, among them:

Anti-Tactical Missile (ATM) Options

Deployment of an anti-tactical missile (ATM) system is an intermediate option that might be available relatively early. The system might combine some advanced mid-course and terminal components identified by the Defensive Technologies Study with a terminal underlay. The advanced components, though developed initially in an ATM mode, might later play a role in continental United States (CONUS) defence. Such an option addresses the pressing need *to protect allied forces as well as our own*, in theaters of operations, from either non-nuclear or nuclear attack. *It would directly benefit our allies* as well as ourselves. Inclusion of such an option in our long-range R&D programme on ballistic missile defences *should reduce allied anxieties* (my emphasis) that our increased emphasis on defences might indicate a weakening of our commitment to the defence of Europe. We can pursue such a programme option *within ABM Treaty constraints*. Such a course is therefore consistent with a policy of deferring decisions on modifying or withdrawing from the treaty.[11]

Lt General James A. Abrahamson, in his first appearance as director of the SDIO before Congress on 25 April 1984, emphasised:

> that our research is focusing on defence against ballistic missiles of all ranges, including tactical and theater range systems as well as ICBMs and SLBMs. We are *not* seeking only to build defences for the United States. As Secretary Weinberger has indicated, our concept of an 'effective' defence is one which *protects our Allies as well as the United States*. (my emphasis)[12]

In March 1985, United States Defence Secretary Caspar Weinberger in his '60-day letter' to eighteen allied nations inviting them to join the SDI stressed:

> Because our security is inextricably linked to that of our friends and Allies, we will work closely over the next several years with our Allies to ensure that, in the event of any future decision to deploy defensive systems (a decision in which consultation with our Allies would play an important part), Allied, as well as United States, security against aggression would be enhanced. Moreover, the SDI program *will not confine* itself solely to an exploitation of technologies with potential against ICBMs and SLBMs, *but will also*

carefully examine technologies with potential against shorter-range ballistic missiles (my emphasis).[13]

These general references by major officials and advisers of the Reagan Administration may be interpreted as a direct reaction to publicly-expressed concerns – in April 1984 West German Defence Minister Manfred Wörner was quoted as saying: 'I can't see that it [the SDI] would provide greater protection or stability'[14] – and as an effort to rally support behind the SDI research programme for political, strategic, economic and tactical reasons. In the view of several West European governments, industry officials and conservative defence experts, however, a strictly European version of SDI, an EDI appears to offer:

> a boost for European aerospace companies, much as SDI has been in the United States, but under the direct sponsorship and therefore control of European governments; a complementary strategy to SDI, meeting the criticism that anti-ICBM technologies could leave Europe vulnerable to other weapons (an argument partially accepted in the Pentagon, which is already studying many of the same ideas); and finally, a step toward the political integration of Europe, building it up to a third superpower through a joint defence strategy, a move viewed with somewhat less enthusiasm from Washington than from Paris.[15]

Since the Spring of 1985 leading retired and active military leaders in Western Europe such as the French Air Force General Pierre Gallois, the German Army General Gerd Schmückle and his active Dutch colleague, General G. C. Berkhof, have become increasingly outspoken in their support for an EDI.[16] Both the former French Defence Minister Charles Hernu[17] and his West German counterpart, Manfred Wörner,[18] came out in support of this during the Summer of 1985, as did many conservative and liberal politicians in the Federal Republic of Germany[19] and leading spokesmen of the parliamentary opposition in France.[20] According to Hernu the European governments should 'launch a European research initiative in the domain of space-based defence', and in view of the president of the French armament company Matra, Jean-Luc Lagardère, 'without a military programme in space, neither France nor Europe can expect a seat in the front row'.[21]

Nevertheless, major differences existed both within the West German government (as far as the legitimation for EDI is concerned) and between the French and the West German Defence Ministers (as to potential co-operation with the United States). While Hernu suggested that the West Europeans, or even France, should launch a programme of their own, Wörner and the leader of the parliamentary CDU/CSU pleaded for close transatlantic co-operation in this respect.[22] While Wörner's director of the defence planning staff, Hans Rühle,[23] stressed the Soviet ballistic missile threat for Western Europe as the primary reason for EDI, Konrad Seitz, the director of the planning staff in the West German Foreign Office, emphasised the technological challenge posed by the SDI for Western Europe:

> What Europe needs, independently of the question of participation in the SDI research programme, is a European initiative to create a common sphere for research and technology . . . Space is the new dimension of human history . . . Only if Europe becomes the third power in space besides the United States and the Soviet Union will it continue to have a voice and influence on earth.
>
> EUREKA is directed at civil projects. At the same time there should be European projects in defence co-operation . . . Here, too, it is a question of finally overcoming national egoisms and – within the framework of NATO strategy – agreeing upon a European defence initiative which would use the new technologies for three major objectives:
>
> 1. *The construction of an integrated air defence system (against aircraft, cruise missiles, ballistic missiles) together with the Americans*. This would have the aim in particular of protecting the system of defences in Europe against a pre-emptive strike (for which in the next decade the other side may have at its disposal precisely targeted missiles with conventional warheads). An integrated air defence system would be *ground-based* and would have a *different role*, both militarily and politically, from that of the *space-based strategic defence system* of the United States, which is aimed at *area and population defence*. It would mean the strengthening of NATO's existing strategy of deterrence, not a strategic revolution. Technologically, however, there are many overlaps with the SDI research programme which could be made use of for a *comprehensive exchange of technology*.
> 2. *The development of 'intelligent' weapons*, which seek their own targets, and development of information and delivery

systems for NATO's *Follow-on Forces-Attack (FOFA) concept* for long range against enemy second-echelon attack.
3. *The development of a European multisensor surveillance satellite* (my emphasis) which is at present under discussion by France and West Germany.

A combination of Eureka and a European defence initiative as suggested here should enable Europe to make the same advances in technology which can be expected for America as a result of SDI and NASA's project of a manned space station.[24]

While the application of SDI technologies for the defence against Soviet TBMs (top-down approach) prevailed in the European debate in 1985, the focus shifted to 'extended air defence' (bottom-up approach) during 1986. Somewhat representative of this conceptual and rhetorical shift has been Dr Wörner who in February 1986 distinguished among three countermeasures against conventionally-armed Soviet TBMs: (a) passive measures, (b) pre-launch destruction of Soviet missiles, (c) interception of missiles by anti-tactical missiles. For the latter, Mr. Enders (a young researcher with the Adenauer Foundation) suggested using the term ATM (*Anti-Tactical Missiles*) or *Extended Air Defence* (EAD) as the general term comprising both the defence against missiles with ballistic missile trajectories or *ATBM (Anti-Tactical Ballistic Missiles)* as the defence against missiles with non-ballistic missile trajectories or *ATAM (Anti-Tactical Aerodynamic Missiles)*.[25]

During 1986 the threat posed by both Soviet TBMs and NATO's AT(B)M became a major topic of studies conducted within NATO and of contract awards by SDIO. In February 1986, NATO's *Advisory Group for Aerospace Research and Development* submitted a secret 'Aerospace Application Study' (AAS-20) and on 22 May 1986, NATO's Defence Planning Committee endorsed additional studies by NATO's Military Committee and its Air Defence Subcommittee on an agreed threat assessment, on the identification of possible actions and countermeasures as well as on mission analysis, system architectures and systems effectiveness.

Independently of NATO, SDIO awarded the first contract outside the United States to Britain for a preliminary 'architecture study' of an ATBM system in Europe ($10 million) in early 1986 and in December 1986, Defence Secretary Weinberger announced that $14 million in SDI funds would go to seven multinational teams, including 51 firms from European countries, that are to study for a six-month period general requirements for European theatre defence architectures.[26] In 1987, two or three of these consortia will be awarded a one-year contract for phase 2 amounting to less than $7 million each.

II SOVIET THEATRE AND TACTICAL NUCLEAR FORCES TARGETED AGAINST WESTERN EUROPE AND POTENTIAL SOVIET COUNTERMEASURES TO SDI AND EDI

An EDI would have to neutralise both the present Soviet short- and medium-range ballistic missiles and the forseeable future offensive (e.g., penetration aids) and defensive countermeasures. An EDA would have to deal in addition with the Soviet airborne threat (bomber, fighter, Cruise missiles) and with the prospective countermeasures (e.g., stealth technology). This section will focus only on one aspect: the Soviet ballistic missiles presently targeted against Western Europe, present Soviet ATBM systems and potential Soviet countermeasures to SDI and EDI.

(a) Soviet nuclear forces targeted against Western Europe in general and against West Germany in particular

In the late 1970s Soviet regional forces had approximately 1,400–1,525 targets for nuclear weapons in NATO Europe, among them 650–675 targets against the Western nuclear threat, 600–650 targets against the conventional threat and 150–200 targets against administrative and economic centres.[27]

Arkin, von Hippel and Levi,[28] in their fictitious scenario of a limited nuclear war in both German states, assumed 676 Soviet nuclear targets on the territory of the Federal Republic of Germany. Of these, 88 targets were against the *nuclear threat* 23 SSBM sites, 9 nuclear air bases, 38 nuclear artillery battalions and 78 nuclear storage sites and the remaining 588 were *other military targets*: 72 national or international headquarters, 31 command/communication centres, 31 Army/Corps/Division headquarters, 37 non-nuclear air bases, 7 naval bases, 256 ground forces bases, 21 radar/early warning sites, 90 SAM sites, 75 munitions/petroleum storage areas, 27 logistic installations and 1 chemical storage site.

According to Western sources, Soviet variable-range missiles (e.g., the 120 SS-11 ICBMs, which had been introduced into the MRBM fields in the late 1960s and replaced in the 1970s by the SS-19);[29] Soviet sea-based strategic reserve forces (SS-N-5 and SS-N-6);

Figure 14.1 Soviet tactical ballistic missile threat to Western Europe

IRBMs; MRBMs – or, in the terminology of the 1979 German Defence White Paper, LRTNF (with a range between 1,000 and 5,500 kilometres, MRTNF (with a range between 100 and 1,000 kilometres) and SRTNF (with a range below 100 kilometres) – as well as nuclear-capable aircraft and Cruise missiles have been targeted against Western Europe. Within the INF context (according to

NATO criteria) two terms were used: INF (intermediate range nuclear forces) with a range from 150 to 5,500 kilometres and SNF (short-range nuclear forces) with a range below 150 kilometres.

According to Berman and Baker, by 1980 the overall Soviet regional-range weapons consisted of 1,031 land-based missiles (SS-12, SS-4, SS-5, SS-20, SS-11 and SS-19), 445 sea-based missiles and 655 bombers which could deliver a total of 3,497 nuclear warheads.[30]

According to an IISS analysis of July 1985 of the potential nuclear weapons systems of the Warsaw Treaty Organisation in Europe or targeted against Western Europe, a portion of (520) of the SS-11 and of the SS-19 ICBMs (300 with six warheads each), and a total of 120 SS-4 MRBMs and 261 SS-20 IRBMs, as well as approximately 450 SS-21/FROG, 516 SS-23/SCUD A/B and 90 SS-12/-22 of the Soviet Union and 208 FROG-3/-5/-7 and 132 SCUD B/C of their Allies were targeted against Western Europe. In addition, the IISS analysis referred to 24 SS-N-5 Soviet SLBMs and 446 Soviet SLCMs confronting NATO Europe.[31] Furthermore, according to the IISS analysis 2,055 medium-range bombers and fighters of the Soviet Union and their Allies could drop nuclear bombs over Western Europe.

An EDA or an integrated or 'extended' European air defence would have to protect either the major military targets in Western Europe – approximately some 700 in West Germany alone – (*minimal requirement for a point defence system*), or the whole spectrum of potential Soviet military, industrial and civilian targets (*maximal requirement for a territorial or area defence system*).

Has there been a major increase in the number of Soviet ballistic missiles targeted against Western Europe since 1970? In 1985, according to the IISS, the number of the ICBMs with variable range (SS-11 and SS-19) dropped below the level of 1972. On the LRTNF level (SS-4, SS-5), the number of the launchers targeted against Western Europe has reached the lowest level while the number of the warheads has steadily increased due to the introduction of MIRVs for the SS-11 and SS-20. The number of the Soviet SLBMs (SS-N-5 and SS-N-6) that could reach Western Europe has also dropped. As far as the overall number of the Soviet MRTNF (SCUD, SS-12, SS-22 and SS-23) slight variations have been observed since 1981, which may be due partly to counting problems. And on the level of the Soviet nuclear battlefield launchers, the overall number has remained stable since 1981 (see Table 14.1).

The IISS analysis does not suggest a major quantitative increase of the Soviet ballistic missile launchers since 1978, but rather a qualita-

Table 14.1 Changes in Soviet ballistic missile launchers aimed at targets in NATO Europe from 1972–86 according to data from the IISS[32]

Category	First year deployed	Range (km)	CEP (m)	Warhead	Yield (KT)
Strategic systems					
ICBM SS-11 mod 1	1966	9,600	1,400	1	950–1,000
mod 2/3	1973/5	10,600–13,000	1,100	1–3	100–(1,000)
SS-19 mod 2	1975–9	10,000	300	1	5,000
mod 3	1982	10,000	300	6 MIRV	6 × 550
SLBM SS-N-5 SERB	1963–4	1,120–1,700	2,800	1	1,000
SS-N-6 mod 1	1968	2,400	1,300	1	500–1,000
mod 2	1973	2,400–3,000	900	1	650–1,000
mod 3	1973–4	3,000	1,300	2–3 MRV	2 × ?500
Theatre systems (LRTNF)					
SS-4 Sandal/SS-5	1958–9	2,000	2,000	1	1,000–(2,000)
SS-20 mod 1	1976–7	5,000	?	1	1,500
mod 2		5,000	400	3	3 × 150
Medium range TNF					
SS-1b SCUD A	1957	150	n.a.	1	KT range
SS-1c SCUD B	1965	300	900		KT range
SS-23	1979–80	500	350	1	KT range
SS-12 Scaleboard	1969	490–900	900	1	200–1,000
SS-12 mod (SS-22)	1979	900	300	1	500–1,000
SCUD B/C with WTO-units	1965	160–300	900	1	KT range
Short-range TNF					
FROG-7	1965	70	400	1	200
SS-21	1978	120	300	1	100
FROG-3/-5/-7 with WTO units	1957	40–70	400	1	200

tive improvement in replacing SS-4s with SS-20s, FROGs with SS-21s, SCUDs with SS-23s and SS-12s with SS-22s or SS-12 modified. According to Arkin and Fieldhouse in 1984 approximately 76 SS-21 launchers of the Soviet Union, 54 SCUD-B launchers of the Soviet and the East German Army and 20 FROG-7 launchers of the East German Army are at present forwardly deployed on the territory of the German Democratic Republic. It is assumed that additional nuclear battlefield systems are being deployed in Poland, Czechoslovakia and Hungary.[33] Some of these forward deployments have been announced – or rather, publicly admitted – by the Soviet

Table 14.1 cont.

1972	74	76	78	79	80	81	83	85	86	Deployment area (1986)
1,464	1,570	1,598	1,562	1,526	1,409	1,390	1,312	1,195	1,151	
970	1,018	900	780	638	580	580	550	(100)	28	USSR } replaced
—	—	—	—	—	—	—	some	(420)	420	USSR } by SS-25
—	—	100	200	300	300	300	few	300	360	USSR
—	—	—	—	—	—	—	330			USSR
30	24	54	54	60	60	57	48	39(24)	39(24)	Golf II, Hotel II
464	528	544	528	528	469	165				Yankee I
							384	(336)	(304)	Golf IV
						288				(no theatre role)
500	600	600	690	710	600	610	599	543	553	
500	600	600	590	590	440	380	239	120	112	USSR
—	—	—	100	120	160	230	360	(423)	441	270 targeted against Europe
300	300	400	1,638	1,638	1,671	618	757	857	908	
						+	+	+	—	GDR, CSSR, USSR
						+410	+440	+420	395	Bulgaria
—	—	—	—	—	—	—	10	180	(240)	USSR
300	300	300	1,300	1,300	1,300	65	70	80	—	USSR
—	—	—	—				100	45	130	USSR, GDR
			100	132	132	163	143	137	132	143	GDR, CSSR
		800	—	—	—	687	700	958	1,014	
		600				482	440	530	(500)	USSR, GDR, CSSR
						n.a.	62	220	(300)	Poland & Hungary
			200	206	206	208	205	198	208	214	

Union as a direct reaction to the implementation of the NATO double decision on behalf of NATO.

(b) Soviet countermeasures against SDI, EDI and extended air defence

Both American[34] and Soviet[35] analysts have referred to various conceivable countermeasures against a multilayered, space- and ground-based BMD system as outlined in the context of the SDI research programme.

A report of a Working Group of the Committee of Soviet Scientists for Peace, Against Nuclear Threat headed by R. Z. Sagdeyev and A. A. Kokoshin distinguished among active and passive countermeasures against a space-based anti-missile system. 'The former include various ground (sea)-, air- or space-based weapons using either missiles or lasers', for example (1) ground (sea)-, air- or space-based ballistic missiles, (2) space mines, (3) high-power ground-based lasers, (4) obstacles put up on the trajectories of the combat stations, and (5) false missile launches. Among the passive countermeasures the Soviet scientists include:

> camouflaging missile launchings in the laser-working optical range (all types of smoke screens), multilayer missile casings and ablating coating. Coatings with a high reflection factor at the working wavelength (including retro-reflectors) may also have a part to play as countermeasures against a number of types of lasers (operating in the visible and infrared spectrum) ... An effective network of the means of countermeasures can be set up within a very short time with the use of the already available technology. As indicated in Western sources, corresponding units and components are much better tested and more reliable than the elements and subsystems to be used for SBAMS based on directed-energy weapons. Estimates show that a system of the means of counter-measures may also be much cheaper than a large-scale SBAMS. The cost of a highly efficient countermeasures system (counter-SBAMS) with regard to the SBAMS itself is most likely to make up 1 or 2 per cent.[36]

As far as a European-based AT(B)M as part of an EDI is concerned, the Soviets could initiate the following additional countermeasures:

1. increase the number of forward deployed SRINF and MRINF (e.g., the SS-21, SS-22 and SS-23);
2. shift the nuclear-capable aircraft to the European theatre in case of a conflict;
3. shift from ballistic to theatre-wide land-, air- and sea-based cruise missiles.

In order to counter this potential Soviet offensive countermeasures against an EDI, European defensive system would require a capability to destroy Soviet

1. short- and medium-range ballistic missiles by an *AT(B)M system*;

2. nuclear- and dual-capable aircraft by an *improved air defence*;
3. dual-capable ground-, air- and sea-based cruise missiles by an *anti-cruise missile defence system*.[37]

(c) Soviet anti-tactical ballistic missiles (ATBMs)

Based on a long-term analysis of IISS data on Soviet substrategic ballistic missiles for the years 1970–85, the initiation of an EDI could hardly be justified. However, since 1983 a major shift in the evaluation of Soviet BMD activities in general and of the Soviet ATBM capabilities could be observed[38] in official United States defence publications and public statements. A joint State Defence Department report on 'Soviet Strategic Defence Programs' of 4 October 1984 stated

> Soviet strategic surface-to-air missiles provide low-to-high-altitude barrier, area, and terminal defences under all weather conditions. Five systems are now operational: The SA-1, SA-2, and SA-3, and the more capable SA-5 and SA-10 . . .
> The SA-10 can defend against low-altitude targets with small radar cross-sections, like cruise missiles. The first SA-10 site was operational in 1980. Over 60 sites are now operational and work is progressing on at least another 30. More than half these sites are located near Moscow; this emphasis on Moscow and the patterns noted for the other SA-10 sites suggest a first priority on terminal defence of command and control, military and key industrial complexes . . .
> The Soviets are also flight-testing another important mobile SAM system, the SA-X-12, which is able to intercept aircraft at all altitudes, cruise missiles, and short-range ballistic missiles. *The SA-10 and SA-X-12 may have the potential to intercept some types of strategic ballistic missiles as well* . . . They could, if properly supported, add a significant point-target defence coverage to a nationwide Soviet ABM deployment.[39]

David Yost, a conservative defence expert who has been working during 1985 as a guest scholar in the Pentagon's Office of Net Assessment supported this official line in several contributions published simultaneously in the United States, in France and in the Federal Republic of Germany:[40]

Three surface-to-air missiles (SAMs) may have some BMD capability: the SA-5, SA-10, and SA-X-12 ...
The SA-10 is deployed in about 800 launchers, more than half near Moscow, with mobile SA-10s expected to be operational in 1985. According to the US Department of Defense, both the SA-10 and the extensively tested experimental SA-X-12 'may have the potential to intercept some types of US strategic ballistic missiles'.[41] All three SAMs are more likely to be able to intercept SLBM warheads than ICBM warheads, because the former are generally slower and offer larger radar cross sections. This is especially significant because SLBMs at sea are, unlike US ICBMs not subject to Soviet counterforce attack.[42]
The SA-X-12 may also be an anti-tactical ballistic missile (ATBM), in that the US Department of Defense judges that the SA-X-12 'may have the capability to engage the Lance and both the Pershing I and Pershing II ballistic missiles'.[43] The SA-X-12 has reportedly been successfully tested against Soviet intermediate-range missiles such as the Scaleboard.[44] The SA-X-12 is also notable because its mobility could enable the Soviets to build thousands of them and conceal them in storage until ready to deploy them relatively rapidly.[44]

However, Yost's claims are obviously in conflict with the 1985–86 issue of *Military Balance*, which states that some 520 SA-10 launchers are deployed in 60 complexes in the Soviet Union, with 30 having a strategic role. However, the 1986–87 issue of *Military Balance* claims that some 735 SA-10 launchers are deployed in 70 complexes, 40 with a strategic role near Moscow.[46]

Being highly sceptical of any arms control solution to the BMD issue, Yost pleads for unilateral Western responses, for example 'improved US abilities to penetrate Soviet air and ballistic missile defences against US intercontinental, intermediate and shorter-range systems [which] would help the Soviets from gaining a great advantage in escalation control capabilities'.[47] Yost bases his call for the deployment of BMD capabilities in NATO countries and the introduction of penetration aids for United States offensive forces on the implications of a worst case assumption of a massive Soviet BMD and ATBM deployment for NATO's security:

First, the credibility of NATO's strategy of 'flexible response' could be reduced. NATO's flexible response strategy includes the threat

of selective employment of nuclear weapons against the Soviet Union, but elements of this threat could be directly countered by Soviet BMD...
Second, Soviet prospects for victory in conventional operations could be improved...
Prospects for victory in such operations would be enhanced if the Soviets could use BMD to neutralise some of the various 'emerging technologies' initiatives of NATO, which are intended to raise the nuclear threshold through enhanced conventional strength. Follow-on-Forces Attack (FOFA) and the Joint Tactical Missile System (JTACMs) and related concepts call for using ballistic missiles such as Lance and Pershing II to deliver advanced conventional submunitions against Warsaw Pact airfields and other military targets.[48] But the SA-X-12 may already be capable of intercepting these missiles, and may thus help the Soviet Union in its attempts to gain and hold air superiority – an indispensable key to victory.
Third, potential Soviet control over the escalation process could be enhanced...
Fourth, the credibility of the British and French nuclear deterrents could be lowered. (my emphasis)[49]

Uwe Nerlich,[50] a defence expert at the *Stiftung Wissenschaft und Politik*, and a supporter of the 'damage limitation school' of Albert Wohlstetter, Fred Hoffman and David Yost,[51] has claimed that in the short run NATO's strategy of flexible response may be undercut by three military developments in the Soviet Union:

1. The most obvious is the combination of improved CEP and improved conventional warheads on some of the Soviet ballistic missiles in the theatre...
2. A second development is the possible Soviet deployment of long-range ground-launched cruise missiles. This may in part be a political counter to NATO's ground-launched cruise missiles...
3. A third possible development would be the deployment of an active ballistic missile defence West of Moscow. In addition to the SAM-10, which is deployed throughout the Soviet Union and supposedly has an anti-cruise missile capability, ... the Soviet Union is developing the SAM-X-

12 missile, which is designed to intercept shorter-range ballistic missiles.[52]

In Nerlich's view the Soviet Union is presently fielding all four components envisaged also in NATO's *Counterair '90 Study:*

1. air defence;
2. offensive counter air (OCA), i.e. neutralisation of airfields;
3. attacks on missile launchers (pre-emption) and
4. active tactical ballistic missile defence.[53]

In full agreement with Fred Hoffman and David Yost, Nerlich tries to legitimate the components of an EDI as a response to similar Soviet developments. In Nerlich's view, the ATBM option suggested in the 'Hoffman Report' would require an appropriate conceptual NATO framework:

> The NATO Counterair '90 Study provides such a framework, although it must not be viewed in isolation from a broader review of what is required to develop NATO strategy and the means to implement it so as to stay in the game.
> Regarding an ATBM capability against conventional missile threats as a candidate for 'early options for the deployment of intermediate systems'[54] may be a way to put realism into SDI, but SDI is hardly providing the conceptual framework for further study of this ATBM option. In fact, it may well create unhelpful perspectives on both accounts because it invites considerations of this option as part of a strategic defence system and because it may suggest a momentum within the SDI, although it is by and large unrelated to what may become an SDI programme.[55]

Components of an integrated (K. Seitz) or extended air defence or of an EDA are already being discussed within NATO circles and among experts in different conceptual frameworks. (See Table 14.2).

III CONCEPTUAL FRAMEWORKS FOR COMPONENTS OF A EUROPEAN DEFENCE ARCHITECTURE (EDA) OR FOR AN INTEGRATED OR EXTENDED AIR DEFENCE SYSTEM

While the political debate in Western Europe until February 1986 has focused primarily on whether to join the SDI research programme and on the terms of such participation, a second controversy on a 'smaller SDI for Europe' or an EDI has gradually emerged, a concept stemming from two different political orientations – an

Atlanticist perspective (in Bonn) calling for a close transatlantic co-operation and technology transfer and a *Europeanist* or *neo-Gaullist perspective* (in Paris) calling for closer technological, military, and strategic co-operation in Western Europe deliberately avoiding the role of a 'subcontractor' to the United States.[56]

While the political debate in Western Europe is continuing, the United States Defense Department has ordered detailed architectural studies for the application of SDI technologies for the defence of Western Europe against Soviet ballistic missiles, cruise missiles, aircraft and Soviet penetration aids.[57] Components of a comprehensive EDA are presently being considered in four different conceptual frameworks:

1. ballistic missile defence (BMD) in the SDI-context;
2. air defence in the context of the United States study Counterair '90, of AGARD's AAS-20 and of more recent NATO studies;
3. new operative concepts dealing with the second and third echelon of Soviet forces (Deep Strike Concepts, such as AirLand Battle, AirLand Battle 2000, Army 21 of the US Army and related NATO concepts such as Follow-on-Forces-Attack (FOFA) concept) or Rogers Plan;
4. Emergent Technologies (ET).

(a) The SDI context for a European defensive system against Soviet ballistic missiles

Will SDI be able to provide an area or only a point defence, for the United States alone or also for their European Allies? While the American political leadership has repeatedly stressed the long-term goal of protecting people, many defence experts have stressed the short-term goal of protecting the offensive nuclear capabilities and thereby strengthening deterrence. On 29 March 1985 George A. Keyworth II, then science advisor to President Reagan, commented on this debate:

> Is SDI to protect people or to protect weapons? Protecting weapons represents no change in present policy. It simply strengthens – entrenches – the doctrine of Mutual Assured Destruction. Protecting people, on the other hand, holds out the promise of dramatic change. This clear purpose of the President has been repeated time and time again by Cap Weinberger, Bud McFarlane, and myself. But the ambiguity over SDI's real goal remains. It is

Table 14.2 Survey of conceptual frameworks and of systems components for an EDA

Overall Concept	*EUROPEAN DEFENCE ARCHITECTURE* (Integrated or extended air defence against ballistic missiles, cruise missiles, and aircraft bombers and fighters)			
Partial goals & concepts	Defence against ballistic missiles		Cruise missile defence	Air defence
Military planning of the US Defense Department	Strategic Defence Initiative (SDI)	Anti-tactical ballistic missile system (ATBM)	Anti-tactical aerodynamic missile (ATAM)	Air Defence plans
	(Research programme)	(Patriot–ATM programme)	(Improved Hawk programme)	Patriot, Roland *et. al.*
Military concepts	BMD	Extended Air Defence	Deep-strike systems	Extended Air Defence
			AirLand Battle FOFA concept NATO's CMF	
		US study: C O U N T E R A I R ' 9 0 of 1982		
Weapons systems	Not yet decided	Patriot–ATM	(Offensive counterair) –JSTARS –JTACMs or Army TACMs & Air Force TACMs	Improved Hawk	Patriot, Roland, Improved Hawk AMRAAM, AIM-9L ASRAAM Fighter aircraft

AGARD: AAS-20 study of early 1986

Example of conceptual study	Military Committee's task (Autumn 1983)	Study's focus on	Study's results
	– investigate Soviet TBM up to the year 2000 MC guidelines to AGARD: (a) passive defence improvements (excluded) (b) emphasis on non-nuclear ATBM systems (c) defence capabilities against TBM with a range of less than 1000 km (SS-21, -12(22), -23)	space-based, airborne and ground-based *weapons* (High Energy Laser (HEL), Particle Beam Weapons; guided missiles electromagnetic guns (EML), SAMs, anti-aircraft artillery) and *sensors* (radar, IR- & UV-sensors, radar)	HEL & EML eliminated as primary ATBM systems; Patriot ATBM suitable for defence of vital installations in FRG and Benelux against Soviet TBM; ground-based radars & airborne IR sensors or space-based 'pop-up' sensors

Organisational units & supporters in the US	SDIO	US Army	Defence Research & Engineering	US Army US Air Force
Defence against Soviet systems	ICBM (SS-11, -19) SLBM (SS-N-6) SS-20 SS-4	SS-12 mod (SS-22) SS-N-5 SS-23 SCUD B/C SS-12	SS-23 SS-21 SCUD/BC FROG	Soviet GLCM ALCM & SLCM SS-NX-21 AS-15 SSC-X-4 SS-NX-24	Backfire Fencer Su-24 Flogger MIG-23 Flogger MIG-27 Fitter Su-17 Foxbat MIG-25 Fishbed MIG-21 Fulcrum MIG-29 Frogfoot Su-25
Requirements and costs	*SDI* Research (1984–93) 50–70 bn $ Deployment: 500–1,000 bn $	1,000–2,000 Patriot–ATM	Army TACMs USAF TACMs	Improved Hawk	ca 6,200 Patriot for 103 fire units

fostered by three main tenets: First is the assertion, embraced by those anxious to protect both past strategic doctrine and future nuclear systems, that 'strengthening deterrence' must be the primary goal for SDI. Second is that protecting weapons, especially ICBM silos, is the near-term most likely goal for SDI. And third is that defence of European military targets against tactical ballistic missiles is the most politically attractive near-term goal for SDI.
If these arguments continue to be used as the basis to achieve Congressional and Allied support, I believe the opportunity for strategic change – and the President's objective – is lost ... Terminal defences within the SDI also can play a very real part in the overall 'layered' defence. But attempts to make terminal defence our first move, within the SDI, does not start us in the direction of the President's objective.[58]

While many American political and military leaders have emphasised President Reagan's long-term vision of overcoming deterrence,[59] many European leaders have stressed the short-term goal of strengthening deterrence.[60] Many European and American experts share the view that a multilayered space-based BMD system can neither provide an area defence for Western Europe nor be able to defend against Soviet SRBMs. If a multilayered space-based BMD should be technically feasible, survivable, and cost-effective,[61] it could protect Western Europe only against Soviet variable range ICBMs (SS-11 and SS-19) and against IRBMs (SS-20) but certainly not against Soviet short-range missiles (SS-21 and SS-23). It is highly doubtful that SDI could cope with the SS-22, given its boost-phase of approximately 90 seconds (compared with 5 minutes for the SS-18 ICBM), its mid-course phase of 6 instead of 30 minutes and its terminal phase of 60 seconds.[62]

Terminal defence of military targets based on available technologies (ATBM options) or on kinetic- and directed-energy kill mechanisms appears to be the most likely conceptual context that SDI could provide for a European-based BMD as one component of a more comprehensive EDA.

(b) The defence against and the destruction of Soviet ballistic missiles in the conceptual framework of 'Counterair '90'

Of the four military components mentioned in the study with the title 'Counterair '90' – air defence, offensive counterair, attacks on missile launchers (pre-emption) and active tactical BMD – the latter two components deal with Soviet SRBMs in the pre-boost phase (pre-emption) and in the terminal phase (ATBM option). Clarence A. Robinson offers the following description of this concept:

> The Counterair '90 system for application in Europe is considered a total systems engineering approach to air and missile defence. It blends offence and defence into an integrated system.
> The program is predicated on defending against a first-strike attack by the USSR with a variety of weapons including conventionally-armed ballistic missiles aimed at US and allied air bases and at air defence sites.
> One of the prime concepts is that the Alliance would respond by using ballistic missiles with conventional warheads to suppress Soviet missiles before they can be reloaded and fired in a second strike.[63]

In the same article Robinson mentioned three related programmes that would directly support the Patriot anti-tactical missile and Counterair '90 programme:

1. joint surveillance and target acquisition radar system (JSTARS);
2. joint tactical missile system designed to provide surface and air-launched stand-off capability – a weapon compatible with the precision location strike system and JSTARS for defence suppression and interdiction, respectively. The program is based on the [JTACMs] ...
3. joint tactical fusion program to exploit on a near real-time basis time-sensitive and high-volume multisensor information.[64]

According to a report of the Military Committee of the North Atlantic Assembly of November 1984:

> European NATO members have been generally sceptical of the Counterair '90 concept on two grounds: cost-effectiveness and crisis stability ... The eastward launch of conventionally-armed ballistic missiles could easily result in a Soviet nuclear response, as they would likely not be able to distinguish between a conventional ballistic missile and a nuclear ballistic missile (such as Pershing or Lance) ... The compressed decision time involved in some of the

new concepts being advocated is also a serious concern for crisis stability and political control. The decision to strike deep at Warsaw Pact forces may have to be made at an early – and ambiguous – stage of the conflict. The proposed use of ballistic missiles for counterair would require a decision to launch within 15–30 minutes of an impending Warsaw Pact air attack, according to one account. One former high-level SHAPE official has expressed concern that real-time response could well border on pre-emption. The implications which technological options may have for political control and crisis decision-making must be more carefully considered in assessing the various military choices facing the Alliance.[65]

However, according to Nerlich, offensive countermeasures would be preferable to a defensive ATBM system:

> A NATO capability for conventional missile attacks on missile launchers and support units does appear to be an option that NATO may have to consider in the future. Either stand-off weapons or ground-launched ballistic missiles could be used ... Unlike attacks on launchers, currently considered active ATBM defence capabilities could not be employed confidently against nuclear shorter-range missile attacks ... The dual mode Patriot is supposed to provide a potentially reliable defence only against conventionally armed shorter-range missiles.[66]

While the ATBM option would overlap with the ground-based terminal defence layer of the SDI, the early or the pre-emptive use of ballistic missiles against second and third-echelon forces and targets is primarily being discussed in the framework of deep-strike concepts (FOFA, AirLand Battle *et al*.).

(c) Deep strike concepts of the United States armed forces and of NATO and pre-launch destruction of Soviet ballistic missile launchers

According to official United States Defense Department testimony in the United States Congress:

> JTACMs and JSTARS are extremely significant for the purposes of Seeing Deep and Striking Deep under AirLand Battle doctrine.[67]

At least two major deep-strike concepts have to be distinguished: the AirLand Battle Doctrine of the United States Army, which foresees an extended battlefield in which Corps level commanders may strike at targets at a distance of 150 kilometres in order to facilitate front-line engagement of opposing forces and the concept of the Allied Command in Europe of Warsaw Pact 'Follow-on-Forces-Attack' (FOFA). The latter, according to the study of the North Atlantic Assembly, 'envisages theatre-wide targeting of massed armour and choke-points from 25 kilometres to more than 400 kilometres beyond the FEBA in order to disrupt and destroy Warsaw Pact reinforcing offensive echelons'.[68]

When General Merryman was asked during congressional testimony in Spring 1983 which range of missions the former JTACMs – in the meantime they have been renamed Army TACMs and Air Force TACMs – are to undertake, he responded:

> The Joint Tactical Missile System will be employed against high-value land and sea targets in the high threat environment. Land targets include area targets such as *surface-to-surface missile sites, air defence sites, airfield facilities, missile refuelling/storage assembly sites, logistic concentrations, choke points, manoeuvre force, and point targets and facilities* that are hardened and reinforced such as *hangeretts, C^3 sites and bridges* (my emphasis).[69]

The technological components of an extended or integrated air defence or of an EDA have also been discussed in the NATO framework in the context of the so-called 'emergent technologies'.

(d) Emergent technologies – the technological linkage to an extended or integrated air defence or to an EDA

At the Bonn summit meeting in June 1982 United States Secretary of Defence Weinberger proposed for the first time that NATO examine emerging technologies to improve conventional defence, and in December 1982 the United States tabled a paper on the application of emerging technologies in Central Europe and in 1983 additional papers were introduced by the United States relating to ETs for the Northern and Southern Regions and for maritime defence. According to a study of the North Atlantic Assembly, Weinberger's paper of December 1982

> noted several areas of technology with the potential to enhance significantly NATO's conventional forces, including increased

target acquisition capability; greater weapon system accuracy and lethality; and command, control and communications (C^3). Four areas of the application of ET were also identified in the December 1982 paper:

1. defence against Warsaw Pact first-echelon forces;
2. NATO's counterair capability;
3. interdiction of Warsaw Pact follow-on forces;
4. command, control, communications and intelligence.[70]

In December 1983 the United States provided a list of United States candidates for inclusion in the emerging technologies effort. The Independent European Programme Group (IEPG) responded in April 1984 with a list of over 200 European projects for United States consideration. At the initiative of West German Defence Minister Wörner, NATO's Military Committee was subsequently asked to develop a 'conceptual military framework' for the application of ET which has not yet been completed. According to the Voigt report to the North Atlantic Assembly:

> The most dramatic new potential offered by ET is the capacity for deep attack of mobile targets with conventional munitions ... The United States has been developing the technology to look and strike well behind the FEBA in the 'Assault Breaker' programme begun in 1978 ... Second-echelon targeting would be accomplished with Pave Mover [now JSTARs], a stand-off airborne radar system for long-range surveillance of fixed and mobile targets beyond the FEBA. Target information would be relayed to ground stations which could assign surface-to-surface and air-to-surface missiles loaded with submunitions to the target. Improved Patriot and Lance missiles are being considered for the surface-to-surface [JTACMs] role while medium and long-range air-delivered cruise missiles (MRASM/LRASM) are being considered for the air-to-surface role.[71])

All four conceptual frameworks – SDI, Counterair '90, Deep Strike and ET – contribute to the development of weapons components which could be incorporated into an EDA or an integrated air defence system:

1. SDI-related weapons systems for the terminal phase using kinetic- and directed-energy kill-mechanism (top-down approach);

2. ATM options based on an application of SAM-D technologies (bottom-up approach);
3. Deep Strike systems exploiting the potential of emerging technologies.

IV AMERICAN WEAPONS PROJECTS AS POTENTIAL COMPONENTS OF AN EDI AGAINST SRBMs

This survey of American weapons projects which are presently in the research and development stage will focus only on systems with a potential to defend against or to destroy Soviet SRBMs. All potential systems that are presently being introduced or developed with a potential against the Soviet airborne threat (aircraft, cruise missiles) will be excluded, as will be potential American countermeasures to Soviet penetration aids to overcome American tactical BMD systems in Europe or the impact of potential Soviet stealth technologies on present and future Western air defences in Europe. The implications of the potential introduction of MIRVs – for example for the SS-22 or possibly the SS-23 – will also not be considered.

(a) SDI-related programmes for defence against Soviet ballistic missiles in Europe

In the SDI context, systems for defence against strategic ballistic missiles are distinguished in technological terms according to: (a) their operative *range* (exoatmospheric and endoatmospheric), (b) the type of the *warhead* (nuclear or non-nuclear) and (c) the *kill mechanism* (warhead or kinetic energy and directed energy).[72] Three or four stages in the flight of a ballistic missile have also been distinguished: the *boost*, the *post-boost*, the *mid-course* and the *terminal* phase. In November 1985 the SDIO planners in the United States Defense Department disclosed that – based on a one-year study – they were considering a complex, seven-layered BMD system.[73] For defence against strategic ballistic missiles, two layers would apply to the *boost phase* (1st layer: directed-energy weapons based on tens of large battle stations; 2nd layer: kinetic-energy weapons based on thousands of smaller satellites), three to the *mid-course phase* – that is, after re-entry vehicles and decoys have been dispersed – (3rd layer: directed-energy weapons; 4th layer: kinetic-energy weapons and 5th layer: clouds of small attack pellets while the decoys and warheads are still closely bunched) and two to the *terminal phase* (6th layer: exoatmospheric ground-based rocket interceptors and 7th layer: endoatmospheric ground-based rocket interceptors).[74] While for the defence against the Soviet SS-20 MRBM all seven layers may be relevant, the defence against Soviet SRBM (SS-12 mod, SS-23, SS-21) could rely only on the latter two. As to the technological requirements for a tactical BMD the scientific literature is divided. According to Sorensen:

> Short horizons are required for detection/intercept (D/I) given especially the Euro-theatre flight distance of times which may not exceed 6–12 minutes from the point of launch. The D/I problem is further complicated by the launch mode of the mobile SS-20 IRBM. The calculations of both target arcs and trajectory 'windows' from fixed launch point to target arrays (such as can be done with fixed-site-land-based missiles) are different with mobile systems, complicating the BMD target acquisition problem. The problem might be slightly less severe for point defence problems, as opposed to area defence.[75]

The technological requirements for a tactical BMD are even more demanding: all Soviet SRBMs are mobile and have a flight time of approximately three–nine minutes.

Canavan, in his discussion of theatre applications of strategic defence concepts, concluded to the contrary:

> A multilayer concept would provide significant, preferential defence at every tier, making a meaningful attack on military targets impossible and a non-military attack on value targets meaningless. Such theatre defences could essentially negate both theatre and extra-theatre threats, particularly when augmented by strategic assets. Interestingly, they have a strong role against air breathing threats of either a nuclear or conventional nature, providing a continuum of deterrence and defence at every level from conventional through strategic.[76]

The official SDI Report of the United States Defense Department has specified the functional needs for the terminal phase:

> A terminal-defence element of a total strategic defence system could serve three separate but similar functions. It could provide the final layer in a defence-in-depth system, stand-alone defense against depressed trajectory SLBMs, and stand-alone capability for

defence of Allies. It is assumed in this discussion that terminal defence needs are defined to exploit the major increase in terminal defence capability possible from the attrition and discrimination in the boost and mid-course elements of the system.[77]

Five weapons concepts based on directed-energy and kinetic-energy kill mechanisms have been listed in the April 1985 SDI report:

1. ground-based laser concepts (GBL);
2. endoatmospheric non-nuclear kill (ENDO-NNK) technology;
3. exoatmospheric non-nuclear kill (EXO-NNK) technology;
4. ENDO-NNK test bed: high endoatmospheric defence interceptor (HEDI);
5. EXO-NNK test bed.

GBL concepts have been repeatedly suggested by Edward Teller[78] as the ideal system to counter Soviet short- and medium-range ballistic systems. The SDI Report offered the following project description:

> The seven tasks of this project provide the technology base, demonstrations, and designs required to provide a firm basis for deciding whether or not to pursue a ground-based laser weapon for boost-phase intercept. This project will establish and demonstrate the major subsystems for this ground-based concept of visible ultraviolet (V/UV) wavelengths. There are three main thrusts: *Technology Development*, *Major Experiments*, and *Concept and Development Definition*.[79]

The remaining four weapons concepts are based on kinetic energy: two for exoatmospheric and two for endoatmospheric interception. The EXO-NNK technology project includes, according to the SDI Report:

> technology development for ground-launched (missile) and space-launched (rocket and hypervelocity gun) non-nuclear kill vehicles to intercept boosters, post-boost vehicles, and re-entry vehicles above the atmosphere.[80]

This technology programme is to support three major BMD concepts:

1. exoatmospheric re-entry Vehicle Interception System (ERIS);

2. space-based hypervelocity gun experiment (Sagittar experiment);
3. space-based kinetic kill vehicle experiment.[81]

The ERIS programme was initiated by the United States Army following the successful conclusion of the HOE (Homing Overlay Experiment),[82] which in 1984 demonstrated the potential for SDI of ground-launched missiles which intercept and destroy re-entry vehicles with a kinetic-energy warhead before they re-enter the atmosphere. ERIS is planned to perform interceptions at heights of roughly between 100 – and 160 kilometres. It will probably be relevant only against the Soviet SS-20.

The second EXO-NNK Test Bed project was established in FY84:

> with a concept definition phase to explore alternative approaches to the design of interceptors that could accomplish non-nuclear kill of re-entry vehicles in the exoatmosphere. The test assembly will be comprised of the off-the-shelf components where applicable. The second phase will be used to validate and demonstrate the solutions of critical issues associated with the preferred interceptor concepts.[83]

For the ENDO-NNK Technology project the SDI Report provided this project description:

> This project is comprised of five tasks and contains research into key technologies associated with the destruction of re-entry bodies within the atmosphere (endoatmospheric) with non-nuclear kill devices. The focus of this programme at this time is the resolution of critical technical issues and the validation of component technologies. Upon validation of the key technologies, integration with supporting elements will subsequently lead to a complete experiment.[84]

Both the ENDO-NNK Technology project and the HEDI[85] project may have some potential for the defence against Soviet SRBMs. The latter has been described in the SDI Report as follows:

> This project consists of a concept definition study initiated in FY84 and a follow-on experimental study of a non-nuclear high endoatmospheric interceptor. The first phase investigated alternative experiments and the required components for each experiment were identified. A functional demonstration will be carried out to validate and demonstrate the critical technology issues.

The critical technology development and functional demonstration will include sensors, seekers, guidance, and warheads. These technologies will be integrated and used for the demonstration of non-nuclear target destruction. This demonstration will include an assessment of vulnerability to countermeasures und survivability in a simulated battle environment.[86]

However, the SDI Report does not discuss the specific demands for the defence of Western Europe on these potentially relevant technology programmes. An additional project – not listed in the SDI Report – of relevance for the defence against Soviet MRBM and SRBM has been referred to in military journals: the SRHIT (small radar-homing intercept technology) project or the FLAGE (flexible lightweight agile guided guided experiment).

SRHIT was originally conceived as a missile to counter the Soviet strategic SS-17 in the context of the non-nuclear ground-based LoADS (low altitude defence system) for the protection of the MX: In January 1983 the Ballistic Missile Advanced Technology Center of the US Army awarded the first contract for development and testing of the SRHIT missile to the Vought Corporation. In December 1984, after the first successful directed SRHIT flight test, *Aviation Week* commented:

> The small radar homing intercept technology programme is a technology demonstration effort aimed at assessing guidance and control technologies with potential applications on future non-nuclear ballistic missile defence systems. Operating at hypersonic speeds, the small radar-homing intercept missile uses kinetic energy from high-closing velocities to destroy without a warhead incoming strategic nuclear missiles. The information derived from the tests will be used to update computer simulations of strategic defence programmes and eventually to define anti-ballistic missile systems possibly using this technology.[87]

On 27 June 1986, in the sixth of nine planned test flights, a ground-based, radar-guided missile intercepted for the first time a moving target, a simulated ballistic missile re-entry vehicle travelling at Mach 5 within the atmosphere. This flexible lightweight agile experiment (FLAGE), as the SR-HIT technology programme was renamed, was carried out by SDIO and the US Army. The main difference between FLAGE (or SR-HIT) and the Army's HOE or ERIS programme is the altitude of interception. While HOE and ERIS are part of an exoatmospheric programme with intercepts at an altitude of more than 100 miles, FLAGE 'is being used to develop technologies applicable of endoatmospheric defense interceptors such as the Army/McDonnell Douglas high endoatmospheric defense interceptor (HEDI)'.[88]

While SRHIT or FLAGE is to acquire a capability to destroy approaching missiles up to ten miles above the target, the HEDI would intercept Soviet ballistic missiles up to a height of 60 miles or 100 kilometres. Hardly any information has been published as to the potential relevance of the SRHIT, FLAGE, HEDI, ERIS AND HOE projects for a tactical BMD in the context of an EDI.[89]

In an effort to apply concepts of strategic defence to the European theatre, Knoth distinguishes two, and Canavan four, layers. For Knoth these two layers of a EUROBMD – a European application of BMD concepts – 'would be almost comparable to the last two layers in the US BMD scheme' which he defined as:

1. *Trans-stratum*: This layer would be the last section of the American mid-course phase plus the very high altitude section of the terminal phase. This layer is relevant for anything higher than 50 or 100 kilometres in altitude.
2. *Cis-stratum*: This layer represents the engagement of targets below the trans-stratum layer extending to the surface of the earth.[90]

Only the SS-20 could be engaged in the trans-stratum layer while the cis-stratum layer would be available to combat the Soviet SS-21, SS-23 and SS-12 mod (SS-22).

Although Knoth considers 'space-based weapons platforms for an EUROBMD ineffective', vital battle management functions would have to be performed by satellites, including early warning, reconnaissance, communication and battle management support:

These satellites will control the trans-stratum battle and assist the ground-based systems in cis-stratum engagements. In addition to the space-based one needs a whole spate of ground-based sensors, everything from large radars to optical telescopes and in between. Even these two would not be enough ... Due to the general weather conditions of central and northern Europe, many ground sensors (mainly the optical and infra-red) will be badly impeded if not completely blocked. Thus as an adjunct (comparable to AWACS with its radar), one needs a plane flying above the weather equipped with those sensors made inoperable through the weather conditions. This LEP (long endurance platform) is comparable to the AOA (Airborne Optical Adjunct) proposed by the US as a part of its SDI research.[91]

Canavan, an assistant physics division leader at the Los Alamos Laboratory, discusses the 'Theater Applications of Strategic Defence Concepts' for four phases: 'low endoatmospheric (below 15 kilometres), high endoatmospheric (below 30 kilometres), mid-course (from 30 kilometres back to the end of MIRVing) and boost phase (MIRVing down to launch)'.[92] Canavan chose these altitude regimes based on physical discriminants: 'in low endo decoys have been discriminated fully by re-entry, in high endo the decoys have been slowed enough for discrimination to be performed, in mid-course decoys are highly credible, and in boost phase they have not yet been deployed'.[93]

In his discussion of the *low endoatmospheric intercept* Canavan points to the 'exploitable weaknesses' of interceptors carrying NNK warheads, 'while nuclear warheads have obvious barriers to achieving the mobility required for survivability and the releasibility required for utility'. Therefore, he suggests a mix of warheads: 'NNK for early defence – including that of the defences themselves – and nuclear concepts for the protracted, precise engagement phase'.[94]

In summary, Canavan emphasises the problems associated with all four intercept regimes:

In the low endo regime there appear to be adequate, capable sensors but interceptor performance must be traded against survivability. In high endo the sensors are the key issue, implying a shift towards airborne IR concepts, which could be particularly effective against lightly decoyed threats. In mid-course the interceptors are relatively simple ... Finally, for boost phase all of the concepts for strategic interaction appear to be applicable to the theatre, where they should work better with even smaller deployments.[95]

Nevertheless, Canavan concludes that 'essentially all of the concepts for strategic defence can be applied in the theatre with little modification and often with greatly improved performance'.[96] Table 1.1 provides a systems architecture for an overall strategic and tactical kinetic energy defensive system based on Canavan.[97] However, instead of relying on highly complex 'exotic' technology programmes, more near-term solutions have been developed by the

United States Army by upgrading existing air defence systems with limited capability against Soviet SRBM.

(b) Antitactical systems (ATM or ATBM) against Soviet SRBMs

Since the 1950s the United States Department of Defence has funded various research projects to study the possibility of surface air missiles (SAMs) to counter tactical ballistic missiles (TBMs). In the late 1960s and early 1970s, in the context of the ABM Treaty negotiations and debate, United States defence officials were concerned that Soviet SAMs might be used as tactical ABMs (TABMs or more recently ATBMs).[98] In 1965 the United States army revived its TABM concept study of the 1950s under the project name 'SAM-D'. During the SALT I hearings officials of the Nixon Administration offered conflicting assessments of the Soviet SAM-upgrading Problem. While the SALT I negotiator, Ambassador G. Smith, downgraded the Soviet SAM, the director of Defence Research and Engineering in the Department of Defense, John Foster, emphasised the Soviet SAM threat and insisted on the need for the American SAM-D project. In the mid-1970s SAM-D was renamed 'Patriot' and since 1980 the debate in the United States Congress has often touched on the issue of an 'anti-tactical defence' (ATM).[99]

During the NPG meeting in Cesme (Turkey) in April 1984 United States Secretary of Defence Weinberger informed his colleagues that the Patriot could be available as an anti-tactical system to extend the SDI umbrella to NATO Europe. According to *Aviation Week*:

> The anti-tactical missile programme is considered a priority effort by both the US and its Allies, according to Pentagon officials. They view the programme as part of another longer-term programme known as Counterair '90 ... The Counterair '90 system for application in Europe is considered a total system engineering approach to air and missile defense. It blends offence and defense into an integrated system ... The Raytheon product improvement programmes for Patriot and Improved Hawk are the cornerstones of an early capability to defend air defence sites against attack by Soviet missiles.[100]

What was offered at Cesme to the European NATO Allies as a tactical version of the SDI had been discussed in the United States Congress since Spring 1980 as a supplementary measure to enhance the survivability of the INF in Europe. Responding to Senator

Warner during the Defence Authorisation Hearings for FY81, James Wade, the then principal Deputy Undersecretary of Defence Research and Engineering, remarked:

> The question of active defence for theater nuclear forces is being looked at quite carefully ... It is reasonably clear that such a course could have merit ... Both GLCM and P-II are designed to achieve survivability against a number of threats through covert field deployment, frequent relocation in the field, and the reduction of signatures associated with field deployment. This mode of operation assumes enough warning to dispense to covert field sites prior to attack. An ATM could reduce the importance of warning time.[101]

Two years later it was announced that the United States army would initiate an ATM programme in FY83 'to counter the long-term postulated tactical missile threat'.[102] In a report to the House Armed Services Committee of 19 April 1984 the following details were offered:

> *The ATM programme is a tri-service programme with the Army as the lead service.* ATM includes defense against all tactical missiles including low-flying cruise missiles, air-launched cruise missiles and tactical ballistic missiles. Active intercept of the missiles in flight as well as detection and destruction of missile launchers and launch platforms fall under the scope of this programme. In order to accomplish this critical mission, extensive use must be made of existing missile systems, fire control systems and radars. In some instances, new systems must be developed to meet the changing threat (my emphasis).[103]

As far as the specific ATM plans of the United States Army were concerned, General Merryman supplied the House committee with this additional information:

> *Near-term anti-tactical missile systems*:
> (a) *Anti-tactical missile programme*: A thorough engineering analysis has been made of the proposed – Anti-Tactical Missile Programme and development efforts were initiated in FY83. Technical risk for this programme is considered to be low with the exception of fuze design which is considered low to medium. Cost risk is also considered to be low.

(b) *Anti-tactical missile upgrade*. At this juncture only preliminary assessments have been made of the technical and cost risks associated with this programme. Technical risk is considered low, since potential enhancements identified to date require engineering tasks which are well understood. The risk associated with RDTE cost is also considered to be low. Risk assessment will continue as the – Upgrade Programme is being developed.

Far-term system:
Programme definition of the far-term total Anti-Tactical Missile Programme is ongoing. As this programme becomes more clearly defined, the associated risks will be assessed (my emphasis).[104]

In an earlier Department of Defense study of the late 1970s onwards: 'ATBM Systems for Defence of Theatre and Naval Forces' it had been recommended:

1. Task the Army to define a programme that would lead to the development and deployment of an ATBM system based on Patriot that does not interfere with the current Patriot programme . . .
2. Task the Navy to define the improvements and programmes required to give AEGIS/SM-2 air defence system an ATBM capability. This system should be developed to sufficient maturity to permit quick deployment in the fleet in a time period approximately equal to that required for widespread deployment of a Soviet ballistic anti-ship missile.
3. A co-operative warhead programme should be initiated to develop nuclear and non-nuclear warhead technology capable of destroying TBMs armed with conventional, chemical, and nuclear warheads. The first goal of this technology programme should be the development of common non-nuclear and nuclear warheads for the Patriot and AEGIS/SM-2 ATBM systems.
4. A comprehensive advanced system concept and technology programme should be implemented. This programme is necessary to develop technology and define system concepts to counter Soviet reaction to first-generation US ATBM systems . . .[105]

According to an account in *Defense Week* in October 1983 the ATBM issue was studied by the 'Army Missile Intelligence Agency,

Armament Research and Development Command, Harry Diamond Laboratories, Ballistic Research Laboratory, Missile Command, and Ballistic Missile Defence Office; the Naval Surface Weapons Center; and other researchers. Both the Army Science Board and the Defence Science Board have studied the issue'.[106] On 22 May 1984 the United States Army and the United States Air Force announced their joint position on ATM:

> The Army and Air Force will complete the tactical missile threat assessment, to include evaluation of the operational impact of anticipated threat technical capabilities ... Using this threat assessment as a baseline [they] will establish a joint Anti-Tactical Missile programme.[107]

The most likely and short-term military component for an EDI has been the Patriot ATM programme of the United States army, which has been discussed in Congressional hearings since Spring 1980, especially in the hearings since 1983. The ATM programme, however, has become neither a part of the BMD organisation of the Army or of the SDIO, but is managed by a small Programme Management Office under the United States Missile Command at Huntsville.[108]

In Congressional testimony two different tasks of the ATM programme have been distinguished: (a) the short-term protection of the air defence system, the 'engineering development' of which was to be initiated in 1984; and (b) the longer-term ATM programme, the development phase of which was to be initiated in FY89. As to its short-term goal, General Merryman remarked:

> It improves the Patriot, and looks into the Hawk upgrade. And the only thing we do as far as preparing for that sophisticated ... is to look into things like non-nuclear kill. In other words, how do you kill one of these things without a nuclear round? We don't have all the answers ... When you have a target going at hundreds of miles an hour, you can destroy that with a nuclear round. But when you have to do it with just high explosives, well, we don't have the answers yet.[109]

In April 1983 *Defense Daily* outlined the longer-term planning:

> The Army proposes to initiate ATM procurement in FY86, involving upgrade of the Patriot and Hawk air defence systems.
> The ATM defence technology study will cover three major areas:

1. *Passive countermeasures*: This includes hardening of the ATM system, movement, dispersal, deception, signature reduction and early warning.
2. *Counterforce*: This includes surveillance, C³I and kill systems to nullify tactical missile threats prior to their launch.
3. *Terminal defence*: This includes surveillance/search, acquisition of low observable targets, C³I, launch detection, acquisition, tracking, discrimination, interdiction and homing functions to destroy or disable after launch each type of missile threat – cruise, short-range ballistic and medium-to-long range ballistic.

The platforms from which the technologies can be used to perform the functions, MiCom said, 'include satellites, probes, aircraft and ground systems'.[110]

According to *International Defense Review*, the Patriot ATM is to acquire a short-term capability against Soviet SS-21, SS-23 and SS-12 mod (SS-22) and after 1989 it is supposed to be able to counter the SS-20.[111] However, in Spring 1985 some officials of the United States Department of Defense were highly sceptical whether the Patriot ATM would ever become an effective counter to the SS-22. In the United States Congress, the ATM plans were deeply cut during the hearings for FY83–FY85.[112] However, on 11 September 1986, as one of several major SDI test experiments, 'a Patriot air defence missile flying at Mach 3 after 8 miles intercepted a Lance tactical missile at Mach 2 at 26,000 feet over White Sands'.[113] This successful test of a Patriot air defence missile in an ATBM mode allayed some of the severe doubts about the effectiveness of the ATM programmes that had existed among members of the House Armed Services Committee who had previously opposed any extension of ATMs that could conflict with the ABM Treaty.[114]

In the academic and in the political debate on the AT(B)M, the conformity of an ATM option with the ABM treaty has been rather controversial. Fred Hoffman, David Yost and Uwe Nerlich[115] have stressed that 'a theatre ballistic missile defence would not constitute an ABM defence in violation of the Treaty unless it were given "capabilities to counter strategic ballistic missiles or their flight trajectory" or it had been tested "in an ABM mode"' (article VIa).[116] As early as April 1983, a few days after President Reagan's 'Star Wars' speech, William E. Jackson, a former high official in the United

States Arms Control and Disarmament Agency, had pointed to legal infringements posed by the ATM issue:

> Surely a European-based ABM system which can shoot down, say, Soviet SLBMs fired from the Baltic or American SLBMs assigned to NATO, is a system that could be used against 'strategic' ballistic missiles. The same can be said of ASBMs (air-to-surface ballistic missiles) which are also counted in the SALT II agreement. Further, the ranges of SLBMs and ASBMs can be shorter than Pershing or SS-20 intermediate-range missiles. For all practical purposes, 'anti-tactical' missiles are strategic ABMs.[117]

Pike and Longstreth have countered the positive evaluation of the Hoffman report that ATMs were not constrained by the ABM Treaty:

> The unconstrained development and deployment of ATMs by both the US and the USSR threatens to circumvent both the letter and purpose of the ABM Treaty. The application of such systems to a strategic ABM role over the long term appears quite likely if the limits of their development and deployment are not clarified in the near future.[118]

According to Stephen Weiner:

> Since the long-range TBM trajectory is virtually identical to a short-range SLBM trajectory, an ATBM system would, almost by necessity, be capable of countering SLBMs ... As yet, no one has developed an unambiguous scheme of defining an effective SAM system without a BMD capability.[119]

According to the SALT definitions strategic systems comprise all ICBMs with a range beyond 5,500 kilometres and SLBMs deployed on nuclear-powered submarines. According to this definition, the Soviet SLBM with the shortest range is the SS-N-5 Serb with a range between 1,000–1,400 kilometres, while the United States SLBM with the shortest range is the Poseidon C-3 which can target over a distance from 4,000–4,600 kilometres. A Soviet SLBM launched from the Baltic or the Black Sea does have a trajectory nearly identical with that of the SS-22. While no legal prohibitions exist that would prevent defence against the SS-22, defending against the SS-N-5 would be a violation. Given the indistinguishability of the missile trajectories, a Soviet ATBM capability against the Pershing II would be permitted while an equivalent United States capability against the SS-20 would be prohibited (and would be controversial against the SS-22). No legal restraints would exist, however, for an AT(B)M system whose capability were limited to countering Soviet SS-21 and SS-23 SRBMs.[120]

(c) Antitactical missile defence by pre-emptive strikes: Army and Air Force TACMs – central components of deep-strike concepts

From a military point of view, a preferable option to a terminal defence against Soviet SRBMs and IRBMs would be destruction of the SS-21, SS-23 and possibly also the SS-22 launchers before they launch their first missile or immediately thereafter, before reloading. This option has been supported by officials of Defense Research and Engineering within the United States Department of Defense. Both within the context of Counterair '90 and the AirLand Battle, the system that would enable pre-emptive strikes against Soviet SRBM launchers forward-deployed in East Germany, Czechoslovakia Poland, or Bulgaria has been referred to as Lance II, Corps Support Weapons System, which was later renamed Joint Tactical Missile System (JTACMs) and more recently Army TACMs and Air Force TACMs.[121] The JTACMs programme was initiated in 1978 in the framework of the Assault Breaker with the aim 'to develop stand-off weapons for second-echelon armour strikes'. In this context the multiple-launch rocket system (MLRS), the Patriot and the Lance II were considered. In FY82, the Army and the Air Force programmes were combined as JTACMs until the agreement between both services was terminated in June 1984. As a consequence each service 'will go its own way in the development of a deep-strike missile; freed of the requirements of co-operation, the concepts for these are developing along more traditional service lines'.[122] The United States Army has been thinking of extending the range of the MLRS to 70 kilometres and to develop a Lance II with a range of 180 kilometres. The United States Air Force, on the other hand, is developing an air-launched cruise missile that could be used to target bridges, petroleum and weapons depots and C^3I installations in the rear, thereby reducing the vulnerability of the bombers and the fighters. The Army TACMs may acquire only limited capability against forward-deployed Soviet launchers for SS-21 and SS-23, while the Air Force TACMs would require highly sophisticated reconnaissance and battle management systems to perform pre-emptive strikes against Soviet SRBM launchers.[123] However, the Air Force TACMs that could be retargeted via NAVSTAR against moving Soviet targets could acquire such a capability during the 1990s. In the FY86 Report of Secretary of Defence Caspar W. Weinberger, the following reference was included:

> JTACMS-Army will be a replacement for a modification of the existing Lance or Patriot Missile System. As such, it will be used to attack targets of importance to the corps at ranges beyond the capability of cannons and rockets. This missile, or similar missiles with a high degree of commonality, will be used by both the Army and the Air Force. The Air Force missile will be launched from both fighter and bomber aircraft. The Army missile will be transported and launched with a mobile ground launcher.[124]

In testimony before the United States Congress both civilian and military officials representing the Department of Defense repeatedly referred to JTACMs as a vital weapons system in the context of AirLand Battle, FOFA and Counterair '90. In his prepared statement to the House Appropriations Committee, General Mahaffay stressed in April 1983:

> Joint Tactical missile system is the culmination of efforts to improve the range, accuracy, and effectiveness of mid-range missile systems, optimise force structure, and reduce the complexity of operations and maintenance. Army studies defined the operational concept for AirLand Battle and Corps Operations – 1986, in which impetus was given to a weapon system that the corps commanders would use to attack targets within their area of influence. A concomitant Air Force requirement to expand the range and efficiency of air-launched weapons pointed out the need for the Army and Air Force to consolidate their efforts and resources to satisfy similar requirements.[125]

In November 1983 *Aerospace Daily* extensively quoted James Wade representing the views of the Defense Research and Engineering division within the Pentagon on the close linkage between improved surveillance and deep strike options:

> The Pentagon is considering satellite surveillance as a way of getting real-time warning of Soviet missile attacks on NATO airbases and air defence sites. Soviet tactical ballistic missiles and what would seem to be the Soviet equivalent of the US's prospec-

tive Joint Tactical Missile System (JTACMS) will be accurate enough by the end of the 1980s to be effective against airfield and air defense targets using only conventional munitions, the Pentagon believes ... Wade gave no details about the satellite surveillance idea except to say that downlinks to NATO would provide a few minutes of warning in real time ... Rockwell International is to study combat intelligence application of its Navstar global positioning system satellites, and Europe will be in view of more than one GPS satellite at all times ... late this decade. And the Milstar ... communications system ... will include a satellite at synchronous altitude in view of Europe. Deciding precisely how to share warning intelligence with the NATO allies will be a 'more difficult problem' than getting the intelligence itself, Wade told the [House Armed Services] Committee. And overall counterair systems operations will have to be carried out through 'a NATO counterair command center, something in terms of command and control that does not necessarily exist today'. Counterair systems will have to be 'affordable and coproducible', he said. Even if NATO can scramble on warning and *take out Soviet missile launchers once an attack has started*, Wade said, *it still needs a tactical anti-ballistic missile, an objective of the Army's Anti-Tactical Missile program. This won't do the job by itself, however: 'I don't think we can afford the numbers required. But we need the capability. One response by the Soviet Union would be to MIRV the front end of their heavy tactical missile systems, which would stress any TABM system'* (my emphasis).[126]

Elements of such considerations could gradually be realised in the context of 'Emergent Technologies', which were adopted during the meeting of NATO's Defence Planning Committee in Brussels in May 1984. According to a report in *Defense Week*:

The ministers gave the highest and most specific endorsement to date of a US-backed plan to bolster NATO's conventional defences with radar, communications and munitions technologies that are either still in technological gestation or just now birthing. The US has pressed its European allies to support the incorporation of these 'emerging technologies' by the early 1990s ...

The list includes a new identification friend or foe (IFF) system for NATO aircraft; the multiple-launch rocket system (MLRS) with-precision-guided submunitions; an electronic support mission (ESM) system ... and a stand-off surveillance and acquisition system.[127]

Besides these general SDI-related research projects and specific weapons concepts and development programmes, the United States Department of Defense awarded various contracts for architecture studies both for a strategic and for a tactical BMD system.

(d) American architecture studies for a European BMD

According to press accounts at least two architecture studies dealing with a European-based BMD against Soviet short- and medium range ballistic missiles have been awarded in 1985 to:

1. *R & D Associates* in Marina del Rey;
2. *Martin Marietta* in Orlando.

The first architecture study will be carried out by Pan Heuristics, an autonomous division of R & D Associates, headed by Fred S. Hoffman. In October 1983 he completed the FSSS, the so-called 'Hoffman Report', which had a major impact on the SDI. Pan Heuristics will 'assess the usefulness of ballistic missile defences that might be deployed in the mid-1990's in countering the rapidly growing Soviet TBM threat to NATO's military forces'. Pan Heuristics will address the following questions:

1. Which NATO targets are most threatened by Soviet TBMs and which of these appear to be cost-effectively defensible with an ATBM system, taking account also of other forms of defence and other threats?
2. How would an ATBM deployment affect crisis stability in Europe? How would NATO's ability to accomplish its military objectives be affected if both NATO and the Soviets deployed ATBMs?
3. On a preliminary basis, which types of technical systems would be most useful for an ATBM?
4. What is the 'cost-effectiveness at the margin' of ATBM systems? Can the defence continue to accomplish its objective even against an increased offensive threat?
5. What are possible and useful roles for the European NATO nations in the development of an ATBM system?[128]

Martin Marietta has been studying – in the framework of an SDIO architecture study – passive and active possibilities in order to enhance the survivability of NATO's armed forces against a Soviet attack with tactical nuclear missiles. The following measures have been considered:

1. *Physical hardening* of the Soviet ballistic missile targets in Western Europe (INF, C³I installations, tactical air, depots) appears to be technically infeasible.
2. *Higher alert rates* could offer significant benefits. However, it would involve high costs and it requires a willingness to give more emphasis to strategic warning indicators.
3. In the context of *active defences* Martin Marietta is considering the possibility of various terminal defence options including Patriot ATBMs, which could deal with the Soviet SRBMs but probably not with the SS-20 and HEDI, which might achieve initial operational capability within five years.

Active defence systems could provide a point defence for military targets but no area or population defence. If 50–100 such targets were to be defended, Martin Marietta experts assume that 10–20 interceptors would be required for each side, or 500–2,000 interceptors altogether.[129]

These preliminary architectural studies for a theatre missile defence system against Soviet TBMs, funded by SDIO and the US Army, contributed to a better understanding of some of the technical problems associated with a tactical BMD for Europe. During 1986, while the conceptual planning for ATM system concepts continued by SDIO and in the context of an extended air defence within NATO, the public debate among defence experts gradually emerged.[130]

In early 1986, SDIO proposed to spend $50 million of its 5.7 billion budget request for FY 1987 to encourage West European governments and defence contractors to counter the growing threat of Soviet SRINF. In June 1986, bids from European and American companies for theatre defence architecture studies were invited in order to 'provide the SDIO with a better understanding of the requirements for a credible, robust defence – a defence which is developed in an evolutionary manner and capitalizes the evaluations of those who confront directly the shorter-range threat'.[131] The Army Strategic Defense Command asked for proposals for a two-phased programme till July 1986: 'a six-month effort to define general requirements and a set of candidate architectures, followed by an option for a second phase that would define a European theatre defence concept in greater detail . . . The first part of the contract will define technical requirements and identify critical technologies required for each concept. The most promising theatre defence architectures will be further defined by selected defence contractors during the second six months of the effort. Life-cycle costs, operation and maintenance plans and basing and deployment estimates will be investigated during Phase 2'.[132]

In the summer of 1986, the US Department of Defense set up an *Anti-Tactical Missile Steering Committee* headed jointly by Fred Iklé, undersecretary of defence for policy, and Donald Hicks, undersecretary of defence for research and engineering, that was to draft a plan till November 1986 for a specific missile defence for Europe based on the results of three separate study groups. One was to study near-term options, such as the prospects for upgrading the Patriot long-range anti-aircraft missile system deployed in Europe during the last two years, while the long-term study group was to analyse the potential applications of SDI technologies for a tactical BMD in Europe and a third study group was to co-ordinate the overall results.[133]

In September 1986, twelve teams representing more than 50 US and European defence companies had submitted bids for the SDIO's theatre defence architecture study, five of which were headed by European firms: Aerospatiale (France), MBB (West Germany), Standard Elektrik Lorenz (West Germany), CITES (Italian consortium) and SNIA BPD (Italy). Seven of these teams were awarded $14 million in SDI contracts in December 1986.[134]

As a reaction to President Reagan's speech of 23 March 1983, NATO's Military Committee tasked its Advisory Group for Aerospace Research and Development (AGARD) in late 1983 to investigate military and technological possibilities for a defence against Soviet TBMs up to the year 2000 based on active engagement methods with non-nuclear systems against Soviet TBMs with ranges less than 1000 km. The AGARD panel, comprising 13 experts from five NATO nations, concluded in its AAS-20 study submitted in early 1986 that High Energy Laser and electromagnetic guns would not contribute to realistic ATM options. Any European ATM system would have to rely on SAM systems and on powerful ground-based radars, airborne infrared sensors and on space-based sensors for early warning purposes.[135]

The first major supporter of a European Defence Initiative (in 1985) or of an extended air defence (in 1986) within NATO was the West German Defence Minister Wörner who had initially been the most outspoken SDI critic at NATO's NPG meeting in Cesme in April 1984. While he did not formally propose EDI during NATO's Defence Planning Committee meeting on 2–3 December 1985, he discussed the problem of an upgraded air defence in a series of bilateral talks with other defence ministers 'on the fringe of the meetings'.[136]

Wörner's EDI initiative reflected the increasing pressure by the leader of the Christian Social Union, Franz Josef Strauss and of the leader of the parliamentary party of the CDU/CSU in the Bundestag, as well as the concern by the former Inspector General of the Bundeswehr, General Altenburg, about the threefold threat posed by Soviet conventional, chemical and nuclear TBMs and the detailed suggestions that had already been tabled by major West German defence contractors. With several million marks from the West German Defence Department, MBB and Diehl (two major defence contractors) had submitted detailed proposals for an 'high energy laser'.[137]

Wörner published his call for 'A Missile Defence for NATO Europe' in the Winter issue of *Strategic Review* in which he argued that the Soviet Union could employ their SRINF armed with conventional warheads against prime NATO targets beneath the nuclear threshold and in combination with active defences, the Warsaw Pact could blunt nuclear options. For Wörner 'the only politically and strategically acceptable alternative for NATO is a direct defence against Soviet missiles' being part of a 'common Alliance initiative'. Such a defence could be accomplished through 'passive measures', 'through the destruction of Soviet missiles before their launch', and 'through the interception of the oncoming missiles before they reach their targets'. It would have to fulfil the following criteria:

1. The anti-missile defense must be non-nuclear. It will be directed primarily against conventionally armed missiles . . .
2. The objective must be, in the first instance, a point defence of priority targets on NATO territory . . .
3. The overall defense need neither be impenetrable, nor cover Western Europe comprehensively in order to have strategic effect . . .
4. The anti-missile defenses must possess high survivability. They must be tied into the NATO air defences . . .
5. The anti-missile defenses must be configured in such a way that the opponent cannot saturate them with only a part of his missile forces.[138]

Wörner's publicly expressed concern about a new Soviet TBM threat was shared by SACEUR, General Rogers. Both were instrumental in initiating various studies in the NATO framework. After the

AGARD study had been submitted to NATO's Military Committee in early 1986, NATO's Defence Ministers meeting at the DPC on 21 and 22 May 1986 for the first time publicly endorsed studies on the ATM system by NATO's military committee and its air defence sub-committee.[139] On 11 July 1986, NATO's Long Term Planning Committee on TBM was asked to draft long-term planning guidelines approving development of a Europe-based antitactical ballistic missile (ATBM) defence system. The guideline, according to General Rogers, once completed will 'set forth the threat and what we need to do to meet it', by laying out the types of weapons systems that would be needed. While the United States was taking the lead on developing ATBM technology, Rogers believes 'that the Europeans will have to lead the political and financial effort for the European system'.[140]

During 1986, three different studies were conducted at NATO: a report on the TBM threat that was tabled to the DPC's ministerial meeting on 4 and 5 December 1986, a study on the identification of possible actions and countermeasures and a mission analysis of the systems architecture and the systems effectiveness. The latter study done by AGARD is due for submittal to NATO's Military Committee on 1 June 1987.[141]

From these NATO studies and the unrelated parallel and complementary SDIO theatre architecture studies the outlines of a European ATM system may emerge during 1987 that may offer a point defence against Soviet TBMs, standoff missiles, cruise missiles and against the airborne threat posed by aircraft. A successful conclusion of a zero-zero option combined with a qualitative freeze and a non-circumvention clause for SRINF within the Reykjavik framework may be the most cost-effective response to the concerns initially expressed by Mr Wörner and Mr Rogers. Given the uncertainties both with respect to the arms control prospects for INF and SRINF and the lack of an agreed systems architecture for a tactical BMD in Europe, any discussion on the military and political implications of ATM options in the framework of either an EDI (top-down approach) or a EAD (bottom-up approach) will have to be of a preliminary and fragmentary nature.

V IMPLICATIONS OF AN EDI FOR EUROPEAN SECURITY – A PRELIMINARY EVALUATION

It is still uncertain what the specific strategic goals of SDI are. Depending on the audience, officials of the Reagan Administration have attempted both to *expound the vision of overcoming nuclear deterrence* and making obsolete the United States military doctrine of MAD by providing a failsafe defence both for the peoples of the United States and their allies and *to reassure the European allies that the credibility of American nuclear deterrent will be enhanced.*[142] The specific purposes of an EDI have remained even more nebulous: is EDI a technical necessity to ensure the credibility of the doctrine of coupling, or will it be a hedge against an assumed massive buildup of Soviet SRBMs armed with conventional, nuclear, biological and chemical warheads? Instead of speculating about future EDI selling strategies,[143] the strategic rationalisations, economic legitimisations and the European 'technical options' which will emerge during the debate on an EDI will have to be discussed in the context of the following fundamental questions:

1. *What is to be protected* by EDI systems: high value military targets (point defence), all major Soviet military targets in Western Europe (extended point defence), the major populations centres (limited population defence) or the peoples of Western Europe (population defence)?
2. Will it be *technically feasible* to achieve a population defence for Western Europe as President Reagan has claimed and General Abrahamson repeated as late as November 1985 in Brussels?
3. Will such a system be *affordable* for the European countries and how is it to be funded: by a major increase in the overall size of defence expenditure in Western Europe while the United States defence budget is frozen, or by a shift from other portions of the defence budgets?
4. What will be the implications on present *force planning* and on future *strategic stability* in general and on crisis, arms race and diplomatic stability in particular?
5. What will be the implications of different EDI options on East–West relations and on the *relationship* between the *United States and its European allies*?
6. What will be the implications on *existing arms control treaties* – for example, on the ABM Treaty – and on *ongoing arms control negotiations* affecting European security – for example, on the bilateral Geneva talks, on the Vienna talks on MBFR and on the Conference on Disarmament and on Confidence- and Security-Building Measures in Europe?
7. What are the possible implications for the *West European*

public: will EDI reassure the Europeans by either overcoming or enhancing deterrence?

Many of the questions raised by Sir Geoffrey Howe relating to the SDI are also of relevance with reference to the EDI options.[144] The West European governments and the international parliamentary bodies, the WEU and the North Atlantic Assembly, as well as the national parliaments have formulated a set of conditions for their support of the SDI research programme.[145] In addition, analytical criteria, as formulated by the Office of Technology Assessment of the United States Congress in its evaluation of BMD and ASAT technologies and weapons systems, may also be applied.[146] The preliminary evaluation of EDI options is based on the assumption that neither MAD nor MAS – the strategic rationalisation of the SDI supporters – will be able to protect both the West European territory and its peoples if deterrence fails. The implications SDI and EDI will have on a truly conventional non-provocative defence posture for Western Europe will therefore be briefly discussed in the concluding remarks.

(a) What is to be protected: military targets or populations?

Except for the SDI and EDI ideologues, all serious military and technical experts in Western Europe and in the United States agree: a perfect defence of the West European territory and its peoples against Soviet short- and medium-range ballistic missiles, cruise missiles and aircraft is neither technically feasible nor economically affordable. Konrad Seitz, Hans Rühle and Lother Ruehl[147] – three major proponents of EDI or of an extended and integrated air defence – have therefore all called for a *limited point defence* of high-value military targets. In its architecture study, Martin Marietta has obviously been considering a protection of the 50–100 most vulnerable high-value military targets in Western Europe, including the Pershing II and GLCM sites, special ammunition sites and major C^3I installations. In this respect EDI or an ATBM option could neither overcome deterrence nor protect the European population but only increase the survivability of major military instruments.

(b) Technical feasibility of ATBM and deep-strike options

The American military literature and civilian and military experts in the United States Department of Defense are divided both as to the military attractiveness and the technical feasibility of the Patriot AT(B)M against Soviet SS-22 and SS-20 ballistic missiles. The technical feasibility of the deep-strike systems to destroy SRBM launchers cannot be evaluated at this stage because it is still unclear what kind of system will finally emerge from the research and development. It would have to rely on all-weather surveillance and battle management systems not yet available. It is also unclear whether the C^3I systems that are supposed to provide near real-time intelligence will be technically feasible, survivable and economically affordable.

(c) Potential costs of SDI and EDI for West Europeans

During the first BMD debate in 1967 and 1968, the cost estimates offered for a BMD for Western Europe ranged from $US 5,000–80,000 million. No cost estimates have been offered by the EDI proponents for an integrated air defence system against Soviet aircraft, ballistic and cruise missiles. The cost figures offered by Martin Marietta lobbyists for 50–100 point defence sites with 500–2,000 ATBM missiles can hardly be taken seriously: $1,000–2,000 million for the missile system and as much again for the radars for the whole system.

All long-term cost estimates have to consider whether, and in what ways, the Europeans will have to contribute financially in case a space- and ground-based BMD should ever be built – either by paying a part of the bill, by offset military purchases or by paying for the withdrawal of American troops. In my own worst case preliminary cost estimate for a European participation in the construction of the SDI – a percentage contribution of 20 per cent by NATO Europe and one-third of this amount alone by the Federal Republic of Germany are assumed – and for an extensive EDI the additional costs per year over a time span of fifteen years would amount to 10,000–20,000 million marks or approximately 10–20 per cent of the West German defence budget by the end of the century.[148]

If the present defence plan for the Bundeswehr with the goals outlined by Defence Minister Wörner should ever be implemented, the percentage for social welfare and defence combined would increase from fifty per cent of the federal budget in 1985 to some sixty per cent by 1997. If both the 1978 long-term NATO defence guidelines and SACEUR Rogers's demands for major expenditure increases to finance FOFA should be realised, the combined percentage for social welfare and defence for the federal budget of West Germany would increase to as much as 80 per cent by the year 1997.[149] No West German and no West European government could be expected to increase its defence expenditures to realise both FOFA and an integrated or extended air defence system or a comprehensive EDA during a time period when the United States defence budget may either be frozen or decreased. Fiscal constraints will more likely decide on the future of SDI and EDI (or the more ambitious SDA and EDA) than the strategic rationalisations which were offered in the late 1960s and the new ones which are now being offered in support of a BMD for Western Europe.

(d) Potential implications of EDI options for strategic stability in Europe

Any debate on the strategic implications of EDI options for NATO will not only be influenced by what the West European countries will be able to afford, but also by how the Soviet Union can be expected to react in the next ten years to NATO's initiatives. The OTA study distinguished four levels and six relevant criteria in its analysis of BMD technologies.[150] The postulated levels are:

1. No additional strategic defence beyond those permitted by the ABM Treaty,
2. protection of some ICBMs (thin point defence),
3. defence of most ICBMs or 'a high degree of urban survival',
4. extremely capable BMD both for ICBMs and cities.

For each level the following six criteria are being discussed:

1. Potential role in United States nuclear strategy,
2. crisis stability effects,
3. arms race stability effects,
4. diplomatic stability effects,
5. feasibility,
6. costs.

This preliminary discussion of the stability effects of EDI options will focus on the two land-based components – ATBM for terminal defence and deep-strike systems for pre-launch destruction of SRBM launchers – and their potential effect on crisis and arms race stability. At least four different scenarios of the behaviour of the two sides may be distinguished:

Scenario 1: Only the Soviet Union and the Warsaw Treaty countries will deploy ATBM and/or deep-strike systems.

Scenario 2: Only the United States and NATO countries will deploy ATBM and/or deep-strike systems while the Soviet Union (WTO countries) will increase their offensive potential.

Scenario 3: Both the United States and the Soviet Union will develop and deploy BMD systems against strategic and tactical ballistic missiles without an arms control regime.

Scenario 4: Both sides will strengthen the ABM Treaty regime by foregoing both BMD and ATBM options and by drastically reducing their deep-strike options.

While both the Soviet Union and the United States are developing and have even tested ATBMs, up till now no deployments have been claimed in Europe outside the Soviet Union. At present the C^3I component of deep-strike missiles appears to be insufficient to permit pre-emptive strikes against mobile SRBMs of either side. But with the increasing accuracy of the SRBMs it is foreseeable that SS-21s, SS-23s or SS-12 (SS-22s mod) could hit stationary and non-hardened high-value targets with conventional warheads. The argument that new systems are needed as a hedge against Soviet systems or treaty violations (anticipatory reaction) has been a major stimulus to bilateral arms competition. If either side announced deployment of an ATBM system it may be assumed that an anticipatory research programme would manifest itself in a subsequent deployment decision. It may be assumed that either Scenario 1 or Scenario 2 would soon lead to Scenario 3, with each Alliance deciding on the specific mixture of offensive and defensive components in the framework of available fiscal resources.

In the context of Scenario 3, what are the potential implications of the deployment of a thin point defence (for example, of 50–100 ATBM sites in both Western and Eastern Europe), of a thick point defence (for example, of up to 1,000 ATBM sites on either side in Europe), of an effective protection of all military and some civilian targets (combined point and limited area defence), and finally of an effective area defence for both crisis and arms race stability?

At present the fourth level of an effective area defence for Western Europe appears to be neither technically feasible nor fiscally affordable. The third level of a combined point and limited area defence would require massive expenditures for civil defence which appear to be unlikely both for Western and Eastern Europe. A thick point defence for only military targets may run into major public acceptance problems, especially in the West European democracies. What would be the implications of a thin ATBM defence for 50–100 high-value military targets in Western and Eastern Europe for crisis and arms race stability?

Will the deployment of a thin ATBM defence by both NATO and the WTO countries increase or decrease incentives to launch a conventional or a theatre nuclear strike in a crisis situation? This alternative may lead to a new level of stability at higher cost. If such a mutual ATBM defensive system is not coupled with a deep-strike component, it may reduce the pressure to pre-empt, and it could therefore be stabilising in a crisis. However, once deterrence has failed the probability of an escalation to the nuclear level may increase for the losing side, the target's being highly vulnerable civilian and economic targets (shift from counterforce to countervalue targets). It is to be assumed that – in the absence of an arms control regime for SRBMs, IRBMs and MRBMs – both the number of the offensive forces may be increased and their penetration potential upgraded. Without an arms control regime for all short- and medium-range theatre nuclear forces, the findings of the OTA study for the mutual deployment of BMD may also apply to ATBM: 'Each side might easily suspect the other of attempting to gain military advantage by seeking the ability to destroy most of the opponent's land-based missiles and then use the defences to keep retaliatory damage to a very low level. If either side feared that its retaliatory capabilities were about to be lost or greatly reduced relative to those of the other side, there would be an incentive to add offensive capabilities and defensive capabilities at the same time'.[151] A direct consequence of this development would be an intensified arms competition in Europe – that is, an increase in arms race instability. It appears unlikely that détente will flourish in such a political environment: diplomatic stability is likely to decrease.

If the deployment of ATBMs is accompanied (or subsequently supplemented) by the deployment of a large number of ballistic missiles for the destruction of the other side's SRBM launchers – for example the JTACMs or the SS-23 and SS-22 option – in a political and military crisis in Europe or spilling over to Europe the pressure to pre-empt may increase, and as a direct consequence crisis stability is likely to decrease. The combination of a thin ATBM defence system and of the massive deployment of deep-strike systems could lead to the worst possible situation in Europe: intensified tensions, an intensified arms race and – in a crisis – increasing pressures to pre-empt or a decrease in diplomatic, arms race and crisis stability.[152] Therefore, strategic stability is to be assumed to be the primary victim of the development and a subsequent deployment of EDI options.[153]

(e) Potential implications of EDI options for arms control treaties and for arms control negotiations affecting Europe

With the limited relevance of space-based BMD components, even of X-ray lasers, at least one arms control Treaty will be affected by various ATBM options. It has already been emphasised that an ATBM with a capability to defend against the SS-20 would certainly violate the ABM Treaty and that the defence against the SS-22 may violate this Treaty if it could defend against the Soviet SLBM SS-N-5 as well. No legal constraints would exist for an ATBM limited to the defence against SS-23 and SS-21 and for a completely independent ATBM defence of Western Europe without an integration of American components.

As far as the deep-strike pre-launch destruction option is concerned, it would not violate any existing arms control Treaty but it would undercut any negotiations for confidence and security-building measures in Europe. The extension of the period for the announcement of manoeuvres would become completely irrelevant if the major offensive and/or defensive components of the other side could be destroyed by stand-off forces without any strategic and tactical warning.[154]

The bargaining chip theorem is highly questionable.[155] The bargaining chips of past arms control talks have often gained an unstoppable momentum of their own, for example the cruise missile option during the SALT I and SALT II negotiations. By the turn of the century there might be 10,000 land-, sea- and air-based cruise missiles.[156] According to the Reagan Administration, SDI cannot be bargained away. If EDI options should become affordable, in an era of increased tensions between East and West, they may develop a momentum of their own once they have crossed the point of no return sometime in the early or mid-1990s.

(f) Potential political implications of EDI options for Western Europe

Although the Soviet Union has failed so far to drive a wedge into the North Atlantic Alliance, SDI has been quite instrumental in splitting the West European countries and the members of the WEU in general and France and the Federal Republic of Germany in particular.[157] Dominique Moisi has pointed to a major 'Star Wars' casualty:

> Politically, the plan for a space-based defence against Soviet missiles had one clear effect: it served to catalyse tensions between two of America's most important allies, France and West Germany. At the diplomatic level, the SDI has had the effect of cancelling years of more positive US diplomacy aimed at cultivating a more united approach to the defence of Europe.[158]

Pierre Lellouche, with Moisi an associate director of the Institute Français des Relations Internationales (IFRI), has pointed to other SDI casualties:

> France and Britain have invested the most effort during the past twenty-five to thirty years in nuclear deterrence as the basis for their security policy, and they were the primary beneficiaries of the previous offensive deterrence regime – established through the Anti-Ballistic Missile Treaty. The reintroduction of defence, particularly on the massive scale contemplated in the SDI programme, directly threatens the very lifeline of their defence policy as well as the existing defence consensus at home (particularly in France) . . . Thus a direct casualty of SDI has been – once again – the goal of European unity. European divisions have appeared not only in the political realm, but also in the technical/industrial front, where in recent months French–German co-operation has suffered two serious setbacks in important space-related instances: . . . SAMRO and Hermès.[159]

Many European observers, both deterrence supporters and critics alike, will agree with Lellouche's outspoken assessment:

> SDI-related technologies are largely irrelevant in dealing with the larger part of the military threat to Europe . . . Europe will remain vulnerable to a whole array of other threats such as shorter-range ballistic missiles (SS-22s in particular), not to mention airbreathing delivery vehicles and a vast panoply of tactical nuclear and non-nuclear short-range delivery means, including artillery. Against all this, SDI is and will remain powerless. What will happen, though, is that once both superpowers embark on any offense–defence race in strategic weaponry, their competition will not stop, but instead will be channelled elsewhere, such as into non-ballistic short-range nuclear and non-nuclear systems in

Europe. Thus the likely outcome of strategic defences will be to augment, rather than reduce, military threats in regional theatres, and particularly in Europe.[160]

From a somewhat different perspective, the Danish defence expert Hans-Henrik Holm has pointed to negative implications of SDI for European scurity:

> Relationships among different domestic groups in the different countries involved will be affected: for example, among the various branches of the US armed forces. Similar cleavages will develop in Western Europe. Indeed, there are already costs in terms of tensions of the domestic consensus of West European countries that was slowly being rebuilt, and between the United States and Western Europe. The potential benefits are presently more elusive than real. It is possible that the SDI proposal as such induced the Soviet Union to return to the negotiations table in Geneva, but it will not induce them to stick to any arms control agreement when SDI is non-negotiable. Overall, SDI made Western Europe more dependent on the United States, and that in and of itself is part of the problem with the proposal.[161]

Holm has pointed to the major alternative confronting West European governments as a reaction to the major challenge posed by both SDI and EDI:

> What Western Europe can do is to extend pressure to reduce superpower confrontations, and such a possibility may exist in firmly rejecting the SDI offer. The rejection itself will be difficult because it runs counter to a long tradition. But acceptance will, as pointed out, increase their dependence and overall strategic stability. The offensive–defensive race is not on yet, and could be stopped. SDI at present is a stumbling block for arms control and confidence-building discussions, and if Western Europe removed itself from the SDI, the enthusiasm of both superpowers for an enormously expensive and probably ineffective space defence system would probably wane.
>
> What Europe should not do is to choose dependence once again. Dependence does not assure security, but neither does isolated independence in a strategically interdependent world.[162]

What could the West European countries do in the framework of the fourth scenario to strengthen the ABM regime and to avoid long-term stabilisation of its present dependent position *vis-à-vis* the United States?

VI CONCLUSION – A PLEA FOR AN ARMS CONTROL AND A POLITICAL ALTERNATIVE TO BOTH SDI AND EDI

During the Reykjavik summit in October 1986 both President Reagan and Secretary General Gorbachev had agreed on the framework of a zero-zero option for INF and on drastic reductions of the number of strategic launchers and warheads. However, President Reagan's insistence on SDI and on the new interpretation of the ABM Treaty – most specifically of the term 'development' – were instrumental for its failure. The acceptance of a zero-zero option for INF would have implied the total elimination of all land-based long-range INF from Europe and a freeze on existing TBMs. It would have removed many of the military concerns that have been put forward as justifications for the development of AT(B)M options.

However, the reaction to the zero-zero option by several European governments was rather critical. Some Conservative politicians called for a balance with respect to TBMs as a precondition for the acceptance of the zero-zero option.

In December 1986, Soviet TBMs had acquired three functions:

(a) to legitimate the call for a *missile defence of Western Europe* by defensive (ATBMs) and offensive ('counterair') measures;
(b) to *undercut a zero-zero option on INF* by adding new conditions;
(c) to justify the need for a *modernisation of NATO's* own TBMs (Lance II or Army TACMs) or for the arming of the F-111 with ALCM.

The realisation of these three demands could lead to the worst possible future for Europe: a competition between offensive systems (TBMs, cruise missiles in the context of deep strike concepts) and defensive options (AT(B)Ms) and to the search for counter-measures and counter-counter-measures. Is there an alternative future? *Scenario 4* could contribute to a more optimistic future: all West European governments have repeatedly expressed their support for the ABM Treaty in its traditional and narrow interpretation. It should be in their interest to strengthen (and not to undercut) the ABM Treaty regime by closing the three grey areas of the Treaty that do presently exist: the relationship between BMD and ASAT technologies, the relationship between ABM and ATBM options, and the problem of discriminating ABM radars from other functions.

As to the first grey area, both the Soviet and the French governments have tabled specific proposals for a ban on ASAT testing and for a subsequent prohibition of ASAT. As to the second grey area – the relationship between ABM and ATBM systems – neither the United States nor the Soviet Union have tabled any proposal for prohibiting or limiting ATBM systems in order to avoid these systems acquiring an ABM capability. However, both the USSR and the US are presently testing ATBM systems, the SA-10 and the SA-X-12 or the Patriot ATBM respectively. Nevertheless, the point of no return may not yet have been crossed. A *moratorium on developing, testing and deployment of dual capable air defence systems* may still be able to foreclose a development that may contribute to a further weakening of (or even to the collapse of) the ABM Treaty regime. A *zero-zero option* as proposed by NATO in November 1981 and agreed upon by both Reagan and Gorbachev at Reykjavik, if linked with a *qualitative freeze* and with a non-circumvention clause for all tactical ballistic missiles, could neutralise many of the military concerns about a future Soviet conventional TBM option in the context of a surprise attack. A *nuclear-free corridor* in Central Europe, as originally proposed by the Palme Commission in June 1982 and agreed upon in principle between the SPD and the SED in October 1986, would eliminate the forward deployment (150 km from the border) of Soviet FROGs and SS-21 in East Germany and Czechoslovakia and of the Lance in the Federal Republic of Germany.

In subsequent arms control negotiations, the number of the remaining TBMs (SS-23, SCUD and SS-12 mod (SS-22) and the backward deployed SS-21 and FROGs should be reduced drastically. The mutual abandonment of AT(B)M systems would be the more credible and stable the sooner the existing TBMs could be removed.

If the TBMs of the United States and of the Soviet Union would be removed from the territory of third countries in Europe, the European governments would as a direct consequence not remain under the gun of the ballistic missiles of the other superpower: a major element of dependence would be removed – the fear of becoming the first casualties if deterrence should fail, if a conventional war outside of Europe spilled over to Europe and escalated to the nuclear level. Crisis stability would be enhanced, arms race stability would be strengthened and the longer-term vision of NATO's Harmel Report of 1967 – 'to create a just and lasting order of peace and security in Europe' – could become a reality. A Europeanisation of European affairs – at present a longer-term vision – could become real.

Notes

1. See chapters 4–9 in this volume. Hans Günter Brauch, 'From Strategic to Tactical Defence? European political, strategic, technological reactions to the "Star Wars" vision and to the Strategic Defence Initiative: Eureka and/or EDI', in John McIntyre (ed.), *International Space Policy* (New York: American Astronautical Society, 1986).
2. David Dickson, 'A European Defence Initiative – The idea that European nations band together for a strictly European version of SDI is gaining support', *Science* (20 September 1985) pp. 1243–5.
3. This term was coined by the West German Christian Democratic MP Wimmer.
4. Richard L. Hudson, 'West Europe Debates Controversial Push to Develop Its Own "Star Wars" Program', *Wall Street Journal*, 22 October 1985; Hans Günter Brauch, *Antitactical Missile Defence – Will the European Version of SDI undermine the ABM-Treaty?* AFES Papier 1 (Stuttgart: AG Friedensforschung und Europäische Sicherheitspolitik, July 1985).
5. Konrad Seitz, 'SDI: the technological challenge for Europe', in *The World Today* (August/September 1985) pp. 154–7; Lothar Rühl, 'Eine Raketenabwehr auch für Europa – Voraussetzungen und Folgen für Strategie und Rüstungskontrolle im Ost–West-Verhältnis', *Frankfurter Allgemeine Zeitung*, 17 January 1986. A shift in the evaluation can be observed with Hans Rühle, 'Löcher im Drahtverhau der Sicherheitsdoktrin–Das Defensivkonzept der Zukunft löst Europas Probleme kaum', *Rheinischer Merkur*, 1 April 1983 and his more favourable pro SDI article 'An die Grenzen der Technologie', *Der Spiegel*, 48/1985, 25 November 1985.
6. 'Paris lädt zu Weltraumprojekt ein – Hernu schlägt enge militärische Zusammenarbeit in Europa vor', *Frankfurter Rundschau*, 22 July 1985; 'Wörner für europäische Weltraumwaffen. Anregung Hernus begrüsst/Union fordert Frankreich zu gemeinsamer Militärplanung auf', *Süddeutsche Zeitung*, 23 July 1985; 'Wörner will Europäische Verteidigungsinitiative', *Süddeutsche Zeitung*, 5 September 1985; 'Wintertagung der Nato in Brüssel', *Neue Zürcher Zeitung*, 5 December 1985.
7. Representative for this school has been Albert Wohlstetter and the writings of members of the European–American Institute for Security Research in Los Angeles.
8. For a recent high-level criticism of MAD, see Fred Iklé, 'Nuclear Strategy: Can There Be a Happy Ending?', *Foreign Affairs* 63 (Spring 1985) pp. 810–26; Fred S. Hoffman, 'Nukleare Bedrohung, Sowjetische Macht und SDI', *Europa-Archiv* 40/21 (10 November 1985) pp. 643–52.
9. See chapter 8. For Wörner's reactions see chapter 6.
10. See 'Excerpts from Statements on BMD by Reagan Administration Officials', in US Congress, Office of Technology Assessment, *Ballistic Missile Defense Technologies*, OTA-ISC-254 (Washington, United States Government Printing Office, September 1985) pp. 297–298.

11. Fred S. Hoffman, Study Director, *Ballistic Missile Defenses and US National Security*, Summary Report, prepared for the Future Security Strategy Study (Washington: October 1983), reprinted in US Congress, Senate, Committee on Foreign Relations, Hearings, *Strategic Defense and Anti-Satellite Weapons* (Washington, DC: United States Government Printing Office, 1984) pp. 125–40.
12. Abrahamson, in *Strategic Defense and Anti-Satellite Weapons*, p. 17.
13. 'Weinberger Invites NATO Allies to Participate in SDI', *Wireless Bulletin from Washington* (Bonn: US Embassy, 59, 29 March 1985) pp. 13–14; Don Cook, 'Reagan Seeking Allied Support on "Star Wars"', *Los Angeles Times*, 25 March 1985; Michael Weiskopf, 'Interim "Star Wars" Defense Eyed for European Protection', *Washington Post*, 26 March 1985; Barry James, 'U.S. Asks Allies to Help in "Star Wars" Research – Weinberger Says Ministers Show Interest', *International Herald Tribune*, 27 March 1985.
14. See chapters 6 and 8.
15. See Dickson 'A European Defence Initiative', p. 1243.
16. See Dickson; Gerd Schmückle, 'Wir dürfen die Zukunft nicht verschlafen', *Rheinischer Merkur*, 30 March 1985; G. C. Berkhof, *Duel om de ruimte. Aspecten van Westeuropese veiligheid* (The Hague: Clingendael, 1985).
17. See note 6 and chapter 5 in this volume.
18. See note 6 and chapter 6 in this volume.
19. See chapter 6 in this volume.
20. John Vinocur, '"Star Wars" Plan for Europe Urged', New York Times, 8 March 1985; 'M. Giscard D'Estaing prône une initiative de défense Européenne en association avec les Etats-Unis', *Le Monde*, 4 December 1985.
21. See Dickson 'A European Defence Initiative, p. 1243.
22. See note 6.
23. Hans Rühle et al., *Standpunkte zu SDI in Ost und West* (Melle: Verlag Ernst Knoth, 1985).
24. Seitz, 'SDI', p. 157.
25. Manfred Wörner, 'A Missile Defense for NATO Europe', *Strategic Review*, vol. 14, no. 1 (Winter 1986), p. 13; Thomas Enders, *Missile Defense as Part of an Extended NATO Air Defense* (Sankt Augustin: Konrad Adenauer Stiftung, May 1986) Interne Studien, Nr. 2/1986; see also bibliography.
26. See Enders, *Missile Defense* for the results of the AGARD study. The developments till December 1987 will be covered in Hans Günter Brauch, *Evaluation of Anti-tactical Ballistic Missile Defence Technologies and Concepts for Western Europe in terms of strategic stability*, a study to be submitted to NATO in fulfilment of the author's NATO Defence Research Fellowship in early 1988.
27. Robert P. Berman and John C. Baker, *Soviet Strategic Forces – Requirements and Responses* (Washington, DC: Brookings Institution, 1982) p. 135.
28. William Arkin, Frank von Hippel and Barbara Levi, 'The Consequences of a "Limited" Nuclear War in East and West Germany', *Ambio — A Journal of the Human Environment*, 11, 2–3 (1982) pp. 163–73.

29. Berman and Baker *Soviet Strategic Forces*, p. 111.
30. Berman and Baker, p. 136.
31. IISS, *The Military Balance 1985–1986* (London: International Institute for Strategic Studies, 1985) pp. 165–6.
32. See Hans Günter Brauch, *Militärische Komponenten einer europäischen Verteidigungsinitiative (EVI). Amerikanische militärische Planungen zur Abwehr sowjetischer ballistischer Raketen in Europa*, AFES Papier 3 (Stuttgart: AG Friedensforschung und Europäische Sicherheitspolitik, February 1986) pp. 10–11. The data have been updated to July 1986, based on IISS, *Military Balance 1986–1987* (London: International Institute for Strategic Studies, 1986) pp. 200–8.
33. William Arkin and Richard W. Fieldhouse, *Nuclear Battlefields. Global Links in the Arms Race*, (Cambridge, Mass. Ballinger, 1985) pp. 265–7.
34. John Tirman (ed.), *The Fallacy of Star Wars* (New York: Vintage Books, 1984).
35. R. Z. Sagdeyev, A. A. Kokoshin for the Committee of Soviet Scientists for Peace Against Nuclear Threat: *Strategic and International Political Consequences of Creating a Space-Based Anti-Missile System Using Directed Energy Weapons* (Moscow: Institute of Space Research, USSR Academy of Sciences, 1984).
36. Sagdeyev, *et al.*, pp. 22–3.
37. Brauch, Militärische Komponenten.
38. See chapter 3 in this volume.
39. 'Soviet Strategic Defence Projects – Text of report by State and Defence Departments', *Wireless Bulletin from Washington* (Bonn: US Embassy, 4 October 1985) p. 13.
40. David Yost, 'Ballistic Missile Defence and the Atlantic Alliance', *International Security* 7 (Fall 1982) pp. 143–74; David Yost, 'Soviet Ballistic Missile Defence and NATO', *Orbis* (Summer 1985) pp. 281–92.
41. Department of Defense, *Soviet Military Power, 1984* (Washington, DC: United States Government Printing Office, 1984) p. 34.
42. Sayre Stevens, 'The Soviet BMD Program', in Ashton B. Carter and David N. Schwartz (eds), *Ballistic Missile Defense* (Washington, DC: Brookings Institution, 1984) pp. 215–16.
43. Department of Defense, *Soviet Military Power 1985*, p. 48.
44. Hubertus G. Hoffmann, 'A Missile Defense for Europe?' *Strategic Review* (Summer 1984) p. 53.
45. Robert Cooper, Director of the Defense Advanced Projects Agency, cited in *Washington Times*, 9 March 1984.
46. See *The Military Balance 1986–1987*, p. 38.
47. Yost, 'Soviet Ballistic Missile Defence', pp. 291–2.
48. Yost, pp. 286–9; see for discussion of FOFA and JCTAMs by General Bernard Rogers, *Air Force Magazine*, February 1985, pp. 20; 23 and Donald R. Cotter, 'Potential Future Roles in Conventional and Nuclear Forces of Western Europe', *Strengthening Conventional Deterrence in Europe: Proposals for the 1980s, Report of the European Security Study* (London: Macmillan, 1983) pp. 224; 238.

49. Yost, 'Soviet Ballistic Missile Defence', pp. 286–9.
50. Uwe Nerlich (ed.), *Sowjetische Macht und westliche Verhandlungspolitik im Wandel militärischer Kräfteverhältnisse* (Baden-Baden: Nomos Verlagsgesellschaft, 1982); Uwe Nerlich, *Die Einhegung sow- jetischer Macht* (Baden-Baden: Nomos Verlagsgesellschaft, 1982).
51. See the forthcoming book by Fred Hoffman, Albert Wohlstetter and David Yost on SDI sponsored by the European American Institute for Security Research (1987).
52. Uwe Nerlich, 'Missile Defences: Strategic and Tactical', *Survival* 27 (May/June 1985) pp. 119–27 especially pp. 120–21.
53. Nerlich, p. 122.
54. Nerlich, p. 123.
55. Nerlich, p. 123.
56. See European Institute for Security, *EIS Journal*, 4–5/85.
57. See chapter 8 in this volume.
58. Keyworth quoted in: US Congress, OTA-ISC-254, p. 302.
59. Iklé, 'Nuclear strategy'.
60. See chapters 4–6 in this volume.
61. For an extensive quote, see US Congress, OTA/ISC-254, pp. 300–2.
62. See chapter 1, table 1.1.
63. Clarence A. Robinson, Jr, 'US Develops Antitactical Weapon for Europe role', *Aviation Week & Space Technology*, 9 April 1984, pp. 45–9.
64. Robinson, p. 46.
65. Karsten Voigt, *Interim Report of the Sub-Committee on Conventional Defence in Europe*, Military Committee, NAA, (Brussels: North Atlantic Assembly, November 1984) pp. 21; 25.
66. Nerlich 'Missile Defences', p. 122.
67. Testimony of a spokesman of the United States Department of Defense before a Congressional Committee, without date and year, in the private files of the author.
68. Voigt, *Interim Report*, p. 17.
69. US Congress, House, Committee on Appropriations, Hearings, Department of Defense Appropriations for 1984, part 4 (Washington, DC: 1983) pp. 335–6.
70. Voigt, *Interim Report*, p. 22.
71. Voigt, *Interim Report*, p. 16.
72. David S. Sorenson, 'Ballistic Missile Defence for Europe', *Comparative Strategy* 5, 1985, pp. 159–78 especially p. 161.
73. Charles Mohr, 'Antimissile Plan Seeks Thousands of Space Weapons-Design After a Year's Study Has 7 Layers of Defenses to Protect US Targets', *New York Times*, 3 November 1985.
74. See also: *The Arms Control Reporter*, 11–85, 575E5.
75. Sorenson, 'Ballistic Missile Defence', pp. 161–2.
76. Gregory H. Canavan, *Theater Applications of Strategic Defense Concepts* (Los Alamos: Los Alamos National Laboratory, June 1985) LA-UR-85-2117 (P/AC:85-149) pp. 19–20.
77. Department of Defense, *Report to the Congress on the Strategic Defense Initiative* (Washington, DC: Department of Defence, April 1985) p. 20.

78. 'Plädoyer für Laser-Systeme in Europa', *Stuttgarter Nachrichten*, 2 July 1985; Helmut Berndt, 'Bodengestützte Laser-Raketen zur Abwehr', *Saarbrücker Zeitung*, 4 July 1985; Eduard Neumaier, 'Edward Teller glaubt an die Grenzenlosigkeit des Machbaren', *Stuttgarter Zeitung*, 1 July 1985.
79. See *Report to the Congress on the Strategic Defense Initiative*, p. 43.
80. *Report to the Congress on the Strategic Defense Initiative*, p. 51.
81. *Report to the Congress on the Strategic Defense Initiative*.
82. 'Homing overlay explained', *Flight International*, 15 December 1984, p. 1642ff.
83. *Report to the Congress on the Strategic Defense Initiative*, p. 56.
84. *Report to the Congress on the Strategic Defense Initiative*, p. 51.
85. 'Army plans early 1990s Ballistic Missile Defence Demonstration', *Aerospace Daily*, 19 October 1984, pp. 249–50; Eugene Kozicharow, 'Army Doubles Missile Defence Fund Bid to Support SDI Effort', *Aviation Week & Space Technology*, 8 April 1985, pp. 62–3.
86. *Report to the Congress on the Strategic Defense Initiative*, pp. 55–6.
87. 'Army Completes First Flight Test of Guided Radar Homing Missile', *Aviation Week & Space Technology*, 10 December 1984, p. 28.
88. 'Army/LTV Missile Intercepts Reentry Vehicle', *Aviation Week & Space Technology*, 14 July 1986, p. 119.
89. For additional details see Hans Günter Brauch, *Militärische Komponenten einer Europäischen Verteidigungsinitiative (EVI). Amerikanische militärische Planungen zur Abwehr sowjetischer ballistischer Raketen in Europa*. AFES Papier 3 (Stuttgart: AG Friedensforschung und Europäische Sicherheitspolitik, February 1986).
90. Arthur Knoth, 'EDI – A European Defence Initiative?', *Military Technology* 12 (1985) p. 23.
91. Knoth, p. 23.
92. Canavan, *Theater Applications*, p. 3.
93. *Theater Applications*, p. 3.
94. *Theater Applications*, p. 6.
95. *Theater Applications*, p. 12.
96. *Theater Applications*, p. 23.
97. See Simon Peter Worden, chapter 1 in this volume.
98. See Brauch, *Antitactical Missile Defence*, p. 33ff.
99. See Brauch, *Antitactical Missile Defence*, p. 25 and *Militärische Komponenten*, p. 35ff.
100. See Brauch *Militärische Komponenten*, p. 47ff; Clarence A. Robinson, 'US Develops Antitactical Weapon for Europe Role' *Aviation Week & Space Technology*, 9 April 1984, pp. 46–9.
101. See Brauch, *Antitactical Missile Defence*, pp. 38–9.
102. US Congress, Senate, Armed Services Committee, *Hearings, Department of Defense, Authorization for Appropriations for Fiscal Year 1981, part 4* (Washington, DC: United States Government Printing Office, 1980) p. 2200.
103. US Congress, House, Committee on Armed Services, *Report on Department of Defense Authorization Act 1985* (Washington, DC: United States Government Printing Office, 1984) p. 152.
104. US Congress, House of Representatives, Committee on Appropriations, *Hearings, Department of Defence Appropriations for Fiscal Year*

1985, part 5 (Washington, DC: United States Government Printing Office, 1984) p. 117.
105. Excerpts referring to this document are in the files of the author.
106. 'Army Scrambles To Devise New Defenses', *Defense Week*, 31 October 1983, pp. 1; 12–13.
107. 'AF, Army agree on ATM', *Military Space*, 11 June 1984, p. 7.
108. US Congress, House, Appropriations Committee, *Hearings, Department of Defense Authorization and Oversight, Fiscal Year 1984, part 5* (Washinton, DC: United States Government Printing Office, 1983) p. 709.
109. US Congress, House, Committee on Appropriations, *Department of Defense Appropriations for Fiscal Year 1984, part 4* (Washington, DC: United States Government Printing Office, 1983) p. 273.
110. 'Army Seeking Anti-Tactical Missile Defense Technology', *Defense Daily*, 7 April 1983, p. 181.
111. 'I-Hawk and Patriot to get antimissile capabilities', *International Defence Review* (11/1983) p. 1535.
112. See Brauch *Militärische Komponenten*, p. 55; US Congress, House, Appropriations Committee, *Hearings, Department of Defense Appropriations, Fiscal Year 1984, part 4* (Washington, DC: United States Government Printing Office, 1983) p. 274; Department of Defense, *Congressional Action on Fiscal Year 1985 Appropriation Request, OASD (Comptroller)* (Washington, DC: United States Government Printing Office, 12 October 1984).
113. 'SDI and Emerging Trends', *Military Technology*, vol. X, no. 12, 1986, p. 50; 'Patriot Air Defense System Intercepts Lance Surface-to-Surface Missile', *Aviation Week & Space Technology*, vol. 125, no. 12, 22 September 1986, pp. 22–3.
114. 'Tussle Over ATM', *Armed Forces Journal International* (August 1984) p. 18.
115. For Hoffman Report see note 11; Fred Hoffman, 'The "Star Wars" Debate: The Western Alliance and Strategic Defence: part I', *Adelphi papers* 199 (London: International Institute for Strategic Studies, 1985) pp. 25–33; Yost, 'Ballistic Missile Defense'; Nerlich, 'Missile Defences', H. Hoffmann, 'A Missile Defense'.
116. See Hoffman, *Adelphi papers*.
117. William E. Jackson, 'Shooting down the ABM Treaty', *Christian Science Monitor*, 11 April 1983, p. 23.
118. Thomas Longstreth and John Pike, *A Report on the Impact of the US and Soviet Ballistic Missile Defense Programs on the ABM Treaty for the National Campaign to Save the ABM Treaty* (Washington, DC: National Campaign, June 1984) p. 24; See also chapter 12 in this volume. For my own criticism, see *Antitactical Missile Defence*, pp. 50–3; Hans Günter Brauch, *30 Thesen und 10 Bewertungen zur Strategischen Verteidigungsinitiative (SDI) und zur Europäischen Verteidigungsinitiative (EVI)* (Stuttgart: AG Friedensforschung und Europäische Sicherheitspolitik, January 1986); Brauch, 'Zur Militarisierung des Weltraums. Auswirkungen auf die Bundesrepublik Deutschland', *Die Friedens-Warte* 65 (1982–1985) pp. 134–48; Hans Günter Brauch, *Angriff aus dem All. Der Rüstungswettlauf im Weltraum* (Berlin, Bonn: Dietz Verlag, 1984).
119. Stephen Weiner, 'Systems and Technology', in Ashton B. Carter and

David N. Schwartz (eds), *Ballistic Missile Defense* (Washinton: Brookings Institution, 1984) pp. 73–5.
120. See note 11.
121. See Brauch, *Militärische Komponenten*, pp. 58–65.
122. *Militärische Komponenten*; Fred Hiatt, 'Hope for a Joint Missile Abandoned by Pentagon', *Army* (October 1984) pp. 401ff; Eric C. Ludvigsen, 'Light Forces Reshaping Modernization Program', *Army* (October 1984) pp. 401ff.
123. Ludvigsen, 'Light Forces'; Melissa Healy, 'JTACMS, Split Asunder, Now Black', *Defense Week*, 25 June 1984, p. 30; 'Senate Wants Army JTACMs Based on Patriot', *Defense Daily*, 11 June 1984, p. 218; James P. Wade, Jr, 'New Strategies and Technologies', Netherlands Institute of International Relations, Clingendael (ed.), *Conventional Balance in Europe: Problems, Strategies and Technologies* (The Hague, Clingendael: 1984).
124. 'Pentagon is said to increase JSTARS, JTACMs funding', *Aerospace Daily* 28 December 1983, p. 298; 'FY 1984 RDTE Congressional Descriptive Summary on JTACMS' in files of the author; *Program Acquisition Costs by Weapon System – Department of Defense Budget for Fiscal Year 1986* (Washington, DC: Department of Defense, 4 February 1985); 'Army Seeking Powerful Battlefield Missile System', *New York Times*, 17 February 1985.
125. US Congress, House, Committee on Armed Services, *Hearings, Defense Department Authorization and Oversight, on HR 2287, Department of Defense Authorization of Appropriations for FY 1984, part 3* (Washington, DC: United States Government Printing Office, 1983) p. 452.
126. 'DoD Eyes Satellites to Warn NATO of Tactical Missile Attack', *Aerospace Daily* 124, 18 November 1983, pp. 97–8.
127. 'NATO' Chief Approves Eleven High-Tech Trade Projects', *Defense Week*, 21 May 1984, p. 5.
128. See letter by Fred S. Hoffman to the author 7 November, 1985.
129. These remarks are based on an informal talk of a Martin Marietta manager.
130. During 1985, and especially in 1986, an increasing number of conferences and workshops have focused on EDI, extended air defence and on AT(B)M options. In late autumn of 1986 the Royal Institute of International Affairs in London (Chatham House) in co-operation with the American Academy of Arts and Sciences in Boston organised an ATBM conference in London, while the Free University of Amsterdam held a two-day workshop on ATBM in late November 1986. Two German–American conferences in December 1986 in Bad Homburg and in Tutzing partly focused on the ATBM issue. A preliminary proposal of a military proponent of a European Aerospace Defence Initiative was offered by the Dutch Brigadier General G. C. Berkhof, 'The American Strategic Defence Initiative and West European Security: A Dutch view', in The Netherlands Institute of International Relations, Clingendael (ed.), *The American Strategic Defence Initiative: Implications for the West European Security. Report on a Workshop, The Hague, 26–27 April 1985* (The Hague: Clingendael, 1986) pp. 215–60.
131. Michael White, 'Europeans to get Star Wars carrot', *Guardian*, 26

April 1986; 'SDI Office Will Sponsor Studies by Allies – Text: Abrahamson Statement to Senate', (London: USIS, 28 April 1986); Theresa M. Foley, 'SDI Organization Plans to Fund Theater Defense Architecture', *Aviation Week & Space Technology*, 19 May 1986, pp. 24–6; Statements by James A. Abrahamson and Fred C. Iklé before the US House of Representatives Committee on Armed Services Defense Policy Panel, 4 June 1986.
132. 'US, Allied Companies to Study European Missile Defense Concepts', *Aviation Week & Space Technology*, 14 July 1986.
133. Walter Andrews, 'Anti-missile system for Europe studied', *Washington Times*, 20 August 1985, p. 5.
134. Brendan M. Greeley, Jr, 'Army Missile Intercept Success Spurs SDI Theater Defense Study', *Aviation Week & Space Technology*, 29 September 1986, pp. 22–3; '12 Teams of US, European Contractors Submit Bids for Missile Defense Study', *Aviation Week & Space Technology*, 29 September 1986, p. 22; 'Wire News Highlights', *Current News*, 5 December 1986, p. 3.
135. Thomas Enders, *Missile Defense as Part of an Extended NATO Air Defense* (Sankt Augustin: Konrad Adenauer Stiftung, May 1986) Interne Studien Nr. 2/1986, pp. 14–16.
136. See note 6 above; 'Wörner mit EVI allein', *Der Spiegel*, 50/1985, 9 December 1985; 'Abwehrpläne der Verteidigungsminister. Zusätzliche Vorsorge gegen Kurz- und Mittelstreckenraketen – Kritik an den Niederlanden', *Rhein-Neckar Zeitung*, 4 December 1985; 'Bemühungen Bonns um verbesserten Schutz gegen Angriffe aus der Luft', *Frankfurter Allgemeine Zeitung*, 4 December 1985; Michael Feazel, 'German Minister Proposes Initiative To Improve European Defenses', *Aviation Week & Space Technology*, 9 December 1985.
137. 'Setzen auf EVI', *Der Spiegel*, 49/1985, 2 December 1985, pp. 25–6; Erhard Heckmann, 'Air defence by means of high-energy lasers – Demonstration by MBB and Diehl', *Military Technology*, 1/86, pp. 60–2.
138. Manfred Wörner, 'A Missile Defense for NATO Europe', *Strategic Review*, vol. 14, no. 1 (Winter 1986), p. 13; Manfred Wörner, 'Europa braucht Raketenabwehr', *Die Zeit*, Nr. 10, 28 February 1986, pp. 45–6; Manfred Wörner, 'Strategie im Wandel, Grundrichtungen und Eckwerte der Strategie aus dem Blickwinkel der Bundesrepublik Deutschland', speech to the Wehrkunde Conference in Munich, 1 March 1986.
139. 'NATO study ties SDI to European air defence', *Jane's Defence Weekly*, 31 May 1986, p. 970; 'NATO Accelerates Antitactical Missile Defense Research', *Aviation Week & Space Technology*, 2 June 1986, pp. 69–70.
140. Michael Feazel, 'NATO Planners Drafting Guidelines For Europe-Based ATBM Development', *Aviation Week & Space Technology*, 14 July 1986, pp. 30–1; 'Move for European missile defence system', *Jane's Defence Weekly*, 16 August 1986, p. 229.
141. See Hans Günter Brauch, *Evaluation of Anti-tactical Ballistic Missile Defence Technologies*, note 26 above.
142. Brauch, see note 118. (*30 Thesen*) pp. 42–4.

143. See note 5 above.
144. See chapter 4 by Trevor Taylor in this volume.
145. See chapters 4–8 in this volume.
146. US Congress, Office of Technology Assessment, *Ballistic Missile Defense Technologies*, OTA-ISC-254 (Washington, DC: United States Government Printing Office, September 1985).
147. See the statements by Seitz, Rühl and Rühle in note 5 above.
148. See Brauch (note 118) p. 32.
149. Hartmut Bebermeyer, 'The Fiscal Crisis of the Bundeswehr', in Hans Günter Brauch and Robert Kennedy (eds), *Alternative Conventional Defense Postures in the European Theater. The Future of the Military Balance and Domestic Constraints*, forthcoming.
150. See US Congress, OTA-ISC-254, pp. 16ff.
151. US Congress, OTA-ISC-254, pp. 22–3.
152. For a detailed discussion see Brauch, note 118 (*30 Thesen*).
153. See also Arnold Kanter, 'Thinking about the strategic defence initiative: an alliance perspective, *International Affairs* (London) 61 (1985) pp. 449–64; Canavan *Theater Applications*; Berkhof 'The American Strategic Defence Initiative'; Ian Bellany, 'Ballistic Missile Defence and Strategic Stability', in Clingendael (ed.), pp. 39–49; Dean Wilkening, 'Strategic Defences and Stability', in Clingendael (ed.) pp. 127–58.
154. See Brauch, note 118 (*30 Thesen*), pp. 62–5.
155. G. W. Rathjens, Abram Chayes, J. P. Ruina, *Nuclear Arms Control Agreements: Process and Impact* (Washington, DC: Carnegie Endowment for International Peace, 1974); Hans Günter Brauch, *Entwicklungen und Ergebnisse der Friedensforschung (1969–1978). Eine Zwischenbilanz und konkrete Vorschläge für das zweite Jahrzehnt* (Frankfurt, 1979).
156. Hans Günter Brauch, 'Der NATO-Doppelbeschluss und die Strategie der Allianz: Ein Plädoyer für die Seestützung', *Die Friedens-Warte* 64 (November 1983) pp. 37–109.
157. Pierre Lellouche, 'SDI and the Atlantic Alliance', *SAIS Review*, Summer–Fall 1985, pp. 67–80; William Drozdiak, 'Kohl, Mitterrand Agree to Push Joint Research on Space Technology. But Leaders Fail to Resolve Split on Joining "Star Wars"', *Washington Post*, 29 May 1985, p. A26; 'Mitterands Fragen an Kohl werden drängender. Die deutsche Beteiligung am Raumtransporter "Hermes"/Deutsch-französische Konsultation in Baden-Baden', *Frankfurter Allgemeine Zeitung*, 16 January 1986.
158. Dominique Moisi, 'A "Star Wars" Casualty: French–German Cooperation', *International Herald Tribune*, 10 January 1986.
159. Lellouche 'SDI and the Atlantic Alliance', p. 71.
160. Lellouche, pp. 74–5.
161. Hans-Henrik Holm, 'SDI and European Security', *Alternatives* 10 (1985) p. 529.
162. Holm, p. 530.

15 Compliance With the ABM Treaty – Questions Related to Soviet Missile Defence Activities and to the United States SDI

Thomas K. Longstreth, John E. Pike and John B. Rhinelander

I INTRODUCTION

Compliance with any treaty is a major indicator of trust and confidence in the relations among states. Challenging the compliance of a treaty partner in public may serve either as a severe warning based on firm evidence or as a public posture to justify a potential future abandonment of a specific treaty. The verification of the ABM Treaty is based on national technical means, for example on the analysis of satellite intelligence. In February 1978 in a report to Congress Secretary of State Vance referred to the following compliance issues relating to the ABM Treaty: (1) installation of a new ABM test range in Kamchatka; (2) SA-5 radar tests 'in an ABM mode' and (3) mobile ABMs. Obviously all three compliance issues could be discussed and solved in the context of the Standing Consultative Comission (SCC).[1]

Renewed allegations about Soviet arms control treaty violations were published on 2 April 1983, ten days after President Reagan's 'Star Wars' speech,[2] as a consequence of deliberate leaks of the existence of secret compliance reports, an element of suspicion and speculation was introduced to the debate. By mid-April President Reagan had rejected proposals to accuse in public the Soviets of treaty violations,[3] instead preferring private diplomatic channels for conveying the United States concerns to the Soviets[4]. In August 1983 the United States called upon the SCC to consider those complaints, and, after an initial Soviet rebuff, bilateral SCC talks discussed the

United States concerns in October 1983.[5] In accordance with congressional requirements in FY84 and FY85 authorisations for the United States ACDA, the Reagan Administration prepared two compliance reports which were transmitted to Congress in January 1984 and February 1985. In October 1984 a declassified version of a report on Soviet compliance since 1958 was prepared by the General Advisory Committee on Arms Control and Disarmament (GAC). As far as the ABM Treaty was concerned, these three reports claimed the following Soviet treaty violations:

1. *Krasnoyarsk ABM radar*: the deployment of a large phased-array radar (LPAR) near Krasnoyarsk *'almost certainly* constitutes a violation' of the ABM Treaty, given its siting, orientation, and capability (January 1984);
2. development and deployment of *mobile ABM radars* (GAC, October 1984);
3. construction of *additional ABM radars* (GAC, October 1984);
4. *Krasnoyarsk ABM radar*: this is now judged to be a violation of the ABM Treaty (February 1985);
5. *mobile ABM systems*: Soviet activities regarding ABM component mobility are 'ambiguous', but the development of such components represents a 'potential violation' of the ABM Treaty (February 1985);
6. *concurrent testing of ABM and SAM components*: evidence of this testing, banned under the ABM treaty, is 'insufficient', although some incidents of such testing are judged 'probable' or highly 'probable';
7. *ABM Territorial defence*: the report 'judges that the aggregate of the Soviet Union's ABM-related actions suggest that the USSR may be preparing an ABM defence of its national territory'. This is prohibited by the ABM Treaty.[6]

These three reports provoked an intensive debate within the United States executive, in the United States Congress and in the media. On 16 January 1985 Richard Perle, one of the most outspoken détente and arms control critics in the Pentagon, cautioned against too high expectations relating to the Geneva talks, stating: 'Soviet violations of virtually all ... important agreements ... will undoubtedly influence the nature of the proposals that we will make'.[7] On 20 February 1985 Perle stated before the Senate Armed Services Committee: 'We must now create penalties that deny the benefits of the violations of the USSR... This could involve research, development or deployment which at least offset the advantage obtained by the Soviets'.[8]

Relating to the most contested issue of the Krasnoyarsk radar, the United States government charged that it could serve as an ABM battle management radar, while the Soviet Union maintained that the radar would be for space tracking and verification of arms agreements. At the same time the United States government stated in an arms control compliance report for the SDI programme published on 15 April 1985 that tests planned for 1989 'will come into conflict with the [ABM] treaty limits'.[9] In late May 1985 both Paul Nitze and Kenneth Adelman pointed to the possibility of adapting the ABM Treaty to accommodate for the new SDI technologies.[10]

In the United States the compliance debate has become an important aspect of the controversial assessment of Soviet BMD activities and of the SDI controversy.[11] While the hawks have claimed that Moscow was cheating,[12] writers of the arms control community, especially of the Federation of American Scientists,[13] of the Arms Control Association[14] and of the National Campaign to Save the ABM Treaty[15] have discussed the official charges in detail. The Administration has used 'alleged Soviet non-compliance as an excuse to explain away a flawed arms control approach and justify the multibillion dollar SDI programme, [which] ... poses one of the most direct threats to the ABM treaty'.[16]

Two political developments in the United States immediately prior to the first Reagan–Gorbachev summit in November 1985 have had a direct impact on the compliance debate:

1. The effort of the Department of Defense to shift from a 'restrictive interpretation' of the ABM Treaty to a 'broader interpretation' that would permit testing and development, but not deployment, of advanced space-based defences against ballistic missiles.[17]
2. The endorsement of the United States claim 'of obvious Soviet Treaty violations' in the communiqué of the NATO Nuclear Planning Group in Brussels on 29 and 30 October 1985.[18]

The latter claim may serve as a future pretext either for a renewed attempt to shift from the restrictive interpretation to a broader interpretation which has been claimed by both Nitze and the legal advisor of the State Department, Mr Abraham Sofaer, as 'fully justified'.[19]

In this chapter we shall focus on the compliance questions related to Soviet missile defence activities and the questions related to the United States Strategic Defence Initiative. In the final section we shall turn to specific proposals as to how the 'grey areas' in the ABM Treaty may be closed by new arms control initiatives.

II COMPLIANCE QUESTIONS RELATED TO SOVIET MISSILE DEFENCE ACTIVITIES

Since the signing of the ABM Treaty in 1972, certain Soviet activities have raised questions regarding their strict compliance with the terms of the ABM Treaty. Past compliance questions have included the possible testing of an air defence (SA-5) radar in an ABM mode; failure adequately to report the dismantling of excess ABM test launchers; and use of an ABM test range on the Kamchatka peninsula.[20] All of these issues were discussed in the SCC and either resolved to the satisfaction of the United States or not further challenged.

Despite the resolution of past issues, the dynamic pace of Soviet ABM, air defence and other programmes continues to create instances of possible conflict with the Treaty. The Reagan Administration has tried to deflect criticism of the SDI and build support for the programme by emphasising areas of possible or probable Soviet non-compliance with the ABM Treaty. The Administration has also alleged that individual Soviet ABM activities, when viewed *in toto*, may indicate a more comprehensive Soviet plan to break out of the ABM Treaty. In summarising an Administration report on Soviet non-compliance released in February 1985, ACDA Director Kenneth Adelman stated 'the Report expresses our concern about Soviet ABM and ABM-related actions which in the aggregate suggest that the Soviet Union may be preparing an ABM defence of its national territory'.[21]

Several current issues related to possible Soviet non-compliance are outlined and summarised below. Some have been mentioned in two United States government reports released in January 1984 and February 1985. Others have not been cited publicly by the Reagan Administration, but are believed to be still under review.

(a) LPAR at Abalakova near Krasnoyarsk

In 1982, the Soviets began construction of a large radar at Abalakova, near Krasnoyarsk,[22] in central Siberia, approximately 400 miles north of Mongolia. The radar is similar to several other Soviet radars designed for early warning of missile attack. It is a LPAR complex with separate transmitter and receiver buildings facing toward the north-east and is expected to become operational by 1988.

In a démarche to the Soviets in the summer of 1983, and during the fall 1983 session of the SCC,[23] the United States questioned the radar's conformity with Article VI(b) of the ABM Treaty, which states, 'Each Party undertakes ... not to deploy in the future radars for early warning of strategic ballistic missile attack except at locations along the periphery of its national territory oriented outward. In a report the President submitted to Congress in January 1984, the Administration charged that the Soviet radar 'almost certainly constitutes a violation of legal obligations under the Anti-Ballistic Missile Treaty'. In the Administration's February 1985 compliance report, this assessment was changed to an unqualified 'violation'.[24]

The United States believes the LPAR[25] being built near Krasnoyarsk is primarily designed for early warning, since its technical characteristics and appearance are very similar to other Pechora-type radars the Soviets have acknowledged are for early warning, and unlike those used for satellite tracking. If it is for early warning, the radar is inconsistent with Article VI(b) because it is not located 'along the periphery' nor 'oriented outward', since it covers a large part of Siberia.

During the course of the ratification hearings on the ABM Treaty, Senator Percy posed the following written question, which was answered for the record by Ambassador Smith:[26]

Question: On ABMs, is there any possibility that potential difficulties in identifying just what the specific purpose of a radar may be, by national [technical] means, could prohibit the construction of space activity or air defense radars?

Answer: We believe that national technical means of verification will be adequate to distinguish between ABM radars and air defense and space tracking radars. There are a number of parameters, such as location, orientation, size, power and signal characteristics, which taken together, should provide sufficient information to make this distinction. If an ambiguous situation regarding radars should nevertheless arise, the situation could be clarified through discussions in the Standing Consultative Commission.

The Soviets have responded at the SCC that the radar is designed for tracking satellites and other space objects and for verification of United States compliance with the Outer Space Treaty, and that it is located and oriented for those purposes.[27] Therefore, the Soviets have argued, pursuant to Agreed Statement F attached to the ABM Treaty, the radar is not subject to the restrictions of Article VI(b). Agreed Statement F states, 'The Parties agree not to deploy phased-array radars . . . except as provided for in Articles III, IV, and VI of the Treaty, or except for the purposes of tracking objects in outer space or for use as national technical means of verification'.

Although the Krasnoyarsk radar would not seem to add much to existing Soviet satellite-tracking capabilities, its location and capabilities may be suitable to support an advanced ASAT weapon system (similar to the new American system) which would engage and destroy target satellites while they were over the Soviet Union. However, the Soviets have not suggested that the radar would perform this function. While it may be able to monitor United States compliance with the Outer Space Treaty, it does not appear to be primarily designed or suited for that purpose.

The United States intelligence community believes the Krasnoyarsk radar is much better suited for early warning than for space tracking. Its location and orientation fills a gap in warning of attack from SLBMs in the Northern Pacific (although a radar located further out on the periphery in a manner fully consistent with the Treaty would provide more warning time than is provided by the radar at Krasnoyarsk). The assertion that the Krasnoyarsk LPAR will serve an ABM battle management role is widely discounted, since the radar is not well-positioned to track United States ICBMs attacking along polar trajectories.[28] The Reagan Administration has not accepted the Soviet explanation, but has been unable to agree upon a remedy to propose to the USSR other than to suggest that construction be stopped.

There have been other instances in which the United States raised questions within the SCC about the locations and purposes of various Soviet Pechora-type LPARs. The United States has periodically expressed its concern that a large number of such radars could provide the Soviets with a base for a nationwide missile defence of their territory, inconsistent with Article I. It has also questioned the Pechora radar's adherence to Article VI(b). The Soviets have responded that all their LPARs of the Pechora type have been deployed in a manner consistent with ABM Treaty obligations while taking account of 'technical and practical considerations.' This seemed to imply that the Pechora radar, for example, was sited further inland because of transportation and geographic impediments. The Soviets have not attempted to explain the location of the Krasnoyarsk radar in this manner, however.[29]

(b) The SA-12 SAM

The SA-12 is a new high-performance SAM system that the Soviets have been testing for the past several years. This system has been code-named 'Gladiator' by NATO. It is still undergoing tests, but is expected to become operational within the next year or two. The SA-12 is a hypersonic missile that is capable of executing very high-speed manoeuvres to engage targets at altitudes in excess of 100,000 feet. It is reported to have a range in excess of 50 miles, while the missile's truck-mounted radar is reported to have a range of over 150 miles. If, as is expected, the SA-12 were deployed with a nuclear warhead (like the SA-5, and unlike other Soviet anti-aircraft missiles), this system could have the capability to intercept some types of long-range ballistic missiles such as SLBMs.[30] The Reagan Administration's February 1986 Report on Soviet Non-Compliance states that:

> 'The SA-X-12 can engage tactical ballistic missiles in flight. Such a system . . . could have many of the features one would expect to see designed into an ABM system, possibly giving it capabilities to intercept some types of strategic ballistic missiles'.[31]

Reportedly, on several occasions during 1983 and 1984, a modified version of the SA-12, called the Giant, was tested against a target vehicle that had a trajectory and flight characteristics similar to the SS-12 tactical ballistic missile. This would be inconsistent with the ABM Treaty *only* if the SA-12 system were 'tested in an ABM mode', which is banned by Article VI(a) of the Treaty.[32]

On 7 April 1972 the United States SALT delegation made a unilateral statement on this issue which stated that:

> we would consider a launcher, missile, or radar to be 'tested in an ABM mode' if, for example . . . an interceptor missile is flight

tested against a target vehicle which has a flight trajectory with characteristics of a strategic ballistic missile.[33]

No definition of a 'flight trajectory with characteristics of a strategic ballistic missile' was agreed to in the ABM Treaty. Within the United States government the Department of Defense has in the past proposed demanding criteria for defining such a trajectory, that is if a target reached an altitude above 40 kilometres and a velocity greater than 2–4 kilometres per second. Although a common position between various US agencies on these criteria was not reached, let alone one with the Soviets, it has been the Pentagon's position that, if a SAM is tested against a target vehicle with such a trajectory, it should be considered 'tested in an ABM mode.' In this regard, it is important to note that the SS-12 does not have a range comparable to United States strategic systems, nor is it likely that the SS-12's warhead (and thus its re-entry characteristics) is comparably designed to those on United States ICBMs and SLBMs.[34]

Also relevant to this issue is the fact that the SS-12's flight trajectory is unlike that of even the older Soviet SLBMs, such as the SS-N-5 and SS-N-6. The SS-N-5 is indirectly limited by the SALT I Interim Agreement, and the SS-N-6 by both SALT I and SALT II, thus some argue they are the minimum standard of a 'strategic ballistic missile'. The SS-N-5 and the SS-N-6, however, have the shortest ranges of any missiles limited by the SALT agreements.[35]

While the SS-12 target reportedly did reach an altitude above 40 kilometres and a velocity greater than 2–4 kilometres, it is difficult to conclude that a test against a target based on the SS-12 constitutes a test against a target 'with characteristics of a strategic ballistic missile', and is therefore inconsistent with the restrictions in Article VI(a). The United States has not yet discussed SA-12 testing activities with the Soviets in the SCC.

Apparently, while the classified version of the Administration's February 1985 non-compliance report referred to the SA-12 testing, the report concluded that this activity was not inconsistent with the ABM Treaty.[36]

Several senior Administration defence officials have also commented on the relation of the SA-12 to the ABM Treaty. Franklin Miller, Director of Strategic Forces Policy for Department of Defense stated in 1984 testimony before the Senate Armed Services Committee that:

> The Soviet System, SA-X-12, . . . is an air defence system which has an anti-tactical ballistic missile capability. There is nothing in

its development that contravenes the ABM Treaty because that treaty deals with strategic anti-ballistic missile systems.[37]

On 9 March 1983, in testimony before the House Armed Services Committee, T. K. Jones stated that:

> the Soviets are developing that [the SA-12] to counter shorter-range ballistic missiles and the ABM Treaty was drafted so that it is a legal development and they could deploy it fully.[38]

These conclusions are not necessarily consistent with Article VI(a) of the Treaty under which the issue is whether an ATBM has the 'capability to counter strategic ballistic missiles'.

(c) Testing of a rapid reload launcher capability

The parties also agreed in Article V(2) 'not to develop, test, or deploy automatic or semi-automatic or other similar systems for rapid reload of ABM launchers'. Press reports indicate that in a test of the SH-08 short-range ABM interceptor (which is one of the new ABM-X-3 components), two interceptors were fired from a single launcher in an interval of two hours. However, no reloading equipment was observed in the launch area, leading some Department of Defense officials to speculate that SH-08 launchers have an underground automatic reload system. The United States reportedly questioned the Soviets at the SCC on this activity. The 1986 ACDA Report on Non-Compliance acknowledged Soviet testing of a reload capability for both the SH-08 and the SH-04 'Galosh' interceptors at Sary Shagan, demonstrating a reload capability of 'much less than a day'.[39]

While the United States and the USSR never reached an agreed statement on what would constitute 'rapid reload' of an ABM launcher, some discussions were held both within the United States government and between United States and Soviet representatives during the drafting of the Treaty. At the time, the chief negotiator, Gerard Smith, informed the Soviets that the United States would consider an ABM launcher tested for rapid reload if it was reloaded in a 'strategically significant' period of time. (It should be noted that United States SLBM warheads could arrive at Soviet targets in less than 15–20 minutes; United States ICBMs could arrive approximately 15 minutes later.) He further indicated that no changes in the existing Soviet Galosh system would be required. United States intelligence believed in 1972 that the Galosh ABM launchers could be reloaded in a period of about 15 minutes. Therefore, it could be inferred that 'rapid reload' would mean a period of less than 15 minutes. Subsequently, estimates of the time required to fire and reload the Galosh launchers were revised upwards.

If the press reports are accurate and the activity in question actually did take place over a period of hours, it would not seem to provide the Soviet Union with a 'rapid reload' capability. The Administration concluded in ACDA's Non-Compliance Report 'that the USSR's actions with respect to the rapid reload of ABM launchers constitute an ambiguous situation as concerns its legal obligations under the ABM Treaty'.[40]

(d) Mobility of new ABM system – testing of a modular radar

Various Soviet practices with respect to deployment of the new ABM-X-3 components led the United States to examine whether the Soviets were developing a mobile, land-based ABM system, and/or components contrary to commitments under Article V of the ABM Treaty. According to the 1986 Non-Compliance Report, 'The Soviets have tested ABM components that are apparently designed so that they could be relocated in months rather than in terms of years required to deploy fixed land-based systems'.[41]

The Soviets have developed and tested a modular radar (Flat Twin) that can be erected on a prepared site in a matter of 6–8 weeks as part of the ABM-X-3. They are also developing another ABM radar (Pawn Shop), that is housed in a van-sized container. Testing of one or both of these radars was initially observed at the Sary Shagan and Kamchatka test ranges in the late 1970s. Reports suggest that a prototype *Flat Twin* radar was initially fielded at Sary Shagan, and subsequently moved to Kamchatka, for testing against longer-range missiles. The quickness of the construction of the radar at Kamchatka was the initial basis for concern about this system. Several of the Flat Twin radars are under construction as part of the upgrade of the Moscow system from the ABM-1B to the ABM-X-3 configuration.[42]

To clarify the Article V(1) prohibition on mobile ABM radars, Common Understanding C provides that this ban 'would rule out the deployment of ABM launchers and radars which were not of permanent fixed types'. The Flat Twin radar is both transportable and modular in the sense that it can be disassembled, moved in component stages, and reassembled in a period of months. This, however, assumes that extensive advanced preparation of the site upon which it is being relocated has occurred. At issue is whether being 'transportable' in this instance means the radar is 'mobile'.

Again, there is no Agreed Statement between the United States and the Soviet Union explicitly defining 'mobile' and/or 'not of a permanent fixed type'. At the time the Treaty was being drafted, the United States was concerned about the mobility of the Soviet SA-2 air defence system, components of which could be disassembled, transported and reassembled in a period of 48–72 hours. The United States interpretation of 'transportable', as it applied to component mobility, was therefore that if such activity occurred in a week or less, it would be considered inconsistent with the Treaty. Flat Twin, which can be transported only in a period of months, even with extensive advanced site preparation, would not appear to conform to that interpretation.

The issue with Pawn Shop is whether the radar's van-like container could be relocated quickly if it were given wheels (which have *not* been observed). Some United States officials believe that Pawn Shop could conceivably be moved and is therefore not of a permanent fixed type and banned by the Treaty. Others disagree.

In its February 1985 Non-Compliance Report the Administration concluded that:

> Soviet actions with respect to ABM component mobility are ambiguous, but the USSR's development of components of a new ABM system, which apparently are designed to be deployable at sites requiring relatively little or no preparation, represent a potential violation of its legal obligation under the ABM Treaty.[43]

The February 1986 Non-Compliance Report contains almost identical language.

(e) Concurrent testing of ABM and SAM components

On occasion, while testing ABM systems and components at their Sary Shagan ABM test range, the Soviets have apparently simultaneously operated certain mechanically steered SAM radars located near the test range. If such SAM radars were tested 'in an ABM mode,' this would be inconsistent with Article VI(a).

There is considerable ambiguity as to what would constitute such activity. During negotiation of the ABM Treaty the two delegations reached a Common Understanding (B) that non-phased-array SAM radars 'used for range safety and instrumentation purposes may be located outside of ABM test ranges'. At the time the Soviets emphasised that the use of non-ABM radars for range safety or instrumentation was not limited under the treaty.

The 7 April 1972 United States Unilateral Statement B attached to the ABM Treaty included an interpretation of 'tested in an ABM mode' as it applied to radars. According to this interpretation, a radar would be 'tested in an ABM mode' if it:

> makes measurements on a co-operative target vehicle (i.e., one with a flight trajectory of a strategic ballistic missile) . . . during the re-entry portion of its trajectory or makes measurements in conjunction with the test of an ABM interceptor missile or an ABM radar at the same test range.[44]

Subsequently, in a 1978 SCC session, the two sides agreed that there would be no concurrent *testing* of air defence and ABM systems or components co-located at the same test range, and that air defence radars used for instrumentation would not make measurements on strategic ballistic missiles or their RVs.

At SCC sessions in 1982–3, the two sides sought to clarify further permitted and prohibited ABM activities at ABM test ranges. A draft understanding between the United States and the Soviet Union concerning concurrent operation (as opposed to testing) of ABM and SAM components was reached in the SCC in April 1982. Subsequently, certain Reagan Administration officials proposed changes to the draft common understanding that had been agreed to with the Soviets.

The discussions on this common understanding continued at the SCC through 1984 and 1985, while the Reagan Administration cited Soviet concurrent testing of ABM and SAM components as a probable violation of the ABM Treaty in its February 1985 Non-Compliance Report. Finally, a Common Understanding clarifying permitted and prohibited operations of SAM radars at ABM test ranges (beyond the clarifications contained in the 1978 Agreed Statement) was reached in June 1985. While its text, like that of the 1978 Agreed Statement, remains classified, it apparently prohibits the operation of SAM radars while ABM tests are taking place unless unusual circumstances (e.g., a hostile aircraft enters the test range) warrant. If such simultaneous operations do take place, the Soviets have to notify the US within thirty days and provide a detailed explanation of the event at the next SCC session.

Despite the completion of this additional US–Soviet understanding on what would constitute 'tested in an ABM mode', certain Administration officials continue to complain publicly and privately that prohibited concurrent operations of SAM radars by the Soviet Union at the Sary Shagan test range continue. Moreover, the Administration's February 1986 Report on Soviet Non-Compliance neglected to mention the satisfactory conclusion of this formal Common Understanding, but *did* state that:

> Soviet actions with respect to concurrent operations ... is [*sic*] insufficient fully to assess compliance with Soviet obligations under the ABM Treaty ... The large number and consistency over time, of incidents of concurrent operation of ABM and SAM components, plus Soviet failure to accommodate fully US concerns, indicate the USSR probably has violated the prohibition on testing SAM components in an ABM mode. In several cases this may be highly probable.[45]

(f) ABM territorial defence

One of the principal purposes of the ABM Treaty was to constrain the ABM capabilities of the parties to such low levels that neither would have to make compensating and offsetting increases in offensive forces. This was codified in the Article I undertaking 'not to deploy ABM systems for a defence of the territory of its country and not to provide a base for such a defence'.

The Administration's February 1985 Non-compliance Report concludes that:

> the aggregate of the Soviet Union's ABM and ABM-related actions suggest that the USSR may be preparing an ABM defence of its national territory.[46]

The Administration reached a similar conclusion in its December 1985 Soviet Non-Compliance Report.

The Administration is attempting to buttress its case against the Soviet Union by asserting that the whole is more significant than the sum of its parts. Legally, the argument would be that the Soviets could violate Article I(2) even though they did not violate any of the specific articles, such as Articles III, V, and VI. A more logical interpretation would be that a violation of either Article III, Article V or Article VI would have to be demonstrated *before* any conclusion that the activities constitute a base for a nationwide ABM system.

Further the facts do not support the case that the Soviets are on the verge of a breakout from the ABM Treaty. While the Soviets have the potential for such a breakout, the Administration has failed to provide any convincing evidence that they are actually making such a move. The Administration's concern does, however, point out areas where agreed statements could provide specific guidelines between the prohibited and the permitted.

There is little doubt that the new ABM-X-3 components *could* be deployed as part of a nationwide defence. Nor is there any dispute over the fact that the Flat Twin modular radar could be deployed more expeditiously than could the much larger phased-array radars of the Pechora type. But these activities – no more than similar American activities such as the Site Defence effort of the early 1970s – are not necessarily preparations for breaking out of the Treaty. In isolation, any ABM component could be regarded as part of preparations for a defence of national territory.

There is no reason to fear, at present, that the Soviets are actually moving to prepare for such an ABM deployment. These preparations could consist of stockpiling ABM interceptors and radars, and preliminary preparations of nationwide ABM sites. Observation of these activities would provide some indication of Soviet intentions, but no such activities have been reported.

In general, possible Soviet motivations for abandoning the ABM Treaty are obscure, unless convinced the United States intended to do so. The technology of the new components is analogous to that of the American Sentinel/Safeguard of the 1960s, and such a system would be of questionable utility when faced with the United States missile threat of the 1990s. Thus, while the Soviets have a hedge against uncertainty in their development programme, the fruits of the ABM-X-3 programme do not provide incentive for breaking out of the ABM Treaty regime. If they became convinced that the United States intends to abrogate the Treaty at a time of its choosing, starting in 1989, the Soviets might decide to match the American 'high-tech' space-based initiative with a deployment of their current 'low-tech' system.

III COMPLIANCE QUESTIONS RELATED TO THE UNITED STATES SDI

From what is publicly known, the following activities within the SDI related to sensor and interceptor development create the greatest cause for concern and raise the necessity for clarification regarding their consistency with the ABM Treaty.[47]

(a) Sensors

The SDI includes work on sensors that would be capable of detecting and tracking ballistic missiles in the initial, boost, mid-course, and terminal phases of flight. Sensors under development for each of these phases raise compliance concerns.

The Boost Surveillance and Tracking System (BSTS),[48] previously known as the Advanced Warning System (AWS), is a follow-on to the present generation of early warning satellites. Initial versions of this satellite are scheduled for testing in space in the early 1990s. BSTS incorporates greatly enhanced infrared sensors which provide high resolution and precision for tracking missiles in their boost phase. The fact that MIRVed warheads are released and individually targeted in the post-boost phase limits the applicability of this system to the early warning mission, since its greater tracking precision does not translate into improved impact prediction or attack characterisation. As part of a layered ABM system, however, BSTS could provide initial target tracking information which would be relayed for use by boost-phase interceptors. Although the BSTS is not intrinsically ABM-related, its inclusion in the SDI does raise questions as to its consistency with the Article V(1) provisions banning the development, testing or deployment of space-based ABM components.

The Space Surveillance and Tracking System (SSTS)[49] will use cryogenically cooled infrared sensors to detect and track warheads and decoys during the mid-course of their flight. This system was previously under development as part of an upgrade to the ground-based Spacetrack satellite tracking network, and would have been used in support of the new air-launched ASAT weapon. As with the BSTS, initial versions of the SSTS will be tested in space in the early 1990s.

In a layered defence, SSTS along with other sensors would provide target tracking and identification information which would be relayed for use by mid-course interceptors. If tested in an ABM mode, SSTS would be inconsistent with Article V(1). Testing SSTS against satellite targets might give it an ABM capability, which is prohibited by Article VI(a).

The Airborne Optical System (AOS),[50] also known as the Airborne Optical Adjunct, has been under development for several years. The first flight of AOS was scheduled for 1987, prior to the advent of SDI. A variety of factors – including technical problems and changing performance requirements – have delayed the first AOS flight test to 1988. Originally, AOS was designed to carry infra-red (IR) sensors. With the programme restructuring, AOS will now carry only a single IR tracking sensor.

AOS is an outgrowth of earlier work on range instrumentation aircraft, such as the C-135 Optical Aircraft Measurement Programme (OAMP),[51] and is intended solely for ABM-related applications. The advanced development and flight testing of AOS would be inconsistent with the provision in Article V(1) banning the development, testing or deployment of air-based ABM components.

The Terminal Imaging Radar (TIR)[52] will be part of a ground-based terminal defence system to defend both cities and hardened military targets. Like the Defence Unit radar of the earlier Low Altitude Defence System (LoADS), it would probably be deployed in a mobile mode to enhance its survivability. The advanced development or testing of the TIR in other than a fixed, ground-based mode would be inconsistent with Article V(1), which bans the development, testing or deployment of mobile, ground-based ABM components.

The Space-Based Imaging Radar and Imaging Laser[53] are relatively new initiatives that provide a greatly improved ability to distinguish actual re-entry vehicles from decoys. These technologies were previously under development for missions other than missile defence, and under the SDI they would be used for air-based and space-based applications. Prior to the accident in January 1986, it was planned to use the Space Shuttle for a late 1980s flight demonstration of some components of a space-based radar. In the early 1990s, either the Imaging Radar or Imaging Laser may be selected for a full-scale demonstration in space.

The advanced development or testing of either of these sensors in other than a fixed, ground-based mode would raise questions as to their consistency with Article V(1).

A new rocket test range is under construction at Shemya Island. It is part of an effort to develop infrared sensors for mid-course and terminal phase interceptors. Rockets will be used to launch test vehicles from the Aleutian Island site into outer space to observe Soviet ballistic missile tests. Tests will include at least two flights under the new Queen Match programme, previously known as the Designating Optical Tracker (DOT), which incorporates an infrared sensor similar to that used in the Homing Overlay Experiment (HOE). DOT has already been tested on several occasions at the Kwajalein Missile Range. In addition, the OAMP[54] C-135, which is a predecessor of the AOS, will be based at Shemya. These projects will obtain data on Soviet systems for use in designing United States missile defences, as well as providing an opportunity to test sensor prototypes against realistic targets. Tests of DOT and OAMP are scheduled over the next several years.

It is not clear whether the DOT or the OAMP should be considered ABM components. If the Shemya range is used to test ABM systems or components, it would become subject to the limits of Article IV, Common Understanding B, and the 1978 Agreed Statement. Article IV allows each party to maintain ABM components for development and testing purposes at 'current or additionally agreed test ranges'. Common Understanding B points out that the only current United States ABM test ranges are at Kwajalein Atoll and White Sands, New Mexico, and that ABM components cannot be located or tested at any other test ranges without prior agreement between the two governments. The 1978 Agreed Statement sets forth procedures of notifying the other party when a new test range is established. The Administration has not indicated an intention to seek agreement that Shemya now be considered an ABM test range.

(b) Interceptors

Interceptors that will be developed and tested under the SDI fall into three general categories: ground-based rockets; space-based and other mobile kinetic-energy weapons; and directed-energy weapons.

Ground-based rockets
SDI work on rocket interceptors will build on the Homing Overlay Experiment (HOE),[55] which was initiated in 1977 and completed in June 1984. HOE consisted of four tests of a ground-based, exoatmospheric non-nuclear kill vehicle. The HOE was designed to deploy its kill vehicle – a large aluminum net carrying metal weights – when it detected, located and converged on its target. The kill vehicle destroyed an incoming warhead by colliding with it at high speed and disintegrating it. The kill vehicle was equipped with sensors to detect the long-wavelength infrared emissions given off by missile warheads as they travel through space prior to re-entry.

In the first three tests, all conducted in 1983, the kill vehicle failed to intercept the target warhead. In the fourth test in June 1984 the

HOE successfully locked onto and destroyed a target warhead that had been launched aboard a *Minuteman* ICBM from Vandenberg Air Force Base. With the completion of the fourth test, information learned from the *HOE* series will now be utilised in designing the ERIS project, described below.

The HOE payload, including the kill vehicle, sensor and signal processor, is carried aboard a modified Minuteman I ICBM. This may be inconsistent with the undertaking in Article VI(a) 'not to give missiles... other than ABM interceptor missiles,... capabilities to counter strategic ballistic missiles or their elements in flight trajectory, and not to test them in an ABM mode'.

In January 1984 the Soviets protested that using Minuteman to test HOE gave it an ABM capability. The Soviet allegation was contained in an *aide memoire* they presented to the United States which listed several dozen alleged United States violations of arms agreements.

In response to the Soviet charge, a bulletin published by ACDA stated that, 'The test missile in question was observably different from Minuteman I, as were its performance characteristics. In any case, the Minuteman I is no longer deployed by the US'.

Another compliance issue pertains to the fact that, although the HOE experiments have been conducted using a single intercept vehicle per launcher, the programme was originally designed to investigate the use of multiple kill vehicles on each launcher. (The Minuteman ICBM used in the tests could certainly accommodate such a payload.) The Soviets protested that HOE was inconsistent with the undertaking in Agreed Statement E to the ABM Treaty 'not to develop, test, or deploy ABM interceptor missiles for the delivery of more than one independently-guided warhead'. The ACDA Bulletin responded that 'the United States is not developing ABM interceptors with multiple warheads and has never pursued such a program'.[56] The Exoatmospheric Reentry Vehicle Interception System (ERIS) is an advanced follow-on to HOE. ERIS is presently in an early definitional phase, with tests scheduled to begin in the late 1980s. ERIS will use a much smaller interceptor kill vehicle than HOE, which might permit the use of multiple warheads on ERIS. When interceptors of this type were first evaluated in the late 1960s under the Homing Intercept Technology programme the use of multiple warheads on a single interceptor was found to enhance the performance of the defence under some circumstances. There may thus be some incentives to incorporate multiple warheads on ERIS. The design of the Braduskill early-midcourse interceptor does currently call for multiple kill vehicles.

Use of multiple warheads could improve the utility of a mid-course ABM interceptor like ERIS and Braduskill. The co-ordination of the release of multiple warheads is a challenging task and, at some point in the testing programme of this procedure, it would have to be either tested or simulated. Any such testing of ERIS and Braduskill would be inconsistent with the undertaking in Agreed Statement E of the ABM Treaty 'not to develop, test, or deploy ABM interceptor missiles for the delivery ... of more than one independently-guided warhead'. However, the Administration has indicated that there are presently no plans to develop a multiple warhead capability for ERIS.[57]

The High Endoatmospheric Defence Interceptor (HEDI) will use a heat seeking hit-to-kill warhead to intercept targets as soon as they enter the atmosphere. HEDI will be used both as the terminal layer of a defence against ICBMs, and as a defence against short-range ballistic missiles. In this latter role, HEDI will be applicable to the ATBM defence of Europe against Soviet theatre nuclear forces. HEDI is one of a number of SDI-related technologies currently being explored for this role. The ATBM issue is discussed more fully in other parts of this volume.

Since HEDI will probably have both a tactical and strategic ABM capability, the transfer of HEDI to Europe may be inconsistent with the undertaking by the United States in Article IX of the ABM Treaty 'not to transfer to other states, and not to deploy outside its national territory, ABM systems or their components limited by this treaty', and with Agreed Statement G, which prohibits the transfer of 'technical descriptions or blueprints especially worked out for the construction of ABM systems and their components'.

Advanced kinetic-energy weapons
The SDI also includes work on a variety of more advanced kinetic-energy weapons. The Hypervelocity Launcher will use an electro-magnetic accelerator, analogous in concept to a particle beam accelerator, to propel projectiles to very high velocities that may be significantly greater than those achieved by conventional rocket interceptors. These projectiles will be comparable in design to the hit-to-kill warheads used by rocket interceptors. The Hypervelocity Launcher offers the prospect of very high rates of fire and is, in a sense, an 'anti-missile gatling gun'.[58]

This concept is applicable to space-based boost-phase and mid-course defence, as well as to ground-based terminal defence. Initial demonstrations will focus on ground-based systems, with space-based demonstrations against satellite targets simulating strategic missile components possible in the early 1990s.

Although the advanced development or field testing of the Hypervelocity Launcher in other than a fixed, ground-based mode would appear to be inconsistent with Article V(1), testing of a space-based version is scheduled for the early 1990s. Testing against orbiting satellite targets would be inconsistent with the Treaty if it demonstrated ABM capabilities. Furthermore, the rapid rate of fire possible with this system (on the order of one shot per second) would appear to be inconsistent with the undertaking in Article V(2) 'not to develop, test, or deploy automatic or semi-automatic or other similar systems for rapid reload of ABM launchers'.

The SLBM Boost Phase Engagement Project will develop and test a sea-based or air-based system for intercepting SLBMs during their boost phase. The potentially short flight times of SLBMs make them more difficult to engage with space-based defence systems. However, sea-based or air-based ABM launcher platforms could move to within a few miles of ballistic missile submarine patrol areas. These systems could intercept SLBMs inside or just above the atmosphere during their boost phase. In this way, SLBMs may prove easier to intercept in boost phase than ICBMs.

Testing of components of this system could be inconsistent with the provision in Article V(1) banning the development, testing or deployment of sea-based and air-based ABM systems and components.

The Space-Based Kinetic Kill Vehicle[59] project is a space-based rocket interceptor system for boost-phase and mid-course defence. A large number of satellites would be deployed in low Earth orbits, with each satellite carrying a number of interceptor rockets similar to the American miniature homing vehicle ASAT system that is presently under development. Testing against orbiting satellite targets simulating missile components is scheduled for the early 1990s. Such testing would demonstrate an ABM capability and would therefore appear to be inconsistent with Article VI(a).

The advanced development or testing in space of this system would also be inconsistent with Article V(1).

Directed-Energy Weapons

The Defense Advanced Research Projects Agency (DARPA) has conducted work on space-based laser development for several years under the so-called laser 'Triad' programme, which has been incorporated into the Space-Based Laser Project of the SDI. Although the revised schedule for this project has not been made public, it can be assumed to parallel that of the Triad programme, which called for an integrated, on-orbit demonstration of a space-based laser in the early to mid 1990s.

The Space-Based Laser project consists of the Talon Gold pointing and tracking component, Large Optics Demonstration Experiment (LODE) mirror system, and the ALPHA hydrogen-fluoride chemical infrared laser.[60]

The large Talon Gold telescope would be attached to the space-based laser and used to ensure that the laser was properly aimed at the target. The testing schedule for Talon Gold initially called for two in-space demonstrations of the system aboard the Space Shuttle in mid-1987 and mid-1988. With the initiation of the SDI these tests were delayed until 1988–9 to permit the inclusion of a second telescope to provide additional surveillance and target acquisition capabilities.

As a result of decisions by the SDIO, the Talon Gold programme has been further restructured. The initial tests of the Talon Gold-derived hardware will be conducted, with the first flight test in space now apparently scheduled for 1989 or later. A full-scale integrated on-orbit demonstration of this system is possible in the early 1990s.[61]

Such advanced development or testing in space would be inconsistent with the provision in Article V(1) banning the development, testing or deployment of space-based ABM components.

Some Reagan Administration officials have argued that Talon Gold is only a generic experiment investigating certain pointing and tracking technologies applicable to many roles, and will not be capable of substituting for an ABM component. Although the technology being demonstrated in Talon Gold is not applicable solely to missile defence, that is the main purpose for which it is intended, as evidenced by Talon Gold's inclusion in the SDI. This argument might have had some merit when applied to the initial Talon Gold configuration. It is clear, that the follow-on to Talon Gold that will be demonstrated in space in the early 1990s will be ABM-capable, and thus inconsistent with the Treaty.

Ground-Based Lasers (GBLs)[62] under development by the SDI would consist of a large GBL that would direct their beam of energy to a target by means of a series of space-based mirrors. Testing of these GBLs at agreed ranges would not be inconsistent with the provisions of the Treaty. However, the inclusion of space-based mirrors could raise concerns about compliance with Article V(1).

The Directed-Energy Programme of the SDI also includes work on Space-Based Neutral Particle Beam Weapon.[63] During the 1980s, work in this area will focus on laboratory demonstrations which are permitted by the Treaty. The 1990 space-based demonstrations of this device would face the restrictions contained in Article V(1).

IV CONCLUSION: RECOMMENDATIONS FOR CLOSING THE 'GREY AREAS' OF THE ABM TREATY

The ABM Treaty has made the strategic arms competition more predictable. The restraints on offensive weapons which have been achieved were possible only because of the restraints on defensive systems agreed to in the ABM Treaty. The absence of a large Soviet ABM system throughout the 1970s gave the United States confidence in its ability to retaliate and reduced the need for more strategic nuclear weapons.

The ABM Treaty curtailed what otherwise would have been a prohibitively expensive race in anti-missile weapons and a more rapid qualitative and quantitative buildup in offensive weapons in the 1970s. The end result of the race would have been billions of dollars thrown away on an ABM system that even missile defence proponents now admit was technologically inadequate to defend against a sophisticated and determined adversary.

The ABM Treaty has enhanced strategic and crisis stability. The Treaty limits Soviet and United States BMDs to such low levels that both nations are unable to protect themselves against missile attack. Each nation could thus not contemplate launching a first strike with the hope that it could survive a retaliatory attack.

The ABM Treaty is now threatened by near-term and far-term United States and Soviet missile defence programmes (see Table 15.1). Any decision to abandon its comprehensive limits on ABM systems invites a number of unfavourable consequences.

The possibility of reductions in strategic offensive nuclear forces will become far more remote. Attempts to achieve such reductions through the SALT process have been difficult enough in the absence of large ABM deployments on each side. In fact, the main impediment to large reductions in offensive weapons during the SALT process has been the presence of MIRVs. The decision in the late 1960s to forge ahead with the development of MIRVs was based on the perceived need to penetrate ABM systems then under develop-

Table 15.1 SDI compliance issues
Some of the scheduled tests and demonstrations within these SDI projects would raise questions of compliance with the ABM Treaty.

Project	Testing mode	Testing period	Treaty Article affected
Sensors			
Boost Surveillance and Tracking System (BSTS)	Space-based	early 1990s	V(1)
Space Surveillance and Tracking System (SSTS)	Space-based	early 1990s	V(1) VI(a)
Airborne Optical System (AOS)	Air-based	late 1980s	V(1)
Terminal Imaging Radar (TIR)	Ground-based (mobile?)	late 1980s	V(1)
Imaging Radar or Laser Demonstration	Space-based	mid 1990s?	V(1)
Directed-energy weapons			
Space-based laser	Space-based	1990s? (integrated demonstration)	V(1)
Ground-based Laser	Space-based (some aspects)	early 1990s	V(1)
Space-based Particle Beam	Space-based	1990s	V(1)
Nuclear-driven Directed Energy	Space-based	early 1990s	V(1)
Kinetic-energy Weapons			
Hypervelocity Launcher	Ground-based	late 1980s	V(2)
SLBM Boost Phase Engagement	Air/sea based	early 1990s?	V(1)
Space-based Hypervelocity Launcher	Space-based	early 1990s?	V(1)
Kinetic Kill Vehicle	Space-based	early 1990s	V(1)

ment. A decision today to build an ABM system would result in a similar scramble by the superpowers to develop additional techniques to penetrate it.

As the White House publication *The President's Strategic Defense Initiative* noted in January 1985, the SDI programme will not eliminate the need for offensive forces. But contrary to that report's suggestion, the SDI will only reduce confidence that United States present retaliatory forces are adequate to penetrate enemy defences and will enhance the need for more offensive weapons.[64]

Preserving the ABM Treaty will require political decisions during this decade by both governments recognising the importance of this objective. The United States and the USSR have failed to make a collective and concerted effort to this end.[65] Instead, each has spent its energies proceeding with new anti-missile weapons and generating rationalisations for why these programmes are consistent with the Treaty.[66]

(a) The Standing Consultative Commission (SCC)

The SCC was established by the ABM Treaty. It was intended to be (and could still become) the main avenue for resolving compliance issues in order to preserve and strengthen the Treaty. Both President Reagan and Secretary of State Shultz have emphasised the need to 'reverse the erosion of the ABM Treaty'. If this is a sincere United States objective, reversing Treaty erosion should take place in the SCC–the forum established specifically for that purpose. Instead, the Reagan Administration has taken actions over the past six years that have severely impaired the usefulness of the SCC and made the task of clarifying and strengthening the ABM Treaty's limits more formidable.

In addressing the problems related to maintaining the integrity of the Treaty, it should be understood that if either side erodes the Treaty by pressing its limits or seeking freedom of action, the other side will seek equal rights. The present approach of the Reagan Administration is to insist on strict Soviet compliance with the Treaty while strenuously avoiding resolution of matters that might impinge upon United States programmes.

Agreed interpretations of the ABM Treaty reached in the SCC may be needed to provide greater clarity and to prevent exploitation of perceived ambiguities which could undermine the Treaty. It is useful, in this respect, to compare the ABM Treaty with SALT II.

Whereas the ABM Treaty has only 12 Agreed Statements and Common Understandings, SALT II includes 98. The degree of detail in SALT II's clarifying terms and limits was not possible when the ABM Treaty was signed in 1972, but may be possible today. The mechanism for 'updating' the ABM Treaty is the SCC. Any agreed interpretations, however, should focus on preserving and strengthening the Treaty, not abandoning it.

(b) Definitional issues

Obviously the most pressing concern in strengthening the ABM Treaty is the resolution of the controversy over the 'old' or 'restrictive' interpretation of the Treaty's limitations on development and testing versus the 'new' and 'permissive' interpretation first made public by the Reagan Administration in October 1985. The recommendations assume that the restrictive interpretation is the operative one.

Definition of ABM development and testing
The parties could reach an agreed statement in the SCC on an explicit definition of development and testing which would clarify that the Article V restrictions on ABM system or component development and testing are applicable to that stage of development *which follows laboratory testing*. These restrictions would apply to that part of the development process where field testing is initiated on either a prototype or breadboard model of a system or its components.

Definition of ABM components
The parties could reach on agreed statement in the SCC on an explicit definition of ABM components which would clarify that the Treaty restrictions on components include restrictions on those devices that are capable of working in conjunction with or substituting for existing types of ABM systems and components. The SALT II numerical limits on the characteristics of permitted tests may serve as a precedent for such a definition. The definition might also specify that prior notification and data exchanges would be required on any testing of a component judged to be similar to but outside of the agreed parameters of an ABM component.

In addition to the SDI, the most important issue that should be addressed in the near future is the problem of 'grey-area' weapon systems and technologies, such as LPARs, ASATs, and ATBMs.

Restrict tests of kill vehicles in space
Another possible means for limiting space-based development and

testing of ABMs, while acknowledging the problems of distinguishing an ABM-capable sensor from one which is not, would be to expand Article V's limits to include *all* testing of any type of kill vehicle in space, while permitting testing of all space-based sensors (which might substitute for or augment ABM radars).

(c) Restrictions on LPARs

There are several approaches to resolving the LPAR issue that might be included in a new Protocol to the Treaty. The following options could be considered together, or separately.

Standstill at the present situation
The parties could agree not to construct any new LPAR for any purpose, without prior consultation and agreement with the other party. The stand-still could either permit or prohibit the completion of radars currently under construction, such as the Soviet Krasnoyarsk and other Pechora-type radars, and the American Pave Paws and BMEWS radars. This consultation process could be extended to include an agreement similar to that covering ABM test ranges, so that the construction of new LPARs would require the agreement of both parties.

Numerical limits on deployed radars
The parties could agree that each country would be permitted no more than a certain number of LPAR transmitter faces. Alternatively, this limit could be figured by aggregating the potential power/aperture product (the product of mean emitted power in watts and antenna area in square metres) of each of these radars. This limit could perhaps take into account ABM test range radars and the 'small' radars at the one permitted ABM site.

The current situation is one that may favour the Soviets slightly, although this will change in coming years. At present, the United States has seven such faces operational (one PARCS, one FPS-85, one Cobra Dane and four Pave Paws), and an additional nine under construction (four Pave Paws and five BMEWS), for a total eventual deployment of fifteen faces (the FPS-85 radar will be replaced by a Pave Paws radar). The Soviets, in contrast, have six currently operational faces (1 Dog House, 1 Cat House, and perhaps 4 Pechora-type), with another 4 Pechora-type radars and the Pushkino LPAR with four faces under construction, for a total ultimate deployment of perhaps 14 faces. (Hen House radars are not modern phased-array radars and would therefore not count toward these limits.) However, Soviet radars typically have a potential that is several times larger than that of comparable American radars although the technology of United States LPARs is more advanced.

A more restrictive type rule for permitted radars
The parties could agree not to deploy in the future any additional LPARs, except as permitted early warning or ABM radars. No new deployments would be permitted for the purposes of space tracking or as national technical means of verification, except to the extent that such deployments were consistent with the limitations on early warning and ABM radars.

Clarification of permitted deployments
The parties could agree not to deploy in the future any LPARs:

   (a) except at locations along the periphery of its national territory that are less than, for example, 150 kilometres from its border, and;
   (b) except oriented outward, with not more than, for example, 5 per cent of the total coverage of the radar (the area described by a section 60 degrees to either side of the bore-sight of each radar face to a range of 2,500 kilometres) covering its national territory;
   (c) including any radar that may be used for early warning, for tracking of space objects, or as a national technical means of verification;
   (d) except for radars located at previously designated ABM test ranges.

Implementation of type rule
The parties could further agree to dismantle or modify the construction of any existing radar that is not:

   (a) located within and along the periphery of its national territory and less than, for example, 150 kilometres from its border, and;
   (b) oriented outward, with not more than 5 per cent of the total coverage of the radar (the area described by a section 60 degrees to either side of the bore-sight of each radar face to a range of 2,500 kilometres) covering its national territory;
   (c) except for existing radars located at previously designated ABM test ranges. This would require the dismantling of the Krasnoyarsk, Thule and Fylingdales radars.

(d) ATBMs

Limit on testing
The parties could agree, the other provisions of the ABM Treaty

notwithstanding, not to test interceptor missiles of any type at altitudes above, for example, 40 kilometres and at velocities in excess of, for example, 2 kilometres per second.

Ban on large mobile radars
The parties could agree not to deploy land-based mobile radars or radars with a potential (the product of mean emitted power in watts and antenna rea in square metres) in excess of 1 million for any purposes or to test such radars against targets which have the characteristics of strategic ballistic missile targets or their components in flight trajectory.

(e) Treaty limiting ASAT weapons

Unless ASAT negotiations are undertaken and are successful in concluding an agreement, the further development, testing, and deployment of ASAT systems will seriously undermine the ABM Treaty. Because directed- and kinetic-energy weapons now under development by both parties could be used for both ASAT and ABM purposes, they should be subjected to stringent limitations.

An agreement on ASAT weapons would avert an arms competition that would result in a mutual lessening of the national security of states. Such a treaty would enhance international security by preserving outer space for civilian applications and military activities such as reconnaissance and early warning.

An ASAT treaty could also help resolve some of the ambiguities that have arisen under the ABM Treaty in the area of large radars. For instance, the Soviet 'space-track' radar near Krasnoyarsk might be dismantled as part of an ASAT agreement limiting ASAT battle management capabilities.

To the extent that limits on ABM-capable, space-based sensors, such as advanced early warning satellites with missile and warhead tracking capability, could be verified, clear differentiation between permitted early warning sensors and prohibited battle management sensors are needed. In addition, and in parallel with an ASAT treaty, a protocol to the ABM Treaty should be agreed on to prohibit the advanced development and testing of fixed, ground-based, exo-atmospheric interceptors using kinetic- or directed-energy weapons. Otherwise, programmes that were labelled as ABM could undermine the ASAT treaty regime, as ASAT activities could today be used to undermine the ABM Treaty.

Ban on ASAT development and testing
The most important limitation in an ASAT treaty would be a prohibition on advanced development and testing of ASAT weapons or

their components by destroying, damaging, disturbing the normal functioning or changing the normal flight trajectory of objects in space. This prohibition could apply either to all types of ASATs, or just to new types, beyond existing systems, and could be formulated in parallel with the provisions of Article V of the ABM Treaty pertaining to limits on space-based and other mobile systems and components.

Ban on ASAT deployment
A prohibition on the deployment of any dedicated system which has been tested by destroying, damaging, disturbing the normal functioning or changing the normal flight trajectory satellites would provide further confidence in the limitation of these capabilities. While such an agreement might pose verification difficulties, it is in the net security interests of the United States. Moreover, both sides would retain residual capabilities against satellites, and the United States would be in a position to recover a dedicated ASAT capability quickly in the event of a Soviet 'breakout'.

(f) Far-term ABM issues

The SDI contemplates a number of activities that pose challenges to the ABM Treaty in the late 1980s and early 1990s, particularly space-based anti-missile systems and related technologies. Other issues of this type would include possible additional restrictions on the testing of long-range exoatmospheric interceptors and of sensors other than radars.

Ban on testing of exoatmospheric interceptors
The parties could agree not to test ABM interceptors or their components against strategic ballistic missiles or their elements in flight trajectory at an altitude in excess of 30 kilometres. This would effectively preclude the further advanced development by either side of exoatmospheric heatseeking interceptors.

The testing and limited deployment of fixed long-range exoatmospheric interceptors is presently permitted under the ABM Treaty. However these interceptors are an essential component of a large-scale BMD system. They could constitute, along with other programmes, a 'base for the defence of territory' which Article I of the ABM Treaty prohibits.

To the extent that limits on certain ABM-capable sensors, such as advanced early-warning satellites, are becoming infeasible due to the blurring between their permitted early warning and prohibited ABM functions, more stringent restraints on other ABM systems and components are called for. In contrast, short-range endoatmospheric interceptors are primarily of interest for defence of single hard targets, such as missile silos or command centres. Such interceptors pose less serious threats to stability.

Ban on development of mobile sensor components
The parties could agree that the ban on mobile ABM systems and components includes a prohibition on components based on new physical principles that are capable of being used in conjunction with or substituting for ABM systems of an existing type.

In recent years there have been a number of advances made in the development of new types of laser and infrared sensors that are capable of substituting for or acting in conjunction with ABM radars. In some instances these systems are air-based or mobile land-based. There may be some ambiguity as to whether these systems are 'components' which are limited by the Treaty, or whether they are permitted 'adjuncts' to components.

Limitation on space-based particle beam devices
The parties could agree to ban or severely limit the testing and deployment in space of particle beam accelerators.

Limitation on space-based lasers
The parties could agree to prohibit or severely limit the placing into space of any directed-energy system which has an aggregate mirror aperture in excess of, for example, 5 square metres.

Limitation on ground-based lasers
The parties could agree to prohibit testing against objects in space or to deploy any ground-based, sea-based or air-based directed-energy system which has an aggregate mirror aperture in excess of, for example, 5 square metres.

General limitation on the brightness of directed-energy systems
'Brightness' is a generic figure of merit that is used to measure the weapons capabilities of all types of directed-energy systems. It is analogous in concept to the 'power-aperture product' that is used to define radars in the ABM Treaty. A brightness of 10^{19} is a useful threshold for significant military capability, and brightness levels of over 10^{21} is the long-term goal of the SDI.

The parties could agree to prohibit the testing of any directed-energy device – regardless of whether it is a laser or particle beam weapon, and regardless of where or how it is based – if the device has a potential brightness, measured in watts per steradian, in excess of 10^{19}. This would preclude the testing of lasers significantly brighter

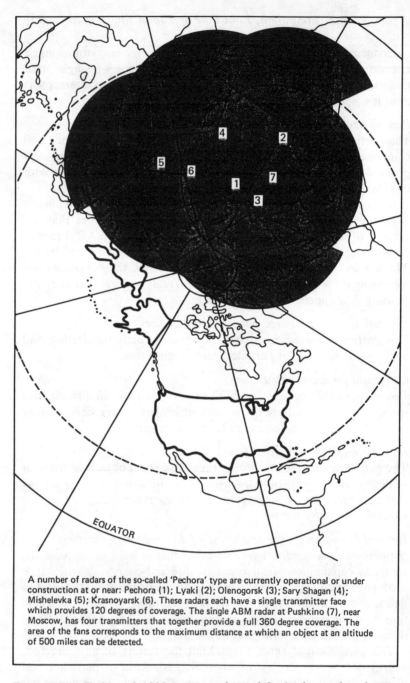

Figure 15.1 Estimated 1990 coverage fans of Soviet large phased-array Radars

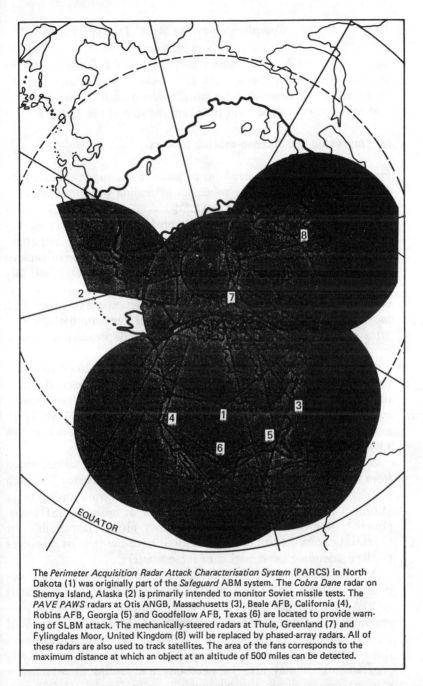

The *Perimeter Acquisition Radar Attack Characterisation System* (PARCS) in North Dakota (1) was originally part of the *Safeguard* ABM system. The *Cobra Dane* radar on Shemya Island, Alaska (2) is primarily intended to monitor Soviet missile tests. The *PAVE PAWS* radars at Otis ANGB, Massachusetts (3), Beale AFB, California (4), Robins AFB, Georgia (5) and Goodfellow AFB, Texas (6) are located to provide warning of SLBM attack. The mechanically-steered radars at Thule, Greenland (7) and Fylingdales Moor, United Kingdom (8) will be replaced by phased-array radars. All of these radars are also used to track satellites. The area of the fans corresponds to the maximum distance at which an object at an altitude of 500 miles can be detected.

Figure 15.2 1990 coverage fans of American large phased-array Radars

than the DARPA space laser triad. 'Potential brightness' can be calculated based on the observable wavelength of a laser and the diameter of the beam director mirror, along with estimates of the maximum power of the laser based on the size of the mirror.

(g) Improving the decision-making process

Finally special attention needs to be given to the organisation of the Soviet and American policy processes in dealing with these issues.

Although the details of the Soviet decision-making process in this field are obscure, General Brent Scowcroft has offered the observation that over-compartmentalisation and the lack of early and effective civilian review are at least in part responsible for the difficulties that the Soviets have experienced in achieving a completely satisfactory record of compliance. While the United States cannot have any direct influence on the Soviet arms control decision-making process, a thorough understanding of that process is useful in formulating future arms control positions and strengthening and preserving existing agreements.

Unfortunately the American process continues to need improvement as well. While some individuals have been assigned the task, there is no permanent body within the United States government advising decision-makers authoritatively on how military programmes affect treaty obligations. Nowhere is this more evident than with respect to the SDI.

Only an internal Defense Department review team, co-ordinated from within the office of the Undersecretary of Defense for Research and Engineering, appears to have access to information allowing informed judgement as to the exact nature of individual SDI programme. Nor do senior level inter-agency groups responsible for overseeing arms control impact questions have the time or resources to allow adequate examination of these matters.

ACDA would seem to be the appropriate location for co-ordinating this review process but (especially in the case of the SDI) ACDA is often uninformed about the details of individual programmes. The General Counsel offices at ACDA and Department of Defense and the Legal Advisor's office at the State Department should play important roles in the early review of United States R&D programmes. It may also be necessary and useful to establish a working group within the National Security Council (NSC) with representatives from the appropriate governmental bodies to co-ordinate this ongoing compliance review process.

The Congress has become increasingly attentive to reviewing the SDI programme and, with its control and oversight of the budget, could limit research and development efforts to what is prudent and necessary, and restrict or deny funding to those projects which would undercut the ABM Treaty.

An important corollary to this is increasing the amount of publicly available information on these issues. Although missile defence has traditionally been regarded as a sensitive area, the currenty available public information on the SDI is at a historic low. Protection of certain information about military programmes is essential, but a delicate balance must be struck.

Recent moves by the Reagan Administration have drastically reduced the level of detail that is available on the SDI, as well as the Defense Department budget as a whole. This has the effect of limiting, if not eliminating, the public debate on this fundamental national security issue. This undermines the process of developing an informed and enduring national consensus, and is inimical to the functioning of a democratic society.

Notes

1. Mark M. Lowenthal, *Verification: Soviet Compliance With Arms Control Agreements* (Washington, DC: Congressional Research Service, 1985) p. 2.
2. Hedrick Smith, 'U.S. Report Said to Accuse Soviet on Arms Treaty', *New York Times*, 3 April 1983; George C. Wilson and Walter Pincus, 'U.S. Weighs Accusing Soviets of Arms Violations', *Washington Post*, 5 April 1985; Hedrick Smith, 'Panel Says Soviet Violated Arms Pact. Conservative US Senators Press Administration to Reveal Findings', *International Herald Tribune*, 22 April 1983.
3. Lou Cannon and Walter Pincus, 'US Said to Shift Track on Arms Treaty Issue', *Washington Post*, 24 April 1985.
4. Hedrick Smith, 'US Sees New Soviet Arms Violation', *New York Times*, 12 May 1983.
5. Hedrick Smith, 'US Seeking Soviet Parley on Arms Violation Issues', *New York Times* 12 August 1985; Michael Getler, 'US Asks Soviets for Special Meeting', *Washington Post*, 13 August 1985; Michael Getler, 'Soviets Veto Bid for Talks On SALT Pact', *Washington Post*, 4 September 1983; Charles Mohr, 'Talks Begin in Possible Arms Violations', *International Herald Tribune*, 6 October 1983.
6. Lowenthal, *Verification* pp. 4–8.
7. 'Soviet Violations Will Affect US Stand at Arms Talks (Transcript: Perle briefing, Foreign Press Center)', *Wireless Bulletin from Washington*, 18 January 1985.
8. Walter Pincus, '"Options" Studied in Arms Violations', *Washington Post*, 21 February 1985, p. 1.

9. Walter Pincus, 'Pentagon Acts to Obey ABM Treaty', *Washington Post*, 26 March 1985, p. 12; Peter Grier, 'As "Star Wars" moves ahead, conflict with ABM treaty looms', *Christian Science Monitor*, 27 March 1985; Bill Keller, 'US Interprets ABM Treaty as Allowing Tests Of Space Arms', *International Herald Tribune*, 22 April 1985.
10. Wayne Biddle, 'Amending of 1972 ABM Pact is Urged', *New York Times*, 31 May 1985, p. 3; 'Nitze: ABM Treaty Adaptable to New Circumstances', *Wireless Bulletin from Washington*, 31 May 1985, pp. 11–16.
11. Jeanette Voas, *The President's Report on Soviet Non-compliance With Arms Control Agreements: A Discussion of the Charges*, Report 84-160F (Washington, DC: Congressional Research Service, 10 September 1984).
12. Colin S. Gray, 'Moscow is Cheating', *Foreign Policy* (Fall 1984) pp. 141–52.
13. John Pike and Jonathan Rich, 'Charges of Treaty Violations Much Less Than Meets the Eye', *FAS Public Interest Report* 37 (March 1984) pp. 1–20.
14. John B. Rhinelander, 'How to Save the ABM Treaty' *Arms Control Today* 15 (May 1985) pp. 1; 5–9.
15. Thomas K. Longstreth, John E. Pike and John B. Rhinelander, *The Impact of US and Soviet Ballistic Missile Defense Programs on the ABM Treaty* (Washington, DC: National Campaign to Save the ABM Treaty, March 1985).
16. Thomas K. Longstreth and John E. Pike, 'US, Soviet programs threaten ABM Treaty', *Bulletin of the Atomic Scientists* 41 (April 1985) p. 11.
17. See John B. Rhinelander, 'Reagan's "Exotic" Interpretation of the ABM Treaty – Legally, Historically, and Factually Wrong', *Arms Control Today* 15 (October 1985) pp. 3–6; Ambassador Nitze and Mr Sofaer, 'The ABM Treaty and the SDI Program', *Current Policy* 755 (Washington DC: United States Department of State, Bureau of Public Affairs, October 1985).
18. See Communiqué of the NPG of NATO of its 38th Ministerial Meeting on 29–30 October 1985 in Brussels, *NATO Review* 6 (1985).
19. See note 1; Frank Greve, '"Star Wars" shift stirs dispute', *Philadelphia Inquirer*, 23 October 1985, p. 4; Walter Pincus, 'US Drafts Response on SDI', *Washington Post*, 23 October 1985, p. 1.
20. See Jeanette Voas, 'The arms-control compliance debate', *Survival* 28, (January/February 1986) pp. 8–31; *The President's Report to the Congress on Soviet Non-Compliance With Arms Control Agreements*, 23 January 1984, US House Doc. 98-158 (Washington DC: United States Government Printing Office, 1984); 'The President's Unclassified Report to the Congress on Soviet Non-compliance with Arms Control Agreements', 1 February 1985, in *Congressional Record*, 5 March 1985, vol. 131, pp. S2530–2534; 'The President's Unclassified Report on Soviet Non-compliance with Arms Control Agreements', *White House Press Release*, 23 December 1985; General Advisory Committee on Arms Control and Disarmament, 'A Quarter Century of Soviet Compliance Practices Under Arms Control Commitments, 1958–1983', *Congressional Record*, 11 October 1984, vol. 130, pp. S14526–S14529.

21. See note 4; Walter Pincus, '"Options" Studied in Arms Violations', *Washington Post*, 21 February 1985, p. 1; Michael R. Gordon, 'CIA Is Skeptical that New Soviet Radar Is Part of an ABM Defense System', *National Journal*, 9 March 1985, pp. 523–6.
22. See Gordon 'CIA is Skeptical'; US Department of Defense, *Soviet Military Power* (Washington, DC: United States Government Printing Office, 1984; 1985).
23. See note 5.
24. See note 20.
25. Large Phased Array Radar (LPAR). A phased array radar (PAR) is a radar that points its beam in different directions without moving the antenna mechanically. PARs can track many targets simultaneously, a critical function for any radar that would be used in an ABM role. Merrill I. Skolnik, *Introduction to Radar Systems* (New York: McGraw-Hill, 1980); Eli Brookner, 'Phased Array Radar', *Scientific American* 252 (1984) pp. 94–102.
26. US Congress, Senate, Committee on Foreign Relations, Hearings on *Strategic Arms Limitation Agreements* (Washington, DC: United States Government Printing Office, 1972) pp. 197–204; US Congress, Senate, Committee on Armed Services, Hearings on *Military Implications of the Treaty on the Limitations of Anti-Ballistic Missile Systems and the Interim Agreement on Limitation of Strategic Offensive Arms* (Washington DC: United States Government Printing Office, 1972).
27. Charles Mohr, 'US and Soviet Discuss Whether Moscow Violated Terms of 2 Arms Pacts', *New York Times*, 5 October 1983; John F. Burns, 'Moscow Accuses US of Violating Arms Agreements', *New York Times*, 30 January 1984; Dusko Doder, 'Tass Again Alleges American Breaches', *Washington Post*, 21 October 1984.
28. See Gordon, 'CIA is Skeptical'.
29. See Lowenthal, *Verification*; Voas, *The President's Report*, 'The arms-control compliance debate'; Gordon, 'CIA is skeptical'.
30. For the contradictory official United States claims see chapter 3 by H. G. Brauch in this volume.
31. See *Soviet Non-Compliance* (Washington, DC: US Arms Control and Disarmament Agency, 1 February 1986), p. 5.
32. See 'Unilateral Statement B' in: Appendix B.
33. Quoted from *Arms Control and Disarmament Agreements. Texts and Histories of Negotiations* (Washington, DC: United States Arms Control and Disarmament Agency, 1980) p. 147.
34. See for details on the problem of distinguishing the trajectories of SLBM and tactical ballistic missiles, Stephen Weiner, 'Systems and Technology', in Ashton B. Carter and David N. Schwartz (eds), *Ballistic Missile Defense* (Washington, DC: Brookings Institution, 1984) pp. 73–5.
35. For details see: *The Military Balance 1985–1986* (London: International Institute for Strategic Studies, 1985).
36. See note 20.
37. US Congress, Senate, Committee on Armed Services, *Hearings Department of Defense Authorization for Appropriations for Fiscal Year 1985*, Part 6 (Washington, DC: United States Government Printing Office, 1984), p. 2956.

38. US Congress, House, Committee on Armed Services, Hearings on *Department of Defense Authorization for Appropriations for Fiscal Year 1984*, Part 5 (Washington, DC: United States Government Printing Office, 1983), p. 242.
39. See 'Soviets Test Defense Missile Reload', *Aviation Week & Space Technology*, 29 August 1983, p. 19.
40. See *Soviet Non-Compliance*.
41. See note 31.
42. See Gordon 'CIA is Skeptical'.
43. For source see note 20.
44. For source see *Arms Control and Disarmament Agreements* pp. 146–7.
45. See note 31.
46. For source see note 20.
47. *Report to the Congress on the Strategic Defense Initiative* (Washington, DC: Department of Defense, April 1985) Appendix B.
48. *Report to the Congress*, p. 34.
49. *Report to the Congress*, p. 35.
50. *Report to the Congress*, pp. 30, 35.
51. *Report to the Congress*, pp. 30, 35.
52. *Report to the Congress*, p. 36.
53. *Report to the Congress*, p. 36.
54. *Report to the Congress*, p. 36.
55. 'RV is Struck in Successful Ballistic Missile Defense Test', *Aerospace Daily*, 12 June 1984, p. 235; Walter Andrews 'Missile Interceptor called capable of 85% protection', *Washington Times*, 24 October 1984; 'Overlay BMD Could Defend ICBMs, Lockheed Says', *Aerospace Daily*, 24 October 1984, pp. 282–3.
56. Eugene Kozicharow, 'Army Doubles Missile Defense Fund Bid to Support SDI Effort', *Aviation Week & Space Technology*, 8 April 1985, p. 62–3; 'Army Plans Early 1990s Ballistic Missile Defense Demonstration', *Aerospace Daily*, 19 October 1984, p. 260.
57. See *Report to the Congress*, p. 55; 'The US Army's role in "Star Wars"', *Flight International*, 9 February 1985, pp. 10ff; Walter Pincus, 'Land-Based Missile Defense May Be Ready in Late '90s – Army System Called Basis for "Star Wars"', *Washington Post*, 30 January 1985; 'State-of-the-art sensor sought for Heds', *Defense Daily*, 27 July 1984.
58. See *Report to the Congress*, p. 54.
59. *Report to the Congress*, p. 58.
60. *Report to the Congress*, p. 40.
61. *Report to the Congress*, pp. 39–43.
62. *Report to the Congress*, pp. 43–6.
63. *Report to the Congress*, pp. 47–9.
64. *The President's Strategic Defense Initiative* (Washington, DC: United States Government Printing Office, 1985).
65. Thomas K. Longstreth, John E. Pike and John B. Rhinelander, *The Impact of US and Soviet Ballistic Missile Defense Programs on the ABM Treaty* (Washington DC: National Campaign to Save the ABM Treaty, March 1985) p. 74.
66. Longstreth, Pike and Rhinelander, p. 73.

16 Closing the Window of Vulnerability
John E. Pike

On 23 March 1983 Reagan delivered his 'Star Wars' speech, calling for the development of an anti-ballistic missile (ABM) system to defend the American population from Soviet missiles. The vision that President Reagan presented as the basis of his Strategic Defence Initiative (SDI) is a world in which nuclear weapons are 'impotent and obsolete'. This is generally taken to mean that the SDI would lead to a virtually perfect defence of populations. Certainly the exuberant rhetoric that has been used in support of the programme would be difficult to sustain in support of less exalted goals, such as defence of retaliatory forces.

While many SDI supporters are trying to bridge the distance between Reagan and reality by advocating an ABM system to defend ICBM silos, the case for such defences is difficult to make. Furthermore, the SDI includes no work on components specifically designed for this mission, and those components that could be applied to the point defence mission are poorly suited to the task.

I PRIOR AMERICAN EFFORTS TO DEFEND ICBM SILOS

The United States has a long history of work on ABM systems.[1] In 1968 President Johnson proposed the deployment of the Sentinel ABM system, which evolved from the Nike-X development effort. Sentinel sought initially to defend American cities from a Chinese attack, with the potential for later expansion to defend against a Soviet attack. In 1969 President Nixon announced a similar but more modest programme, Safeguard, to replace Sentinel.

These systems utilised two types of large phased-array radars (LPARs), the Perimeter Acquisition Radar and the Missile Site Radar, which permitted greatly improved tracking and targeting compared with their predecessors. A long-range rocket, the Spartan, was designed to use high-yield nuclear warheads to destroy incoming re-entry vehicles beyond the atmosphere. A high-acceleration short-range rocket, the Sprint, was to be employed to intercept incoming targets during re-entry into the atmosphere (that is, an endo-atmospheric interceptor), defeating such simple countermeasures as balloon decoys.

The decision to deploy Sentinel and Safeguard was very controversial. While most experts doubted any ABM system's ability to protect populations against a determined attacker, some nevertheless favoured deployment of a limited missile defence to protect land-based missiles and other military targets.

However, the Safeguard components, which had been developed to defend cities, were poorly suited to the defence of military targets. The large Spartan and Sprint interceptors had ranges far in excess of that required to defend hardened ICBM silos. The large radars of the Safeguard system were very expensive and highly vulnerable to direct attack and to blinding by the explosions of the long-range interceptor's warheads. The cost of the defence was thus greater than the cost of the offensive buildup needed to overcome it.

After the signing of the ABM Treaty in 1972 American missile defence research focused on improving techniques for defending ICBM silos. The United States Army's Site Defence Programme developed a rapidly deployable radar and a short-range interceptor missile for this mission. Both of these components were considerably smaller than their Safeguard predecessors, and the long-range Spartan was dispensed with entirely. Site Defence formed the basis for the Low Altitude Defence System (LoADS) which was considered in the early 1980s for defence of the MX missile.

Longer-range non-nuclear interceptors were under development in the Overlay programme, which in conjunction with LoADS would form a layered defence. These systems would use air-based and space-based infrared sensors in addition to radars, and non-nuclear kinetic-energy interceptors along with nuclear-tipped interceptors. A layered defence of this sort was under evaluation during 1982 for possible use to defend the MX ICBM, but it was rejected due to high cost and uncertain performance.

For more than a decade American ABM research has thus concentrated on the development of systems to intercept enemy warheads as they re-enter the atmosphere. These terminal phase components were optimised for the defence of missile silos.

II POTENTIAL SDI COMPONENTS FOR SITE DEFENCE

Now the SDI proposes to move away from this traditional area of concentration and instead focus on more exotic boost-phase, post-boost phase and mid-course interception systems. These systems are much more costly, have higher technical risk and uncertainty and are not required for traditional ABM missions such as silo defence.

The budgetary emphasis of the SDI^2 makes it clear that traditional ABM missions will receive decreasing emphasis in coming years. The SDI is not focused on missions such as silo defence, but on a futile effort to protect the American population.

These priorities were confirmed in the allocation by the SDIO the cuts made in the FY85 budget request. Most of the reduction was made in the projects that comprise the Terminal System Demonstration, that part of the SDI that is most applicable to traditional ABM missions. These components include the Airborne Optical System (AOS) and Terminal Imaging Radar (TIR) sensors, and the Low and High EndoAtmospheric Defence Interceptors (LEDI and HEDI) and the Exoatmospheric Re-entry Vehicle Interception System.

AOS, also known as the Airborne Optical Adjunct (AOA), is a modified Boeing 767 that will carry two mid-wavelength infrared telescopes for tracking and identification of targets for mid-course and terminal defence. This system has been under study for several years, and the first test flight of AOS is planned for 1988, with operational demonstrations in conjunction with other ABM systems in the early 1990s.

TIR is a long range X-band radar that will provide enhanced capabilities for the discrimination of warheads from decoys. TIR has a much longer operating range than the radar of the earlier LoADS. TIR will be demonstrated by the end of the decade and is intended to support the HEDI as part of a terminal defence system.

LEDI, previously known as the Short-Range Homing Interceptor Technology (SR-Hit), is the SDI interceptor with greatest relevance to the silo defence mission. LEDI is a small high-velocity interceptor with a high-power conformal array millimetre-wave radar antenna for a terminal homing guidance. Terminal propulsion is provided by a number of small rockets that provide lateral thrust to guide the missile to its target. In January 1983 Vought Corporation was awarded a $69,000,000 thirty-month contract for development and testing of SR-Hit. In January 1984 the first flight of the SR-Hit test vehicle was conducted against an air-launched simulated re-entry vehicle target. In 1986 ER-HIT was renamed the Flexible Lightweight Agile Guided (FLAG) Experiment, and succeeded in intercepting a simulated target. Further flights are planned to test high-particle velocity non-nuclear warheads. The initial Vought contract was supplemented in late November 1983 by an award of $15,700,000 for development and ground-based experiments on an extended range version of SR-Hit, known as ER-Hit, which has recently been refocused into the LEDI effort. LEDI could also be used for the ATBM mission.

HEDI will demonstrate by the end of the decade a large, long-range ground-based rocket interceptor using a heat-seeking explosive warhead to engage re-entry vehicles as soon as they enter the atmosphere. However, in the event that an opponent deployed manoeuvring re-entry vehicles, HEDI could require a nuclear warhead. HEDI is a two-stage solid-fuel rocket weighing about 7,000 kilogrammes. Each stage would burn for only a few seconds, providing very high acceleration and a range of between 100 and 200 kilometres. HEDI would be committed to intercept targets once they reached an altitude of between 80 and 120 kilometres, and would intercept the target altitudes between 15 and 50 kilometres. HEDI will enforce a keep-out range of about 15 kilometres, mandated by the vulnerability of soft urban targets to the 2 psi overpressure of a 5 megaton detonation. A single battery could defend about 30,000 square kilometres, and 100 batteries the entire United States.

Exoatmospheric Re-entry Vehicle Interception System (ERIS) will demonstrate by the end of the decade an exoatmospheric heat-seeking hit-to-kill interceptor. ERIS is a follow-on to the HOE, although it will use a significantly smaller interceptor kill vehicle, smaller than the miniature homing vehicle of the new F-15 launched ASAT. Development of this new kill vehicle is required in order to reduce the cost of the defence relative to the offence and will represent a major technological challenge. The entire force of interceptors would be located at a single base in the central United States, from which it would be fired to engage targets at ranges from 500 to over 2,000 kilometres, at altitudes of between 100 and 1000 kilometres. Flight times from launch to intercept of between 100 and 600 seconds are envisioned. The single interceptor base, like the MX 'Dense Pack' basing mode, is intended to complicate attacks on the defence itself.

With the exception of LEDI, the capabilities of these components,

Table 16.1 Comparison of endoatmospheric interceptors

Interceptor programme	Sprint Safeguard	Sentry Loads	HEDI SDI	LEDI SDI
Date	1965	1987	1989	1984
Stages	2	1	2	2
Length metres	8.3	5.0	12.0	8.0
Diameter metres	1.4	1.4	1.0	1.0
Weight tons	3.4	1.8	7.0	3.5
Range km	40	10	200	50?
Warhead	10 kt	1 kt	NNK	NNK
Guidance	Command	Inertial	IR Homing	MMW Homing

designed as part of a population defence, are clearly far beyond those required for defence of ICBM silos. Performance is expensive, and the cost of these systems render them very poorly suited to ICBM defence. If the SDI is to be used solely to protect retaliatory forces, it would have to be significantly reoriented, and the recent formation of the LEDI effort is the only evidence that such a reorientation is in prospect.

III GOALS FOR LESS-THAN-PERFECT INTERMEDIATE DEFENCES

Strategic policy faces the dilemma of attempting to maximise the utility of one's own forces while at the same time minimising the utility of the opponent's forces. Enhancements to American strategic capabilities could include denying the Soviets the ability significantly to damage United States society, as well as achieving escalation dominance at lower levels of nuclear conflict. During the first two decades of the nuclear era the United States tended to emphasise the first goal of damage limitation, primarily through numerical strategic superiority. In the more recent period of rough parity, the United States has emphasised efforts to maintain escalation dominance, which has led to such improvements as increases in the targeting flexibility of ICBMs.

At the same time United States policy has sought to maintain an assured destruction capability against Soviet society, as well as to reduce or eliminate Soviet incentives and capabilities for escalation in time of crisis.

However, these four goals are in conflict. Strategic stability requires that both sides retain a retaliatory assured destruction capability against the other side's urban–industrial centres. Efforts to maintain American assured destruction capabilities, such as the ABM Treaty, have reduced the prospects for United States damage limitation.

Crisis stability requires that the characteristics of nuclear forces should provide incentives to escalate to higher levels of violence, including a pre-emptive attack on the other side's nuclear forces, in time of crisis. However, the United States would like to retain the option of deliberate escalation.

Each of these four goals have been used as possible missions that have been discussed for the less-than-perfect defences that are the most likely technological product of the SDI. Missions related to strategic stability include significantly limiting damage to the United States, and thereby achieving United States strategic superiority and preserving American ICBM assured destruction retaliatory capabilities. Missions related to crisis stability include defending ICBMs to reduce post-attack force asymmetries and defending ICBMs so as to preserve a flexible response capability.

IV US DAMAGE LIMITATION/DENIAL

Much of the public rhetoric surrounding the SDI is cast in terms of Mutual Assured Survival (MAS)[3] – a perfect defence of populations. Less exuberant advocates speak in terms of protecting ICBMs. However, as the two subsequent sections conclude, it is difficult to make a coherent military case for defending ICBMs, even if the SDI were working on technologies relevant to this mission, which it is not.

This debate has been obscured by an Aesopian dialogue in which some Star Wars advocates have attempted to mask their desire for American strategic superiority and its putative political advantages. Given the weakness of the case for less-than-perfect defences of ICBMs, it is difficult to avoid the impression that those who continue to make this case in support of the SDI either have not thought the matter through or are too embarrassed or discrete forthrightly to state their real objectives (this of course does not include those who are inclined toward technologies not included in the SDI that would be strictly limited to point defence). The terms of the SDI debate would be greatly clarified, and intellectual honesty served, if more would follow the example of Colin Gray and Keith Payne in candidly making the case for strategic superiority[4] rather than hoping to achieve that end by deception and dissimulation.

Although many would deny any such intentions, it is a little difficult to believe that a weapons development programme that otherwise makes little sense would be pursued in a fit of absentmindedness. Analysis of capabilities properly takes precedence over analysis of intentions, but indications of intent are not difficult to find. Paul Nitze's 'New Strategic Concept' speaks of 'replacing deterrence based on offensive forces with deterrence based on defences and other means'.[5] Fred Iklé has initiated an effort to unite offensive and defensive forces into an integrated warfighting capability. And the Have Temp and Master programmes, initiated in the 1986 budget, seem designed to give life to such aspirations.

It is thus not surprising that the Soviets publicly complain that the primary significance of the SDI is its potential for giving the United States a 'splendid' first-strike capability, which would greatly limit or deny the Soviets the ability effectively to retaliate.

An assessment of American strategic capabilities planned for the 1990s leads to a similar conclusion. The United States plans to field an impressive array of counterforce systems.[6] These can be thought of as the 'pre-boost phase' layer of a strategic defence system. The addition of boost-phase and subsequent layers to the defence would greatly add to the damage-limiting potential of the 'pre-boost phase'.

Soviet ICBM silos will be vulnerable to destruction by the very capable Trident II. Mobile ICBMs would be attacked by the Stealth bomber. Ballistic missile submarines will be threatened by the new SSN-21 Seawolf-class attack submarines, which are specifically designed to operate under the Arctic ice pack, where they can seek out and destroy even the Typhoon-class submarines.[7] The small force of Soviet bombers would be subject to destruction on the ground by the Trident II and interception over the Arctic by an upgraded North American Air Defence system. Soviet command and control facilities could be attacked by the Pershing II and ground-launched and sea-launched cruise missiles, further inhibiting a retaliatory response.

But there are some limitations to these capabilities. Although the relatively small number of mobile ICBMs could be targeted by the Stealth bomber using the new KH-12 photographic reconnaissance satellite, there are limits to the survivability of this satellite. Many of these mobile ICBMs would escape destruction, as would at least a small fraction of the silo-based missiles. The Seawolf might prove highly effective in an attrition campaign during a conventional theatre conflict prior to the initiation of the use of central forces, but some Soviet missile submarines would surely remain at sea. Together, these surviving forces could clearly destroy the United States.

In the absence of an anti-missile defence, the military and political utility of these American systems is difficult to identify. The damage limitation they could provide would be so slight as to be insignificant. But with the addition of the sort of less-than-perfect defence that the SDI could provide in the 1990s, the United States could hope to achieve a meaningful damage limitation capability.

The United States might hope to maintain this position by virtue of its current lead in strategic defence technologies, magnified by the massive SDI effort. At present the United States holds a five to ten year lead over the Soviets in this area,[8] and the SDI might open this lead to several decades. SDI advocates have characterised the Soviet programme in advanced anti-missile research as a 'technology limited' effort, suggesting that there is little opportunity for the massive expansion contemplated by the SDI.

Although the cost of deploying an SDI system (as well as the additional costs for supplementary offensive forces) would not be trivial, it might not exceed the level of effort for strategic forces that the United States experienced under the Eisenhower Administration. During the 1950s the United States spent almost 3 per cent of its Gross National Product (GNP) on strategic forces, about equally divided between offence and defence. The Reagan Administration, while almost doubling the strategic forces budget to over $40 thousand million, is only devoting about 1 per cent of the GNP to this mission. A strategic offence and defence budget of $150 thousand million each year in the 1990s would probably pay for all that would be required for a damage-limiting/denying capability. Although this number may seem high by recent standards, it would represent a levy on the United States economy no greater than that exacted in the 1950s. Some would further argue that the Soviet economy would be unable to bear a similar burden, particularly given the vast technological lead that the United States might enjoy.

Unfortunately the public reticence of many of the advocates of this approach leaves the supposed political or military value of this capability somewhat unclear. In general terms, it could represent a return to the situation of the first decade or so of the nuclear era, when American strategic superiority enabled the United States to use nuclear forces for compellance as well as deterrence. The United States could thus be in a superior position to threaten to escalate a conventional conflict to the nuclear level or to threaten to escalate a tactical nuclear exchange to the strategic level. This advantage could apply both to a European conflict as well as to conflicts in other areas. In this sense, the SDI could enhance extended deterrence (and compellance) through the greater readiness of the United States to threaten the use of nuclear weapons. The level of conflict at which this nuclear compellance could operate would of course depend on the margin of superiority that the United States enjoyed and the political significance of the conflict. A very great asymmetry in strategic capabilities could enable the United States to use nuclear threats even in fairly low-level conflicts in Third World countries.

This is an attractive prospect for those in and around the Reagan Administration who view the past several decades as a time of general American retreat and Soviet expansion and who see the hand of Moscow in virtually every Third World conflict. A usable margin of strategic superiority could enable the United States to confine Moscow's influence to the borders of the Soviet Union, and perhaps place sufficient pressure on the Soviet regime to lead to its ultimate transformation.

However attractive this prospect for some Americans or fearful for some Soviets, it is likely to prove a mirage. The United States is unlikely to achieve a really usable margin of strategic superiority. The Soviets have given notice that they will not acquiesce in such an effort. And the experience of the first two decades of the nuclear era suggests that even if the United States did achieve a major strategic advantage, it would probably find it difficult (if not impossible) to translate this nuclear edge into political gains.

Furthermore, there is a fundamental difference between damage limitation and damage denial. While an integrated offensive and defensive warfighting capability may provide an American damage limitation capability, it is no more likely completely to protect American society than is President Reagan's vision of a perfect astrodome defence. It makes little difference whether the initial phase of the defence is boost-phase or pre-boost phase, as both are vulnerable to the same countermeasures of offensive proliferation, decoys and direct attack, as well as circumvention through such things as the delivery of nuclear weapons by covert means.

Although the United States might be able to steal a technological march over the Soviets for a period of several years, there is little precedent for believing that a condition of marked asymmetry would endure for very long. While the United States has generally led the technological arms race and has maintained an edge in the technical sophistication and 'sweetness' of some of its deployed systems, the Soviets have consistently matched the American capability within a few years. It is difficult to believe that the political promises of a strategy of superiority could be realised during the fleeting window of opportunity that might briefly open, or that any such gains could be maintained once the Soviets managed to close the window.

From 1945 through the mid-1950s the United States enjoyed a virtual nuclear monopoly, and yet found it impossible to derive any tangible political advantage from it. Until the early 1970s the United States maintained a large margin of superiority, which was similarly impotent. Most of the political and military crises of the period were resolved without reference to nuclear weapons. The actual use of nuclear weapons was contemplated only rarely and never seriously. The threatened use of nuclear weapons was rarely a decisive factor in any conflict or crisis.[9]

There is little reason to believe that nuclear weapons would become more usable should the United States regain strategic superiority through the SDI. Although the SDI could provide an enhanced damage-limiting capability, American society would remain at risk. Since the risk from acting would have to be proportional to the risk from not acting, American nuclear compellance would continue to be credible only in the gravest of circumstances. It is unclear that the United States would be any more inclined to use nuclear threats in relatively peripheral Third World situations than it is today.

Even if the United States were inclined to make such threats, their effectiveness is open to question. If the Reagan Administration's view of the Soviet Union as the source of the world's turmoil is correct, then perhaps the Soviets could be compelled to induce their clients to bend to America's will. Recent history suggests, however, that the Soviets have little more control over their clients than does the United States, and that in any event, Soviet involvement in Third World conflicts is more opportunistic than causative. And less developed countries and guerrilla forces offer notoriously poor targets for nuclear weapons. The SDI is essentially irrelevant to the resolution of conflicts such as the war in Central America.

The technological potential for damage limitation and the political will to implement such a strategy are two different things. The political and financial costs, as well as the military risks, would seem to outweigh any transitory gain that might be realised. The reluctance of many of the advocates of this strategy publicly to articulate their position suggests at a minimum that it is highly unlikely that the United States would be able to achieve and maintain the political consensus needed to implement this strategy over the long haul.

V PRESERVING UNITED STATES ASSURED DESTRUCTION

If the SDI would not provide the United States with a splendid first-strike capability against the Soviet Union, perhaps it can ensure that the Soviets do not achieve such a capability against the United States. The Soviets have demonstrated a continued fascination with damage limitation, calling into question the American potential for assured destruction. The 'Window of Vulnerability' debate is generally cast in these terms. The most widely publicised version of this threat is that the Soviets could destroy the bulk of United States bombers and land-based and sea-based missiles and suffer only 'acceptable' damage in retaliation.[10] 'Window' theorists argue that this gives the Soviets the political advantage of greater willingness to resort or threaten to resort to the use of force in Europe and around the world. Some polemicists of this persuasion attribute putative Soviet gains in Asia and Africa in the 1970s to the vulnerability of American strategic forces to Soviet attack. They further argue that a key to reversing Soviet geopolitical momentum is assuring the survival of American ICBMs through deployment of an anti-missile system.

The current and foreseeable Soviet potential for damage limitation is not overly impressive. Soviet counterforce capabilities are confined to 308 deployed SS-18 Mod 4 heavy ICBMs, each with ten accurate 500 kiloton warheads. Although the SS-17 and SS-19 also carry warheads with similar yields, neither has the accuracy needed successfully to attack American ICBM silos (the United States intelligence community downgraded its estimate of the accuracy of the SS-19 in early 1985). The newer solid-fuelled SS-24, which carries ten warheads, will have neither the yield nor the accuracy to threaten ICBM silos.

The SS-18s could probably destroy many, but not all, of the current force of about 1,000 American ICBMs. The exact number is a matter of controversy in the United States and probably of uncertainty in the Soviet Union. Both sides would have some difficulty assessing the reliability of the SS-18s and their warheads, as well as the accuracy of the warheads and the regularity with which they detonate with their planned yield. There are also a number of factors that preclude a precise estimate of the number of silos that would survive an attack, including the precise hardness of each silo against blast, which is difficult to determine with precision, as well as the vulnerability of the silos and their missiles to nuclear effects such as electromagnetic pulse (EMP). The number of surviving American ICBMs could range from a few dozen to several hundred, and in practice the actual value could diverge significantly from the expected value predicted by either side.

Furthermore, ICBM vulnerability would present a problem for strategic stability only if the other legs of the Triad were also vulnerable. But the robust survivability of American SLBMs and bombers should discourage the Soviets from hoping to limit the scope of United States retaliation by attacking missile silos.

'Window' advocates argue that the American SLBM submarines either are or soon will be vulnerable to Soviet strategic anti-submarine warfare (ASW). While the Soviets may have a SSBN vulnerability problem in the form of the American SSN-21 attack submarine, the United States does not seem to have a similar problem. American submarines continue to be significantly quieter than their Soviet counterparts. In addition, American SSBNs can hide in an open ocean patrol area of over 40,000,000 square miles, while Soviet SSBNs are concentrated in a much smaller area in bastions in the Sea of Okhotsk, the Norwegian Sea and the Arctic Ocean. The Soviet bastions can be monitored by United States undersea acoustic sensors and can be penetrated by United States surface ships and land-based ASW aircraft. In contrast, the Soviets lack an ASW acoustic sensor network beyond the vicinity of their shores, and American SSBNs are effectively beyond the reach of Soviet surface and airborne ASW platforms.

Recently the survivability of American SSBNs has been questioned on the basis of Soviet experiments using space-based radars to detect submerged submarines.[11] It is argued that once the SSBNs are localised, they could be destroyed by an open-ocean barrage of Soviet ICBMs. It is rather difficult to give much credence to this threat. Submarines do produce a wake on the surface of the ocean that can be detected by radar, but only if the submarine is moving at high speed near the water's surface. Since American SSBNs typically patrol at low speed and great depth, they are undetectable by this technique. Other non-acoustic detection methods have proved even less promising. At present the United States spends about $10 million each year on submarine survivability research. This paltry budget is indicative of the absence of credible threats to American SSBNs.

The American bomber force is also highly survivable. About one-third of the force is kept on 15-minute alert on a day-to-day basis. In a time of crisis a higher percentage of the force (up to 90 per cent) could be placed on ground alert, and some planes would be kept on airborne alert. In addition, bombers are moved to shorter alert times in response to movements of Soviet submarines closer to the American coast to ensure that Soviet SLBMs would not catch them on the ground.

Although the Soviets have continued to improve their strategic air defences, it is generally acknowledged that most American bombers and cruise missiles would reach their targets.

In the face of the high survivability of American SLBMs and bombers, attacking American ICBMs would not enable the Soviets to limit the damage that could be inflicted by an American retaliatory strike. Consequently, anti-missile systems to protect ICBM sites are not required to preserve US assured destruction capabilities.

VI ESCALATION STABILITY

A more subtle version of the 'Window' argument looks to the special technical and military attributes of American ICBMs and the contribution that they make to the United States strategy of extended deterrence and flexible response.[12] Although the NATO Alliance has never succeeded in formally working out the operational implications of its flexible response strategy, several levels of escalation seem to be contemplated. In the event that deterrence failed and a war started, conventional force would be used in an attempt to prevent Soviet forces from crossing the inter-German border. Long-range conventional munitions (including advanced 'emerging technology' munitions) would be used to disrupt the rear area of the Warsaw Pact. If this was unsuccessful, tactical nuclear weapons would be used on the battlefield and against Warsaw Pact airfields and other military facilities to retard and break up the Soviet advance. Should this fail, American ICBMs would be used to execute Limited Nuclear Options (LNO) against airfields and other military sites in the Soviet Union.

The number of targets attacked by ICBM warheads would probably number between several dozen and a few hundred.

ICBMs are the weapon of choice for executing the LNO component of flexible response for several reasons.[13] Unlike bombers, they can be launched in relatively small numbers with high confidence that their assigned targets will be destroyed. Although a mass bomber raid would successfully penetrate Soviet air defences, a small raid might not fare as well. SLBMs share the ICBMs' certainty of penetration, but they face several shortcomings. Communications with submerged submarines is much more difficult than with ICBM silos. In a protracted conflict the National Command Authority might be unsure whether the submarine chosen to execute the attack was in a position to do so, or even whether the submarine had been sunk by the Soviets. It might also prove difficult to relay targeting data for SLBMs. The submarine would betray its position when it launched its missiles, and those missiles that were not part of the limited attack would be subject to a Soviet counterattack.

If the Soviets were able to destroy the American ICBM force, they could knock several rungs out of the extended deterrence escalation ladder. The United States could then be faced with the choice of either backing down or escalating to large countervalue strikes which would provoke a retaliation in kind.

This is probably the most credible version of the Window of Vulnerability. Although the number of people who profess concern about ICBM vulnerability greatly exceeds the number of people who are believers in limited nuclear warfighting, the degradation of American LNO capabilities is probably the worst consequence of silo vulnerability. One has to make a number of assumptions in order to reach this point of concern over silo vulnerability. Having reached this point, it becomes clear that anti-missile protection of ICBMs is a cure that is worse than the disease.[14]

To begin with, it is far from clear that LNOs are the most pressing question for the defence of NATO, since they are only relevant at very intense levels of conflict. Rectifying potential deficiencies relevant at lower levels of conflict is clearly more important. If NATO has a robust conventional defence, with a tactical nuclear weapons doctrine that makes sense, then there is little likelihood that the need would arise to contemplate executing LNOs. And there is clearly a need for improving NATO's conventional capabilities and clearing up the present mess in tactical nuclear weapons.

Political questions are of even greater importance. An enduring

Alliance consensus on political and strategic questions is a fundamental precondition for a successful and credible military posture, but the perennial crisis in NATO on these matters is becoming increasingly acute. In addition, the fundamental political aim of NATO must be the avoidance of war, which should be of equal if not greater priority than preparations for war. Confidence-building measures and other initiatives to reduce international tension and reduce the likelihood of conflict should be a matter of greatest concern.

Unfortunately, these priorities have been reversed in recent years. Preparation for war has taken precedence over preventing war. And preparations for war at the highest levels of destruction have been given greater attention than measures that would terminate a conflict with the least possible damage to the contestants.

Interest in LNOs also assumes a certain optimism that a limited nuclear war can be kept limited. There is little cause for such optimism.[15] The Second World War quickly evolved into a total conflict, despite efforts on both sides to avoid escalation. The enormous destructive potential of nuclear weapons makes it unlikely that militarily significant attacks could be conducted without major inadvertant collateral damage, including disruption in damage assessment and command and control capabilities. Nuclear weapons are a very blunt instrument to convey political messages, and intentions are difficult to discern in mushroom clouds.

This version of the 'Window' argument also assumes that ICBMs are the unique instrument for executing LNOs and that their destruction would deprive the United States of limited nuclear options. This is clearly not the case. Present SLBM forces are only slightly less capable of being used for LNOs, and with some effort any deficiencies could be rectified. Sea-launched Cruise missiles might be equally useful.

It is further assumed that the Soviets would be prepared to initiate a massive escalation in the scope of a conflict which has been thus far confined to conventional weapons and the limited use of small nuclear weapons against military forces. The jump from dozens of kilotons to thousands of megatons would pose difficult choices, and political leaders would be highly motivated to temporise and prevaricate. No leader would be eager to wager his country on such a 'cosmic roll of the dice'.

This also assumes that the United States would continue to be interested in executing limited attacks following a massive Soviet counterforce strike. But the very extensive collateral damage that an attack of several thousand megatons would produce would surely reduce American interest in LNOs. The Soviets are unlikely to strike United States ICBMs if they know (or fear) that the level of damage to American society would provoke a retaliatory attack on Soviet value targets.

The Midgetman mobile small ICBM (SICBM) can close the Window of Vulnerability sooner than can the SDI, and without degrading American flexible response capabilities. A force of perhaps 200 of these missiles would provide an adequate LNO force. Larger strikes could be executed using the Trident II, since each submarine carries 24 missiles with a total of 192 warheads. Deployment of these mobile missiles over a territory of 10,000 square miles would force the Soviets to strike with several thousand megatons to destroy the entire force. The Soviets would have little confidence that the extensive collateral damage from such a massive attack would not provoke a major American retaliation. Midgetman does not have to be absolutely survivable, as long as its basing mode exacts a sufficiently high attack price that the LNO capabilities of the ICBM are rendered irrelevant.

Midgetman mobile SICBMs can also close the window sooner than can the SDI. Midgetman will begin flight testing in 1987, and initial deployment is scheduled for 1992. The SDI components that could be used to defend ICBM silos (HEDI, ERIS, etc.) will not begin flight testing until 1989, and deployments could not begin until 1996.

Recent concern about the limited payload of Midgetman ignore the underlying military rationale of the system. Current plans call for Midgetman to carry the same 350 kiloton Mk-21 warhead as the MX. The weight of the warhead might limit the range of the missile or prevent the inclusion of simple penetration aids. But for limited attacks with low collateral damage the Mk-21 has an excessive yield. Replacing the Mk-21 with the Mk-4 warhead carried by the Trident I or some other smaller warhead would solve any shortcomings in the performance of Midgetman and make it better suited to its primary mission.

However, one does not necessarily have to agree with the premises or details of the LNO aspects of current United States strategy[16] to recognise that the SDI and potential Soviet BMD deployments could have significant and negative implications for extended deterrence and flexible response. Although it is unlikely that any United States anti-missile system could ever succeed in protecting the

American population from Soviet attack, it is not difficult to imagine that the Soviets could deploy an anti-missile defence that would significantly degrade American extended deterrence and flexible response capabilities.

It would not be difficult for the Soviets to deploy a thin nationwide defence that would, at a minimum, force the United States to take actions to negate the defence (such as increasing the size of the attack to saturate the defence or using chaff and decoys to mask the attack). This would, moreover, reduce or entirely negate the strategic utility of LNO. American and NATO nuclear flexible response strategy depends on the limited and selective application of force, yet the countermeasures required to overwhelm even a thin Soviet ABM system would result in attack that was either too large to be regarded as limited or too completely masked by chaff and decoys to be interpreted as selective.

The silo vulnerability problem is not one of how many warheads survive in the United States, but rather how many warheads reach their targets in the Soviet Union. While an American ABM system could increase the number of surviving warheads, a Soviet ABM system could entirely eliminate the number of warheads that could reach their targets in a small attack. The cure is worse than the disease.

VII CONCLUSION

Although the technical challenges of erecting an 'effective defence' are not trivial, these modest strategic goals for the SDI certainly fall far short of public expectations that the SDI will provide a permanent and perfect shield from the nuclear threat. The SDI now seems to be little more than an effort to perfect 'weapons to defend weapons' rather than 'weapons to defend people'. The question remains, however, whether the cure of defence will be preferable to the malady of vulnerability.

Former Defence Secretary James Schlesinger is frequently quoted to the effect that 'defences that might be desirable are not feasible, and those that are feasible are not desirable'. Although the President's goal of an impermeable shield over Western Civilisation may be an attractive one, there is little reason to expect that it is attainable. The United States has had the technological capacity to erect less capable defences, for instance of ICBM silos, for some time but has chosen not to do so for a variety of political and military reasons. These reasons will continue to be valid for the foreseeable future.

Notes

1. US Army Ballistic Missile Defense Command, *ABM Research and Development at Bell Laboratories Project History* DAHC60-71-C-0005 (Huntsville, Al.: October 1975).
2. John Pike, *The Strategic Defense Initiative Budget and Program*, (Washington, DC: Federation of American Scientists, 10 February 1985).
3. Daniel O. Graham, *High Frontier A New National Strategy* (Washington, DC: Heritage Foundation, 1982).
4. Colin Gray and Keith Payne, 'Victory is Possible', *Foreign Policy*, 39 (1980) p. 14.
5. Paul Nitze, *The Objectives of Arms Control*, Current Policy 677, (Washington, DC: United States State Department, 28 March 1985).
6. Howard Moreland, *US and Soviet 'First Strike' Capabilities*, (Washington, DC: Coalition for a New Foreign and Military Policy, 1985).
7. Walter Andrews, 'Navy Will Seek New Killer Subs to Counter Soviet Missile Fleet,' *Washington Times*, 26 December 1983, p. 2.
8. John Pike, 'Is There an ABM Gap?' *Arms Control Today*, July 1984.
9. Dean Rusk, *et al.*, 'The Lessons of the Cuban Missile Crisis,' *Time*, 27 September 1982, p. 85.
10. Robert Jastrow, 'First Strike,' *The American Legion*, August 1985.
11. Stephen Gibert, 'Our Submarine Fleet is Vulnerable,' *Wall Street Journal*, 19 September 1983, p. 32.
12. Walter Slocombe, 'Extended Deterrence,' *Washington Quarterly* (Fall 1984).
13. James A. Thomson *Planning for NATO's Nuclear Deterrent in the 1980s and 1990s* RAND P-6828 (Santa Monica, Ca.: RAND Corporation, November 1982).
14. Kevin Lewis, *Ballistic Missile Defense, ICBM Modernization, and Small Strategic Attacks: Out of the Frying Pan?*, RAND P-6902 (Santa Monica, Ca.: RAND Corporation, March 1983).
15. Desmond Ball, *Can Nuclear War Be Controlled?* Adelphi Papers 169 (London, International Institute for Strategic Studies, 1981).
16. Herbert Scoville, 'Flexible Madness?' *Foreign Policy* Spring 1974.

Appendix A

Treaty Between the United States of America and the Union of Soviet Socialist Republics on the Limitation of Anti-Ballistic Missile Systems

In the Treaty on the Limitation of Anti-Ballistic Missile Systems the United States and the Soviet Union agree that each may have only two ABM deployment areas,[1] so restricted and so located that they cannot provide a nationwide ABM defense or become the basis for developing one. Each country thus leaves unchallenged the penetration capability of the other's retaliatory missile forces.

The treaty permits each side to have one limited ABM system to protect its capital and another to protect an ICBM launch area. The two sites defended must be at least 1,300 kilometres apart, to prevent the creation of any effective regional defense zone or the beginnings of a nationwide system.

Precise quantitative and qualitative limits are imposed on the ABM systems that may be deployed. At each site there may be no more than 100 interceptor missiles and 100 launchers. Agreement on the number and characteristics of radars to be permitted had required extensive and complex technical negotiations, and the provisions governing these important components of ABM systems are spelled out in very specific detail in the treaty and further clarified in the 'Agreed Statements' accompanying it.

Both parties agreed to limit qualitative improvement of their ABM technology, e.g., not to develop, test, or deploy ABM launchers capable of launching more than one interceptor missile at a time or modify existing launchers to give them this capability, and systems for rapid reload of launchers are similarly barred. These provisions, the Agreed Statements clarify, also ban interceptor missiles with more than one independently guided warhead.

There had been some concern over the possibility that surface-to-air missiles (SAMs) intended for defense against aircraft might be improved, along with their supporting radars, to the point where they could effectively be used against ICBMs and SLBMs, and the treaty prohibits this. While further deployment of radars intended to give early warning of strategic ballistic missile attack is not prohibited, they must be located along the territorial boundaries of each country and oriented outward, so that they do not contribute to an effective ABM defense of points in the interior.

Further, to decrease the pressures of technological change and its unsettling impact on the strategic balance, both sides agree to prohibit development, testing, or deployment of sea-based, air-based, or space-based ABM systems and their components, along with mobile land-based ABM systems. Should future technology bring forth new ABM systems 'based on other physical principles' than those employed in current systems, it was agreed that limiting such systems would be discussed, in accordance with the treaty's provisions for consultation and amendment.

The treaty also provides for a U.S.–Soviet Standing Consultative Commission to promote its objectives and implementation. The commission was established during the first negotiating session of SALT II, by a Memorandum of Understanding dated December 21, 1972. Since then both the United States and the Soviet Union have raised a number of questions in the Commission relating to each side's compliance with the SALT I agreements. In each case raised by the United States, the Soviet activity in question has either ceased or additional information has allayed U.S. concern.

Article XIV of the treaty calls for review of the treaty 5 years after its entry into force, and at 5-year intervals thereafter. The first such review was conducted by the Standing Consultative Commission at its special session in the fall of 1977. At this session, the United States and the Soviet Union agreed that the treaty had operated effectively during its first 5 years, that it had continued to serve national security interests, and that it did not need to be amended at that time.

Note

1. Subsequently reduced to one area (see section on ABM Protocol).

Source

US Arms Control and Disarmament Agency, *Arms Control and Disarmament Agreements: Texts and Histories of Negotiations 1980 Edition* (Washington, DC: US ACDA, August 1980) pp. 137–42.

Treaty Between the United States of America and the Union of Soviet Socialist Republics on the Limitation of Anti-Ballistic Missile Systems

Signed at Moscow May 26, 1972
Ratification advised by U.S. Senate August 3, 1972
Ratified by U.S. President September 30, 1972
Proclaimed by U.S. President October 3, 1972
Instruments of ratification exchanged October 3, 1972
Entered into force October 3, 1972

The United States of America and the Union of Soviet Socialist Republics, hereinafter referred to as the Parties,

Proceeding from the premise that nuclear war would have devastating consequences for all mankind,

Considering that effective measures to limit anti-ballistic missile systems would be a substantial factor in curbing the race in strategic offensive arms and would lead to a decrease in the risk of outbreak of war involving nuclear weapons,

Proceeding from the premise that the limitation of anti-ballistic missile systems, as well as certain agreed measures with respect to the limitation of strategic offensive arms, would contribute to the creation of more favorable conditions for further negotiations on limiting strategic arms,

Mindful of their obligations under Article VI of the Treaty on the Non-Proliferation of Nuclear Weapons,

Declaring their intention to achieve at the earliest possible date the cessation of the nuclear arms race and to take effective measures toward reductions in strategic arms, nuclear disarmament, and general and complete disarmament,

Desiring to contribute to the relaxation of international tension and the strengthening of trust between States,

Have agreed as follows:

Article I

1. Each party undertakes to limit anti-ballistic missile (ABM) systems and to adopt other measures in accordance with the provisions of this Treaty.
2. Each Party undertakes not to deploy ABM systems for a defense of the territory of its country and not to provide a base for such a defense, and not to deploy ABM systems for defense of an individual region except as provided for in Article III of this Treaty.

Article II

1. For the purpose of this Treaty an ABM system is a system to

counter strategic ballistic missiles or their elements in flight trajectory, currently consisting of:

(a) ABM interceptor missiles, which are interceptor missiles constructed and deployed for an ABM role, or of a type tested in an ABM mode;
(b) ABM launchers, which are launchers constructed and deployed for launching ABM interceptor missiles; and
(c) ABM radars, which are radars constructed and deployed for an ABM role, or of a type tested in an ABM mode.

2. The ABM system components listed in paragraph 1 of this Article include those which are:

(a) operational;
(b) under construction;
(c) undergoing testing;
(d) undergoing overhaul, repair or conversion; or
(e) mothballed.

Article III

Each Party undertakes not to deploy ABM systems or their components except that:

(a) within one ABM system deployment area having a radius of one hundred and fifty kilometres and centered on the Party's national capital, a Party may deploy: (1) no more than one hundred ABM launchers and no more than one hundred ABM interceptor missiles at launch sites, and (2) ABM radars within no more than six ABM radar complexes, the area of each complex being circular and having a diameter of no more than three kilometers; and

(b) within one ABM system deployment area having a radius of one hundred and fifty kilometers and containing ICBM silo launchers, a Party may deploy: (1) no more than one hundred ABM launchers and no more than one hundred ABM interceptor missiles at launch sites, (2) two large phased-array ABM radars comparable in potential to corresponding ABM radars operational or under construction on the date of signature of the Treaty in an ABM system deployment area containing ICBM silo launchers, and (3) no more than eighteen ABM radars each having a potential less than the potential of the smaller of the above-mentioned two large phased-array ABM radars.

Article IV

The limitations provided for in Article III shall not apply to ABM systems or their components used for development or testing, and located within current or additionally agreed test ranges. Each Party may have no more than a total of fifteen ABM launchers at test ranges.

Article V

1. Each Party undertakes not to develop, test, or deploy ABM systems or components which are sea-based, air-based, space-based, or mobile land-based.
2. Each Party undertakes not to develop, test, or deploy ABM launchers for launching more than one ABM interceptor missile at a time from each launcher, not to modify deployed launchers to provide them with such a capability, not to develop, test, or deploy automatic or semi-automatic or other similar systems for rapid reload of ABM launchers.

Article VI

To enhance assurance of the effectiveness of the limitations on ABM systems and their components provided by the Treaty, each Party undertakes:

(a) not to give missiles, launchers, or radars, other than ABM interceptor missiles, ABM launchers, or ABM radars, capabilities to counter strategic ballistic missiles or their elements in flight trajectory, and not to test them in an ABM mode; and

(b) not to deploy in the future radars for early warning of strategic ballistic missile attack except at locations along the periphery of its national territory and oriented outward.

Article VII

Subject to the provisions of this Treaty, modernization and replacement of ABM systems or their components may be carried out.

Article VIII

ABM systems or their components in excess of the numbers or outside the areas specified in this Treaty, as well as ABM systems or their components prohibited by this Treaty, shall be destroyed or dismantled under agreed procedures within the shortest possible agreed period of time.

Article IX

To assure the viability and effectiveness of this Treaty, each Party undertakes not to transfer to other States, and not to deploy outside its national territory, ABM systems or their components limited by this Treaty.

Article X

Each Party undertakes not to assume any international obligations which would conflict with this Treaty.

Article XI

The Parties undertake to continue active negotiations for limitations on strategic offensive arms.

Article XII

1. For the purpose of providing assurance of compliance with the provisions of this Treaty, each Party shall use national technical means of verification at its disposal in a manner consistent with generally recognized principles of international law.

2. Each Party undertakes not to interfere with the national technical means of verification of the other Party operating in accordance with paragraph 1 of this Article.

3. Each Party undertakes not to use deliberate concealment measures which impede verification by national technical means of compliance with the provisions of this Treaty. This obligation shall not require changes in current construction, assembly, conversion, or overhaul practices.

Article XIII

1. To promote the objectives and implementation of the provisions of this Treaty, the Parties shall establish promptly a Standing Consultative Commission, within the framework of which they will:

(a) consider questions concerning compliance with the obligations assumed and related situations which may be considered ambiguous;

(b) provide on a voluntary basis such information as either Party considers necessary to assure confidence in compliance with the obligations assumed;

(c) consider questions involving unintended interference with national technical means of verification;

(d) consider possible changes in the strategic situation which have a bearing on the provisions of this Treaty;

(e) agree upon procedures and dates for destruction or dismantling of ABM systems or their components in cases provided for by the provisions of this Treaty;

(f) consider, as appropriate, possible proposals for further increasing the viability of this Treaty; including proposals for amendments in accordance with the provisions of this Treaty;

(g) consider, as appropriate, proposals for further measures aimed at limiting strategic arms.

2. The Parties through consultation shall establish, and may amend as appropriate, Regulations for the Standing Consultative Commission governing procedures, composition and other relevant matters.

Article XIV

1. Each Party may propose amendments to this Treaty. Agreed amendments shall enter into force in accordance with the procedures governing the entry into force of this Treaty.
2. Five years after entry into force of this Treaty, and at five-year intervals thereafter, the Parties shall together conduct a review of this Treaty.

Article XV

1. This Treaty shall be of unlimited duration.
2. Each Party shall, in exercising its national sovereignty, have the right to withdraw from this Treaty if it decides that extraordinary events related to the subject matter of this Treaty have jeopardized its supreme interests. It shall give notice of its decision to the other Party six months prior to withdrawal from the Treaty. Such notice shall include a statement of the extraordinary events the notifying Party regards as having jeopardized its supreme interests.

Article XVI

1. This Treaty shall be subject to ratification in accordance with the constitutional procedures of each Party. The Treaty shall enter into force on the day of the exchange of instruments of ratification.
2. This Treaty shall be registered pursuant to Article 102 of the Charter of the United Nations.

DONE at Moscow on May 26, 1972, in two copies, each in the English and Russian languages, both texts being equally authentic.

FOR THE UNITED STATES OF AMERICA

FOR THE UNION OF SOVIET SOCIALIST REPUBLICS

President of the United States of America

General Secretary of the Central Committee of the CPSU

Appendix B

Agreed Statements, Common Understandings, and Unilateral Statements Regarding the Treaty Between the United States of America and the Union of Soviet Socialist Republics on the Limitation of Anti-Ballistic Missiles

1. Agreed Statements

The document set forth below was agreed upon and initiated by the Heads of the Delegations on May 26, 1972 (letter designations added);

AGREED STATEMENTS REGARDING THE TREATY BETWEEN THE UNITED STATES OF AMERICA AND THE USSR ON THE LIMITATION OF ANTI-BALLISTIC MISSILE SYSTEMS

[A]

The Parties understand that, in addition to the ABM radars which may be deployed in accordance with subparagraph (a) of Article III of the Treaty, those non-phased-array ABM radars operational on the date of signature of the Treaty within the ABM system deployment area for defense of the national capital may be retained.

[B]

The Parties understand that the potential (the product of mean emitted power in watts and antenna area in square meters) of the smaller of the two large phased-array ABM radars referred to in subparagraph (b) of Article III of the Treaty is considered for purposes of the Treaty to be three million.

[C]

The Parties understand that the center of the ABM system deployment area centered on the national capital and the center of the ABM system deployment area containing ICBM silo launchers for each Party shall be separated by no less than thirteen hundred kilometers.

[D]

In order to insure fulfillment of the obligation not to deploy ABM systems and their components except as provided in Article III of the Treaty, the Parties agree that in the event ABM systems based on other physical principles and including components capable of substituting for ABM interceptor missiles, ABM launchers, or ABM radars are created in the future, specific limitations on such systems and their components would be subject to discussion in accordance with Article XIII and agreement in accordance with Article XIV of the Treaty.

[E]

The Parties understand that Article V of the Treaty includes obligations not to develop, test or deploy ABM interceptor missiles for the delivery by each ABM interceptor missile of more than one independently guided warhead.

[F]

The Parties agree not to deploy phased-array radars having a potential (the product of mean emitted power in watts and antenna area in square meters) exceeding three million, except as provided for in Articles III, IV and VI of the Treaty, or except for the purposes of tracking objects in outer space or for use as national technical means of verification.

[G]

The Parties understand that Article IX of the Treaty includes the obligation of the US and the USSR not to provide to other States technical descriptions or blue prints specially worked out for the construction of ABM systems and their components limited by the Treaty.

2. Common Understandings

Common understanding of the Parties on the following matters was reached during the negotiations:

A. Location of ICBM Defenses

The U.S. Delegation made the following statement on May 26, 1972:

Article III of the ABM Treaty provides for each side one ABM system deployment area centered on its national capital and one ABM system deployment area containing ICBM silo launchers. The two sides have registered agreement on the following statement: "The Parties understand that the center of the ABM system deployment area centered on the national capital and the center of the ABM system deployment area containing ICBM silo launchers for each Party shall be separated by no less than thirteen hundred kilometers." In this connection, the U.S. side notes that its ABM system deployment area for defense of ICBM silo launchers, located west of the Mississippi River, will be centered in the Grand Forks ICBM silo launcher deployment area. (See Agreed Statement [C].)

B. ABM Test Ranges

The U.S. Delegation made the following statement on April 26, 1972:

Article IV of the ABM Treaty provides that "the limitations provided for in Article II shall not apply to ABM systems or their components used for development or testing, and located within current or additionally agreed test ranges." We believe it would be useful to assure that there is no misunderstanding as to current ABM test ranges. It is our understanding that ABM test ranges encompass the area within which ABM components are located for test purposes. The current U.S. ABM test ranges are at White Sands, New Mexico, and at Kwajalein Atoll, and the current Soviet ABM test range is near Sary Shagan in Kazakhstan. We consider that non-phased array radars of types used for range safety or instrumentation purposes may be located outside of ABM test ranges. We interpret the reference in Article IV to "additionally agreed test ranges" to mean that ABM components will not be located at any other test ranges without prior agreement between our Governments that there will be such additional ABM test ranges.

On May 5, 1972, the Soviet Delegation stated that there was a common understanding on what ABM test ranges were, that the use of the types of non-ABM radars for range safety or instrumentation was not limited under the Treaty, that the reference in Article IV to "additionally agreed" test ranges was sufficiently clear, and that national means permitted identifying current test ranges.

C. Mobile ABM systems

On January 29, 1972, the U.S. Delegation made the following statement:

Article V(1) of the Joint Draft Text of the ABM Treaty includes an undertaking not to develop, test, or deploy mobile land-based ABM systems and their components. On May 5, 1971, the U.S. side indicated that, in its view, a prohibition on deployment of mobile ABM systems and components would rule out the deployment of ABM launchers and radars which were not permanent fixed types. At that time, we asked for the Soviet view of this interpretation. Does the Soviet side agree with the U.S. side's interpretation put forward on May 5, 1971?

On April 13, 1972, the Soviet Delegation said there is a general common understanding on this matter.

D. Standing Consultative Commission

Ambassador Smith made the following statement on May 22, 1972:

The United States proposes that the sides agree that, with regard to initial implementation of the ABM Treaty's Article XIII on the Standing Consultative Commission (SCC) and of the consultation Articles to the Interim Agreement on offensive arms and the Accidents Agreement,[1] agreement establishing the SCC will be worked out early in the follow-on SALT negotiations; until that is completed, the following arrangements will prevail: When SALT is in session, any consultation desired by either side under these Articles can be carried out by the two SALT Delegations; when SALT is not in session, *ad hoc* arrangements for any desired consultations under these Articles may be made through diplomatic channels.

Minister Semenov replied that, on an *ad referendum* basis, he could agree that the U.S. statement corresponded to the Soviet understanding.

E. Standstill

On May 6, 1972, Minister Semenov made the following statement:

In an effort to accommodate the wishes of the U.S. side, the Soviet Delegation is prepared to proceed on the basis that the two sides will in fact observe the obligations of both the Interim Agreement and the ABM Treaty beginning from the date of signature of these two documents.

In reply, the U.S. Delegation made the following statement on May 20, 1972:

The U.S. agrees in principle with the Soviet statement made on May 6 concerning observance of obligations beginning from date of signature but we would like to make clear our understanding that this means that, pending ratification and acceptance, neither side would take any action prohibited by the agreements after they had entered into force. This understanding would continue to apply in the absence of notification by either signatory of its intention not to proceed with ratification or approval.

The Soviet Delegation indicated agreement with the U.S. statement.

3. Unilateral Statements

The following noteworthy unilateral statements were made during the negotiations by the United States Delegation:

A. Withdrawal from the ABM Treaty

On May 9, 1972, Ambassador Smith made the following statement:

The U.S. Delegation has stressed the importance the U.S. Government attaches to achieving agreement on more complete limitations on strategic offensive arms, following agreement on an ABM Treaty and on an Interim Agreement on certain measures with respect to the limitation of strategic offensive arms. The U.S. Delegation believes that an objective of the follow-on negotiations should be to constrain and reduce on a long-term basis threats to the survivability of our respective strategic retaliatory forces. The USSR Delegation has also indicated that the objectives of SALT would remain unfulfilled without the achievement of an agreement providing for more complete limitations on strategic offensive arms. Both sides recognize that the initial agreements would be steps toward the achievement of more complete limitations on strategic arms. If an agreement providing for more complete strategic offensive arms limitations were not achieved within five years, U.S. supreme interests could be jeopardized. Should that occur, it would constitute a basis for withdrawal from the ABM Treaty. The U.S. does not wish to see such a situation occur, nor do we believe that the USSR does. It is because we wish to prevent such a situation that we emphasize the importance the U.S. Government attaches to achievement of more complete limitations on strategic offensive arms. The U.S. Executive will inform the Congress, in connection with Congressional consideration of the ABM Treaty and the Interim Agreement, of this statement of the U.S. position.

B. Tested in ABM Mode

On April 7, 1972, the U.S. Delegation made the following statement:

Article II of the Joint Text Draft uses the term "tested in an ABM mode," in defining ABM components, and Article VI includes certain obligations concerning such testing. We believe that the sides should have a common understanding of this phrase. First, we would note that the testing provisions of the ABM Treaty are intended to apply to testing which occurs after the date of signature of the Treaty, and not to any testing which may have occurred in the past. Next, we would amplify the remarks we have made on this subject during the previous Helsinki phase by setting forth the objectives which govern the U.S. view on the subject, namely, while prohibiting testing of non-ABM components for ABM purposes: not to prevent testing of ABM components, and not to prevent testing of non-ABM components for non-ABM purposes. To clarify our interpretation of "tested in an ABM mode," we note that we would consider a launcher, missile or radar to be "tested in an ABM mode" if, for example, any of the following events occur: (1) a launcher is used to launch an ABM interceptor missile, (2) an interceptor missile is flight tested against a target vehicle which has a flight trajectory with characteristics of a strategic ballistic missile flight trajectory, or is flight tested in conjunction with the test of an ABM interceptor missile or an ABM radar at the same test range, or is flight tested to an altitude inconsistent with interception of targets against which air defenses are deployed, (3) a radar makes measurements on a cooperative target vehicle of the kind referred to in item (2) above during the reentry portion of its trajectory or makes measurements in conjunction with the test of an ABM interceptor missile or an ABM radar at the same test range. Radars used for purposes such as range safety or instrumentation would be exempt from application of these criteria.

C. No-Transfer Article of ABM Treaty

On April 18, 1972, the U.S. Delegation made the following statement:

In regard to this Article [IX], I have a brief and I believe self-explanatory statement to make. The U.S. side wishes to make clear that the provisions of this Article do not set a precedent for whatever provision may be considered for a Treaty on Limiting Strategic Offensive Arms. The question of transfer of strategic offensive arms is a far more complex issue, which may require a different solution.

D. No Increase in Defense of Early Warning Radars

On July 28, 1970, the U.S. Delegation made the following statement:

Since Hen House radars [Soviet ballistic missile early warning radars] can detect and track ballistic missile warheads at great distances, they have a significant ABM potential. Accordingly, the U.S. would regard any increase in the defenses of such radars by surface-to-air missiles as inconsistent with an agreement.

Note

1. See Article 7 of Agreement to Reduce the Risk of Outbreak of Nuclear War Between the United States of America and the Union of Soviet Socialist Republics, signed Sept. 30, 1971.

Source

US ACDA, *Arms Control and Disarmament Agreements*, 1980 Edition (Washington DC: US ACDA, August 1980) pp. 143–7.

Appendix C

Protocol to the Treaty Between the United States of America and the Union of Soviet Socialist Republics on the Limitation of Anti-Ballistic Missile Systems

At the 1974 Summit meeting, the United States and the Soviet Union signed a protocol that further restrained deployment of strategic defensive armaments. The 1972 ABM Treaty had permitted each side two ABM deployment areas, one to defend its national capital and another to defend an ICBM field. The 1974 ABM Protocol limits each side to one site only.

The Soviet Union had chosen to maintain its ABM defense of Moscow, and the United States chose to maintain defense of its ICBM emplacements near Grand Forks, North Dakota. To allow some flexibility, the protocol allows each side to reverse its original choice of an ABM site. That is, the United States may dismantle or destroy its ABM system at Grand Forks and deploy an ABM defense of Washington. The Soviet Union, similarly, can decide to shift to an ABM defense of a missile field rather than of Moscow. Each side can make such a change only once. Advance notice must be given, and this may be done only during a year in which a review of the ABM Treaty is scheduled. The treaty prescribes reviews every 5 years; the first year for such a review began October 3, 1977.

Upon entry into force, the protocol became an integral part of the 1972 ABM Treaty, of which the verification and other provisions continue to apply. Thus the deployments permitted are governed by the treaty limitations on number and characteristics of interceptor missiles, launchers, and supporting radars.

Source

US ACDA, *Arms Control and Disarmament Agreements* (1980 Edition) (Washington DC: US ACDA, August 1980) pp. 161–63.

Protocol to the Treaty Between the United States of America and the Union of Soviet Socialist Republics on the Limitation of Anti-Ballistic Missile Systems

Signed at Moscow July 3, 1974
Ratification advised by U.S. Senate November 10, 1975
Ratified by U.S. President March 19, 1976
Instruments of ratification exchanged May 24, 1976
Proclaimed by U.S. President July 6, 1976
Entered into force May 24, 1976

The United States of America and the Union of Soviet Socialist Republics, hereinafter referred to as the Parties,

Proceeding from the Basic Principles of Relations between the United States of America and the Union of Soviet Socialist Republics signed on May 29, 1972,

Desiring to further the objectives of the Treaty between the United States of America and the Union of Soviet Socialist Republics on the Limitation of Anti-Ballistic Missile Systems signed on May 26, 1972, hereinafter referred to as the Treaty,

Reaffirming their conviction that the adoption of further measures for the limitation of strategic arms would contribute to strengthening international peace and security,

Proceeding from the premise that further limitation of anti-ballistic missile systems will create more favourable conditions for the completion of work on a permament agreement on more complete measures for the limitation of strategic offensive arms,

Have agreed as follows:

Article I

1. Each Party shall be limited at any one time to a single area out of the two provided in Article III of the Treaty for deployment of anti-ballistic missile (ABM) systems or their components and accordingly shall not exercise its right to deploy an ABM system or its components in the second of the two ABM system deployment areas permitted by Article III of the Treaty, except as an exchange of one permitted area for the other in accordance with Article II of this Protocol.

2. Accordingly, except as permitted by Article II of this Protocol: the United States of America shall not deploy an ABM system or its components in the area centered on its capital, as permitted by Article III(a) of the Treaty, and the Soviet Union shall not deploy an ABM system or its components in the deployment area of intercontinental ballistic missile (ICBM) silo launchers as permitted by Article III(b) of the Treaty.

Article II

1. Each Party shall have the right to dismantle or destroy its ABM system and the components thereof in the area where they are presently deployed and to deploy an ABM system or its components in the alternative area permitted by Article III of the Treaty, provided that prior to initiation of construction, notification is given in accord with the procedure agreed to in the Standing Consultative Commission, during the year beginning October 3, 1977 and ending October 2, 1978, or during any year which commences at five year intervals thereafter, those being the years for periodic review of the Treaty, as provided in Article XIV of the Treaty. This right may be exercised only once.

2. Accordingly, in the vent of such notice, the United States would have the right to dismantle or destroy the ABM system and its components in the deployment area of ICBM silo launchers and to deploy an ABM system or its components in an area centered on its capital, as permitted by Article III(a) of the Treaty, and the Soviet Union would have the right to dismantle or destroy the ABM system and its components in the area centered on its capital and to deploy an ABM system or its components in an area containing ICBM silo launchers, as permitted by Article III(b) of the Treaty.

3. Dismantling or destruction and deployment of ABM systems or their components and the notification thereof shall be carried out in accordance with Article VIII of the ABM Treaty and procedures agreed to in the Standing Consultative Commission.

Article III

The rights and obligations established by the Treaty remain in force and shall be complied with by the Parties except to the extent modified by this Protocol. In particular, the deployment of an ABM system or its components within the area selected shall remain limited by the levels and other requirements established by the Treaty.

Article IV

This Protocol shall be subject to ratification in accordance with the constitutional procedures of each Party. It shall enter into force on the day of the exchange of instruments of ratification and shall thereafter be considered an integral part of the Treaty.

DONE at Moscow on July 3, 1974, in duplicate, in the English and Russian languages, both texts being equally authentic.

For the United States of America:

RICHARD NIXON

President of the United States of America

For the Union of Soviet Socialist Republics:

L. I. BREZHNEV

General Secretary of the Central Committee of the CPSU

Appendix D

Standing Consultative Commission Agreements

1. *Memorandum of Understanding Establishing SCC* (December 21, 1972, Unclassified).
This memorandum implements Article XIII of the ABM Treaty and Article VI of the Interim Agreement by establishing the Standing Consultative Commission.

2. *Protocol Establishing Regulations for the Standing Consultative Commission* (May 30, 1973, Unclassified).
This protocol formally adopts the regulations worked out for the SCC, pursuant to the Article XIII of the ABM Treaty and Article VI of the Interim Agreement.

4. *Protocol on Procedures for ABM Systems* (July 3, 1974, Secret).
This protocol is an implementation of the ABM Treaty and establishes procedures for the replacement and dismantling or destruction of ABM systems.

6. *Supplementary Protocol on ABM Procedures* (October 28, 1976, Secret).
This supplementary protocol is in implementation of the ABM Treaty and establishes procedures for the replacement, dismantling, or destruction of operational ABM systems and for the exchange of ABM deployment areas.

9. *Agreed Statement on ABM Treaty Topics* (November 1, 1978, Secret)
This Agreed Statement defines test ranges within the meaning of Article IV of the ABM Treaty on the basis of the presence of ABM components for testing, identifies the current test ranges for each side, and sets forth procedures of notifying the other Party when a new test range is established. The Agreed Statement also specifies criteria for applying the term "tested in an ABM mode" as used in the ABM Treaty to missiles, launchers, and radars. Finally, the Agreed Statement specifies that the ABM Treaty permits air defense radars located at ABM test ranges to carry out air defense functions, but to avoid ambiguous situations or misunderstandings the sides will refrain from concurrent testing of air defense components and ABM system components co-located at the same test range, and air defense radars utilized as instrumentation equipment will not be used to make measurements on strategic ballistic missiles.

Source

U.S. Senate Committee on Foreign Relations, *SALT II Treaty: Background Documents/*"Miscellaneous Agreements Relating to the Standing Consultation Commission:" forward from J. Brian Atwood, Dept. of State, to Senator Frank Church, November 13, 1979.

Bibliography

This bibligrapy provides not only a first general introduction to *Star Wars and European Defence* but also a supplement to the literature referred to in the individual chapters. It covers a selection of monographs, readers and contributions to scientific journals and magazines – primarily in the English language – that had been published up to November 1986.

For more extensive bibliographical references in both English and German see also:

Brauch, Hans Günter and Fischbach, Rainer, *Military Use of Outer Space: A Research Bibliography* (Stuttgart: AG Friedensforschung und Europäische Sicherheitspolitik, Institut für Politikwissenschaft, 1986) (AFES Papier no. 4). This 78-page bibliography covers entries up to February 1985.

Brauch, Hans Günter and Fischbach, Rainer, *Militärische Nutzung des Weltraums: Eine Forschungsbibliographie* (Berlin: Berlin Verlag, Arno Spitz [Reihe Militärpolitik und Rüstungsbegrenzung, Band 8], forthcoming). (This approximately 250-page bibliography covers English and German entries up to December 1986.)

The American SDI debate has been covered in:

Lawrence, Robert M., and Reynolds, Sally M., *The SDI Bibliography and Reference Guide* (Boulder: Westview, 1986).

For more recent bibliographical references on SDI the reader should check, for example, the bibliographical sections of the following journals:

Arms Control Today, a monthly journal of the Arms Control Association, whose address is 11 Dupont Circle, Washington, DC, 20036, USA.

ADIU Report, a bi-monthly journal of the Armament and Disarmament Information Unit, SPRU, Mantell Building, University of Sussex, Falmer, Brighton BN1 9RF, United Kingdom.

For detailed up-to-date bibliographies on the ABM Treaty see also:

National Campaign to Save the ABM Treaty, *Briefing Book on the ABM Treaty and Related Issues* (Washington, National Campaign, 1601 Connecticut Avenue, NW, # 704, Washington, DC, 20009, USA, 1986).

Military Use of Outer Space – Scientific and Technological Reference Books

Air Command and Staff College, *Space Handbook* (Maxwell Air Force Base: Air University Press, January 1985).

Angelo, Joseph A., Jr, *The Dictionary of Space Technology* (New York: Facts on File, 1982).
Baker, David, *The Rocket: The History and Development of Rocket and Missile Technology* (London: Crown, 1978).
Beckett, Brian, *Weapons of Tomorrow* (London: Orbis, 1982).
Cambridge Encyclopaedia of Astronomy, The (London: Cape, 1979).
Friedman, Richard S.; Gunston, Bill; Hobbs, David; Miller, David; Richardson, Doug, and Walmer, Max, *Advanced Technology Warfare* (New York: Harmony Books, and London: Salamander Books, 1985).
Gatland, Kenneth (ed.) *The Illustrated Encyclopedia of Space Technology* (London: Salamander Books, 1981).
Gunston, Bill, *Jane's Aerospace Dictionary* (London: Jane's Publishing Company, 1980).
Hecht, Jeff, *Beam Weapons: The Next Arms Race* (New York and London: Plenum Press, 1984).
Pardoe, Geoffrey, KC, *The Future for Space Technology* (London, and Dover New Hampshire: Pinter, 1984).
Parmentola, J., and Tsipis, Kosta, 'Particle Beam Weapons', *Scientific American*, vol. 240, no. 4 (April 1979) pp. 54–65.
Schroer, Dietrich, *Science, Technology, and the Nuclear Arms Race* (New York, Chichester, Brisbane, Toronto, Singapore: Wiley, 1984).
Tsipis, Kosta, *Arsenal – Understanding Weapons in the Nuclear Age* (New York: Simon & Schuster, 1983).
Tsipis, Kosta, 'Laser Weapons', *Scientific American*, vol. 245, no. 6 (December 1981) pp. 51–7.
Turnill, Reginald (ed.) *Jane's Spaceflight Directory* (London: Jane's Publishing Company, 1984).
Turnill, Reginald (ed.) *Jane's Spaceflight Directory 1986* (London: Jane's Publishing Company, 1986).
United Nations, Ralph Chipman (ed.) *The World in Space: A Survey of Space Activities and Issues* (Englewood Cliffs, New Jersey: Prentice-Hall, 1982).

History of the ABM/BMD Issues and of the Military Use of Outer Space

Barnaby, Frank and Boserup, Anders (eds), *Implications of the Anti-Ballistic Missile Systems* (London: Souvenir Press, 1969).
Cahn, Anne Hessing, *Eggheads and Warheads: Scientists and the ABM* (Cambridge, Massachusetts: MIT Center for International Studies, 1971).
Chayes, Abram, and Wiesner, Jerome B. (eds) *ABM: An Evaluation of the Decision to Deploy an Antiballistic Missile System* (New York: Harper & Row, 1969).
Garwin, Richard L., and Pike, John, 'Space Weapons History and Current Debate', *Bulletin of the Atomic Scientists*, vol. 40, no. 5 (May 1984) pp. 25–9.
Holst, Johan J., and Schneider, William, Jr, (eds) *Why ABM? Policy Issues in the Missile Defense Controversy* (New York: Pergamon, 1969).

Jayne, Edward Randolph, II, 'The ABM Debate: Strategic Defense and National Security', Ph.D. dissertation, Massachusetts Institute of Technology, 1969.

Jeffers, Harvey Paul, *How the US Senate Works the ABM Debate* (New York: McGraw Hill, 1970).

Killian, James Rhyne, *Sputnik, Scientists and Eisenhower: A Memoir* (Cambridge, Massachusetts, MIT Press, 1972).

Logsdon, John M., *The Decision to Go to the Moon: Project Apollo and the National Interest* (Chicago: University of Chicago Press, 1970).

Manno, Jack, *Arming the Heavens: The Hidden Military Agenda for Space 1945–1995* (New York: Dodd & Mead, 1984).

McDougall, Walter A., . . . *The Heavens and the Earth: A Political History of the Space Age* (New York: Basic Books, 1985).

Oberg, James E., *The New Race for Space: The US and Russia Leap to the Challenge for Unlimited Rewards* (Harrisburg: Stackpole, 1984).

Ordway, Frederick I., III, and Sharpe, Mitchell R., *The Rocket Team: From the V-2 to the Saturn Moon Rocket* (Cambridge, Massachusetts: MIT Press, 1982).

Schichtle, Cass, *The National Space Program: From the Fifties into the Eighties* (Washington, DC: The National Defense University, 1983).

Stein, Jonathan, B., *From H-Bomb to Star Wars: The Politics of Strategic Decision-Making* (Lexington, Massachusetts: Lexington Books, 1984).

Stoffer, Howard, *Congressional Defense Policy-Making and the Arms Control Community: The Case of the Anti-Ballistic Missile*, Ph.D. dissertation, Columbia University, New York, 1980.

US Congress, House Committee on Science and Technology, Subcommittee on Space Science and Applications, *Space Activities of the United States, Soviet Union and Other Launching Countries/Organizations, 1957–1981*, 97th Congress, 2nd session (Washington, DC: US Government Printing Office, 1982).

Yanarella, Ernest J., *The Missile Defense Controversy. Strategy, Technology, and Politics, 1955–1972* (Lexington: University of Kentucky Press, 1977).

Ballistic Missile Defense and Space Warfare: The US Debate

Adelman, Kenneth L., Director, Arms Control and Disarmament Agency, 'What's Next for Strategic Stability and Arms Control', Speech to the International Institute for Strategic Studies in London, 13 February 1985.

Abrahamson, James A., 'The Strategic Defense Initiative', *Defense/84*, August 1984.

Baker, David, *The Shape of Wars to Come* (Feltham: Hamlyn Press, 1981).

Bellamy, Ian, and Blacker, Coit (eds) *Anti-Ballistic Missile Defence in the 1980s* (London: Cass, 1983).

Bethe, Hans A., et al., 'Space-Based Ballistic Missile Defense'. *Scientific American*, vol. 251, no. 4, 1984, pp. 39–49.

Bethe, Hans, Boutwell, Jeffrey and Garwin, Richard L., 'BMD Technologies and Concepts in the 1980s', *Daedalus*, vol. 114, no. 2 (Spring 1985) pp. 53–72.

Bethe, Hans and Garwin, Richard L., 'Appendix A: New BMD Technologies', *Daedalus*, vol. 114, no. 3 (Summer 1985) pp. 331–68.

Blacker, Coit D., 'Defending Missiles, Not People: Hard-Site Defense', *Issues in Science and Technology* (Fall 1985) pp. 30–44.
Boutwell, Jeffrey, and Scribner, Richard A., *The Strategic Defense Initiative: Some Arms Control Implications* (Washington, DC: American Association for the Advancement of Science, May 1985).
Bowman, Robert M., *Star Wars: Defense or Death Star* (Potomac, Maryland: Institute for Space and Security Studies, 1985).
Bowman, Robert M., 'Arms Control in Space: Preserving Critical Strategic Space Systems Without Weapons in Space', *Air University Review* (November/December, 1985) pp. 58–72.
Brauch, Hans Günter, 'Allgemeine Entwicklungslinien der Waffentechnik: Weltraumrüstung der USA', *Die Friedens-Warte*, vol. 65 (1982–5) pp. 13–27.
Brennan, Donald M., 'The Case for Missile Defense', *Foreign Affairs*, vol. 47, no. 3 (1969) pp. 433–48.
Broad, William J., *Star Warriors* (New York: Simon & Schuster, 1985).
Brown, Harold, 'The Strategic Defense Initiative: Defensive Systems and the Strategic Debate', *Survival* vol. 27, no. 2, March/April 1985) pp. 55–64
Brown, Harold, 'Is SDI technically feasible?', *Foreign Affairs*, vol. 64, no. 3 (1986) pp. 435–54.
Brzezinski, Zbigniew, Jastrow, Robert and Kampelman, Max M., 'Defense in Space is Not Star Wars', *New York Times Magazine*, 27 January 1985.
Bundy, McGeorge, *et al*, 'The President's Choice: Star Wars or Arms Control', *Foreign Affairs*, 63 (Winter 1984/5) pp. 264–78.
Burrows, Robert M., 'Ballistic Missile Defence: The Illusion of Security', *Foreign Affairs*, vol. 62, no. 4 (Spring 1984) pp. 843–56.
Canan, James, *War in Space* (New York: Berkeley Publishing Group, 1983).
Carter, Ashton B. and Schwartz, David N. (eds) *Ballistic Missile Defense* (Washington, DC: Brookings Institution, 1984).
Cimbala, Stephen J., 'The Strategic Defense Initiative: Political Risks', *Air University Review*, (November/December 1985) pp. 24–37.
Clausen, Peter A., 'SDI in Search of A Mission', *World Policy Journal*, (Spring 1985) pp. 249–303.
Cooper, Robert S., 'No Sanctuary: A Defense Perspective on Space', *Issues in Science and Technology* (Spring 1986) pp. 38–45.
Council of Economic Priorities, *The Strategic Defense Initiative. Costs, Contractors and Consequences* (New York: Council of Economic Priorities, 1985).
DeLauer, Richard D., *The Strategic Defense Initiative: Defensive Technologies Study* (Washington, DC: US Department of Defense, 1984).
Domenici, Pete V., 'Toward a Decision on Ballistic Missile Defense', *Strategic Review*, vol. 10, no. 1 (1982) pp. 22–7.
Downey, Arthur J., *The Emerging Role of the US Army in Space* (Washington: National Defense University Press, 1985).
Drell, Sidney, Farley, Philip J. and Holloway, David, *The Reagan Strategic Defense Initiative: A Technical, Political, and Arms Control Assessment*, (Stanford Center for International Security and Arms Control, Stanford University, 1984).
Drell, Sidney D. and Panofsky, Wolfgang, 'The Case Against: Technical and Strategic Realities', *Issues in Science and Technology*, vol. 1, no. 1 (Fall 1984) pp. 45–65.

Drell, Sidney, and Johnson, Thomas H. (eds) *Strategic Missile Defense: Necessities, Prospects, and Dangers in the Near Term*, report of a workshop at the Center for International Security and Arms Control, Stanford University, April 1985.

Durch, William J. (ed.) *National Interests and the Military Use of Space* (Cambridge Massachusetts: Harvard University Press, 1984).

Fletcher, James C., 'Technologies for Strategic Defense', *Issues in Science and Technology*, vol. 1, no. 1 (Fall 1984) pp. 15–29.

Fought, Stephen O., 'SDI: A Policy Analysis', *Naval War College Review* (November/December 1985) pp. 59–95.

Freedman, Lawrence, *The Evolution of Nuclear Strategy* (London: Macmillan, 1981).

Gallis, Paul. E., Lowenthal, Mark M., and Smith, Marcia S., *The Strategic Defense Initiative and United States Alliance Strategy*, report no. 85–48-F (Washington: Congressional Research Service, US Congress, 1 February 1985.

Garfinkle, Adam M., 'The Politics of Space Defense', *Orbis*, vol. 28, no. 2 (Summer 1984) pp. 240–57.

Garwin, Richard L., 'Countermeasure: Defeating Space-based Defense', *Arms Control Today*, (May 1985).

Glaser, Charles L., 'Why Even Good Defenses May Be Bad', *International Security*, vol. 9, no. 2 (1984) pp. 92–123.

Gouré, Daniel, 'Strategic Offense and Defense: Enhancing the Effectiveness of US Strategic Forces', *Annals of the American Academy of Political and Social Science*, no. 457 (September 1981) pp. 28–45.

Graham, Daniel, *High Frontier: A New National Strategy* (Washington: High Frontier, 1982).

Graham, Daniel, *High Frontier: A Strategy for National Survival* (New York: Tom Doherty Associates, 1983).

Graham, Daniel O., and Fossedal, Gregory A., *Defense That Defends; Blocking Nuclear Attack* (Greenwich, CT: Devin-Adair, 1983).

Gray, Colin S., *American Military Space Policy: Information Systems, Weapon Systems and Arms Control* (Cambridge, Massachusetts: Abt Books, 1982).

Gray, Colin S., 'Space is Not a Sanctuary', *Survival*, vol. 25, no. 5 (1983) pp. 194–203.

Gray, Colin S., 'Strategic Defenses. A Case for Strategic Defense', *Survival*, vol. 27, no. 2 (March/April 1985) pp. 50–4.

Guertner, Gary L., 'Star Wars: What is "Proof"?', *Foreign Policy*, no. 59 (Summer 1985) pp. 73–84.

Guertner, Gary L. and Snow, Donald M., *The Last Frontier: An Analysis of the Strategic Defense Initiative* (Lexington, Massachusetts: Lexington Books, 1986).

Haley, Edward, and Merrit, Jack, *Strategic Defense, Folly or Future* (Boulder, Colorado: Westview, 1986).

Hartung, William D., *et al.*, *The Strategic Defense Initiative: Costs, Contractors and Consequences*, (New York: Council on Economic Priorities, 1985).

Hoffman, Fred S., *Ballistic Missile Defenses and US National Security: Summary Report*, prepared for the Future Security Strategy Study, Washington, DC: Department of Defense, October 1983.

Hoffman, Fred S., 'The SDI in U.S. Nuclear Strategy. Senate Testimony', *International Security*, vol. 10, no. 1 (Summer 1985) pp. 25–33.
Iklé, Fred C., 'Nuclear Strategy: Can There be a Happy Ending?', *Foreign Affairs*, vol. 63, no. 4 (Spring 1985) pp. 810–26.
Jasani, Bhupendra (ed.) *Outer Space – Battlefield of the Future?* (London: Taylor & Francis, 1978).
Jasani, Bhupendra (ed.) *Outer Space – A New Dimension of the Arms Race* (London: Taylor & Francis, 1982).
Jasani, Bhupendra (ed). *Space Weapons – The Arms Control Dilemma* (London and Philadelphia: Taylor & Francis, 1984).
Jasani, Bhupendra and Lee, Christopher, *Countdown to Space War* (London and Philadelphia: Taylor & Francis, 1984).
Jastrow, Robert, *How to Make Nuclear Weapons Obsolete* (Boston and Toronto: Little, Brown and Co., 1985).
Kalish, Jack, 'The Technologies of Hard-Site Defense', *Issues in Science and Technology*, vol. II, no. 2 (Winter 1986) pp. 122–7.
Karas, Thomas, *The New High Ground: Strategies and Weapons of Space-Age War* (New York: Simon & Schuster, 1983).
Keeney, Spurgeon, Jr., and Panofsky, Wolfgang K. H., 'MAD versus NUTS', *Foreign Affairs*, vol. 60, no. 2, (1982) pp. 287–304.
Keyworth, George, 'Reassessing Strategic Defense' (Washington, DC: Council of Foreign Relations, 1984).
Keyworth, George A., II, 'A Sense of Obligation – the Strategic Defense Initiative', *Aerospace America* (April 1984).
Keyworth, George A., II, 'Strategic Defense Initiative: The Rational Route to Effective Nuclear Arms Control', *Government Executive* (June 1984).
Keyworth, George A., II, 'The Case For: An Option for a World Disarmed', *Issues in Science and Technology*, vol. 1, no. 1 (Fall 1984) pp. 30–44.
Keyworth, George, A., II, 'The Case for Arms Control and the Strategic defense Initiative', *Arms Control Today* (April 1985) pp. 1, 2, 8.
Lebow, Richard Ned, 'Assured Strategic Stupidity: The Quest for Ballistic Missile Defense', *Journal of International Affairs*, vol. 39, no. 1 (Summer 1985) pp. 57–80.
Lin, Herbert, 'The Development of Software for Ballistic-Missile Defense', *Scientific American* (December 1985) pp. 46–53.
Long, Franklin A., Hafner, Donald, Boutwell, Jeffrey (eds) *Weapons in Space* (New York and London: Norton, 1986).
Lupton, David, 'Space Doctrine', *Strategic Review*, vol. 11, no. 4 (1983) pp. 36–47.
McNamara, Robert S., and Bethe, Hans A., 'Reducing the Risk of Nuclear War', *The Atlantic Monthly*, July 1985, pp. 43–51.
Mische, Patricia M., *Star Wars and the State of Our Souls* (East Orange, New Jersey: Global Education Associates, 1984).
Myers, Ware, 'The Star Wars software debate', *Bulletin of the Atomic Scientists*, vol. 42, no. 2 (February 1986) pp. 31–6.
Nitze, Paul H., Special Adviser to the President and the Secretary of State for Arms Reductions, 'On the Road to a More Stable Peace', speech to the World Affairs Council, in Philadelphia, Pennsylvania, 20 February 1985.
Office of Technology Assessment, US Congress, *Ballistic Missile Defense Technologies*, OTA–ISC–254 (Washington, DC: US Government Printing Office, September 1985).

Office of Technology Assessment, US Congress, *Directed Energy Missile Defense in Space* (Washington, DC: US Government Printing Office, April 1984).

Panofsky, Wolfgang, 'SDI: Perceptions versus Reality', *Physics Today* (June 1985) pp. 34–45.

Papp, Dan S., *Ballistic Missile Defense, Space-Based Weapons, and the Defense of the West*, AD–A140 280/9 (Alexandria, Virginia: Department of Commerce, NTIS, 1983).

Parnas, David L., 'Software Wars. Ein offener Brief', *Kursbuch*, No. 83 (March 1986) pp. 49–69.

Payne, Keith B. (ed.) *Laser Weapons in Space: Policy and Doctrine* (Boulder, Colorado: Westview Press, 1983).

Payne, Keith B., *Strategic Defense: 'Star Wars' in Perspective* (Lanham, Massachusetts: Hamilton, 1986).

Payne, Keith B., 'Strategic Defense and Stability', *Orbis*, vol. 28, no. 2 (1984) pp. 215–27.

Payne, Keith B. and Gray, Colin S., 'Nuclear Policy and the Defensive Transition', *Foreign Affairs*, vol. 62, no. 4 (Spring 1984) pp. 820–42.

Perle, Richard N., 'The Strategic Defense Initiative: Addressing Some Misconceptions', *Journal of International Affairs*, vol. 39/no. 1 (Summer 1985) pp. 23–30.

Pike, John, *The Strategic Defense Initiative Budget and Program* (Washington DC: Federation of American Scientists, July 1985).

Poole, Randall A., 'Ballistic Missile Defense and Strategic Deterrence', *National Defense* (November 1985) pp. 39–49.

Pournelle, Jerry E., *Mutual Assured Survival* (New York: Baen, 1984).

Ra'anan, Uri, and Pfaltzgraff, Robert L. Jr. (eds) *International Security Dimensions of Space* (Hamden, Connecticut: Archon, 1984).

Rathjens, George, 'The ABM Debate', in Brodie, Bernard; Intriligator, Michael D. and Kolkowicz, Roman (eds) *National Security and International Stability* (Cambridge Massachusetts: Oelgeschlager, Gunn & Hain, 1983) pp. 379–406.

Rathjens, George and Ruina, Jack, 'BMD and Strategic Instability', *Daedalus*, vol. 114, no. 3 (Summer 1985) pp. 239–56.

Reagan, Ronald, *The President's Strategic Defense Initiative* (Washington DC: White House pamphlet issued 3 January 1985).

Richardson, Robert C. III, 'Conceptual and Political Overview of High Frontier and the Strategic Defence Initiative', *EIS-Journal*, no. 4–5 (1985) pp. 8–28.

Ritchie, David, *Space War* (New York: Atheneum, 1982).

Rosen, Stephen Peter, 'Nuclear Arms and Strategic Defense', *Washington Quarterly*, vol. 4, no. 2 (1981) pp. 82–99.

Ruina, Jack, 'Perspectives on Hard-Site Defense', *Issues in Science and Technology* (Winter 1986) pp. 128–133.

Schlesinger, James R., speech on ballistic missile defense at Symposium on Space, National Security and C-cubed-I (Bedford, Massachusetts: The Mitre Corporation: 25 Oct, 1984. Mitre Document M 85–3) pp. 55–62.

Schlesinger, James R., 'Rhetoric and realities in the Star Wars debate', *International Security*, vol. 10, no. 1 (1985) pp. 3–12.

Schneider, William *et al.*, *U.S. Strategic Nuclear Policy and Ballistic Missile*

Defense: The 1980s and Beyond (Cambridge, Massachusetts: Institute for Foreign Policy Analysis 1980).

Scott, Armstrong and Grier, Peter, *Strategic Defense Initiative: Splendid Defense or Pipe Dream?* (New York: Foreign Policy Association, 1986).

Scowcroft, Brent, et al., *Report of the President's Commission on Strategic Forces* (Washington, DC: US Government Printing Office, 6 April 1983).

Scowcroft, Brent, et al., *Second Report of the President's Commission on Strategic Forces* (Washington, DC: US Government Printing Office, 21 March 1984).

Sloss, Leon, 'The Return of Strategic Defense', *Strategic Review*, vol. 12, no. 2 (1984) pp. 37–44.

Slocombe, Walter, 'An Immediate Agenda for Arms Control (Or what to do until Darth Vader comes)', *Survival*, vol. 27, no. 5 (1985) pp. 204–13.

Snow, Donald M., 'Ballistic Missile Defense and the Strategic Future', *Parameters*, vo. 13, no. 2 (June 1983) pp. 11–22.

Stanford University, *Strategic Missile Defense: Necessities, Prospects, and Dangers in the Near Term* (Stanford: Center for International Security and Arms Control, April 1985).

Star Wars Quotes (Washington, DC: Arms Control Association, 1986).

Stares, Paul B., *The Militarization of Space: US Policy, 1945–1984* (Ithaca, New York: Cornell University Press, 1985).

Stares, Paul B., *Space Weapons and US Strategy: Origins and Development* (London and Sydney: Croom Helm, 1985).

Starsman, Raymond E., *Ballistic Missile Defense and Deceptive Basing: A New Calculus for the Defense of ICBMs* (Washington: National Defense University Press, 1981).

Stine, Harry, G., *Confrontation in Space: Wars of the Future will be Fought in Space* (Englewood Cliffs, New Jersey: Prentice-Hall, 1981).

Sundaram, G. S., 'Military Space Programs: Emphasis on Survivability', *International Defense Review* vol. 17, no. 8 (1984) pp. 1019–30.

Tirman, John (ed.) *The Fallacy of Star Wars: Based on Studies Conducted by the Union of Concerned Scientists* (New York: Vintage Books/Random House, 1984).

Tirman, John (ed.), *Empty Promise – The Growing Case Against Star Wars – The Union of Concerned Scientists* (Boston: Beacon Press, 1986).

US Air Force Academy, *Book of Readings for the USAFA Military Space Doctrine Symposium* (Colorado Springs, Colorado: USAF Academy, 1981).

US Congress, House, Committee on Banking, Finance and Urban Affairs, *The Impact of Strategic Defense Initiative (SDI) on the US industrial base*, Hearings, 10 December 1986 (Washington, DC: US Government Printing Office, 1986) Ser. no. 99–59.

US Congress, Senate Committee on Foreign Relations, *Strategic Defense and Anti-Satellite Weapons* (Washington, DC: US Government Printing Office, 25 April 1984) p. 361.

US Department of Defense, *Overview of Strategic Defense Initiative: Fact Sheet* (Washington, DC: US Department of Defense, 9 March 1984).

US Department of Defense, *The Strategic Defense Initiative, Defensive Technologies Study* (Washington, DC: US Department of Defense, March 1984).

US Department of Defense, *Defense Against Ballistic Missiles: An Assessment of Technologies and Policy Implications*. Washington, DC: US Department of Defense, April 1984).
US Department of Defense, *Report to the Congress on the Strategic Defense Initiative* (Washington: US Department of Defense, 1985).
US Department of Defense, *Report to the Congress on the Strategic Defense Initiative* (Washington: US Department of Defense, 1986).
Weinberg, Alvin and Barkenbus, Jack H., 'Stabilizing Star Wars', *Foreign Policy*, no. 54 (Spring 1984) pp. 164–71.
Weinberger, Casper W., *Report of the Secretary of Defense to the Congress* (Washington DC: Department of Defense, 4 February 1985).
White House, *Fact Sheet on The Strategic Defense Initiative* (Washington, DC: White House, 1 June 1983).
Yonas, Gerold, 'The Strategic Defense Initiative', *Daedalus*, vol. 114, no. 2 (Spring 1985) pp. 73–90.
Yonas, Gerold, 'SDI: The Politics and Science of Weapons in Space', *Physics Today*, June 1985, pp. 24–32.
Yonas, Gerold, 'Research and the Strategic Defense Initiative', *International Security*, vol. 11, no. 2 (Fall 1986) pp. 185–9.
York, Herbert F., 'Nuclear Deterrence and the Military Use of Space', *Daedalus*, vol. 114, no. 2 (Spring 1985) pp. 17–32.
Zimmerman, Peter D., 'Pork bellies and SDI', *Foreign Policy*, no. 63 (1986) pp. 76–87.

Soviet BMD and Military Space Activities

Berman, Robert P., and Baker, John C., *Soviet Strategic Forces: Requirements and Responses* (Washington: The Brookings Institution, 1982).
Davis, Jaquelyn K. *et al.*, *The Soviet Union and Ballistic Missile Defense* (Cambridge, Massachusetts: Institute for Foreign Policy Analysis, 1980).
Deane, Michael J., *The Role of Strategic Defense in Soviet Strategy* (Coral Gables, Florida: Advanced International Studies Institute in association with the University of Miami, 1980).
Ermarth, Fritz W., 'Contrasts in American and Soviet Strategic Policies', *International Security*, vol. 3, no. 2 (Fall 1978) pp. 138–55.
Garthoff, Raymond, L., 'Mutual Deterrence and Strategic Arms Limitation in Soviet Policy', *International Security*, vol. 3, no. 1 (1978) pp. 112–47.
Garthoff, Raymond L., 'BMD and East–West Relations', in Carter, Ashton B., and Schwartz, David N. (eds) *Ballistic Missile Defense* (Washington, DC: The Brookings Institution, 1984).
Garthoff, Raymond L., *Detente and Confrontation: American–Soviet Relations from Nixon to Reagan* (Washington, DC: The Brookings Institution, 1985).
Gouré, Daniel, and McCormick, Gordon H., 'Soviet Strategic Defense: the Neglected Dimension of the US–Soviet Balance', *Orbis*, vol. 24, no. 1, (Spring 1980) pp. 103–27.
Gouré, Leo and Deane, J. J., 'The Soviet Strategic View: Target – the US Strategic Defense Initiative', *Strategic Review*, vol. 13, no. 2 (Spring 1985) pp. 73–6.

Graybeal, Sidney, and Gouré, Daniel, 'Soviet Ballistic Missile Defense Objectives: Past, Present, and Future', in *US Arms Control Objectives and the Implications for Ballistic Missile Defense*, proceedings of a symposium at Harvard University in November, 1979 (Cambridge, Massachusetts: Harvard University Center for Science and International Affairs, 1980).
Green, William C., *Soviet Nuclear Weapons Policy: A Research Guide* (Boulder, Colorado: Westview Press, 1985).
Herspring, Dale, and Laird, Robin, *The Soviet Union and Strategic Arms* (Boulder, Colorado: Westview Press, 1984).
Holloway, David, *The Soviet Union and the Arms Race* (New Haven, Connecticut: Yale University Press, 1983).
Holloway, David, 'The Soviet Union and the SDI', *Daedalus* (Summer 1985) pp. 257–78.
Leebaert, Derek (ed.) *Soviet Military Thinking* (London: George Allen & Unwin, 1981).
Menshikov, S., 'What is behind the "Star Wars" Debate?', *International Affairs* (Moscow) vol. 31, no. 6 (June 1985) pp. 67–77.
Meyer, Stephen M., 'Soviet Military Programs and the New High Ground', *Survival*, vol. 25, no. 5 (September/October 1983) pp. 240–15.
Meyer, Stephen M., 'Soviet Strategic Programs and the US SDI', *Survival* vol. 27, no. 6 (November/December 1985) pp. 274–92.
Nincic, Miroslav, 'Can the US Trust the USSR?', *Scientific American*, vol. 254, no. 4, (April 1986) pp. 33–41.
Ovinnikov, R., ' "Star Wars" programme – A New Phase in Washington's Military Policy', *International Affairs*, vol. 31, no. 8 (1985) pp. 13–22.
Richelson, Jeffrey T., 'US Intelligence and Soviet Star Wars', *Bulletin of the Atomic Scientists*, vol. 42, no. 5 (May 1986) pp. 12–14.
Rivkin, David B., Jr., 'What does Moscow Think?', *Foreign Policy*, no. 59 (Summer 1985) pp. 85–105.
Shenfield, Stephen, 'Soviets May Not Imitate Star Wars', *Bulletin of the Atomic Scientists*, vol. 41, no. 6 (June/July 1985) pp. 38–9.
The Soviet Propaganda Campaign against the US Strategic Defence Initiative (Washington, DC: Office of Public Affairs, US Arms Control and Disarmament Agency, August 1986) ACDA Publication 122.
Velikhov, Yevgeny, 'Space Weapons' – Effect on Strategic Stability', Special Section, *Bulletin of the Atomic Scientists*, vol. 40, no. 3 (March 1984) pp. 12s–15s.
Walt, Stephen M., *Interpreting Soviet Military Statements: A Methodological Analysis*, Report no. CNA-81-0260.10, Alexandria, Va.: Center for Naval Analyses, (December 1983).

SDI, EDI and Extended Air Defence – The European Debate

Afheldt, Horst, 'Der Morgen nach SDI', *Kursbuch* No. 83 (March 1986) pp. 125–48.
Altmann, Jürgen, *Laserwaffen. Gefahren für die strategische Stabilität und Möglichkeiten der vorbeugenden Rüstungsbegrenzung* (Frankfurt: HSFK, September 1986) HSFK-Report 3/1986.

Baer, Alain, ' "ATBM" ', Défense aérienne élargie et concept de dissuasion globale', *Défense National*, vol. 42, no. 8 (August–September 1986) pp. 7–20.
Bardaji, Rafael L., *La 'Guerra de las Galaxias': Problemas y perspectivas de la nueva doctrina militar de las Administración Reagan* (Madrid: INAPPS, 1986).
Barnaby, Frank, *What on Earth is Star Wars?* (London: Fourth Estate, 1986).
Baumel, Jacques (ed.) *La nouvelle stratégie de l'espace* (Paris: Les Cahiers de la Fondation du Futur, 1984).
Berkhof, G. C., *Duel om de ruimte: Aspecten van Westeuropese veiligheid* (The Hague: Instituut Clingendael, 1985).
Berloznik, Robert and de Boosere, Patrick, *Star Wars* (Berchem: Uitgeverij EPO, 1986).
Bertram, Christoph, 'Strategic Defense and the Western Alliance', *Daedalus*, vol. 114, no. 3 (Summer 1985) pp. 279–98.
Bluth, Christoph, 'SDI: The Challenge to West Germany', *International Affairs*, vol. 62, no. 2 (Spring 1986) pp. 247–64.
Boniface, Pascal and Heisbourg, François, *La Puce, Les Hommes et la Bombe: L'Europe face aux nouveaux défis technologiques et militaires* (Paris: Hachette, 1986).
Boyer, Yves, 'Raketenabwehr im Weltraum: Antwort auf eine moralische Frage oder Reform der Strategie?', *Europa-Archiv*, vol. 40, no. 15 (August 10, 1985) pp. 467–74.
Brauch, Hans Günter, *Angriff aus dem All: Der Rüstungswettlauf im Weltraum* (Berlin–Bonn: Dietz, 1984).
Brauch, Hans Günter, 'Zur Militarisierung des Weltraums: Auswirkungen auf die Bundesrepublik Deutschland', *Die Friedens-Warte*, vol. 65 (1982–5) pp. 134–48.
Brauch, Hans Günter, *30 Thesen und 10 Bewertungen zur Strategischen Verteidigungsinitiative (SDI) und zur Europäischen Verteidigungsinitiative (EVI)* (Starnberg: Forschungsinstitut für Friedenspolitik, 1986) (AFES-Papier no. 2).
Brauch, Hans Günter, *Militärische Komponenten einer Europäischen Verteidigungsinitiative* (Stuttgart: AG Friedensforschung und Europäische Sicherheitspolitik, February 1986) (AFES-Papier no. 3).
Brauch, Hans Günter, 'Rüstungsdynamik und Waffentechnik: Ein Versuch der Interpretation der amerikanischen strategischen Raketenabwehrsysteme mit Hilfe von Theoremen aus dem Bereich der Rüstungsdynamik', in Beate Kohler-Koch (ed.), *Technik und internationale Politik* (Baden-Baden: Nomos, 1986) pp. 411–48.
Bulkeley, R., and Spinardi, G., *Space Weapons – Deterrence or Delusion?* (Cambridge: Polity Press, 1986).
Chalfont, A., *Star Wars: Suicide or Survival* (London: Weidenfeld & Nicolson, 1985).
Enders, Thomas (ed.) *Standpunkte zu SDI in West und Ost* (Melle: Ernst Knoth, 1985).
Enders, Thomas, *Raketenabwehr als Teil einer erweiterten NATO–Luftverteidigung* (Sankt Augustin: Konrad-Adenauer-Stiftung, April 1986), (Interne Studien, no. 2/1985).

Engels, Dieter, Scheffran, Jürgen, and Sieker, Ekkehard, *Die Front im All: Weltraumrüstung und atomarer Erstschlag* (Köln: Pahl-Rugenstein, 1984).
European Institute for Peace Research and Security, *Star Wars Strategy: The Implications for the Europeans and the Atlantic Alliance* (Brussels: EIPRS, 1984).
Everts, Philip P. (ed.), *De Droom der onwetsbaarheid. Het Amerikaanse Strategisch Defensie Initiatief en het belang van Europa* (Kampen: Kok Agora, 1986).
Felden, Marceau, *La Guerre dans L'Espace* (Paris: Berger-Levrault/Boréal Express, 1984).
Fenske, John, 'France and the Strategic Defence Initiative: Speeding up or Putting on the Brakes?', *International Affairs*, vol. 62, no. 2 (Spring 1986) pp. 231–46.
Freedman, Lawrence, 'A new strategic revolution?', *Space Policy*, vol. 1, no. 2 (May 1985) pp. 131–4.
Freedman, Lawrence, 'The "Star Wars" Debate: The Western Alliance and Strategic Defense, Part II', *Adelphi Papers*, no. 199 (July 1985) pp. 34–30.
Gallois, Pierre, *La Guerre de Cent Secondes: Les Etats-Unis, l'Europe et la guerre des étoiles* (Paris: Fayard, 1985).
Gorce, Paul-Marie de la, *La guerre et l'atome* (Paris: Plon, 1985).
High Frontier Europe, *The European Defense Initiative (EDI): Some Implications and Consequences* (Rotterdam: High Frontier Europe, 1986).
Hockaday, Arthur, *The Strategic Defence Initiative: New Hope or New Peril*: (London: The Council on Christian Approaches to Defence and Disarmament, 1985).
Hoffman, Fred S., 'The "Star Wars" Debate: The Western Alliance and Strategic Defense', *Adelphi Papers*, no. 199 (July 1985) pp. 25–33.
Hoffman, Hubertus, G., 'A Missile Defense for Europe?', *Strategic Review*, vol. 12, no. 3 (1984) pp. 45–55.
Horn, Erwin, ' "Es ist nur ein europäischer Ableger des SDI-Projekts". Der Sicherheitspolitiker Erwin Horn zur Ablehnung des Raketenabwehrprogramms EVI/Ein Positionspapier für die SPD', *Frankfurter Rundschau*, 5 April 1986.
Howe, Sir Geoffrey 'Defence and Security in the Nuclear Age', speech to the Royal United Services Institute in London, 15 March 1985.
Kaiser, Karl, 'SDI und deutsche Politik', *Europa-Archiv*, vol. 41, no. 19, pp. 569–78.
Kaltefleiter, Werner, *The Strategic Defense Initiative: Some Implications for Europe* (London: Institute for European Defense and Strategic Studies, 1985).
Kennet, Wayland, 'Star Wars: Europe's polite Waffle', *Bulletin of the Atomic Scientists*, vol. 41, no. 9 (September 1985) pp. 7–11.
Labusch, Reiner, Maus, Eckart, and Send, Wolfgang (eds) *Weltraum ohne Waffen: Naturwissenschaftler warnen vor der Militarisierung des Weltraums* (München: Bertelsmann, 1984).
Larrabee, Stephen F., 'Westeuropäische Interessen im amerikanisch-sowjetischen Dialog über Kernwaffen und Waffen zur strategischen Verteidigung', *Europa-Archiv*, vol. 40, no. 6 (25 March 1985) pp. 165–74.
Lellouche, Pierre, *L'avenir de la guerre* (Paris: Mazarine 1985).

Lucas, Michael, 'SDI and Europe. Militarization or Common Security?', *World Policy Journal*, vol. 3, no. 2 (Spring 1986) pp. 219–49.
Menaul, Stewart, 'A European Defense Initiative', *Journal of Defense & Diplomacy* (February 1986) pp. 18–22.
Nerlich, Uwe, 'Missile Defences: Strategic and Tactical', *Survival*, vol. 27, no. 3 (May/June 1985) pp. 119–36.
Nerlich, Uwe, 'Folgerungen aus SDI für Strategie, Rüstungskontrolle und Politik', *Europa-Archiv*, vol. 41, no. 4 (1986) pp. 89–98.
Rühl, Lothar, 'Eine Raketenabwehr auch für Europa', *Frankfurter Allgemeine Zeitung*, 17 January 1986, pp. 10–11.
Rühle, Hans, 'Löcher im Drahtverhau der Sicherheitsdoktrin – Das Defensivkonzept der Zukunft löst Europas Probleme kaum', *Rheinischer Merkur*, 1 April 1983.
Rühle, Hans, *Standpunkte zu SDI in Ost und West* (Melle: Ernst Knoth, 1985).
Rühle, Hans, '"An die Grenzen der Technologie" – Der Planungschef des Verteidigungsministeriums, Hans Rühle, zur wissenschaftlichen Kritik am SDI-Projekt', *Der Spiegel*, vol. 39, no. 48 (November 1985) pp. 155–9.
Rühle, Hans and Rühle, Michael, *Raketenabwehr. Strategic Defense Initiative (SDI)* (Koblenz: Bernard & Graefe, 1985).
Scheffran, Jürgen, *Rüstungskontrolle bei Antisatellitenwaffen. Risiken und Verifikationsmöglichkeiten* (Frankfurt: HSFK, October 1986) HSFK-Report 6/1986.
Taylor, Trevor, 'Britain's response to the Strategic Defence Initiative', *International Affairs*, vol. 62, no. 2 (Spring 1986) pp. 217–30.
Schreiber, Wolfgang, *Die Strategische Verteidigungsinitiative* (Melle: Ernst Knoth, 1985) (Forschungsbericht 45, Konrad-Adenauer-Stiftung).
Schweizersiche Arbeitsgemeinschaft für Demokratie, *SDI – Informationen und Analysen zur strategischen Verteidigungsinitiative der USA* (Zürich: SAD, January 1986) (Schriften der SAD 22).
Seitz, Konrad, 'SDI – die technologische Herausforderung für Europa', *Europa-Archiv*, vol. 40, no. 13 (10 July 1985) pp. 381–90.
Senghaas, Dieter, 'Strategic Defense Initiative – Ansatzpunkte für eine Lagebeurteilung', *Neue Gesellschaft/Frankfurter Hefte*, vol. 32, no. 9 (1985) pp. 842–9.
Sevaistre, O., 'L'Europe face à l'iniciative de défense stratégique', *Défense Nationale* (May 1985).
Thompson, E. P. (ed.) *Star Wars* (Harmondsworth: Penguin, 1985).
Thompson, E. P., and Thompson, Ben., *Star Wars: Self-Destruct Incorporated* (London: Merlin, 1985).
Western European Union, *The Military Use of Space* (Paris: WEU, 1984).
Wilson, Pete, 'A Missile Defense for Europe: We must respond to the Challenge', *Strategic Review*, vol. 14, no. 2 (Spring 1986) pp. 9–15.
Wörner, Manfred, 'A Missile Defense for NATO Europe', *Strategic Review*, vol. 14, no. 1 (Winter 1986) pp. 13–18.
Wörner, Manfred, 'Europa braucht Raketenabwehr', *Die Zeit*, 28 February 1986.
Yost, David S., 'Ballistic Missile Defense and the Atlantic Alliance', *International Security*, vol. 7, no. 2 (1982) pp. 143–74.

Yost, David S., 'European Anxieties about Ballistic Missile Defense', *The Washington Quarterly* vol. 7, no. 4 (1984) pp. 112-29.
Zuckerman, Lord, 'The Wonders of Star Wars', *The New York Review of Books* (30 January 1986) pp. 32-40.

SDI and the ABM Treaty

Boutwell, Jeffrey, and Scribner, Richard A., *The Strategic Defense Initiative: Some Arms Control Implications* (Washington, DC: The American Association for the Advancement of Science, May 1985).
Brauch, Hans Günter, *Anti-Tactical Missile Defense: Will the European Version of SDI Undermine the ABM-Treaty?* (Stuttgart: AG Friedensforschung und Europäische Sicherheitspolitik, July 1985) (AFES-Papier no. 1).
Bulkeley, Rip, *The Anti-Ballistic Missile Treaty 1972-83* (University of Bradford: School of Peace Studies, December 1983).
Bundy, McGeorge, Kennan, George, McNamara, Robert S., and Smith, Gerard C., 'The President's Choice: Star Wars or Arms Control', *Foreign Affairs*, vol. 63, no. 3, (Winter 1984-5) pp. 264-78.
Carnesale, Albert, testimony, 'Department of Defense Authorization for Appropriations for Fiscal Year 1986', *Hearings Before the Committee on Armed Services, US Senate*, 21 February 1985, 20 March 1985, pp. 4007-50.
Chayes, Abram, 'Treaties and Legal Issues', speech on ballistic missile defense at Symposium on Space, National Security and C-cubed-I, Bedford, Massachusetts: The Mitre Corporation, 25 October 1984). Mitre Document M85-3, pp. 29-32.
Chayes, Abram, Chayes, Antonia Handler, and Spitzer, Eliot, 'Space Weapons: The Legal Context', *Daedalus*, vol. 114, no. 3 (Summer 1985) pp. 193-218.
Chayes, Abram and Chayes, Antonia Handler, 'Commentary: Testing and Development of "Exotic" Systems Under The ABM Treaty: The Great Reinterpretation Caper', *Harvard Law Review*, vol. 99, no. 8 (June 1986) pp. 1956-71.
Drell, Sidney D.; Farley, Philip J. and Holloway, David, 'Preserving the ABM Treaty: A Critique of the Reagan Strategic Initiative', *International Security*, vol. 9, no. 2 (Fall 1984) pp. 51-91.
Finkelstein, Mark A., 'Star Wars Meets The ABM Treaty: The Treaty Termination Controversy', *The North Carolina Journal of International Law and Commercial Regulation* (Summer 1985) pp. 701-27.
Fischer, Horst, 'Die Wiener Konvention über das Recht der Verträge und die Auslegung des ABM Vertrages', *S & F*, vol. 4, no. 1 (1986) pp. 14-19.
Gordon, Michael R., 'US-Soviet Arms Control Negotiations: Nuclear and Space Weapons', *AEI Foreign Policy and Defense Review*, vol. 5, no. 2 (1985) pp. 21-44.
Gordon, Michael R., 'CIA is Skeptical that New Soviet Radar is Part of an ABM system', *National Journal* (9 March 1985) pp. 523-6.
Gray, Colin, S., 'Space Arms Control: A Skeptical View', *Air University Review* (November/December 1985) pp. 73-86.

Graybeal, Sidney N., and Krepon, Michael, 'Making Better Use of the Standing Consultative Commission', *International Security*, vol. 10, no. 1 (Fall 1985) pp. 183-99.
Graybeal, Sidney N. and Krepon, Michael, 'SCC: Neglected Arms Control Tools', *Bulletin of the Atomic Scientists* vol. 41, no. 10 (November 1985) pp. 30-3.
Judge, John, 'The Strategic Defense Initiative and Arms Control', *Defense Electronics* (March 1985) pp. 96-106.
Kennedy, Edward M., 'Star Wars v. the ABM Treaty', *Arms Control Today* vol. 14, no. 6 (July/August 1984) pp. 1, 18-19, 24.
Keeny, Spurgeon, M., Jr., 'Uncertain Future of Arms Control', *Arms Control Today*, vol. 15, no. 6 (July/August 1985) pp. 2-5.
Krepon, Michael, *Arms Control Verification and Compliance* (Washington, DC: Foreign Policy Association, 1984).
Krepon, Michael and Peck, D. Geoffrey, 'Another Alarm on Soviet ABMs', *Bulletin of the Atomic Scientists*, vol. 14, no. 6 (June/July 1985) pp. 34-6.
Krepon, Michael, 'Dormant Threat to the ABM Treaty', *Bulletin of Atomic Scientists*, vol. 14, no. 1 (January 1986) pp. 31-4.
Kröger, Herbert, 'Die Weltraumrüstungspläne der USA und das Völkerrecht unserer Zeit', *IPW-Berichte*, vol. 15, no. 3 (March 1986) pp. 11-17, 41.
Longstreth, Thomas K., Pike, John E. and Rhinelander, John B., *The Impact of US and Soviet Ballistic-Missile Defense Programs on the ABM Treaty* (Washington, DC: National Campaign to Save the ABM Treaty, March 1985).
Longstreth, Thomas K., and Pike, John E., 'US, Soviet Programs Threaten ABM Treaty', *Bulletin of the Atomic Scientists*, vol. 41, no. 4 (April 1985).
Meredith, Pamela L., 'The Legality of a High-Technology Missile Defense System: the ABM and Outer Space Treaties', *The American Journal of International Law* (April 1984).
National Academy of Sciences, *Nuclear Arms Control: Background and Issues* (Washington, DC: National Academy Press, 1985).
Newhouse, John, *Cold Dawn: The Story of SALT* (New York: Holt, Rinehart & Winston, 1973).
Nitze, Paul H., *SDI and the Anti-Ballistic Missile Treaty* (Washington, DC: US Department of State, Bureau of Public Affairs, June 1985) Current Policy no. 711.
Nitze, Paul H., "SDI and the Anti-Ballistic Missile Treaty', *US-Soviet Relations* (Washington, DC: Foreign Policy Institute, School of Advanced International Studies, Johns Hopkins University, August 1985) pp. 21-4.
Nunn, Sam, 'Access to the ABM Treaty Record – The Senate's Constitutional Right', *Arms Control Today*, vol. 16, no. 6 (September 1986) pp. 3-7.
Paine, Christopher, 'The ABM Treaty: Looking for the Loopholes', *Bulletin of the Atomic Scientists*, vol. 39, no. 2 (August/September 1983) pp. 13-16.
Payne, Keith and Gray, Colin S., 'Nuclear Policy and the Defensive Transition', *Foreign Affairs*, vol. 62, no. 3, (Spring 1984) pp. 820-42.
Pike, John, 'Is There an ABM Gap?', *Arms Control Today*, vol. 14, no. 6 (July/August 1984) pp. 2-3, 16-17.

Rhinelander, John B., and Willrich, Mason (eds) *SALT: The Moscow Agreements and Beyond* (New York: Macmillan, 1974).
Rhinelander, John B., 'How to Save the ABM Treaty', *Arms Control Today* vol. 15, no. 4 (May 1985) pp. 1, 5–9, 12.
Rhinelander, John B., 'Reagan's "Exotic" Interpretation of the ABM Treaty', *Arms Control Today* vol. 15, no. 8 (October 1985) pp. 3–6.
Schear, James A., 'Arms Control Treaty Compliance: Buildup to a Breakdown?' *International Security*, vol. 10, no. 2 (Fall 1985) pp. 141–82.
Schneiter, George, 'The ABM Treaty Today', in Carter and Schwartz (eds) *Ballistic Missile Defense*, pp. 221–50.
Schneiter, George R., 'Implications of the Strategic Defense Initiative for the Anti-Ballistic Missile Treaty', *Survival*, vol. 27, no. 5 (September/October 1985) pp. 213–25.
Sherr, Alan B., *Legal Issues of the Star Wars Defense Program*, Lawyers Alliance Issues Brief #3 (Boston, Massachusetts: Lawyers Alliance for Nuclear Arms Control, Inc., June 1984).
Sherr, Alan B., *Legal Analysis of the 'New Interpretation' of the Anti-Ballistic Missile Treaty* (Boston, Massachusetts: Lawyers Alliance for Nuclear Arms Control, Inc., March 1986).
Sherr, Alan B., The Languages of Arms Control', *Bulletin of Atomic Scientists* vol. 41, no. 9 (November 1985) pp. 23–9.
Smith, Gerard C., *Doubletalk: The Story of SALT I* (New York: Doubleday & Co., 1980).
Smith, Gerard C., 'Star Wars is Still the Problem', *Arms Control Today*, vol. 16, no. 2 (March 1986) pp. 3–6.
Smith, R. Jeffrey, 'Arms Control Agreement Breathes New Life into SCC', *Science* (9 August 1985) pp. 535–6.
Sofaer, Abraham D., 'The ABM Treaty and the Strategic Defence Initiative', *Harvard Law Review*, vol. 99, no. 8 (June 1986) pp. 1972–85.
US Congress, House, Arms Control, International Security, and Science Subcommittee of the Foreign Affairs Committee, *Implications of the President's Strategic Defense Initiative and Anti-Satellite Weapons Policy* (Washington, DC: US Government Printing Office, 24 April and 1 May 1985).
US Congress, House Committee on Foreign Affairs, *Anti-Ballistic Missile Treaty Dispute* (Washington, DC: US Government Printing Office, 22 October 1985).
US President, 'Report of Soviet Non-compliance With Arms Control Agreements', *Department of State Bulletin* (March 1984) pp. 8–11.
Voas, Jeanette, *ABM Treaty Interpretation: the Soviet View* (Washington, DC: Congressional Research Service, 25 November 1985, pp. 1–17.

Index

Abalakova see Krasnoyarsk
Abrahamson, James A., General 131, 272, 403, 437
Adelman, Kenneth 502-3
Adomeit, Hannes 55
AEG company 211
Aerospatiale 162-3, 188, 272, 477
AGALEV party, Belgium 239-40
AGARD 440, 451, 453, 477-9
  'Aerospace Application Study' (AAS-20) 440, 451, 453, 477
airborne optical system (AOS) 18-19, 399, 405-6, 465, 514, 522, 539
airborne warning and control systems (AWACS) 362
aircraft, bomber 24, 30-3, 549
Akhromeyev, Sergei, Marshal xxv, 425-7
Alford, Jonathan v, 143
Allison, General 377, 416
Altenburg, Wolfgang, General 176, 192, 478
Amerongen, Otto Wolff von 212
Anderson, Jack 93
Anti-Ballistic Missile (ABM) Treaty (1972) xxiii, 64-5, 373-431, 555-71
  activities permitted and prohibited 387-9
  and arms control 64-5
  ASAT weapons 395-7, 526-8
  ATBMs 91-5, 395, 397-400, 470-1, 489-90, 526
  compliance with 36-7, 500-33: Soviet BMD 11, 407, 503-13; US SDI 513-21
  definitional issues 390-5, 414, 524
  European attitudes toward 262-3, 331, 343-4
  evolution of xxxv, xxxvii, 374-83
  'grey areas' 113, 395-403, 521-33
  modifications, possible 355-6, 366
  offensive-defensive system linkages 33-4
  provisions xxxix, 384-7 see also Appendixes
  radars 10-11, 395, 400-3, 524-6
  Reagan Administration and 403-23
  reinterpretation, efforts at 407-21, 524

Soviet attitudes toward 62-3, 67, 84-5, 423-8
  Standing Consultative Committee 386-9, 430, 505, 511, 523-4, 556
  and weapons reduction 9-10, 34
anti-ballistic missiles (ABMs) see ballistic missile defence (BMD)
anti-satellite (ASAT) systems xxiii, 22, 30-1
  and ABM Treaty 113, 395-7, 526-8
  and arms control 45
  and BMD 30-1, 39-41
  Soviet 72, 78-82, 97
anti-submarine warfare (ASW) 548
anti-tactical aerodynamic missiles 440
anti-tactical ballistic missiles (ATBMs/ATMs) 437
  and ABM Treaty 91-5, 395, 397-400, 437, 470-1, 489-90, 526
  European 74, 466-72, 485-6
  feasibility of 470, 481-2
  Soviet 91-5, 447-50, 466, 506-8
  US 466-73, 476
anti-tactical missile steering committee 477
Anureyev, I., Major General 59
APVO Strany (Soviet aviation units) 72
ARIANE launcher 155-6, 201, 315-16, 349
Arkin, William 441, 444
arms control 26-46
  and EDI 486
  offensive-defensive system linkages 30-4
  and SDI 44-5, 365-7
  SDI and negotiating strategy 34-7
  Soviet countermeasures to SDI 37-44, 73-4
  and strategic doctrines 27-9
arms race xxi-iv
art, military, Soviet 57-8
Asia xx, 363 see also China
Atlantic Institute 260

Baekelandt, Ambassador 237
Bahr, Egon 204-5
Baker, John C. 68, 72, 443
balance of power xx-iii, 66, 545-6

ballistic missile defence (BMD) 9–25
  cost-effectiveness 21–2
  countermeasures to 37–44, 73–4, 110–13, 445–7
  and European defence 19–21
  global 15–19
  international parliamentary debates on 257–61
  and radars (*q.v.*) 10–11
  Soviet capabilities 11, 50–1, 57–73, 82–95, 334–9, 547–9
  space-based weapons (*q.v.*) 24
  stability of 23
  survivability of 22–3
  tactical xxxiii, xxxix
  technical feasibility 12–15, 481–2
  see also ABM Treaty; Strategic Defense Initiative (SDI)
Bangemann, Martin 178, 186, 193–4, 202
Banks, Robert 276
Basic Research in Industrial Technology for Europe (BRITE) programme 315–16
Batitsky, General 84
Baum, Gerhard 194
Baumel, Jacques 161
Becker, Kurt 196, 258
Belgium, reactions to SDI 234–43, 251–3
  government position 235–8
  industry 242–3
  political parties 238–41
  scientific community 241–2
Benthem van den Bergh, G. van 249
Bergh, Mr van den 286–90
Berkhof, G. C., General (Report) 249, 438, 478–9
Berloznik, Robert J. xiii, 3–4, 234–53
Berman, Robert P. 68, 72, 443
Berrier, Mr (Report) 288, 290–1, 293
Bertram, Christoph 168, 214–15
Beyme, Klaus von 55
Bierich, Marcus 211
Bingaman, Jeff, Senator 88
Blaauw, Mr 286
Boer, Mr de 246–7
bombers 24, 30–3, 549
boost phase of missile flight, and interception in 13, 16–18, 20–1, 37–40, 459
boost surveillance and tracking system (BSTS) 513–14, 522
Bourg, Willy 250
Braduskill 517–18

Brandt, Willy 203
Brauch, Hans Günter xiii, 1–6, 50–113, 166–216, 256–99, 436–90
Brezhnev, L. I. 66, 571
brightness 529
Broek, Hans van den 244
Brookings study 94
Brown, Harold 377, 381
Brzezinski, Zbigniew 141, 353
Buckel, Werner 213–14
Buckley, James, Senator 382, 420
Bukman, Mr 246
Bundestag debates 207–10
Burch, Mr 236
Burke, Kelly H., Lt General 80, 101
Burt, Richard R. 80, 267
Buteaux, Paul 261

Campaign for Nuclear Disarmament (CND) 142
Canada 271–2
Canavan, Gregory H. 17, 25, 460, 464–5
Carrington, Lord 262
Carter, Jimmy xxxv, 166–7
Carton, Alain xiii–xiv, 3–5, 150–64, 311–26, 341–50
Cartwright, John 139, 275, 278–9, 281
Centre National d'Etudes Spatiales (CNES) 155–6
Chalfont, Lord 141
Charzat, Gisèle 296
Chevaline programme 344–5
Cheysson, Claude 153, 317
China, People's Republic of 257, 354–7, 360, 363–4
Chirac, Jacques 161–2
Christian Democratic Party, Belgium 238, 240
Christian Democratic Party, Netherlands 245–7, 252
Christian Democratic Union (CDU), FRG 168–9, 171, 175, 178, 180, 182, 197–201, 215
Christian Social Union (CSU), FRG 169–71, 175, 180, 182, 197–202, 215
Churchill, Sir Winston xxi
CITES consortium 272, 477
civil defence 154
Close, Robert, General 240–1
COLUMBUS project 201, 316
Committee of Soviet Scientists for Peace 73, 445–6
Communist Party, Belgium 240

Communist Party, France 161
Communist Party, Netherlands 247–8
compliance with ABM Treaty 500–33
  ASAT weapons 526–8
  ATBMs 526
  'grey areas' 521–33
  radars 524–6
  Soviet BMD 503–13
  US SDI 513–21
computers 14–15, 313–14
conflict *see* escalation; theatre
Conservative Party, UK 140–2
  government 130–8, 322–4
consultation
  ABM Treaty 257–63
  SDI xxxv–vi, 132, 272–98; European Parliament 294–8; NATO 263–72; WEU 282–94
Consultative Group Co-ordinating Committee (COCOM) 195–6, 314
conventional forces xxvi–ix, 42, 477, 549–50
Cooper, John Sherman, Senator 420
Cooper, Robert S. 80, 88, 93
costs of BMD 21–2, 26, 480, 482–3, 544
Counterair '90 450, 453–6
countermeasures to BMD 37–44, 73–4, 110–13, 445–7
cruise missiles xxiii–iv, xxx–i, 32–3, 355, 486
  air-launched (ALCMs) 44, 472
  and BMD 24, 338–9
  sea-launched (SLCMs) 44 *see also* submarines
Curien, Hubert 156, 288, 350

Daggett, Stephen 39
damage limitation xxxvii, 27, 57, 69–70, 436, 542–8
Deane, Michael J. 51, 61–2
Debré, Michel 161, 262, 344
decoupling of USA from NATO xxxi, xxxv–viii
decoys 13, 15, 17–21, 23, 39–41
deep-strike options xxviii, 456–7, 481–2
defensive–offensive system linkages *see* offensive–defensive
DeGrasse, Robert 39
De Lauer, Richard 88
Delors, Jacques 318, 322, 324
Demarets, José 236

Denmark 268–9
designating optical tracker (DOT) 515
destruction, assured *see* mutual assured destruction
deterrence
  French policy 157, 342–3
  'gridlock' 363–4
  mutual 61–2, 64–6, 68–9
  and SDI xix–xxxi, 545
  and strategic doctrine 27, 61–2
  'through denial' 27–8
  UK policy 331–9
'development', definition of 392–5, 524
Diehl company 477
Dierickx, Mr 239–40
directed-energy weapons
  and ABM Treaty 382–3, 387, 519–21, 529
  and BMD 14, 22–4, 459–61
  Soviet 95–103
  see also 'exotic' systems; lasers; particle beam weapons
dispersal of forces 41
doctrine, military, Soviet 52–7
  and space 57–71
doctrine, nuclear 27–8
  France 157, 343–5, 364–5
  NATO xix–xx, xxvi–vii, 27, 112, 174, 176, 365, 448–9, 549, 553
  Soviet Union 28–9, 68–9
  USA 28–9, 35, 68–9, 145, 543; *see also* damage limitation; mutual assured destruction
Donnelly, C. N. xxvi
Dornier Aerospace 211, 272
Douglass, Joseph D., Jr 61
Dregger, Alfred 169, 192, 199–200, 208
Drell, Sidney 67–8
Drozdiak, William 178, 181
Dumas, Roland 160, 176, 181, 285, 291, 294, 311, 317–18, 326
Dürr, Hans Peter 213
Dürr, Heinz 211
Dyson, Freeman 46

early warning
  AWACS 362
  BMEWS 401–3, 525
  radars 10–11, 400
Eberle, Sir James, Admiral 144
Ehmke, Horst 203–6, 208–9
Eisenhower, Dwight D. xxii

Ellis, General 389
emergent technologies (ET) 457–9, 474–5
Enders, Thomas 440
endoatmospheric defence 463–4, 517
endoatmospheric non-nuclear kill (ENDO-NNK) technology 461–3
Ermath, Fritz 68–9
escalation 363–4, 549–53
control of 27–8, 167, 191
and flexible response (*q.v.*) xxvi–vii
see also limited nuclear war
ESPRIT project 315
EUREKA *see* European Research Co-ordination Agency
EUROBIOT 321, 346
EUROBOT 321, 346
EUROCOM 321, 346
EUROMAT 321, 346
EUROMATIC 320, 346
EUROPA rocket project 155
Europe
EDI (*q.v.*) 479–89
EUREKA (*q.v.*) 311–26
reactions to SDI *see under individual countries*
and Soviet space weapons 103–13
European Defence Architecture (EDA) 450–9
American studies for 475–9
European Defence Initiative (EDI) 169, 172, 192, 200–1, 349–50, 436–90
and European security 479–89; arms control 486; costs 482–3; and politics 487–9; protection 481; strategic stability 483–6; technical feasibility 481–2
framework, EDA 450–9
origins 436–40
Soviet countermeasures 73–4, 445–7; ATBMs 447–50
and US weapons projects 459–79; architecture studies 475–9; ATBMs 466–72; SDI 459–66; TACMs 472–5
European Economic Community (EEC) 294, 315–25
see also European Parliament
European Parliament 273, 294–8
European Research Co-ordination Agency (EUREKA) 311–27, 342, 349
aims 315–17
Benelux response 237, 244, 252–3
European Parliament debates 294–8
FRG response 169, 172, 175–6, 182, 185–9, 200–1
origins of 158–63, 186, 311–15, 324–7
programme 317–21, 345–6
questions about 321–4
UK response 135–7
European Space Agency (ESA) 141, 154–5, 314
Everts, Philip 253
exoatmospheric non-nuclear kill (EXO-NNK) technology 461–2
exoatmospheric re-entry vehicle interception system (ERIS) 461–2, 464, 516–17, 528, 539–40
'exotic' systems 376–8, 380–1, 383, 409, 413–19, 421–2, 424, 429
extended air defence 436, 440, 443, 452, 457, 481, 483
extended-range homing intercept technology (ER-HIT) 399, 540

Faber, Mient Jan 249
Fabius, Laurent 317
Fack, Fritz Ullrich 215
Farley, Philip 67
Feldmann, Olaf 202, 209
Feldmeyer, Karl 268
Fieldhouse, Richard W. 444
'first strike' xxii, 543
FLAGE 463–4, 540
Flat Twin radar 509–10, 512
Flemish Christian Democratic Party 240
Flemish Socialist Party 236, 239
Fletcher, James C. (Report) 14–15, 264–5
flexible lightweight agile guided experiment (FLAG) 540
flexible response strategy xix–xx, xxvi–xxvii, 27, 112, 174, 176, 343, 365, 448–9, 549, 553
follow-on-forces attack (FOFA)/Rogers Plan 112, 151, 168, 189, 241, 440, 449, 457, 482–3
Forrestal, Michael 273–4
Foster, John 382, 394, 416–19, 466
Fourré, Mr 284
fractional orbital bombardment system (FOBS) 78
France, reaction to SDI 137–8, 150–64, 341–50
defence policy 341–50; and ABM Treaty 343–4; deterrence

342–3, 354; EDI 349–50; space technology 345–8
government reaction 151–60, 341–2
industry 162–3
political debate 160–2
strategy, nuclear 157, 343–4, 361, 364–5
Francke, Klaus 282
Frank, Lewis Allen 61
Freedman, Lawrence 143–4, 147, 331
Freeson, Mr 292
freeze on nuclear weapons xxv
Freyend, Eckard John Von 210
Front Democratique des Francophones 238
Fylingdales radar 526

Gallis, Paul E. 263
Gallois, Pierre, General 349, 438
Galosh missile 83, 85–6, 90, 257, 337
Gandhi, Rajiv 363
Gansel, Norbert 292
Gardner, John 101, 277
Garn, Senator Jack, 100
Garthoff, Raymond L. xiv, xxxiii–ix, 2, 65–7, 84–5, 257, 259, 262, 378–9
Gaullist Party (RPR), France 161–2
Geiger, Michaela 208–9
Geissler, Heiner 171, 199, 201
General Electric Company, UK 187, 320, 323
Genscher, Hans-Dietrich 160, 169–87, 193, 202–3, 208–10, 216, 270, 285–9, 318, 324, 436
George, Bruce 277
geostrategic risks of SDI 352–68
arms control 365–7
BMD policy 364–5
changing military environment 360–3
EDI 483–6
escalation 363–4
Marxist-Leninism 357–8
Soviet attitude 354–7
Soviet economy 358–60
Germany, Federal Republic of, reactions to SDI 137, 166–216
Bundestag debate 207–10
business and trade unions 210–12
government position 170–97: arms control 179–82; economics and technology 183–9; military strategy 189–93; participation terms 193–7
and INF 166–70
media 214–15

political parties 197–207: CDU 197–201; CSU 201–2; FDP 202–3; Greens 206–7; SPD 203–6
public opinion 215
scientific community 212–14
Gershwin, Lawrence K. 89
Glinne, Mr 296
Goldwater, Barry, Senator 415–16, 419
Goodge Report 186
Gorbachev, Mikhail xx, xxiv–v, 2, 182, 424–5, 427, 489
Gray, Colin 542
Graybeal, Sid 377
Grechko, Andrei A., Marshal 84–5
Greece 268–9
Green Party, FRG 196–8, 206–7, 215
Griffon missile 83, 257
Grinevsky, Mr 378
Gromyko, Anatoly A. xxiv, 390
ground-based lasers (GBL) 461, 520, 522, 529
Grüner, Martin 287–8
Guertner, Gary L. xiv–xv, 3–4, 26–46

Hamilton, Andrew xxix
Harmel Report 166–7, 174, 490
Hart, Douglas M. 56–7
Hawk missile system 398, 466, 469, 490
Healey, Denis xv, xix–xxxi, 139, 258
Heath, Edward 140
Hecht, Jeff 101–2
Heckmann, Erhard 477
Hen House radar 83, 87, 337, 400, 525, 530, 567
Hensel, Howard M. 63, 70–1
HERMES shuttle programme 155–6, 181, 201, 210, 316, 349
Hernu, Charles 157–8, 169, 347, 438–9
Herter, Christian xxi
Heseltine, Michael 132, 136, 285, 333
Hicks, Donald 477
high endoatmospheric defence interceptor (HEDI) 463–4, 517, 539–41
high-energy laser (HEL) 477
Hill, Mr 286, 292
Hill, Roger 263–4
Hippel, Frank von 441
Hoeber, Amoretta M. 61
Hoff, Günter 210
Hoffman, Fred S. (Panel Report) 264–5, 399, 436–7, 449–50, 470–1, 475

Holloway, David 52-3, 58, 65, 67, 69, 71, 110-12
Holm, Hans-Henrik 488
Holst, Johan Jørgen 260-1
homing overlay experiment (HOE) 16, 409, 462, 464, 515-17
Houwelingen, Jan van 243
Howe, Sir Geoffrey 133-4, 142, 180, 270, 285, 289, 334, 481
Huntzinger, Jacques 160
Hussein, Farooq 145
hypervelocity launcher 518, 522

Iklé, Fred 43, 167, 176, 411, 477, 543
Independent European Programme Group (IEPG) 138, 458
industry, and SDI 185, 210-12, 242-3
  Belgium 242-3
  France 162-3, 342-3
  FRG 185, 187-8, 210-12
  Japan 312-14
  Netherlands 250
  USA 312-14
Institute for European Defence and Strategic Studies (IEDSS) 143
interception of missiles 12-15, 397-400, 516-21, 537-8
intercontinental ballistic missiles (ICBMs) 550-3
  interception of 13
  site defence xxiv, 43, 353, 451, 537-42, 553, 568
  Soviet 10, 16, 20-1, 32-3
  US 10, 32-3
intermediate-range nuclear forces (INF)
  debate on 166-8
  deployed in Europe xxx-i, xxxiv-v
  negotiations 355, 366-7, 375
International Institute for Strategic Studies (IISS) xxvii, 54, 143, 443-5, 448

Jackson, Senator Henry 382, 393, 415-20
Jackson, William E. 471
Japan 270, 272, 312-14, 359
Jasani, Bhupendra 75, 77, 82
Jastrow, Robert 353
Johnson, Lyndon B. 257, 374, 537
Johnson, Nicholas L. 77, 81
Johnson Smith, Sir Geoffrey 275, 277-8
joint surveillance and target acquisition radar systems (JSTARS) 455-8
joint tactical missile systems (JTACMS) 112, 449, 455-8, 472-5

Jones, T. K. 92-3, 508
Jospin, Lionel 160-1

Kamchatka test range 381, 509
Kampelman, Max M. 353
Kapitza, Pyotr 84
Karpov, Victor 377
Keegan, George J, General 100
Kelly, Petra 206-9
Kennedy, John F. 257
Kennet, Lord 140
Kerr, Donald 80
Keyworth, George A. 130, 281, 397, 404, 451, 454
kill mechanisms 459
  directed-energy 14, 22-4, 459, 461
  kinetic-energy 14, 16-17, 459, 461-3
kinetic energy interception 14, 16-17, 518-19, 522
Kinnock, Neil 138
Kishilov, Nicolai 378-9
Kissinger, Henry xxi, 141, 380-1
Klein, Hans 209
Knoth, Arthur 464-5
Kohl, Helmut 168-75, 178-83, 186, 194-5, 203-11, 276, 326, 436
Köhler, Georges 213
Kokoshin, A. A. 73, 446
Kosygin, Aleksei 84, 375
Krasnoyarsk LPAR 11, 401, 407, 412, 501-6, 524, 526-7, 530
Kreemers, H. P. M. 243-4
Kuznetsov, Vasily V. 423
Kwajalein missile range 16, 515, 564

Labour Party, UK xix, xxv, 138-9
Lagardère, Jean-Luc 162, 438
la Gorce, François de 152
Laird, Melvin 382, 415-16
Lambsdorff, Count 203
Lance missile system 92, 472-3
LANDSAT 53
large phased-array radars (LPARs) 11, 93, 386, 407, 412, 501-6, 524-6
lasers xxiv
  and ABM Treaty 388-9, 519-20
  for BMD 14, 22-4, 89
  ground-based 461, 520, 529
  high-energy 477
  land-based 378, 413, 418, 422-3
  Soviet 95-103
  space-based 382, 413, 419, 422-3, 515, 519-20, 522, 529
  see also 'exotic' systems

layers of defence 12–13, 16–19, 23–4, 459–60
Lefebvre, Thomas-Henri (Report) 273–5, 277
le Guen, René 161
Lellouche, Pierre 487–8
Leningrad defence system 83, 257
Lenzer, Christian 199–201, 288–9, 291, 293–4
Levi, Barbara 441
Lewin, Lord 336
Liberal Party, Belgium 238, 240
Liberal Party (FDP), FRG 168–71, 175, 180, 196–8, 202–3, 215
Liberal Party, Luxembourg 251
Liberal Party, Netherlands 245–6
Liberal Party, UK 139–40
limited defence *see* partial defence
limited nuclear war xxvii, xxxv, 12, 27, 549–53
Linkohr, Rolf 296, 298
Lizin, Ms 296
Lohr, Helmut 210
Longstreth, Thomas K. xv, 2, 5, 471, 500–33
low altitude defence system (LoADS) 538, 541
low endoatmospheric defence interceptor (LEDI) 539–41
Lowenthal, Mark M. 263
Lowi, Theodore, J. 4, 50
Luxembourg, reactions to SDI 234–5, 250–3
 government position 250–1
 opposition parties 251

McFarlane, Robert C. 281, 289, 407–8, 425
Mackensen, Ulrich 214, 268
McNamara, Kevin 281
McNamara, Robert xxi, xxvi, 83, 256–8, 354, 356
Mahaffay, General 473
Malinovsky, R. Y., Marshal 84
manoeuvrable re-entry vehicles (MARVs) 41, 44
Mao Zedong 366
Marietta, Martin, company 476, 482
Martens, Wilfried 235–6
Martin, Lawrence W. 260
Marxism-Leninism 55, 58, 66, 357–8
Mathias, Charles McC., Jr, Senator 278, 281
Matra 162, 188, 272, 438
Matre, Mr 163

Maystadt, Philippe 237
Mecklinger, Roland 211
media, and SDI 142–3, 214–15
Merryman, General 457, 467–9
Mertes, Alois 180
Messerschmitt-Boelkow-Blohm (MBB) 188, 211, 272, 477
Metten, Mr (Report) 296–7
Meyer, Stephen M. 71–2, 75, 77–8, 81, 101–2
mid-course of missile flight, and interception in 13, 18, 20–1, 38–40, 459
Middle East 362–3
Midgetman ICBM 552
Miert, Karel van 239
Miller, Franklin 507
Milstein, Mikhail A., General 59
Minuteman ICBM 10, 516–17
Mischnick, Wolfgang 202
Mitterrand, François xxvi, 2, 135, 151–5, 159, 162, 182, 187, 317, 326, 341
Moisi, Dominique 487
Möllemann, Jürgen 180, 202
Moscow defence system 83, 85–6, 257, 337, 568
Mulroney, Brian 271–2
multiple independently-targetable re-entry vehicles (MIRVs) xxii, 10, 13, 16, 83, 521
multiple orbital bombardment system (MOBS) 78
mutual assured destruction (MAD) 50, 171, 436, 480–1, 547–9
mutual assured survival (MAS) 171, 481, 542
mutual vulnerability 56–9, 64–8, 167, 352

Nakasone, Yasuhiro 313
Narjes, Karl Heinz 297, 324
National Aeronautics and Space Administration (NASA) 312
navigation satellites 78
negotiations
 INF 355, 366–7, 375
 strategy for, and SDI 34–7, 365–7
 see also ABM Treaty; SALT; START
Nerlich, Uwe 210, 449–50, 456, 470
Netherlands, reactions to SDI 234–5, 245–50, 251–3
 government position 243–5
 industry 250

political parties 245-9
scientific community 249-50
'new strategic concept', US 35, 145, 543
Newhouse, John 378
Nike-X ABM system 257, 537
Nike-Zeus ABM system 257, 405
Nitze, Paul H. 35, 75, 79, 83, 145, 280-1, 289, 377, 381, 416, 502-3, 543
Nixon, Richard M. 261, 356, 374, 380, 537, 570
North Atlantic Assembly 272-82, 408, 455, 481
North Atlantic Treaty Organisation (NATO)
AGARD 440, 451, 453, 477, 479
consultations within xxxv-vi, 263-72
conventional forces xxvii-ix, 550-1
Counterair '90 450, 453-6
decoupling of US defence from xxxi, xxxv-viii
France and 342-3, 361
FRG and 168
Nuclear Planning Group (NPG) xxxv-vi, 1, 169, 190, 235-6, 252, 257-61, 265-7
and SDI 147, 263-72, 487
strategy *see* flexible response; follow-on forces attack
US position in 166-7
Norway 268-9
Nuclear-free corridor 490

ocean reconnaissance 77-8
offensive-defensive system linkages xix-xx, xxxvii, 30-3
and arms control 34-7
and strategic doctrine 28-9
treaties 33-4
Ogarkov, N., Chief of Staff 361
Outer Space Treaty (1967) xxiii, 59, 79, 207, 422, 505
'oversell' 4, 50-1, 88-90
Owen, David 139-40

Palme Commission xxviii
Palmer, General 382, 419-20
Pan Heuristics 475
parliaments, international 256-99, 481
on ABM 257-63
NATO 263-72
on SDI/EUREKA 272-98:
European Parliament 294-8;
WEU 282-94
Parsons, Ambassador 377-8

partial defence 43, 146-7, 353, 481
participation in SDI 411, 437-8
Benelux 252
France 158-9
FRG 162, 177-8, 184, 193-7, 210-12
UK 135-7
particle beam weapons
and BMD 14, 22-4, 520-2, 529
Soviet 95-103
space-based 520-2, 529
Patriot ATM 95, 398, 455, 466, 469-70, 472-3, 477, 490
Pave Paws radar 401-2, 524-5
Pawn Shop radar 509-10
Payne, Keith 63, 101, 396-7, 542
peace movements 142, 196-8, 206-7, 241, 248-9
Pechora radar 506, 524-5, 530
penetration aids (penaids) 338, 347
Percy, Charles H., Senator 402, 504
perimeter acquisition radar (PAR) 401, 405, 531, 537
Perle, Richard N. xxi, 195-6, 271, 407, 411, 420, 501-2
Pershing missile xxx-i, 92-3, 355
Philips, N. V., company 187, 320, 323
Pichoud, Daniel 349
Pike, John E. xv, 3, 5, 471, 500-33, 537-54
Pipes, Richard 61
PKO Strany (Soviet ASAT) 72
Polaris missile xxxi, 331, 334-5
Pompidou, Georges 342
Pond, Elizabeth 169, 188
Poniatowski, Mr (Report) 295-7
pre-emption 27-8, 32, 58
PRO Strany (Soviet ABM) 72
public opinion, on SDI 129-30, 215
PVO Strany (Soviet air defence) 72
Pym, Francis 140

Quiles, Paul 345-9
Quinlan, Michael 335

Ra'anan, Uri 64
radars 395, 400-3, 524-6
and ABM Treaty 10-11, 385, 388, 395, 400-3, 405, 509-10, 524-6
early warning 10-11, 400-3, 525
JSTARS 455-8
LPARs 11, 93, 386, 524-6, 530-1
PAR 401, 405, 531, 537
Soviet 11, 83, 85-7, 337, 400-1, 407, 412, 501-6, 524-5, 567

radars (*contd*)
 space-based imaging 515
 TIR 19, 399, 405, 514, 522, 539
 range, of defence systems 459, 461–4
rapid reload launchers 508–9
Reagan, Ronald xix, xxi, xxiii–iv, xxx, xxxv–vii, 1–2, 14, 26, 45, 50–1, 88, 105–7, 130–3, 151, 166, 256, 264, 313, 333–4, 352–3, 375, 390, 403, 408, 412, 436, 489, 500, 523, 537
reconnaissance *see* satellites
reduction, in nuclear weapons xxv, 9–10, 34, 42–4, 180
 see also START
remotely-piloted vehicles (RPVs) 478
Research and Development in Advanced Communication and Technology for Europe (RACE) programme 315–16
Reykjavik summit 147, 216, 299, 429, 489
Rhinelander, John B. xvi, 2, 5, 373–431, 500–33
Riesenhuber, Heinz 184–7, 209, 326
Robinson, Clarence A., Jr 80–1, 94, 455
Rogers, Bernard W., General xxvii, xxix, 478–9, 482
Rogers Plan *see* follow-on-forces attack
Rogers, William, US Secretary 412–14
Roth, William, Senator 276
Roth, Wolfgang 196, 203
Royal Institute of International Affairs (RIIA) 144
Royal United Services Institute (RUSI) 133, 142–3, 334
Rühl, Lothar 179, 193, 258–9, 481
Rühe, Volker 199–200, 208
Rühle, Hans 179, 189–90, 439, 481
Ruiter, Jacob de 244, 266
Rumsfeld, Donald 80

Saby, Mr 296
Safeguard ABM system 261, 356, 401, 405, 537–8, 541
Sagdeyev, R. Z. 73, 446
Sanden, Mr van der 283
Sary Shagan test site 83, 98, 509–10, 530, 564
satellites xxiii, 473–4
 Soviet 74–8, 81–2
 US and European 53, 82, 156
 see also anti-satellite systems
Schäfer, Mr 203, 208
Scheer, Hermann 204, 207, 284–5
Schierholz, Henning 209
Schlesinger, James 553
Schmidt, Helmut 2, 182, 205, 259
Schmückle, Gerd, General 438
Schomerus, Lorenz 195–6
Schoser, Franz 212
Schröder, Dieter 214
Schulze, Franz-Joseph, General 210
science, military, Soviet 57–8, 69
Scienkiewicz, Stanley 68
scientific communities, reaction to SDI 144–5, 212–14, 241–2, 249–50
Scowcroft, Brent, General 529, 532
 Commission Report 45
Seitz, Konrad 182, 188–9, 203, 439, 450, 481
Semenov, Ambassador 84, 565
Senate, US, hearings 382, 393, 415–21, 467–8
sensors, missile 14, 16–18, 21, 39, 367, 513–16, 522, 528–9
Sentinel ABM system 257, 356, 401, 405, 537–8
short-range ballistic missiles (SRBMs) 19
Shulz, George P. 176, 268–9, 280–1, 289, 390, 408, 410, 430, 523
Siemens, AG 187, 211, 320, 323
Skoludek, Horst 211–12
Sloan, Stanley R. 263
small-radar homing intercept technology (SR-HIT) 399, 463–4, 539–40
Smith, Gerard C., Ambassador 34, 375–7, 393–5, 402, 408, 413–16, 466, 504–5, 508, 564–5
Smith, Marcia S. 75, 77, 81, 263
Smith, Margaret Chase, Senator 416–17
SNIA BPD 477
Social Democratic Party (SPD), FRG 168, 196–8, 203–6, 215
Social Democratic Party, Luxembourg 251
Social Democratic Party (SDP), UK 139–40
Socialist Party, Belgium 236, 238–9
Socialist Party (PS), France 160–1
Socialist Party, Netherlands 247–8
Sofaer, Abraham 421, 503
Sokolovskii, V. D., Marshal 58–9, 63, 65
Sommer, Theo 194
Soviet Union (USSR)
 and ABM Treaty: attitude toward 62–3, 67, 84–5, 423–8; compliance with 11, 407, 501–13

Index

air defence missiles, BMD potential 91–5
air defence, BMD and space organisations 71–3
ASATs 78–82, 395–7, 526–8
ATBMs 447–50, 506–8, 526
BMD activities 501–13
BMD capabilities 50–1, 57–73, 82–95, 334–9, 547–9
countermeasures to SDI 37–44, 73–4, 110–13, 445–7
damage limitation 547–8
directed-energy weapons 95–103
economy 358–60
forces targeted against W. Europe 441–5
Marxist–Leninism 55, 58, 66, 357–8
military doctrine 52–71
radars 11, 83, 85–7, 337, 400–1, 407, 412, 501–6, 509–10, 524–5, 530, 567
reaction to SDI 354–7, 424–7
space systems 57–71; passive (satellites) 74–8, 81–2; weapons 78–82, 95–103
SAMs 83, 91–5, 447–8, 466, 506–8, 510–11
SS missiles xxxi, 10, 16–17, 20–1, 443–5, 486, 507, 547–8
space, military activity in
BMD 22–4, 30–1
kinetic energy defensive system 17, 22–4
Soviet 51–113: ASAT systems 78–82; directed-energy weapons 95–103; and military doctrine 57–71; passive (satellite) systems 74–8, 81–2; and security of Europe 103–13
US 17, 21–4, 30–1, 74, 446, 519–22
space-based anti-missile systems (SBAMS) 17, 22–4, 30–1, 74, 446, 519–20
space-based kinetic kill vehicle project 519, 522
space-based laser project 519–20, 522
space-based neutral particle beam weapon 520–2
space surveillance and tracking system (SSTS) 514, 522
Spartan interceptor 537–8
Späth, Lothar 182, 184–5, 187
SPOT satellite 53
Sprint interceptor 538, 541
Staab, Heinz A. 213

Standard Elektrik Lorentz 211, 477
Standing Consultative Commission (SCC) 386–9, 430, 505, 511, 523–4, 556
Stares, Paul B. xvi, 3–4, 81, 130, 134, 331–9
Stevens, Sayre 69–70, 90, 94
Stoltenberg, Gerhard, Minister 182, 187
Stourdze, Yves 317
Strategic Arms Limitation Treaties
SALT I (1972) xxxiii, 33–4, 44, 62, 65, 262–3, 374–5, 380–1
SALT II (1979) 34, 44, 79, 375
Strategic Arms Reduction Talks (START) 32, 35–6, 44, 366–7, 375
strategic defence 26–46
Strategic Defense Initiative (SDI)
and ABM Treaty 373–431, 513–21
and arms control 44–5, 365–7
as BMD (*q.v.*) 9–25
Benelux reactions to 234–53
costs of 21–2, 26, 480–3, 544
defence of ICBM sites xxiv, 43, 353, 451, 537–42
European response to 311–26
French reaction to 150–64, 341–50
geostrategic risks 352–68
German reaction to 166–216
international parliamentary debates on 256–99
and negotiating strategy 34–7
participation in 135–7, 158–62, 177–8, 184, 193–7, 210–12, 252, 411, 437–8
rationale of 50–113
Soviet countermeasures to 37–44, 73–4, 110–13, 445–7
Soviet reaction to 354–7, 424–7
UK reaction to 129–47, 331–9
Strategic Defence Initiative Organisation (SDIO) 106, 131, 211
Strategic Rocket Forces, (RSVN), Soviet Union 72–3
Strauss, Franz Joseph 169–71, 179–87, 192–4, 201–3, 217, 287, 478
Strode, Rebecca V. 89–90
Stroude, Dan 63
Stürmer, Michael 210
Stützle, Walther 169, 197, 214
submarines 548
French 347
offensive–defensive 30–1
UK xxxi, 134, 331–9, 361, 543
surface-to-air missiles (SAMs) 83, 85, 555

surface-to-air missiles (SAMs) (*contd*)
  Soviet 83, 91–5, 447–8, 466, 506–8, 510–11
surface-to-surface (SS) missiles, Soviet xxxi, 10, 16–17, 20–1, 443–5, 486, 507, 547–8
survivability of defence system 15, 22–3
Swaelen, Mr 236, 240
SYRACUSE system 156, 348

tactical anti-ballistic missiles (TABMs) *see* anti-tactical ballistic missiles
tactical ballistic missile defence xxxiii, xxxix, 455, 460
Tactical Defence Initiative (TDI) 245–7, 436
tactical missiles (TACMs) 457, 472–5
Talensky, N., Major General 59, 65, 67, 84
Tallinn system 83, 257
Talon Gold telescope 519–20
targeting 338, 360, 441–5
Taylor, Trevor xvii, 3–4, 129–47
technology xxxiii–iv
  and BMD 11, 12–15, 46, 99–100
  EDI 481–2
  emergent (ET) 457–9, 474–5
  France 162–3
  FRG 183–5
  sharing, Europe and USA xxxviii, 176, 185 *see also* participation
  UK 135–7
  US and Soviet 99–100, 108–10
  see also European Research Co-ordination Agency
Teller, Edward 461
Teltschick, Horst 176–9, 185, 192, 203, 211
terminal imaging radars (TIRs) 19, 399, 405, 514, 522, 539
terminal phase of missile flight, and interception in 12–13, 18–21, 38–40, 459, 538
'testing', definition of 391, 524
Texas Instruments 312
Thatcher, Margaret 130–2, 135, 333–4, 411
theatre war 361
  limited xxvi–vii, xxxv, 12, 27, 441–5
Thomson company 187, 320, 323
Thule, radar 526
Tiedtke, Stephan v
Tindemans, Leo 235–7
Tobback, Mr 236, 239
Todenhöfer, Jürgen 199, 201

Toom, Mr den 259
trade unions 210–12
treaties 421
  offensive–defensive system linkages 33–4
  and SDI 113
  test ban xxv
  see also ABM Treaty; arms control; negotiation; Outer Space Treaty; SALT
Triad laser programme 519–20
Trident missile system xxxi, 134, 331–9, 361, 543
Tummers, Mr 285

Ulburghs, Mr 296
Union of Soviet Socialist Republics (USSR) *see* Soviet Union
Union pour la Democratie Francaise 162
United Kingdom (UK)
  implications of BMD for deterrence 331–9, 354
  reactions to SDI 129–47: academics and defence analysts 142–5; government 130–8, 332–4; political parties 138–42
  Trident 334–9, 361
United States of America (USA)
  and ABM Treaty 403–23; compliance with 513–21; reinterpretation, efforts at 407–21
  assessments of Soviet Union: BMD 85–90; military capabilities 54–7; military doctrine 60–8; space programmes 95–103
  decoupling from NATO xxxi, xxxv–viii
  'new strategic concept' 35, 145, 543
  radars 401, 405, 455–8, 514, 531, 537
  space weapons 17, 21–4, 30–1, 74, 81–2, 446, 519–22
  strategic doctrine 27–9, 68–9, 456–7
  technological challenge to Europe 312–14
  weapons projects for EDI 459–79
'use or lose' dilemma xxviii

Van Cleave, William R. 61
Vannick, Sir Peter 295
Vedrine, Hubert 157
Vogel, Hans-Jochen 196, 203
Voigt, Karsten 169, 204, 276, 282, 458
Voorhoeve, J. C. C. 246

Index

Vosen, Josef 209
Vreven, A. 235-6
Vries, Klaas de 285
vulnerability, mutual 56-9, 64-8, 167, 352
vulnerability, 'window' of 537-54
  assured destruction 547-9
  damage limitation 542-7
  escalation stability 549-53
  ICBM sites, defence of 537-41
  intermediate defence goals 541-2

Wade, James P. 100, 467, 473-4
Wallop, Malcolm, Senator 100
Warner, John W., Senator 467
Warsaw Pact xxvii, 361-2
Wehrkunde Conference 172-3
Weinberger, Caspar W. xix, xxi, xxx, 1, 43, 137, 193-4, 235-6, 243, 259, 265-71, 395, 407, 411, 426, 437, 440, 457-8, 466, 473
Weiner, Stephen 471
Weisweiler, Mr 211
Weizsäcker, Richard von, President 181

Western European Union (WEU) 137, 140-1, 169, 171, 202, 237, 270-3, 282-94, 479, 481, 487
Wilkinson, John (Report) 141, 283-5
Williams, James A., Lt General 63
Williams, Phil 144
Williamsburg Summit (1983) 317
Wilson, Peter A. xvii, 2, 5, 352-68
Wimmer, Willy 199-200, 207-8
Wohlstetter, Albert 449
Worden, Simon Peter xvii, 2-3, 9-25
Wörner, Manfred 168-71, 175-9, 190-3, 203, 215-17, 266, 269, 438, 440, 458, 476-9, 482

York, Herbert 106
Yost, David 62-3, 83, 85, 101, 112, 447-50, 470

Zenith Missile Troops (ZMT), Soviet Union 72
'zero option' xxxi, xxxv, 489-90
Zhang Aiping 355
Zundel, Rolf 194